Sustainable Landscape Planning and Natural Resources Management

IEREK Interdisciplinary Series for Sustainable Development

Sustainable Landscape Planning and Natural Resources Management (SLNR) is a peer-reviewed book series that presents groundbreaking research through the intersection of two main areas that of which are 'Sustainable Landscape Planning' and 'Natural Resources Management'. The series initially covers a wide range of academic disciplines in sustainability, including landscape ecology, green urbanism, urban regeneration, land restoration and renaturation, energy management, and nature conservation. Moreover, it also incorporates the latest developments and research in engineering topics that contribute to more sustainable cities, such as Geosciences, infrastructure, civil engineering, geology, mining, water, and natural resources conservation and management.

By showcasing methods for discovering, managing, and employing natural resources, the series aims to provide approaches that enable the responsible use of these resources. Ultimately, the SLNR series is committed to advancing the United Nations' Sustainable Development Goals for 2030.

The wide range of topics covered in this series are primarily related to sustainable development, involving durable infrastructure projects, civil engineering, urban regeneration, the built, sustainability, climate change mitigation, energy management, geosciences, geology, and the future of cities, all of which depend on the availability of natural resources and effective wealth management. The series also features conceptual and empirical contributions from various interrelated fields of earth science, with an emphasis on the era of new discovery. It focuses on providing practical solutions to ensure sustainable development and provides new case studies that offer concrete examples of how to manage natural resources effectively.

The series encompasses the best research papers from various international conferences organized by IEREK conferences, also accepts proposals from other international conferences, including interdisciplinary contributions. It exhibits innovative approaches and highlights how they can best support both economic and sustainable development through the responsible use of natural resources and global, local, and individual activities for the welfare of all societies on Earth.

The SLNR series is a valuable source for experts in government, international organizations, consultancies, industry, business sectors, and academics. Published in cooperation with IEREK, the Springer SLNR series displays important new research in various sectors, delivering valuable insights and practical solutions to the complex challenges of sustainable development for the readers.

Jyoti Devi
Editor

Biotechnological Intervention in Production of Bioactive Compounds

Biosynthesis, Characterization and Applications

Springer

Editor
Jyoti Devi
Agricultural Biotechnology
Govardhan Agricultural University
Bulandpur, Uttar Pradesh, India

ISSN 2948-1910 ISSN 2948-1929 (electronic)
Sustainable Landscape Planning and Natural Resources Management
IEREK Interdisciplinary Series for Sustainable Development
ISBN 978-3-031-76858-3 ISBN 978-3-031-76859-0 (eBook)
https://doi.org/10.1007/978-3-031-76859-0

This Springer imprint is published by the registered company Springer Nature Switzerland AG
The registered company address is: Gewerbestrasse 11, 6330 Cham, Switzerland

Preface

Bioactive Compounds are substances that are added to or naturally present in food matrices that have the potential to offer health benefits beyond nutritional value. Since bioactive compounds have been found in a variety of sources, their use as functional food ingredients and medications has been supported by their healing advantages, dietetic importance, and defensive properties in the treatment of humans and animals. One of the most exciting advances in contemporary natural medicine is the methodical investigation of naturally active products and the prospective biological actions of these bioactive compounds, including their potential biomolecular interfaces, mode of action, clinical applications, quality control, and standardization.

The current state of information on the biosynthesis of bioactive compounds from plant, microbes, algae, and other biological sources is described in the book *Biotechnological Intervention in Production of Bioactive Compounds: Biosynthesis, Characterization, and Applications*. Furthermore, the molecular method for identifying bioactive substances is examined, along with instances of how these molecules can benefit human health. Information on several sources of bioactive chemicals, such as bacteria, algae, plants, and food items, is covered in the first part of the book. The second part of book examines synthetic methods, a few chosen bioactivities, and the potential for biotechnological and biomedical applications. Together with a summary of the medical uses for chemicals produced from plants, an overview of the use of bioactive substances as stem cell research subjects, anti-inflammatory and antioxidant agents, and medications for allergies is also included. Undergraduate and graduate biomedical chemistry students who are eager to learn more about the possibilities of bioactive natural compounds will find this book to be an informative read. Additionally, it offers helpful information for scientists engaged in a variety of studies where natural products play a major role.

Bulandpur, India | Jyoti Devi

Contents

About the Editor

Dr. Jyoti Devi is an Assistant Professor at Department of Agricultural Biotechnology, Govardhan Agricultural University, Bulandpur, Uttar Pradesh, India. Dr. Devi completed her doctoral degree in Plant Biology from Patiala University, Punjab, India. Her specialization is genetic improvement of plants, Molecular Breeding, and bioactive compounds characterization. She has published more than 11 research articles in leading international and journals having high impact factors. She has received many awards and fellowships from different government organizations. She is involved in teaching and research and her area of research is plant biotechnology. She is an active reviewer for journals Plant Gene, Plant Biotech Reports, 3 Biotech, Industrial Crops and Products, Gene, Current Science, Biologia etc.

Exploring Bioactive Compounds: Biosynthesis, Extraction, Characterization, and Diverse Applications—An Overview

E. Sandhya Rani, Bojjala Sravya, Sagi Sowmika, and Narasimha Golla

Abstract

Bioactive compounds are crucial in various industries, including pharmaceuticals, agriculture, cosmetics, and personal care. They are synthesized through molecular mechanisms, enzymatic reactions, and regulatory pathways. Extraction techniques for separating bioactive compounds from natural sources include solvent extraction, Soxhlet extraction, ultrasound-assisted extraction, microwave-assisted extraction, and enzyme-assisted extraction. Characterization methods like mass spectroscopy, NMR, and FTIR are used for molecular identification and structural information. Chromatography techniques like column chromatography, HPLC, and GC are essential for separating and analyzing bioactive compounds. They are also used in drug development, treatments, cardiovascular health, antioxidants, neurological disorders, and immunomodulation. They promote overall wellness through vitamins, antioxidants, omega-3 fatty acids, probiotics, herbal extracts, polyphenols, collagen, and adaptogens in nutraceuticals and dietary supplements. Agricultural applications include bio-stimulants, biofertilizers, plant growth promoters, biopesticides, and allelopathic agents. Bioactive compounds are also used in cosmetics and personal care products as antioxidants, moisturizers, anti-aging, skin-brightening, anti-inflammatory, UV filters, and hair care ingredients. This chapter emphasizes the versatility and importance of bioactive compounds in diverse industries and their potential to contribute to human health and agricultural sustainability.

E. Sandhya Rani · N. Golla (✉)
Department of Virology, Sri Venkateswara University, Tirupati, Andhra Pradesh 517502, India
e-mail: dr.g.narasimha@gmail.com

B. Sravya · S. Sowmika
Department of Biotechnology, Sri Padmavati Mahila Visvavidyalayam, Tirupati, Andhra Pradesh 517502, India

Keywords

Bioactive compounds • Extraction methods • Characterization • Pharmaceutical • Nutraceuticals and agriculture applications

1 Introduction

One of the essential components of life is biosynthesis, the complex process by which organisms create complex chemicals from smaller predecessors. When it comes to the world of bioactive compounds, biosynthesis is essential to the creation of molecules with a variety of biological functions, from therapeutic effects to ecological interactions. Comprehending the mechanisms that underlie biosynthesis not only illuminates the complex machinery of life, but also has great potential applications in the pharmaceutical, agricultural, and biotechnology industries (Boadella, 2015). Fundamentally, biosynthesis is a sequence of enzyme reactions that are coordinated by an organism's cellular apparatus. These reactions are controlled by a network of genes that encode the enzymes in charge of catalyzing each stage, and they take place inside specialized cellular compartments like organelles or particular cellular areas (Liu et al., 2021). Utilizing successive conversions of precursor molecules, which often commence from basic building blocks such as sugars, amino acids, or fatty acids, organisms produce an astounding variety of structurally varied bioactive compounds. The intrinsic intricacy and adaptability of biosynthesis are one of its most fascinating features (Vinogradov and Suga, 2020). One set of precursor molecules can result in many different end products, all of which have different chemical structures and biological functions. The modular structure of biosynthetic pathways, which allows for the production of distinct chemicals by combining, altering, or even skipping particular enzyme-catalyzed processes, is the source of this diversity (Kour et al., 2019). Further

J. Devi (ed.), *Biotechnological Intervention in Production of Bioactive Compounds*,
Sustainable Landscape Planning and Natural Resources Management, https://doi.org/10.1007/978-3-031-76859-0_1

broadening the range of substances that can be generated is the fact that organisms frequently have numerous biosynthetic routes that can diverge or converge. Biosynthesis is the foundation for the synthesis of natural products with enormous medicinal potential when it comes to bioactive compounds (Zepka et al., 2021). Many of the medications used in clinical practice today, ranging from immunomodulators and analgesics to antibiotics and anticancer medicines, are derived from bioactive substances produced by living things. Researchers can use nature's chemical toolkit to find new medication candidates, improve production methods, and even create designer molecules with specific features by deciphering the complexities of biosynthetic pathways (Salwan and Sharma, 2020). Beyond the domain of human health, biosynthesis is also important in determining ecological relationships and environmental dynamics. As secondary metabolites, microorganisms, plants, and other organisms create bioactive molecules to ward off diseases, and herbivores. Through their interactions with other species, these natural products affect entire ecosystems in addition to the survival and evolution of the producing creatures (Davies and Ryan, 2012). This book chapter delves into the intriguing realm of biosynthesis, examining the fundamental concepts, workings, and control of this vital biological function. We explore the wide range of bioactive substances produced by living things, from microbial natural products to phytochemicals derived from plants, using a multidisciplinary lens. By clarifying the complexities of biosynthesis processes and their uses, we hope to offer a thorough grasp of how nature's chemical diversity can be utilized for agriculture, human health, and other purposes.

2 Biosynthesis of Bioactive Compounds

The process by which living things create complex chemicals from simpler precursors often under the guidance of enzymes- is called biosynthesis (Yang et al., 2016). When it comes to bioactive compounds, biosynthesis is essential to the creation of molecules with a variety of biological activities, from interactions with the environment to medical effects. Clarifying the sources, modes of action, and possible uses of bioactive compounds requires an understanding of the biochemical pathways that underpin their synthesis. The biosynthesis of bioactive compounds begins with precursor molecules, which serve as fundamental components for the creation of more complex structures (Salwan and Sharma, 2020). Precursors include simple molecules such as fatty acids, sugars, amino acids, nucleotides, and intermediate metabolites generated by major metabolic pathways.

Enzymes catalyze biosynthetic pathways, allowing precursor molecules to be converted into bioactive compounds through a sequence of enzymatic processes (Rizzo et al., 2020). Biosynthesis-related enzymes are extremely selective; they can identify certain substrates and catalyze particular chemical processes, including oxidation, reduction, condensation, and rearrangement. The process of biosynthesizing bioactive substances is frequently controlled by a complex system of genes that encode the enzymes that catalyze the various steps in the route (Zepka et al., 2021). The expression of genes involved in biosynthesis is regulated by genetics and is influenced by a range of internal and external factors, including metabolic demands, environmental stressors, and developmental signals. To manipulate the production of compounds in organisms and optimize yields for commercial uses, it is essential to comprehend the genetic regulation of biosynthetic pathways (Alagoz et al., 2016). The biosynthesis of bioactive substances can involve a wide range of complicated pathways, from highly branching networks comprising several intermediates and parallel routes to linear processes with few enzyme steps. The structural diversity of bioactive compounds and the evolutionary adaptations of organisms to various ecological niches are reflected in the complexity of biosynthetic pathways (Salwan and Sharma, 2020). Enzymes and substrates are arranged spatially to support metabolic reactions in specialized cellular compartments, such as organelles or certain areas inside the cell, where biosynthesis frequently takes place. The efficiency and product specificity of biosynthetic pathways is enhanced by the compartmentalization of these processes, which permits the sequestration of intermediates, control over enzyme activity, and synchronization of metabolic fluxes (Bharadwaj et al., 2021).

For example, In Plant Secondary Metabolites, through specific biosynthetic pathways, plants generate a wide range of bioactive molecules, such as flavonoids, terpenoids, phenolics, and alkaloids. For instance, the production of alkaloids, like morphine and caffeine, entails a series of steps that are facilitated by enzymes found in particular cellular compartments, such as plastids and the endoplasmic reticulum (Twaij and Hasan, 2022).

In Microbial Natural Products, A wide range of bioactive substances with potential applications in medicine are produced by microorganisms, such as fungi, bacteria, and actinomycetes. Antibiotics, antifungals, and immunosuppressants are a few examples. Gene clusters encode complex biosynthetic mechanisms used by fungi and bacteria to produce these. The development of structurally varied compounds with strong antibacterial properties is the result of the coordinated action of several enzymes and regulatory proteins during the biosynthesis of antibiotics such as streptomycin and penicillin (Challinor and Bode, 2015).

In Marine Natural Products, Bioactive substances with industrial, medicinal, and agricultural uses can be found in abundance in marine creatures such as algae, sponges, and corals. The biosynthesis of marine natural products frequently takes place via symbiotic connections between species or in reaction to indications from the environment, like the pressure of predators and the availability of nutrients. Marine bacteria, cyanobacteria, and marine invertebrates create polyketides, peptides, and alkaloids as examples. These compounds have a variety of biological functions, such as antiviral, anticancer, and antifouling qualities (Carroll et al., 2023).

3 Overview of Biosynthetic Pathways

Biosynthetic pathways are intricate metabolic mechanisms that enable living organisms to synthesize complex molecules from simpler constituents. The synthesis of numerous distinct kinds of biomolecules, including lipids, proteins, carbohydrates, and nucleic acids, as well as specialized metabolites like pigments, bioactive chemicals, and secondary metabolites, depends on these pathways. Clarifying the molecular underpinnings of life and utilizing organisms' metabolic potential for a variety of purposes need an understanding of the structure, dynamics, and regulation of biosynthetic pathways (Ma et al., 2020). A sequence of enzyme events known as biosynthetic pathways change precursor molecules, also known as metabolic intermediates, into final products through successive biochemical transformations. These intermediaries participate in a variety of metabolic processes and cellular functions as substrates, coenzymes, or regulatory molecules in metabolic pathways (Bajguz et al., 2020). When they are used to catalyze biosynthetic reactions, enzymes are biological catalysts that accelerate chemical reactions without being consumed. Enzymes catalyze particular chemical transformations, such as oxidation–reduction, condensation, hydrolysis, and isomerization events, to produce intermediate and final products (Goyal and Verma, 2023). They are very specific to their substrates. The ability of enzymes to identify particular substrates based on their chemical structures, binding affinities, and active site layouts is known as substrate specificity. Substrate specificity is influenced by the molecular complementarity between the substrate and the enzyme as well as the chemical, steric, and electrostatic interactions that occur during substrate binding and catalysis (Peracchi, 2018). Complex regulatory systems that manage metabolite flux, synchronize metabolic processes, and preserve cellular homeostasis are imposed on biosynthetic pathways. Allostery, suppression of feedback, activation, and repression of enzymes, post-translational changes, and signal transduction pathways that react to cellular signals and environmental stimuli are examples of regulatory processes (Qian-Wen et al., 2022). A complex metabolic web is formed by the convergence, divergence, and intersection of interconnected networks of biological reactions known as metabolic pathways. Through the coordinated synthesis, breakdown, and interconversion of metabolites made possible by pathway integration, organisms can adjust to shifting environmental factors, energy requirements, and nutritional availability (Dhaniaputri et al., 2022). Different pathways are shown in Fig. 1.

1. **Glycolysis**: This essential metabolic process produces ATP and reduces equivalents in the form of NADH by oxidizing glucose to pyruvate. This route is found in the cytoplasm of cells and provides a significant amount of energy for metabolism within the cell. Pyruvate, ATP, and NADH are produced during a sequence of chemical events known as glycolysis, which is catalyzed by enzymes like pyruvate kinase, phosphofructokinase, and hexokinase (Chandel 2021).
2. **Fatty Acid Biosynthesis**: This biosynthetic route converts precursors of acetyl-CoA and malonyl-CoA into fatty acids. This process is found in the cytoplasm of cells and is catalyzed by fatty acid synthase, a multi-enzyme complex. The following steps of fatty acid biosynthesis are condensation, reduction, and dehydration, which result in long-chain fatty acids. These are the fundamental components of signaling molecules, energy storage molecules, and membrane lipids (Harwood, 2020).
3. **Polyketide Biosynthesis**: This biosynthetic process converts basic substrates like acetyl-CoA and malonyl-CoA into polyketide molecules, including antibiotics, antifungals, and anticancer medicines. Polyketide synthase enzymes catalyze this process, which is found in plants, fungi, and bacteria. Polypeptide biosynthesis involves repeated cycles of condensation, reduction, dehydration, and cyclization. The result is a structurally diverse set of molecules with potent biological activity (Weissman, 2015).

Clarifying the molecular underpinnings of cellular metabolism, genetic control, and organism physiology requires understanding biosynthetic pathways. It offers prospects for metabolic engineering, synthetic biology, and biotechnological applications, as well as insights into the molecular mechanisms underpinning disease processes, cellular activities, and evolutionary adaptations. Researchers can create innovative approaches to drug discovery, bioproduction, and sustainable agriculture by deciphering the intricacies of biosynthetic pathways and utilizing the metabolic potential of living things. These efforts will ultimately benefit human health, environmental sustainability, and economic well-being (Wong and Khosla, 2012).

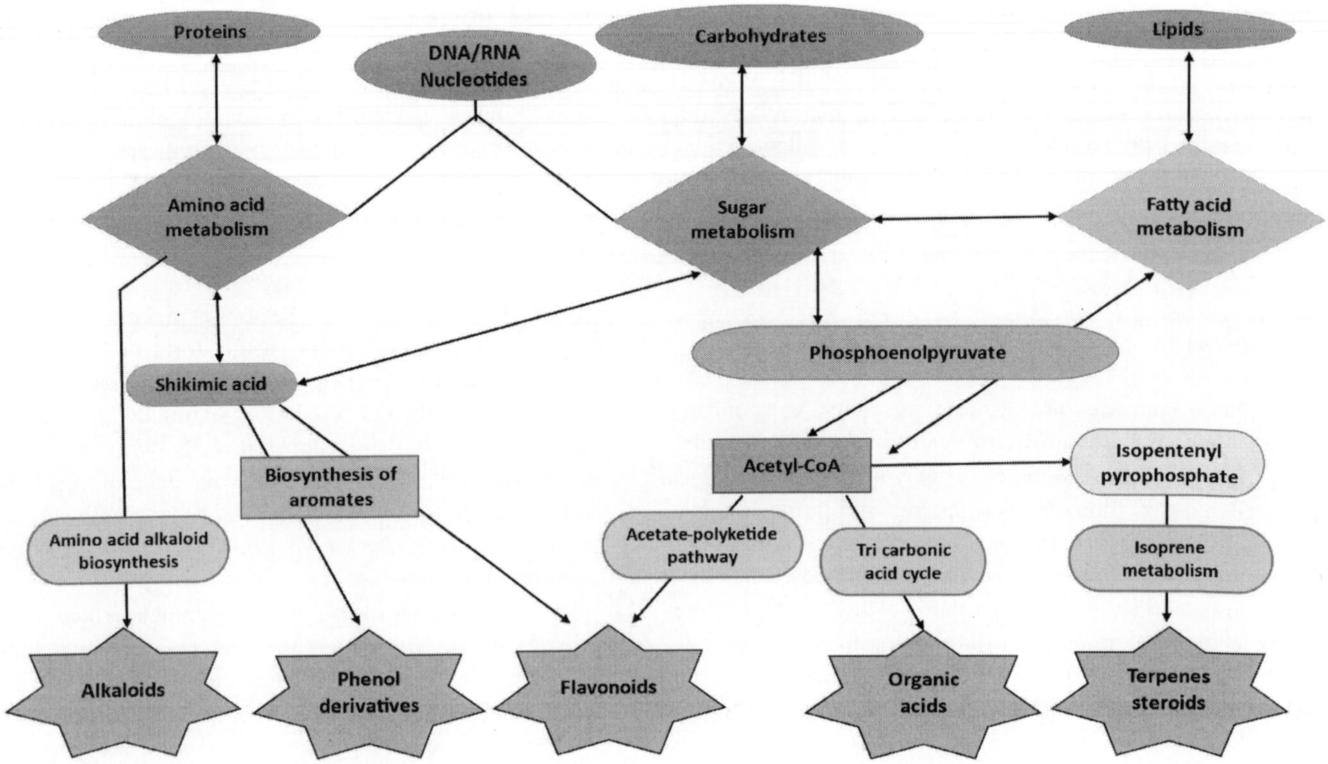

Fig. 1 Different pathways of bioactive compounds

4 Bioactive Compounds from Natural Sources

The biosynthesis of bioactive compounds from various natural sources, including plants, microorganisms, and marine organisms (as shown in Fig. 2), has potential therapeutic applications. In all these organisms, the biosynthesis of bioactive compounds involves complex biochemical pathways and regulatory mechanisms.

4.1 Plants

Alkaloids, phenolics, terpenoids, and flavonoids are just a few of the many specialized metabolites that plants create. These metabolites serve a variety of purposes, including attracting pollinators, protecting plants from diseases and herbivores, and regulating interactions between microbes and plants (Akhtar et al., 2019). Intricate metabolic pathways, frequently divided within certain cellular organelles, including the cytoplasm, chloroplasts, and endoplasmic reticulum, are involved in the biosynthesis of specialized metabolites in plants. Enzymes produced by genes found in the plant genome catalyze these processes, which are regulated by hormonal stimuli, developmental signals, and environmental cues (Jimenez-Garcia et al., 2013).

Plants are subject to strict regulatory control on the manufacture of specialized metabolites at several levels, including transcriptional regulation, post-transcriptional regulation, and post-translational modification (Balawejder et al., 2022). The expression and activity of genes involved in biosynthesis are regulated by regulatory proteins, which include transcription factors, microRNAs, and protein kinases, in reaction to both internal and external stimuli.

Environmental elements that can affect a plant's ability to synthesize specialized metabolites include light, temperature, the availability of water, nutrients, and biotic stresses. For instance, ultraviolet (UV) radiation exposure can stimulate the synthesis of flavonoids and other phenolic chemicals that are important for antioxidant defense and UV protection (Samyn et al., 2023).

4.2 Microorganisms

Actinomycetes, fungi, and bacteria are examples of microorganisms that produce a wide variety of bioactive compounds with different chemical structures and biological activity. These substances have a variety of ecological purposes, such as protection against predators, rivals, and environmental stresses (Janardhan et al., 2014).

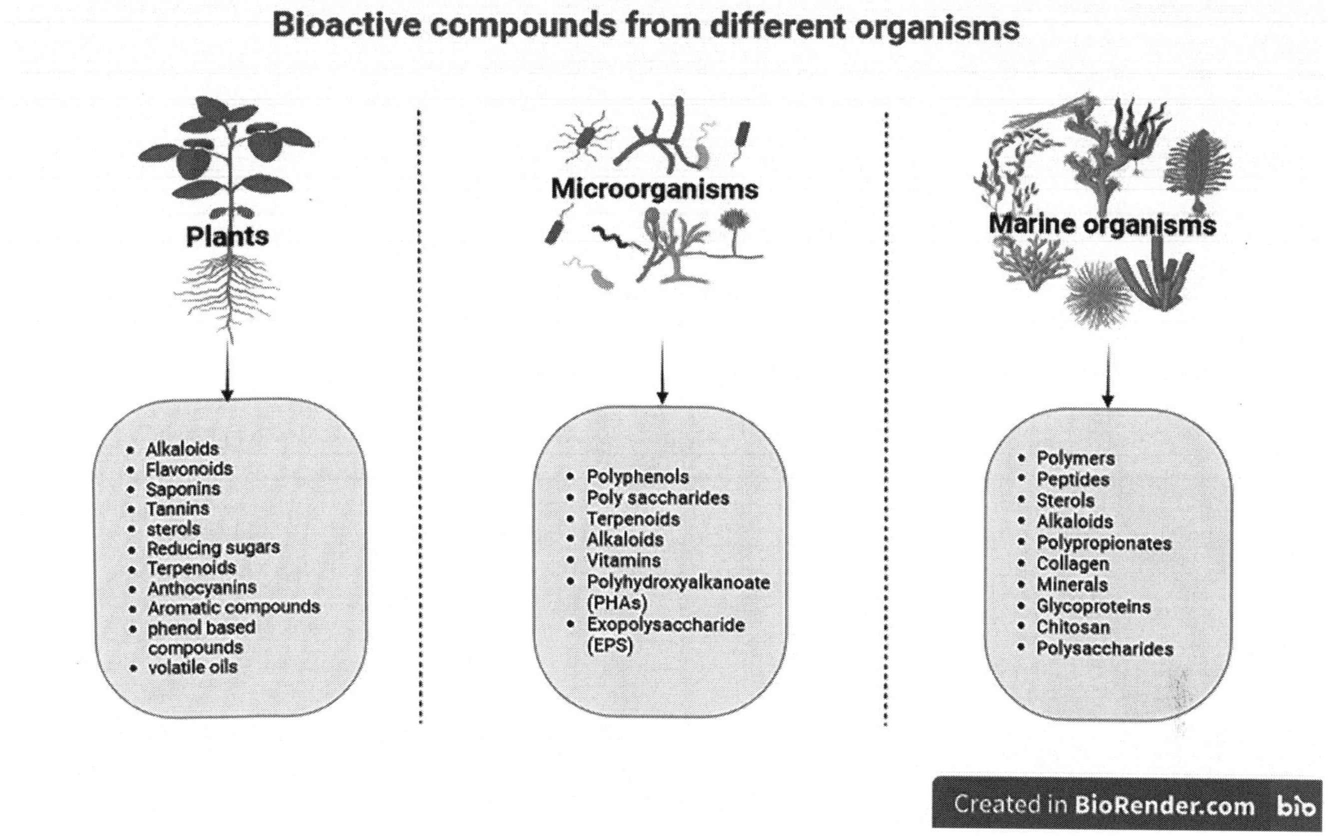

Fig. 2 Bioactive compounds from plants and microorganisms

Microorganisms' biosynthesis is characterized by several important factors.

Microorganisms produce a vast array of natural substances through specialized metabolic pathways, such as immunosuppressants, antifungals, antibiotics, and anticancer drugs. These pathways, which call for the coordinated activity of several different enzymes, are frequently encoded by gene clusters found within the genome of microorganisms (Tran et al., 2019). Natural product biosynthesis in microorganisms is usually linked to secondary metabolism, which takes place in response to certain physiological processes such as quorum sensing, stress, or resource limitation. Separate from core metabolic pathways, specific biosynthetic enzymes produce secondary metabolites, which are controlled by environmental and genetic variables.

Biosynthetic gene clusters (BGCs) found in the genomes of microorganisms encode the regulatory proteins and enzymes required for the synthesis of natural products. To enable the coordinated expression and control of genes involved in biosynthesis, BGCs are frequently arranged into operons or gene clusters (Singh et al., 2017). BGCs in microbial genomes are found and characterized to find natural products using genome mining and bioinformatics techniques. To create bioactive compounds of interest, microbial biosynthetic pathways can be heterologously produced in host species like bacteria, yeast, or plants. Heterologous expression systems facilitate drug discovery, bioproduction, and metabolic engineering endeavors by enabling complex natural products controlled and scalable production (Davies and Ryan, 2012).

4.3 Marine Organisms

Numerous marine species, including bacteria, sponges, corals, and algae, are abundant sources of bioactive substances with a wide range of chemical and biological properties. In marine ecology, these substances are essential for protection against predators, resource competition, and interspecies communication (Janardhan et al., 2014). Numerous natural chemicals with strong biological properties such as anticancer, antiviral, and antibacterial are produced by marine organisms (Ameen et al., 2021). These products include alkaloids, peptides, terpenoids, and polyketides. Because of their distinct pharmacological profiles and chemical structures, marine natural products are desirable targets for drug discovery and development. Natural product biosynthesis in marine animals is frequently linked to ecological processes like defense against microbial diseases, fouling organisms, and predators (Kiuru et al., 2014). In marine ecosystems, secondary metabolites produced by marine animals can influence interspecies interactions, suppress microbial growth, and repel herbivores. Marine products from nature exhibit remarkable biogeographic variety because their chemical profiles differ among different marine environments, such as coral reefs, deep-sea vents, and intertidal zones. Environmental elements that can affect marine species' ability to synthesize natural products include temperature, salinity, the availability of nutrients, and symbiotic relationships. Biotechnological applications of marine natural products are noteworthy in the fields of biomedicine, biotechnology, and drug discovery. Commercial goods for materials science, pharmaceuticals, and cosmetics have been developed using chemicals derived from marine creatures, including antifouling, antibiotics, and anticancer drugs (Parte et al., 2017).

5 Factors Influencing the Biosynthesis of Bioactive Compounds

The process of biosynthesizing bioactive compounds is intricate and dynamic, impacted by several internal and external influences. Comprehending these variables is vital to maximize the yield of bioactive substances and capitalize on their versatility. Numerous important variables that affect biosynthesis are shown in Table 1.

6 Extraction Methods of Bioactive Compounds

Pre-extraction and extraction techniques are essential in isolating bioactive components from natural sources. The method of isolating bioactive compounds from components using a substantial solvent is known as solvent extraction. Several variables, including agitation, temperature, and the kind of solvent employed can impact solid–liquid extraction (SLE). The SLE method's solvents' viscosity can be lowered and the bioactive component's solubility can be increased by temperature. Compounds have been extracted via maceration, Soxhlet, microwave-assisted, and ultrasound-assisted extraction. Extraction methods of bioactive compounds are listed in Table 2.

The bioactive compounds are extracted, separated, purified, and then identified to find out which compounds are present, functional activities of these substances are also classed based on several bioactivity experiments.

7 Characterization Techniques

Many analytical approaches are employed to separate, purify, and identify bioactive compounds as shown in Fig. 3; however, these procedures are screened according to three criteria: specificity, simplicity, and speed (Altemimi et al., 2017).

7.1 Spectroscopy Techniques

Several spectroscopy techniques are employed for molecular identification, composition, and bonding once bio-active compounds. The fundamental idea behind these techniques is that molecules absorb electromagnetic radiation, creating a spectrum in the process. Since each spectrum is specific to a certain kind of bonding inside the molecule, the spectra are intended for identification. Fourier transforms and UV visibility Among those techniques are mass spectroscopy (MS), nuclear magnetic resonance (NMR), and infrared spectroscopy (FTIR) (Ivanović et al., 2020).

7.1.1 Mass Spectroscopy

Using the mass-to-charge ratio (m/z) of chemical compounds as a basis for identification and characterization, mass spectrometry (MS) is a potent analytical method. The procedure involves ionizing the sample molecules to produce charged particles, or ions, which are subsequently separated using an electric or magnetic field based on their mass-to-charge ratio. The mass-to-charge ratio of the ions is used to detect and quantify them, revealing details on the molecular mass and the structure of the molecules (Hameed et al., 2015).

7.1.2 NMR Spectroscopy

Determining the structure of bioactive substances is the main goal of NMR analysis. It is based on the interaction of certain nuclei in an external magnetic field. Radio waves are absorbed and released by the nuclei in the sample, depending on which other atoms or groups are associated with them. A detector is in place to detect these waves. According to a property called molecular change, the detector presents the energy absorbed as a spectrum (Santos and Silva, 2014). To further aid in the molecular comprehension of the mechanisms of action of enzymes, and transporter proteins, The NMR is a powerful tool for studying conformational changes in proteins and chemical kinetic processes. The research revealed that the protein's capacity to dimerize is essential for the creation of cell adhesion in multicellular organisms (Luo et al., 2019).

Table 1 Factors affecting the biosynthesis of bioactive compounds

S. No.	Factor	Description	References
1.	*Genetic factors*		
	Genetic regulation	Regulation of genes encoding regulatory proteins and enzymes controlling biosynthetic pathways	Zhan et al. (2022)
	Gene expression	Regulation of biosynthetic genes' expression in response to environmental cues, developmental signals, and metabolic demands	Gupta et al. (2015)
	Genetic variation	Genetic variation within populations or species affects the production of bioactive compounds through natural selection, genetic drift, and gene flow	Manach et al. (2017)
2.	*Environmental elements*		
	Abiotic factors	Influence of environmental factors such as light, temperature, humidity, water availability, soil nutrients, and pH on biosynthesis	Toscano et al. (2019)
	Biotic interactions	Impact of interactions with other species, including competitors and herbivores, on biosynthesis	Szakiel et al., (2011)
	Ecological context	Influence of habitat type, ecosystem structure, and community composition on the production of bioactive compounds in natural settings	Santi et al. (2023)
3.	*Physiological factors*		
	Developmental stage	Dependency of bioactive chemical production on the organism's developmental stage, including plant growth phases and marine organism lifecycle stages	Toscano et al. (2019)
	Metabolic state	Influence on metabolic activity, energy status, and precursor pool size on biosynthesis	Dijkstra, et al. (2011)
	Cellular differentiation	Cellular specialization and differentiation to the production of cell type-specific bioactive compounds	Dijkstra, et al. (2011)
4.	*Biotechnological factors*		
	Bioprocess engineering	Use of biotechnological methods such as fermentation, cell culture, and metabolic engineering to enhance bioactive compound production	Xu et al. (2018)
	Genetic modification	Alteration of biosynthetic pathways and incorporation of new features into organisms using synthetic biology and genetic engineering	Sharma et al. (2019)
	Bioreactor design	Impact of bioreactor design and operating parameters on biosynthesis in large-scale production processes	Motolinía-Alcántara et al. (2021)
5.	*Metabolic and nutritional factors*		
	Substrate availability	Influence of precursor, cofactor, and energy source availability on bioactive compound production	Septembre-Malaterre et al. (2018)
	Metabolic intermediates	Role of metabolic intermediates and final products of primary metabolism as substrates and regulators of biosynthetic pathways	Aharoni and Galili (2011)

7.1.3 Infrared (IR) Spectroscopy

The measurement of the absorption, and emission of infrared radiation by matter is known as infrared spectroscopy, also referred to as vibrational spectroscopy. It is employed in the study and identification of solid, liquid, and gaseous chemical compounds or functional groups (Johnson et al., 2023). When a molecule absorbs light in the infrared portion of the electromagnetic spectrum, it can be detected using IR spectroscopy. A molecule needs a bond in its structure that can display a "dipole moment," or unequal distribution of electrons within a bond, to be able to absorb light (Daniel, 2017).

7.2 Chromatographic Techniques

7.2.1 Column Chromatography

Column chromatography techniques are used to separate and isolate the individual compounds from the mixture of extracts. Zhang et al. (2018), isolated and purified the bio-active compounds based on their ionic strength, boiling temperatures, molecular size, and adsorption capabilities. With the use of adsorption column chromatography, target molecules, including silica gel, aluminum oxide, polyamides, and silver nitrates, can be separated according to their affinities for these adsorbent surfaces. The liquid–liquid

Table 2 Extraction methods of bioactive compounds and their applications

S. No.	Extraction method	Mechanism	Applications	References
1.	Solvent extraction	Classical analytical method for detecting different inorganic and organic compounds. complexation with organic ligands precedes the extraction of inorganic molecules	Beneficial components in an industry, such as petrochemicals, food beverages, and pharmaceuticals	Eshwari and Kannahi (2018)
2.	Soxhlet extraction	Employed when the component to be extracted has a restricted solubility in the selected solvent and the impurities are insoluble	Several bioactive compounds extracted from herbs, seeds, and plants, such as flavonoids, alkaloids, and essential oils Pharmaceutical industry	López-Bascón and De Castro (2020)
3.	Ultrasound-assisted (UAE) extraction	Bioactive compounds from water and ethanol water are utilized as solvents to extract a range of bioactive compounds utilizing ultrasound frequencies that span from 20 to 2000 kHz	Herbal and oil extraction Phytopharmaceutical extraction. Protein extraction. Bioactive extraction from animal and plant materials	Kumar et al. (2021)
4.	Microwave assisted extraction	(MAE) is the technique of separating analytical substances from a sample aggregate into the solvent by heating solvents while interacting with the sample using microwave energy. The key benefit of this technique is that MAE has an inherent capacity to quickly heat the sample solvent mixture	Analysis of the heavy metals and other contaminants found in the soil samples using a shorter test duration, less power, and less solvent A wide range of naturally occurring resources have been used to extract several biologically active compounds utilizing the MAE method	Tomasi et al. (2023)
5.	Enzyme assisted extraction	Plant cells of biomass are bound to an enzyme's active site, which modifies the enzyme's structure to accommodate the substrate in its active regions. As a result, the chemical compounds in the cells are liberated into the extraction medium	Extraction of phenolic compounds with high stability and high antioxidant activity for pharmaceutical and food applications Extraction of enzymes like cellulase, hemicellulase, pectinase, and proteases	Das et al. (2021)

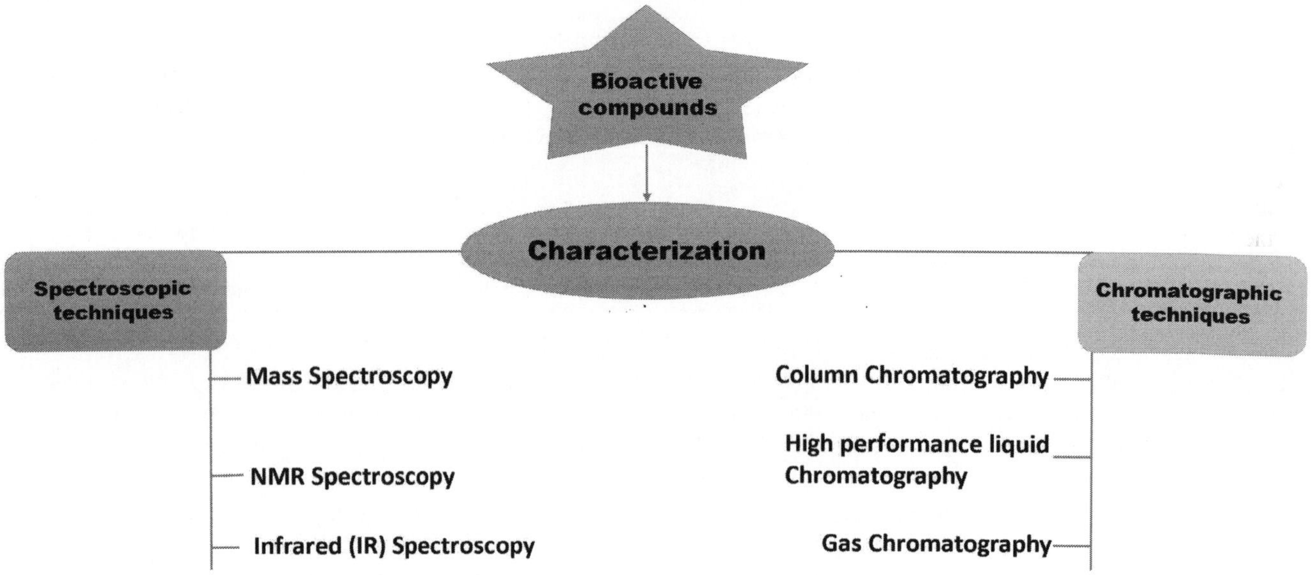

Fig. 3 Characterization techniques of bio-active compounds

extraction, in which one liquid phase is stationary and another is in the mobile phase, forms the basis for partition chromatography and counter-current chromatography. In addition, the bio-active substances are divided according to molecular size in the cases of membrane filtering and gel chromatography. The big molecular particles are held in place while the smaller molecules move through. When using gel chromatography, the bioactive substances are separated based on how long they stay in the solution. Ion-exchange chromatography, in the meantime, divides the bioactive substances according to their net surface charge. By modulating the ionic charge of the mobile phase, ion-exchange resin can either catch or release the molecules in this phase. Distillation, on the other hand, in the same direction, allows for the separation of high molecular weight or thermosensitive bio-active compounds (Mahato et al., 2019).

7.2.2 High-Performance Liquid Chromatography (HPLC)

High-performance liquid chromatography (HPLC) is a chromatographic technique that is robust, versatile, and widely used in analytical chemistry for the isolation of natural substances. It is used to identify, quantify, and purify the individual components of a mixture. This method is more and more popular among other analytical techniques as the primary option for fingerprinting studies for herbal plant quality control (Lozano-Sánchez et al., 2018). After a relatively crude extract is evaluated in a biological assay to completely describe its characteristics, natural components are often separated. Reversed-phase mode is used for 65% of all HPLC separations. The ease of use, adaptability, and breadth of the reversed-phase method which can identify secondary plant metabolites, for instance, and handle compounds with a wide range of polarity and molecular mass are some of the explanations for this (Prathap et al., 2013). Furthermore, “buffer” is the common term for the mobile phases in reversed-phase chromatography. However, because mobile phase solutions typically contain strong acids at low pH levels together with significant quantities of organic solvents, they have little ability to act as buffers. When operating in physiological settings, it is important to maintain an adequate buffering capacity (Prathap et al., 2013).

7.2.3 Gas Chromatography (GC)

Gas chromatography (GC) is a method of separation that has been employed for separating extremely complex mixtures, mostly based on polarity and changes in boiling point and vapor pressure (Falaki 2019). Gas chromatography separation takes place inside the column along with the mobile phase, and the sample containing various substances is loaded into the column. In gas phase separation, the mobile phase is a gas known as the carrier gas. The mobile phase and the sample both move through the column, but the frequency at which they do so varies according to the molecule. The times at which the various compounds arrive at the column. Consequently, there is a separation of each compound (Al-Rubaye et al., 2017).

8 Applications of Bioactive Compounds

8.1 Pharmaceutical Applications

Bioactive compounds found in nature have a biological effect on living organisms. They are often associated with health benefits in various applications, there are some common pharmaceutical applications of bio-active compounds (Lomartire and Gonçalves, 2022).

a. **Drug discovery and development**: Bio-active compounds are the first step in the production of new drugs. It is practical to extract them from their native habitats, which include flora, microorganisms, and marine life, and then study any potential therapeutic benefits. Two examples are the discovery of the anti-cancer drug paclitaxel from the Pacific yew tree and the anti-malarial drug artemisinin from the sweet wormwood plant (Dominguez et al., 2017).
b. **Therapeutic substances**: Bio-active compounds have a wide range of therapeutic applications (Rao and Narasimha, 2021). For example, the cannabis plant’s cannabinoids have been used to create drugs that suppress nausea, stimulate appetite, and control pain. These compounds, which are similar to curcumin found in turmeric, have anti-inflammatory and antioxidant properties that make them prospective treatments for a variety of illnesses (Ortega and Campos, 2019).
c. **Antibiotics**: Bioactive substances from marine microbes have antibacterial properties (Avilala Janardhan et al., 2019). For instance, penicillin, one of the earliest antibiotics ever discovered, comes from the *Penicillium* family. Other examples include tetracycline, which originates from the *Streptomyces* bacteria, and erythromycin, which is derived from the Saccharopolyspora erythraea bacteria (De Simeiset and Serra 2021).
d. **Cardiovascular Health**: Certain bioactive compounds, including the omega-3 fatty acids found in fish oil, have positive impacts on cardiovascular health. They can lessen the chance of developing heart disease by decreasing blood pressure, decreasing inflammation, and increasing cholesterol levels (Sindhu et al., 2021).
e. **Antioxidants**: Fruits and vegetables include flavonoids, which are among the many bioactive compounds with antioxidant qualities. According to Banerjee et al., they

support the body's defense against harmful free radicals that may otherwise damage cells and contribute to several types of diseases, including heart disease and cancer (Palithya et al., 2022).

f. **Neurological disorders**: Bioactive substances like polyphenols are found in large amounts in foods like tea, berries, and nuts show as neuroprotective agents. By defending neurons from damage and lowering inflammation in the brain, they may reduce the likelihood of developing neurodegenerative diseases, including Parkinson's and Alzheimer's (John et al., 2020).
g. **Anti-inflammatory Agents**: A small number of bioactive compounds have lower levels of inflammation, making them useful in the treatment of inflammatory conditions such as arthritis and inflammatory bowel disease. Two examples are quercetin from apples and onions and retractive iron from red wine (Wu et al., 2019).
h. **Immunomodulators**: Several bioactive compounds possess the capacity to either activate or suppress the immune system. For example, research has been done on the immunomodulatory properties and potential use of compounds made from mushrooms, such as beta-glucans, in cancer immunotherapy (Pathak et al., 2022).

8.2 Nutraceutical and Dietary Supplement Applications

Bioactive substances improve health and wellness; often, these products supply concentrated quantities of beneficial molecules present in natural sources. Some common applications of bioactive compounds found in nutraceuticals and dietary supplements are as follows:

a. **Vitamins and minerals**: Essential vitamins and minerals are bioactive compounds that are vital to many physiological functions. Nutraceuticals and nutritional supplements often contain concentrated forms of vitamins (C, D, and E) and minerals (including calcium, magnesium, and iron) to improve overall health and repair deficiencies (Li et al., 2023).
b. **Antioxidants:** The body uses bioactive compounds called antioxidants to counteract free radicals and reduce oxidative stress. Flavonoids, selenium, and vitamins C and E are examples of foods and supplements high in antioxidants that are believed to preserve cellular health, reduce the risk of chronic illnesses, and slow down the aging process.
c. **Omega-3 Fatty Acids**: The primary sources of these beneficial compounds, which comprise EPA (eicosapentaenoic acid) and DHA (docosahexaenoic acid), are fatty fish like salmon, mackerel, and sardines. Nutraceutical supplements containing omega-3 fatty acids are widely used to enhance cardiovascular health, reduce inflammation, improve cognitive function, and preserve healthy skin (Clark and Lands 2015).
d. **Probiotics and Prebiotics**: Prebiotics help the growth of good gut flora, whereas probiotics are live microorganisms that provide health benefits when consumed in large enough quantities. Nutraceuticals and dietary supplements contain probiotics (*lactobacillus* and *bifidobacterium*) and prebiotics (insulin and fructo oligosaccharides), which are used to support digestive health, improve gut microbiota balance, and strengthen the immune system (Reque and Brandelli, 2021).
e. **Herbal extracts**: These are made from medicinal plants and herbs and are high in bioactive substances. Nutraceutical supplements may contain herbal extracts from plants like garlic, echinacea, ginger, ginseng, and turmeric (*curcumin*). Numerous health benefits, such as immune support, anti-inflammatory qualities, and stress reduction, are believed to be offered by these extracts (Williamson et al., 2020).
f. **Polyphenols**: Fruits, vegetables, tea, coffee, chocolate, and other plant-based diets contain a variety of healthful chemicals called polyphenols. Nutraceuticals and dietary supplements, rich in polyphenols such as anthocyanins, resveratrol, and catechins, are well known for their anti-inflammatory, antioxidant, and cardiovascular health-promoting properties (Harun-Ur-Rashid and Krishna, 2024).
g. **Collagen**: Found in connective tissues, skin, and bones, collagen is a structural protein. Collagen peptides sourced from animal sources (marine or collagen) are found in nutraceutical supplements that enhance bone density, joint health, and skin and nail growth (Choi et al., 2019).
h. **Adaptogens**: The body uses adaptogens, which are bioactive substances present in some herbs and mushrooms, to maintain homeostasis and adjust to stress. It is thought that adaptogens such as Reishi mushroom, Rhodiola, holy basil, and ashwagandha, enhance adrenal health and lessen exhaustion from stress, increasing resistance to both physical and mental pressures (Gulati et al., 2016).

8.3 Agricultural Applications

Bioactive compounds have several applications in agriculture, contributing to plant growth, health, and protection against pests and diseases.

a. **Plant Growth Promoters**: Auxins, cytokinins, gibberellins, and abscisic acid are examples of bioactive compounds that are important in controlling the growth and development of plants. To improve agricultural yield and quality, they can be used exogenously as growth regulators to encourage fruit development, flowering, root and shoot growth, and seed germination (Malik et al., 2020).
b. **Biofertilizers**: To improve soil fertility and plant nutrient availability, bioactive substances generated from helpful microorganisms, such as phosphate-solubilizing bacteria (e.g., *Bacillus, Pseudomonas*) and nitrogen-fixing bacteria (e.g., *Rhizobium, Azospirillum*), can be employed as biofertilizers. By enhancing soil health and decreasing the demand for synthetic fertilizers, these biofertilizers support sustainable agriculture (Kadimpati and Golla, 2022).
c. **Bio-stimulants**: By promoting physiological processes in plants, bio-stimulants improve plant growth and stress tolerance. Plant growth-promoting rhizobacteria (PGPR), amino acids, seaweed extracts, and humic acids are examples of bioactive substances that are frequently employed as biostimulants to enhance crop stress resistance, nutrient uptake, root development, and seed germination (Singh et al., 2021).
d. **Biopesticides**: Insect pheromones, microbial metabolites, and plant extracts are examples of naturally occurring bioactive substances that can be utilized as biopesticides to manage diseases and pests in agricultural settings. Biopesticides are less harmful to non-target creatures and leave fewer pesticide residues in food products than synthetic pesticides, making them more environmentally benign than their synthetic counterparts (Rani et al., 2021).
e. **Allelopathy**: Allelopathy is the process by which some plants impede the growth of pests or nearby plants by releasing bioactive substances. To increase crop yield and lessen weed competition in agricultural systems, allelochemicals such as phenolic compounds, terpenoids, and alkaloids can be used for insect control, weed suppression, and allelopathic crop rotation (Jabran et al., 2015).
f. **Phytoremediation**: Using plants to eliminate, break down, or immobilize pollutants in soil and water is a technique that can be made easier by certain bioactive substances, such as Phyto-chelators and enzymes generated by hyperaccumulating plants. For the repair of contaminated agricultural areas' soil and water, phytoremediation provides an economical and environmentally friendly solution (Khan et al., 2020).
g. **Plant Défense Mechanisms**: Key components of plant defense against diseases and herbivores include bioactive substances produced by the plant, such as phytoalexins, pathogenesis-related (PR) proteins, and secondary metabolites (e.g., alkaloids, terpenoids, and phenolics). By comprehending and utilizing plant defense mechanisms, new approaches to crop protection and disease control in agriculture can be developed (de Souza Candido et al., 2011).

8.4 Cosmetics and Personal Care Products

Bioactive compounds are increasingly used in cosmetics and personal care products for their beneficial effects on skin, hair, and overall appearance.

a. **Antioxidants**: These are common ingredients in skincare and cosmetic products and include antioxidant molecules like coenzyme Q10, vitamins C and E, and polyphenols derived from botanical extracts (including pomegranate, grape seed, and green tea). By preventing oxidative stress, protecting the skin from environmental damage, and neutralizing free radicals, they help maintain youthful-looking skin and postpone the onset of aging (Soto et al., 2018).
b. **Moisturizers and hydrating agents**: Hyaluronic acid, glycerine, and ceramides are examples of bioactive substances that are utilized in skincare products as hydrating and moisturizing agents. Softer, smoother, and more supple skin is the result of their assistance in preserving skin hydration, enhancing skin barrier function, and halting moisture loss (Viseras et al., 2021).
c. **Anti-aging ingredients**: Retinoids (retinol, and retinyl palmitate), peptides, and botanical extracts (plant stem cells, and peptides) are examples of bioactive substances with anti-aging qualities that are frequently used for anti-aging creams, serums, and treatments. These components enhance the firmness, elasticity, and general texture of the skin while also lessening the visibility of wrinkles, fine lines, and age spots (Dhalaria et al., 2020).
d. **Skin Brightening agents**: Botanical extracts (kojic acid and licorice extract) and bioactive substances such as vitamin C, alpha hydroxy acids (AHAs), and beta hydroxy acids (BHAs) are utilized in skincare and cosmetic goods as skin brightening agents. They contribute to a more radiant, even-toned complexion by assisting in the exfoliation of dead skin cells, blocking the synthesis of melanin, and reducing hyperpigmentation (Viseras et al., 2021).
e. **Anti-inflammatory agents**: To relieve and calm irritated or sensitive skin, skincare products contain bioactive substances with anti-inflammatory qualities, such as chamomile extract, aloe vera gel, and niacinamide (vitamin B_3). People with acne, rosacea, or other

inflammatory skin diseases can benefit from these substances because they help lessen redness, inflammation, and skin sensitivity (Freitas et al., 2021).

f. **UV Filters and Sunscreens**: Widely used to provide broad-spectrum UV protection and fight against sunburn, premature aging, and skin cancer, sunscreens and sun protection products contain certain bioactive components, such as zinc oxide, titanium dioxide, and organic UV filters (avobenzone and octocrylene). These components assist in dispersing or absorbing dangerous UV radiation, shielding the skin from UV-induced damage (Viseras et al., 2021).
g. **Hair Care Ingredients**: The compounds are used to nourish, strengthen, and condition hair, hair care products contain bioactive substances such as proteins (keratin, silk amino acids), vitamins (biotin, panthenol), and botanical extracts (argan oil, coconut oil, and bamboo extract). They strengthen shine and luster, help repair damaged hair, and improve hair texture and manageability (Kumari et al., 2024).
h. **Natural and Organic Ingredients**: Bioactive compounds derived from botanical sources (plant extracts, and essential oils) are being employed more frequently as active ingredients in cosmetics and personal care products due to the growing consumer demand for natural and organic skincare products. Perceived advantages of these compounds include sustainability, compatibility with delicate skin, and purity (Savary et al., 2016).

9 Future Prospectives

a. **Personalized Nutrition and Health**: Personalized nutrition and health therapies that are catered to each person's unique genetic composition, lifestyle choices, and medical problems are becoming more and more popular because of technological breakthroughs in the omics sciences (genomics, metabolomics, etc.). Researchers are examining the potential of bioactive compounds in customized dietary advice and focused medical interventions.
b. **Microbiome modulation**: Research has proven that the gut microbiota has a significant influence on both health and illness. Prebiotics, probiotics, and postbiotics are among the bioactive substances whose capacity to alter the makeup and activity of the gut microbiota is being studied. This could have implications for immune system support, gut health, and metabolic balance.
c. **Nanotechnology applications**: Nanotechnology provides novel ways to improve the stability, distribution, and bioavailability of bioactive substances. To address issues with low solubility, stability, and bio-accessibility, researchers are investigating nanoencapsulation, nano-emulsions, and nanoscale delivery systems. This will increase the use of bioactive substances in functional foods, medications, and nutraceuticals.
d. **Plant-Based Functional Foods**: Interest in plant-derived bioactive substances with functional qualities is rising as consumers become more aware of the health advantages of plant-based diets. To address certain health issues and dietary preferences, research focuses on finding and characterizing bioactive components in fruits, vegetables, grains, and herbs, as well as creating novel plant-based functional foods and ingredients.
e. **Multi-Targeted Strategies**: A lot of chronic illnesses have intricated pathophysiological processes, necessitating the use of many targeted strategies. Polyphenols, flavonoids, and terpenoids are examples of bioactive compounds with pleiotropic effects that are being researched for their capacity to modify many cellular pathways and targets linked to the pathophysiology of illness, opening the door to novel treatment approaches.

10 Challenges

Bioavailability and Absorption Improving the bioavailability and absorption of bioactive compounds in the body is a significant challenge in the field of bioactive chemical research. Enhancing the solubility, stability, and delivery methods of bioactive compounds is crucial for optimizing their therapeutic benefits and guaranteeing their effectiveness in functional foods and supplements. Standardization and quality control for these bioactive substances to be commercialized and receive regulatory approval is essential to guarantee their uniformity, purity, and safety. For natural extracts and final products to contain bioactive compounds, their identity, potency, and purity must be confirmed using standardization procedures, quality control measures, and authentication techniques. Safety and toxicity although bioactive compounds have the potential to provide health benefits, a detailed assessment of their safety profiles and potential toxicities is necessary. To determine safe consumption levels and recommendations for consumer use, extensive research evaluating the safety, tolerability, and unfavourable impacts of bioactive substances is required. Regulatory hurdles accessing markets and developing new products might be difficult due to the disparate regulatory frameworks that control the use of bioactive compounds throughout the world. Innovation and commercialization in the field require the harmonization of rules, the clarification of recommendations, and the evidence-based validation of health claims. Sustainable Sourcing and production to reduce environmental impact, protect biodiversity, and guarantee ethical standards along the supply chain, sustainable sourcing and manufacture of bioactive compounds are crucial. Promoting fair trade policies, wild-gathering methods, and sustainable agriculture are essential to the long-term survival of the bioactive chemical industry.

11 Conclusion

Bioactive compounds are essential in various fields like medicine, agriculture, food science, and cosmetics due to their antioxidant, antimicrobial, anti-inflammatory, and anticancer properties. They are naturally produced in plants, microbes, and marine organisms. Technological advancements have made extraction, isolation, and characterization of these compounds more efficient and environmentally friendly. Modern extraction methods, such as supercritical fluid extraction, ultrasound-assisted extraction, and microwave-assisted extraction, have improved yield and purity. Advanced characterization techniques have enabled the precise identification of structure–activity relationships. Future research should focus on optimizing extraction and synthesis processes, improving bioavailability, and expanding applications while ensuring sustainable and ethical use of natural resources.

References

Aharoni, A., & Galili, G. (2011). Metabolic engineering of the plant primary–secondary metabolism interface. *Current Opinion in Biotechnology, 22*(2), 239–244.

Akhtar, M. S., Swamy, M. K., & Sinniah, U. R. (2019). Natural bio-active compounds. *Natural Bio-Active Compounds, 1*, 1–608.

Alagoz, Y., Gurkok, T., Zhang, B., & Unver, T. (2016). Manipulating the biosynthesis of bioactive compound alkaloids for next-generation metabolic engineering in opium poppy using CRISPR-Cas 9 genome editing technology. *Scientific Reports, 6* (1), 30910.

Al-Rubaye, A. F., Hameed, I. H., & Kadhim, M. J. (2017). A review: Uses of gas chromatography-mass spectrometry (GC-MS) technique for analysis of bioactive natural compounds of some plants. *International Journal of Toxicological and Pharmacological Research, 9*(1), 81–85.

Altemimi, A., Lakhssassi, N., Baharlouei, A., Watson, D. G., & Lightfoot, D. A. (2017). Phytochemicals: Extraction, isolation, and identification of bioactive compounds from plant extracts. *Plants, 6* (4), 42.

Ameen, F., AlNadhari, S., & Al-Homaidan, A. A. (2021). Marine microorganisms as an untapped source of bioactive compounds. *Saudi Journal of Biological Sciences, 28*(1), 224–231.

Avilala Janardhan, A. J., Kumar, A. P., Buddolla Viswanath, B. V., Saigopal, D. V. R., & Narasimha, G. (2014). *Production of bioactive compounds by actinomycetes and their antioxidant properties.*

Bajguz, A., Chmur, M., & Gruszka, D. (2020). Comprehensive overview of the brassinosteroid biosynthesis pathways: Substrates, products, inhibitors, and connections. *Frontiers in Plant Science, 11*, 535077.

Balawejder, M., Matłok, N., Piechowiak, T., Szostek, M., Kapusta, I., Niemiec, M., et al. (2022). The modification of substrate in the soilless cultivation of raspberries (*Rubus idaeus* L.) as a factor stimulating the biosynthesis of selected bioactive compounds in fruits. *Molecules, 28*(1), 118.

Bharadwaj, R., Kumar, S. R., Sharma, A., & Sathishkumar, R. (2021). Plant metabolic gene clusters: Evolution, organization, and their applications in synthetic biology. *Frontiers in Plant Science, 12*, 697318.

Boadella, D. (2015). *Lifestreams: An introduction to biosynthesis*. Routledge.

Carroll, A. R., Copp, B. R., Davis, R. A., Keyzers, R. A., & Prinsep, M. R. (2023). Marine natural products. *Natural product reports, 40*(2), 275–325.

Challinor, V. L., & Bode, H. B. (2015). Bioactive natural products from novel microbial sources. *Annals of the New York Academy of Sciences, 1354*(1), 82–97.

Chandel, N. S. (2021). Glycolysis. *Cold Spring Harbor Perspectives in Biology, 13*(5), a040535.

Choi, F. D., Sung, C. T., Juhasz, M. L., & Mesinkovsk, N. A. (2019). Oral collagen supplementation: A systematic review of dermatological applications. *Journal of Drugs in Dermatology, 18*(1), 9–16.

Clark, C., & Lands, B. (2015). Creating benefits from omega-3 functional foods and nutraceuticals. *Food and Nutrition Sciences, 6* (17), 1613–1623.

Das, S., Nadar, S. S., & Rathod, V. K. (2021). Integrated strategies for enzyme assisted extraction of bioactive molecules: A review. *International Journal of Biological Macromolecules, 191*, 899–917.

Davies, J., & Ryan, K. S. (2012). Introducing the parvome: Bioactive compounds in the microbial world. *ACS Chemical Biology, 7*(2), 252–259.

De Simeis, D., & Serra, S. (2021). Actinomycetes: A never-ending source of bioactive compounds—an overview on antibiotics production. *Antibiotics, 10*(5), 483.

de Souza Cândido, E., Pinto, M. F. S., Pelegrini, P. B., Lima, T. B., Silva, O. N., Pogue, R., Grossi-de-Sá, M. F., & Franco, O. L. (2011). Plant storage proteins with antimicrobial activity: Novel insights into plant defense mechanisms. *The FASEB Journal, 25* (10), 3290–3305.

Déniel, M., Puspita, M., Douzenel, P., Stiger-Pouvreau, V., Bedoux, G., Bourgougnon, N., & Vandanjon, L. (2017). Seasonal variation of *Sargassum Muticum* biochemical composition determined by Fourier transform infrared spectroscopy. *Journal of Analytical, Bioanalytical and Separation Techniques, 2*(1), 75–84.

Dhalaria, R., Verma, R., Kumar, D., Puri, S., Tapwal, A., Kumar, V., Nepovimova, E., & Kuca, K. (2020). Bioactive compounds of edible fruits with their anti-aging properties: A comprehensive review to prolong human life. *Antioxidants, 9*(11), 1123.

Dhaniaputri, R., Suwono, H., Amin, M., & Lukiati, B. (2022, May). Introduction to plant metabolism, secondary metabolites biosynthetic pathway, and in-silico molecular docking for determination of plant medicinal compounds: An overview. In *7th International Conference on Biological Science (ICBS 2021)* (pp. 373–382). Atlantis Press.

Dijkstra, P., Thomas, S. C., Heinrich, P. L., Koch, G. W., Schwartz, E., & Hungate, B. A. (2011). Effect of temperature on the metabolic activity of intact microbial communities: Evidence for altered metabolic pathway activity but not for increased maintenance respiration and reduced carbon use efficiency. *Soil Biology and Biochemistry, 43*(10), 2023–2031.

Dominguez, H., & Muñoz, M. J. G. (Eds.). (2017). *Water extraction of bioactive compounds: From plants to drug development*. Elsevier.

Eshwari, N. T., & Kannahi, M. (2018). Characterization and partial purification of bioactive compounds produced from marine microbes. *International Journal for Advance Research and Development, 3*(3), 57–67.

Falaki, F. (2019). Sample preparation techniques for gas chromatography. In *Gas chromatography-derivatization, sample preparation, application*. IntechOpen.

Freitas, L. M., Antunes, F. T. T., Obach, E. S., Correa, A. P., Wiiland, E., de Mello Feliciano, L., et al. (2021). Anti-inflammatory effects of a topical emulsion containing Helianthus annuus oil, glycerin, and vitamin B3 in mice. *Journal of Pharmaceutical Investigation, 51*, 223–232.

Goyal, J., & Verma, P. K. (2023). An overview of the biosynthetic pathway and therapeutic potential of rutin. *Mini Reviews in Medicinal Chemistry, 23*(14), 1451–1460.

Gulati, K., Anand, R., Ray, A. (2016). Nutraceuticals as adaptogens: their role in health and disease. In *Nutraceuticals* (pp. 193–205). Academic Press.

Gupta, V., Estrada, A. D., Blakley, I., Reid, R., Patel, K., Meyer, M. D., et al. (2015). RNA-Seq analysis and annotation of a draft blueberry genome assembly identifies candidate genes involved in fruit ripening, biosynthesis of bioactive compounds, and stage-specific alternative splicing. *Gigascience, 4*(1), s13742–s14015.

Hameed, I. H., Hamza, L. F., & Kamal, S. A. (2015). Analysis of bioactive chemical compounds of *Aspergillus niger* by using gas chromatography-mass spectrometry and Fourier-transform infrared spectroscopy. *Journal of Pharmacognosy and Phytotherapy, 7*(8), 132–163.

Harun-Ur-Rashid, M., & Krishna, S. B. N. (2024). Narasimha Golla*, Abu Bin Imran* 12 polyphenol oxidases: The future toward global sustainability. *Polyphenol Oxidases: Function, Wastewater Remediation, and Biosensors*, 309.

Harwood, J. L. (2020). Fatty acid biosynthesis. In *Plant lipids* (pp. 27–66). Blackwell.

Ivanović, M., Islamčević Razboršek, M., & Kolar, M. (2020). Innovative extraction techniques using deep eutectic solvents and analytical methods for the isolation and characterization of natural bioactive compounds from plant material. *Plants, 9*(11), 1428.

Jabran, K., Mahajan, G., Sardana, V., & Chauhan, B. S. (2015). Allelopathy for weed control in agricultural systems. *Crop Protection, 72*, 57–65.

Janardhan, A., Kumar, A. P., Viswanath, B., Saigopal, D. V. R., & Narasimha, G. (2014). Production of bioactive compounds by actinomycetes and their antioxidant properties. *Biotechnology Research International*, 2014.

Jimenez-Garcia, S. N., Vazquez-Cruz, M. A., Guevara-Gonzalez, R. G., Torres-Pacheco, I., Cruz-Hernandez, A., & Feregrino-Perez, A. A. (2013). Current approaches for enhanced expression of secondary metabolites as bioactive compounds in plants for agronomic and human health purposes-A review. *Polish Journal of Food and Nutrition Sciences, 63*(2).

John, T., Samuel, B., Abolaji, O., Folashade, O., Oyetooke, A., & Oluwatosin, F. (2020). Functional foods and bioactive compounds: Roles in the prevention, treatment, and management of neurodegenerative diseases. *GSC Biological and Pharmaceutical Sciences, 11*(2), 297–313.

Johnson, J. B., Walsh, K. B., Naiker, M., & Ameer, K. (2023). The use of infrared spectroscopy for the quantification of bioactive compounds in food: A review. *Molecules, 28*(7), 3215.

Kadimpati, K. K., & Golla, N. (2022). Cost-effective biogenic production of inorganic nanoparticles, characterizations, and their antimicrobial properties. In *Cost effective technologies for solid waste and wastewater treatment* (pp. 265–290). Elsevier.

Khan, M. I., Cheema, S. A., Anum, S., Niazi, N. K., Azam, M., Bashir, S., et al. (2020). Phytoremediation of agricultural pollutants. *Phytoremediation: In-situ applications* (pp. 27–81).

Kiuru, P., D'Auria, M. V., Muller, C. D., Tammela, P., Vuorela, H., & Yli-Kauhaluoma, J. (2014). Exploring marine resources for bioactive compounds. *Planta medica, 80*(14), 1234–1246.

Kour, D., Rana, K. L., Kaur, T., Singh, B., Chauhan, V. S., Kumar, A., et al. (2019). Extremophiles for hydrolytic enzymes productions: Biodiversity and potential biotechnological applications. *Bioprocessing for Biomolecules Production*, 321–372.

Kumar, K., Srivastav, S., & Sharanagat, V. S. (2021). Ultrasound-assisted extraction (UAE) of bioactive compounds from fruit and vegetable processing by-products: A review. *Ultrasonics Sonochemistry, 70*, 105325.

Kumari, B., Rathi, J., Kumar, S., Singh, Y., & Kumar, S. (2024). Medicinal role of cosmetic products in beautifying of skin. *Research Journal of Pharmacy and Technology, 17*(1), 433–438.

Li, Y. O., González, V. P. D., & Diosady, L. L. (2023). Microencapsulation of vitamins, minerals, and nutraceuticals for food applications. In *Microencapsulation in the food industry* (pp. 507–528). Academic Press.

Liu, W., Feng, Y., Yu, S., Fan, Z., Li, X., Li, J., et al. (2021). The flavonoid biosynthesis network in plants. *International Journal of Molecular Sciences, 22*(23), 12824.

Lomartire, S., & Gonçalves, A. M. (2022). An overview of potential seaweed-derived bioactive compounds for pharmaceutical applications. *Marine Drugs, 20*(2), 141.

López-Bascón, M. A., De Castro, M. L. (2020). Soxhlet extraction. In *Liquid-phase extraction* (pp. 327–354). Elsevier.

Lozano-Sánchez, J., Borrás-Linares, I., Sass-Kiss, A., & Segura-Carretero, A. (2018). Chromatographic technique: High-performance liquid chromatography (HPLC). In *Modern techniques for food authentication* (pp. 459–526). Academic Press.

Luo, C., Xu, X., Wei, X., Feng, W., Huang, H., Liu, H., et al. (2019). Natural medicines for the treatment of fatigue: Bioactive components, pharmacology, and mechanisms. *Pharmacological Research, 148*, 104409.

Ma, Y. H., Dou, X. X., & Tian, X. H. (2020). Natural sesquiterpenoids: An overview of their chemical structures, pharmacological activities, and biosynthetic pathways. *Phytochemistry Reviews, 19*, 983–1043.

Mahato, N., Sinha, M., Sharma, K., Koteswararao, R., & Cho, M. H. (2019). Modern extraction and purification techniques for obtaining high-purity food-grade bioactive compounds and value-added co-products from citrus wastes. *Foods, 8*(11), 523.

Malik, A., Mor, V. S., Tokas, J., Punia, H., Malik, S., Malik, K., et al. (2020). Biostimulant-treated seedlings under sustainable agriculture: A global perspective facing climate change. *Agronomy, 11*(1), 14.

Manach, C., Milenkovic, D., Van de Wiele, T., Rodriguez-Mateos, A., de Roos, B., Garcia-Conesa, M. T., et al. (2017). Addressing the inter-individual variation in response to consumption of plant food bio-actives: Towards a better understanding of their role in healthy aging and cardiometabolic risk reduction. *Molecular Nutrition & Food Research, 61*(6), 1600557.

Motolinía-Alcántara, E. A., Castillo-Araiza, C. O., Rodríguez-Monroy, M., Román-Guerrero, A., & Cruz-Sosa, F. (2021). Engineering considerations to produce bioactive compounds from plant cell suspension culture in bioreactors. *Plants, 10*(12), 2762.

Ortega, A. M. M., & Campos, M. R. S. (2019). Bioactive compounds as therapeutic alternatives. In *Bioactive compounds* (pp. 247–264). Woodhead Publishing.

Palithya, S., Gaddam, S. A., Kotakadi, V. S., Penchalaneni, J., Golla, N., Krishna, S. B. N., et al. (2022). Green synthesis of silver

nanoparticles using flower extracts of *Aerva lanata* and their biomedical applications. *Particulate Science and Technology, 40*(1), 84–96.
Parte, S., Sirisha, V. L., & D'Souza, J. S. (2017). Biotechnological applications of marine enzymes from algae, bacteria, fungi, and sponges. In *Advances in food and nutrition research* (Vol. 80, pp. 75–106). Academic Press.
Pathak, M. P., Pathak, K., Saikia, R., Gogoi, U., Ahmad, M. Z., Patowary, P., et al. (2022). Immunomodulatory effect of mushrooms and their bioactive compounds in cancer: A comprehensive review. *Biomedicine & Pharmacotherapy, 149*, 112901.
Peracchi, A. (2018). The limits of enzyme specificity and the evolution of metabolism. *Trends in Biochemical Sciences, 43*(12), 984–996.
Prathap, B., Dey, A., Srinivasa Rao, G. H., Johnson, P., & Arthanariswaran, P. (2013). A review importance of RP-HPLC in analytical method development. *IJNTPS, 3*(1), 15–23.
Qian-Wen, C., Xiao, Y., Xiao-Qian, L., Yao-Hua, L., Wei-Hong, F., Chun, L., et al. (2022). Alkaloids from Pic*rasma quassioides*: An overview of their NMR data, biosynthetic pathways and pharmacological effects. *Phytochemistry, 193*, 112987.
Rani, K., Kaur, H., & Wati, L. (2021). Actinobacteria mediated nanoparticles: A pioneering technology for agriculture. *Journal of Nanoscience and Nanotechnology, 21*(6), 3611–3621.
Rao, Y. J., & Narasimha, G. (2021). Isolation and screening of marine actinobacteria for their antimicrobial compounds. *Pharmacognosy Research, 13*(1).
Reque, P. M., & Brandelli, A. (2021). Encapsulation of probiotics and nutraceuticals: Applications in functional food industry. *Trends in Food Science & Technology, 114*, 1–10.
Rizzo, P., Altschmied, L., Ravindran, B. M., Rutten, T., & D'Auria, J. C. (2020). The biochemical and genetic basis for the biosynthesis of bioactive compounds in Hypericum perforatum L., one of the largest medicinal crops in Europe. *Genes, 11*(10), 1210.
Salwan, R., & Sharma, V. (2020). Bioactive compounds of streptomyces: Biosynthesis to applications. *Studies in Natural Products Chemistry, 64*, 467–491.
Samyn, P., Meftahi, A., Geravand, S. A., Heravi, M. E. M., Najarzadeh, H., Sabery, M. S. K., et al. (2023). Opportunities for bacterial nanocellulose in biomedical applications: Review on biosynthesis, modification and challenges. *International Journal of Biological Macromolecules, 231*, 123316.
Santi, I., Carrari, E., Gasperini, C., Cabrucci, M., & Selvi, F. (2023). Impact of active coppice management on microclimate and understorey vegetation in a Mediterranean oak forest. In *118° Congresso della Società Botanica Italiana IX international plant science conference (ipsc) pisa, 13–16 September 2023* (pp. 1–252). SBI.
Santos, C. M., & Silva, A. M. (2014). Nuclear magnetic resonance spectroscopy for structural characterization of bioactive compounds. In *Comprehensive Analytical Chemistry* (Vol. 65, pp. 149–191). Elsevier.
Savary, G., Grisel, M., & Picard, C. (2016). Cosmetics and personal care products. *Natural Polymers: Industry Techniques and Applications*, 219–261.
Septembre-Malaterre, A., Remize, F., & Poucheret, P. (2018). Fruits and vegetables, as a source of nutritional compounds and phytochemicals: Changes in bioactive compounds during lactic fermentation. *Food Research International, 104*, 86–99.
Sharma, M., Koul, A., Sharma, D., Kaul, S., Swamy, M. K., & Dhar, M. K. (2019). Metabolic engineering strategies for enhancing the production of bio-active compounds from medicinal plants. *Natural Bio-Active Compounds: Volume 3: Biotechnology, Bioengineering, and Molecular Approaches*, 287–316.
Sindhu, R. K., Goyal, A., Algın Yapar, E., & Cavalu, S. (2021). Bioactive compounds and nano delivery perspectives for the treatment of cardiovascular diseases. *Applied Sciences, 11*(22), 11031.
Singh, R., Das, R., Sangwan, S., Rohatgi, B., Khanam, R., Peera, S. P. G., Das, S., et al. (2021). Utilization of agro-industrial waste for sustainable green production: A review. *Environmental Sustainability, 4*(4), 619–636.
Singh, M., Kumar, A., Singh, R., & Pandey, K. D. (2017). Endophytic bacteria: A new source of bioactive compounds. *3 Biotech, 7*, 1–14.
Soto, M. L., Parada, M., Falqué, E., & Domínguez, H. (2018). Personal-care products are formulated with natural antioxidant extracts. *Cosmetics, 5*(1), 13.
Szakiel, A., Pączkowski, C., & Henry, M. (2011). Influence of environmental biotic factors on the content of saponins in plants. *Phytochemistry Reviews, 10*, 493–502.
Tomasi, I. T., Santos, S. C., Boaventura, R. A., & Botelho, C. M. (2023). Optimization of microwave-assisted extraction of phenolic compounds from chestnut processing waste using response surface methodology. *Journal of Cleaner Production, 395*, 136452.
Toscano, S., Trivellini, A., Cocetta, G., Bulgari, R., Francini, A., Romano, D., et al. (2019). Effect of preharvest abiotic stresses on the accumulation of bioactive compounds in horticultural produce. *Frontiers in Plant Science, 10*, 468818.
Tran, P. N., Yen, M. R., Chiang, C. Y., Lin, H. C., & Chen, P. Y. (2019). Detecting and prioritizing biosynthetic gene clusters for bioactive compounds in bacteria and fungi. *Applied Microbiology and Biotechnology, 103*, 3277–3287.
Twaij, B. M., & Hasan, M. N. (2022). Bioactive secondary metabolites from plant sources: Types, synthesis, and their therapeutic uses. *International Journal of Plant Biology, 13*(1), 4–14.
Vinogradov, A. A., & Suga, H. (2020). Introduction to thiopeptides: Biological activity, biosynthesis, and strategies for functional reprogramming. *Cell Chemical Biology, 27*(8), 1032–1051.
Viseras, C., Sánchez-Espejo, R., Palumbo, R., Liccardi, N., García--Villén, F., Borrego-Sánchez, A., et al. (2021). Clays in cosmetics and personal-care products. *Clays and Clay Minerals, 69*(5), 561–575.
Weissman, K. J. (2015). The structural biology of biosynthetic mega enzymes. *Nature Chemical Biology, 11*(9), 660–670.
Williamson, E. M., Liu, X., & Izzo, A. A. (2020). Trends in use, pharmacology, and clinical applications of emerging herbal nutraceuticals. *British Journal of Pharmacology, 177*(6), 1227–1240.
Wong, F. T., & Khosla, C. (2012). Combinatorial biosynthesis of polyketides—a perspective. *Current Opinion in Chemical Biology, 16*(1–2), 117–123.
Wu, M. S., Aquino, L. B. B., Barbaza, M. Y. U., Hsieh, C. L., De Castro-Cruz, K. A., Yang, L. L., et al. (2019). Anti-inflammatory and anticancer properties of bioactive compounds from *Sesamum indicum* L.—A review. *Molecules, 24*(24), 4426.
Xu, Y., Geng, L., & Zhao, S. (2018). Biosynthesis of bioactive ingredients of Salvia miltiorrhiza and advanced biotechnologies for their production. *Biotechnology & Biotechnological Equipment, 32*(6), 1367–1377.
Yang, L., Yang, C., Li, C., Zhao, Q., Liu, L., Fang, X., et al. (2016). Recent advances in biosynthesis of bioactive compounds in traditional Chinese medicinal plants. *Science Bulletin, 61*(1), 3–17.
Zepka, L. Q., Do Nascimento, T. C., & Jacob-Lopes, E. (Eds.). (2021). *Bioactive compounds: Biosynthesis, characterization and applications*. BoD–Books on Demand.
Zhan, X., Chen, Z., Chen, R., & Shen, C. (2022). Environmental and genetic factors involved in plant protection-associated secondary metabolite biosynthesis pathways. *Frontiers in Plant Science, 13*, 877304.
Zhang, Q. W., Lin, L. G., & Ye, W. C. (2018). Techniques for extraction and isolation of natural products: A comprehensive review. *Chinese Medicine, 13*, 1–26.

Bioactive Molecules of Microbial Origin

Devargya Ganguly, Sneha Sengupta, Soumajit Ghosh, Subhojit Sengupta, and Gargi Goswami

Abstract

Microorganisms produce complex biomolecules with unique properties to help them survive in unfavorable environmental conditions. Secondary metabolites are natural products, produced by a wide range of bacteria, cyanobacteria, fungi, and microalgae. These natural products are rich in various biologically active molecules with varying chemical structures and possess antimicrobial, anticancer, and anti-inflammatory properties. Many of these compounds have been developed into potential therapeutics for human diseases and have found applications in relevant industries. This book chapter provides a detailed overview of past and ongoing studies with reports of promising microorganisms which can serve as cell factories to generate biologically active molecules.

Keywords

Microorganisms • Bioactive molecules • Antimicrobial properties • Anticancer compounds • Anti-inflammatory molecules • Carotenoids

1 Introduction

The existence of a diverse microbial population and their ability to produce unique bioactive molecules have innovatively directed novel biotechnological research. In general, biologically active molecules produced by microorganisms are referred to as secondary metabolites. These biomolecules, when introduced into the biological system, may have either a positive or a negative impact. Majority of the microbial products are toxic or have side effects, making them unsuitable for use in clinical practices. Nevertheless, microorganisms have also been reported to produce a wide range of substances that are extensively being used in human and animal medicines (Běhal, 2006).

Overuse of antibiotics have escalated the prevalence of Multi Drug Resistant (MDR) microorganisms. Hence, novel or underexploited biomolecules of microbial origin can be potent against the drug-resistant pathogens due to their restricted usage (Rani et al., 2021a). A greater part of modern-day research focuses on bacteria and fungi as possible sources of new antimicrobial compounds. These bioactive molecules have been reported to be effective against viral infections like HIV, pathogenic bacterial and fungal infections (Chaisuwan et al., 2021; Frediansyah et al., 2022; Kardani and Bolhassani, 2021; Monciardini et al., 2014; Stincone and Brandelli, 2020).

Conversely, antioxidant-rich bioactive molecules are also produced by microorganisms. Antioxidants possess the ability to scavenge free radicals thereby reducing oxidative stress and improving human health by minimizing cellular damage (Chandra et al., 2020) and thus may also be used as a therapeutic alternative as anticancer agents (Kardani and Bolhassani, 2021). The applications of microbiological bioactive molecules are widespread owing to their antioxidant properties, antimicrobial properties, anticancer properties, and pigment production (Demain, 2014; Rani et al., 2021b) and involve the food, cosmetic, and pharmaceutical industries (Rani et al., 2021b). This book chapter aims to deliver a comprehensive overview of the different bioactive molecules, reported from microorganisms, that exhibit health benefitting properties and have extensive industrial applications.

D. Ganguly · S. Sengupta · S. Ghosh · S. Sengupta · G. Goswami (✉)
Department of Life Sciences, GITAM School of Science, Gandhi Institute of Technology and Management (Deemed to be University), Visakhapatnam, Andhra Pradesh, India
e-mail: ggoswami@gitam.edu; gargigoswami.83@gmail.com

J. Devi (ed.), *Biotechnological Intervention in Production of Bioactive Compounds*, Sustainable Landscape Planning and Natural Resources Management, https://doi.org/10.1007/978-3-031-76859-0_2

2 Industrial Applications and Health Benefitting Aspects of Bioactive Molecules

Owing to their health benefitting potential, the bioactive molecules are generally consumed as a nutrient supplement, together with food (Bansal et al., 2023). These bioactive molecules have the ability to influence the metabolic system of the consumer through quenching free radicals by antioxidants or inhibiting infections by anti-microbials (Correia et al., 2012). Also, in the present scenario, the anti-inflammatory, anti-carcinogenic, anti-mutagenic, and anti-allergenic activities demonstrated by these molecules enhance the necessity for their consumption, as a protective measure for disease prevention (Ham et al., 2009; Parvathy et al., 2009).

2.1 Antioxidants

Free radicals are involved in immune responses, cell signaling, and play a major role in cell apoptosis. However, free radicals can also catalyze carcinogenesis, cardiovascular diseases, and aging (Birch-Machin and Bowman, 2016). Oxidation reactions generate free radicals that may trigger a chain reaction eventually leading to cell damage. Antioxidants slow down or prevent the oxidation reactions by eliminating or stabilizing free radicals and eventually avoiding chain reactions (Chandra et al., 2020; Valentão et al., 2002). Prolonged storage and processing of packaged food can generate free radicals thereby compromising with the food quality (Bansal et al., 2023). Antioxidants find potential applications in the food and packaging industry as they have been reported to quench reactive oxygen species (ROS), chelate molecules, and break chain reactions (Frankel and Meyer, 2000).

2.2 Anti-microbials

Increase in the abundance of antimicrobial resistant microorganisms have rendered treatment of microbial infections difficult. The effectualness of the existing antimicrobial molecules are questionable owing to the copiousness of such antimicrobial resistant organisms (Laws et al., 2019). Microbes have been reported to produce antimicrobials such as antibacterial, antifungal, antiviral, and cytotoxic molecules (Amaning Danquah et al., 2022) and thus serve as a reservoir of antimicrobial agents which are yet to be exploited to their full potential.

2.3 Carotenoids

Carotenoids are produced by microorganisms such as bacteria, fungi, and algae. Lutein and zeaxanthin are produced and accumulated in humans as well (Cezare-Gomes et al., 2019). Microbial pigments of the colour red to yellow are supplemented as nutrients for human consumption. Microorganisms generally produce and accumulate carotenoids as a counter to stress response (Kirti et al., 2014). Due to their antioxidant and protective properties against cellular damage, these high-value metabolites have found widespread use in cosmetic, pharmaceutical, and nutraceutical industries (Galasso et al., 2017).

2.4 Fatty Acids

The fatty acid composition of a cell membrane determines its functionality and fluidity. The fluidity of the cell membrane influences the proteins present in the cell membrane and their movement (Calder, 2015). Saturated fatty acids like palmitic acid and myristic acid are involved in the acylation of proteins. Polyunsaturated fatty acids like arachidonic acid are a precursor for eicosanoids. Essential fatty acids, like docosahexaenoic acid (DHA), are produced in very limited quantities and are mainly supplemented through diet. DHA helps in cell signaling and development of neural cells (Sun et al., 2018). Fatty acids are also involved in regulating the expression and activity of several transcription factors (Calder, 2015).

3 Bioactive Compounds from Bacteria

Bacteria act as indigenous sources of various secondary metabolites, known to have a significant degree of bioactivity against other microorganisms, thus enhancing their chances of survival (Bhatnagar and Kim, 2010). These metabolites, commonly referred to as bioactive molecules, may have antimicrobial, anticancer, and antioxidant properties. They might also act as immuno-suppressants, toxins, or enzymes.

3.1 Antimicrobial Properties

Increase in the number of drug-resistant pathogens such as multi-drug resistant strains of *Staphylococcus aureus* (MRSA) and recurring mycobacterial infections as well as persistent diseases caused by viral agents such as HIV, Ebola, and Hanta virus have necessitated the need for novel drugs (Bérdy, 2005). Bacteria produce secondary metabolites to protect themselves against other microorganisms. These molecules can potentially be used owing to their antimicrobial properties (Srinivasan et al., 2021). Clinically significant bioactive molecules with antimicrobial activity generally belong to the classes β-Lactam, Glycopeptide, Macrolide, Lincosamide, Streptogramin, Lipopeptide, and Cationic peptides (Wright, 2014).

3.1.1 Antibacterial Compounds

Polyketide compounds with antibacterial activity such as Pikromycin from *Streptomyces venezuelae* (Xue and Sherman, 2001), Erythromycin A from *Saccharopolyspora erythraea* (Chng et al., 2008), Tylosin A from *Streptomyces fradiae* (Cundliffe et al., 2001), and Spiramycin from *Streptomyces ambofaciens* (Karray et al., 2007) have been reported to work against both gram-positive and gram-negative bacteria like *Haemophilus influenzae*, *Staphylococcus aureus*, *Streptococcus pyogenes*, *Streptococcus pneumoniae*, *Legionella pneumophila*, *Neisseria gonorrhoeae*, and *Mycoplasma pneumonia*. The compounds belonging to the polyketide class inhibit protein biosynthesis by interacting with the larger 50S ribosomal subunit (Robertsen and Musiol-Kroll, 2019). Another bioactive antibacterial polyketide molecule, Anthracimycin obtained from *Streptomyces* sp. strain CNH365, has been reported to effectively inhibit Gram-negative bacteria like *Haemophilus influenzae*, *Burkholderia thailandensis*, *Escherichia coli*, *Pseudomonas aeruginosa* and against Gram-positive Methicillin-Resistant *Staphylococcus aureus* (MRSA) and *Bacillus anthracis*. Metabolic labeling experiments have revealed Anthracimycin to be involved in the inhibition of DNA/RNA synthesis (Jang et al., 2013). However, the smaller ribosomal subunit (30S) is inhibited by the action of Aminoglycosides (Długosz and Trylska, 2009) such as Streptomycin, Kanamycin, Gentamicin, Sisomicin, and Lividomycin. These aminoglycosides are used against Gram-negative bacterial infections (Pham et al., 2019; Ryeol Park et al., 2013).

Allocyclinones or hyperchlorinated angucyclinones have reportedly been isolated from *Actinoallomurus* sp. ID145698 and have demonstrated to be effective against *Staphylococcus aureus*, *Streptococcus pyogenes*, and *Enterococcus faecalis*. Four different types of allocyclinones have been reported till date (Cruz et al., 2017). These allocyclinones differ in the number of chlorine atoms.

3.1.2 Antifungal Compounds

Prevalence of multiple drug-resistant fungal pathogens like *Candida albicans* has encouraged the search for novel antifungal agents (Lee et al., 2020). Rustmicin, an antifungal agent obtained from *Micromonospora* sp., inhibits the biosynthesis of sphingolipids (Wang et al., 2023) thereby inhibiting the growth of fungal species like *Candida albicans*, *Candida tropicalis*, *Cryptococcus neoformans*, and *Aspergillus fumigatus* (Cruz and Wuest, 2023).

Bacillus licheniformis 09IDYM23-derived Glycolipids, Ieodoglucomide C, and Ieodoglycolipid have the ability to form pores and channels on the membrane thus compromising with the membrane integrity and permeability and eventually resulting in the death of pathogenic fungal strains (Inès and Dhouha, 2015). Their inhibitory effect on the growth of fungal species like *Candida albicans*, *Rhizoctonia solani*, *Aspergillus niger*, *Botrytis cenerea* and *Colletotrichum acutatum* has been reported by Tareq et al. (2015).

3.1.3 Antiviral Compounds

Abyssomicin, a polyketide produced by the marine bacterium *Verrucosispora* sp. MS100137 isolated from the South China Sea, can inhibit the para-aminobenzoic acid synthesis pathway—an essential cofactor in folic acid biosynthesis. Four variants of the polyketide have been reported to exhibit inhibitory effects against the influenza A virus. Three of the four abyssomicins may serve as prospective candidates for combating influenza A (H1N1) virus as suggested by the study conducted by Zhang et al. (2020).

3.2 Anticancer Compounds

Anthraquinone derivatives—Marmycins A and B, obtained from the marine *Streptomyces* strain CHN990, have demonstrated cytotoxicity against the human colon carcinoma cell line, HCT-116. Marmycin A has been reported to be more effective compared to Marmycin B (Martin et al., 2007). The cytotoxic potential of the pentacyclic C-Glycosides was also tested against 12 human tumor cell lines including prostate, colon, lung, breast, and leukemia. Upon further study on the action mechanism of Marmycin A, it was revealed that the anthraquinone derivative accumulates in the lysosome, triggering an abrupt release of metal ions and acidic contents. This causes the activation of the caspase-dependent cell death pathway (Cañeque et al., 2015). Salinosporamide A, a γ-lactam-β-lactone product of *Salinispora tropica* strain CNB-392, has also demonstrated cytotoxic activity against human colon carcinoma cell lines. Tests against Non-Small Cell Lung Cancer (NCI-H226), Melanoma (SK-MEL-28), Breast Cancer (MDA-MB-435), and CNS Cancer (SF-539) cell lines have also confirmed notable efficacy (Feling et al., 2003). Salinosporamide is a potent inhibitor of NF-κβ, which in turn inhibits various transcription factors crucial for the survival of cancer cells. Also, salinosporamide is a proteasome inhibitor and can induce apoptosis in multiple myeloma cells (Chauhan et al., 2005). These promising anticancer characteristics of Salinosporamide A have encouraged Nereus Pharmaceuticals towards the development of a novel drug NPI-0052 (Gulder and Moore, 2010).

A cyclic lipopeptide with biosurfactant properties, Pseudofactin II, was obtained from *Pseudomonas fluorescens* BD5 (Janek et al., 2010). Psudofactin II has been reported to promote caspase-3-dependent apoptotic activity in melanoma A375 cells. On interaction with the lipid bilayer, PFII induces the formation of micelles thereby destabilizing the membrane integrity and leading to a sudden surge in the Ca^{2+} influx, which eventually activates apoptosis (Janek et al., 2013). Anti-cancer properties have also been demonstrated by Chalcomycins, a macrolide collected from different *Streptomyces* sp. These molecules have been reported to show activity against human cervical carcinoma cell line owing to their ability to inhibit protein synthesis (Wu et al., 2007).

3.3 Toxins

Bacterial toxins, also known as bacteriocins, can be divided into two types: endotoxins and exotoxins. Endotoxins are lipopolysaccharides of Gram-negative bacteria while exotoxins are peptides synthesized by ribosomes (Baindara and Mandal, 2020). These toxins may serve as antimicrobial agents and anticancer molecules. Toxins function in three distinct pathways: (i) some toxins cause damage to cellular membranes; (ii) others target cells by altering or modifying their secondary messengers; and (iii) some toxins may also inhibit protein synthesis (Oswald et al., 2005).

Streptococcus bovis HC5 synthesizes bovicin HC5 which belongs to the class of lantibiotics and is synthesized by ribosomes. This antimicrobial peptide exhibits efficacy against foodborne pathogens and prevents infectious diseases in humans. The cytotoxic potential of bovicin has been tested against human breast adenocarcinoma cell line MCF-7, Vero cells (monkey kidney epithelial cell line), and human liver hepatocellular carcinoma cell line HepG2. Generally, lantibiotics function by targeting the cell membranes (Paiva et al., 2012). Examples of lantibiotics include Epidermin—collected from *Staphylococcus epidermidis* Tu3298 and Variacin 8—obtained from *Micrococcus varians* MCV8 (McAuliffe et al., 2001). Laterosporulin is a defensin-type bacteriocin produced by *Brevibacillus* sp. strain SKDU10. This peptide-based toxin exhibits antimicrobial properties and demonstrates selective antagonism towards *Mycobacterium tuberculosis* while preserving mammalian cell integrity. Laterosporulin exerts its antimicrobial effects by targeting the cell membrane and perturbing ATP, NAD/NADH, and NADP/NADPH homeostasis (Baindara et al., 2016). Another bacteriocin of 7887 daltons with channel-forming properties was reported to be synthesized by a *Klebsiella pneumoniae* strain. This bacteriocin, Microcin E492, is capable of inducing cell shrinkage and reducing mitochondrial membrane potential. Additionally, it facilitates the release of intracellular calcium ions (Ca^{2+}) and consequently activates the apoptotic pathway. The cytotoxic effects of microcin E492 have been documented across RJ2.2.5, HeLa, and Jurkat cell lines (Hetz et al., 2002). The Lipid A analogue Eritoran, synthesized by *Rhodobacter sphaeroides*, functions as an endotoxin antagonist and targets the signaling pathway of TLR4 (Toll-like receptor 4). It is currently undergoing pharmaceutical development and is aimed at suppressing cytokine storms in patients (Iwai et al., 2023).

3.4 Enzymes

Enzymes are biocatalysts composed of either protein or RNA and are known to have functionalities in detergent, paper, agro-food, textile, animal feed, petroleum, leather, chemical, and biochemical industries. Enzymes provide a more sustainable approach compared to other chemical processes.

3.4.1 Food Industry

In the food industry, enzymes are employed for a variety of processes including bread manufacturing, brewing, cheese production, and for enhancing the quality of juices. The bacterial species *Bacillus licheniformis*, *Bacillus stearothermophilus*, or *Bacillus amyloliquefaciencs* serve as sources of heat stable α-Amylase to cleave the α-1,4 glycosidic polysaccharide bonds (van der Maarel et al., 2002). Lipases obtained from *Pseudomonas* sp. play significant roles in organic synthesis (Rios et al., 2018). Xu et al. (2017a, b) demonstrated the production of Ferulic acid, a precursor of vanillin, through enzymatic activity of feruloyl esterase derived from various strains of *Lactobacillus* sp. *Bacillus licheniformis* can also serve as a source of thermostable esterase enzymes which find use in the production of short-length flavoring agents (Alvarez-Macarie and Baratti, 2000). *Bacillus licheniformis* strain P11(C) synthesizes xylanases with molecular weights of 17.5 and 23 kDa that have potential applications in fruit juice processing and bakery operations. These xylanases are stable across a pH range of 5–11 and over temperatures ranging from 40–100 °C (Bajaj and Manhas, 2012). Also, catalases that play a role in food preservation, have been identified from the bacterial species *Bacillus halodurans* LBK 261 (Raymond Oluoch et al., 2006) and *Rhizobium radiobacter* strain 2–1 (Nakayama et al., 2008).

3.4.2 Detergent Industry

Detergents are surface-active agents that contain a variety of chemicals such as bleach factors, foam agents, perfume agents, and other materials responsible for their cleansing action (Chaturvedi and Tiwari, 2013). Most of these chemicals pose a threat to the ecosystem. Hence, naturally available enzymes can undoubtedly serve as a better alternative.

Lipases are used in the detergent industry due to their stability and long shelf life (Javed et al., 2018). Commercially available lipases have been isolated from *Pseudomonas mendocina*, *Pseudomonas glummae*, and *Pseudomonas alcaligenes*. Lipase isolated from *Pseudomonas mendocina* has been developed under the name Lumafast by Genencor International and lipase obtained from *Pseudomonas alcaligenes* has been developed by the name Lipomax by Gist Brocades (Li et al., 2014).

Alkaline serine proteases are predominantly employed for the removal of protein-based stains. In 1960s, the first protease detergent was developed by Novozyme. It was isolated from *Bacillus licheniformis* and was labelled as Alcalase (Singh and Bajaj, 2017). Subtilisin, an approximately 27 kDa protease, was initially isolated from *Bacillus amyloliquefaciens* and *Bacillus licheniformis*. Presently, it can also be obtained from the bacterial species *Bacillus clausii*, *Bacillus halodurans*, *Bacillus lentus*, and *Bacillus alkaloophilus* and do serve as a prominent candidate in detergent formulations (Maurer, 2004). Examples of other protease detergents developed are Esperase, Savinase, and Everlase (Singh and Bajaj, 2017).

Bacillus species have been reported to be potent producers of detergent-compatible amylases. Commercially used amylases have been isolated from *Bacillus amyloliquefaciens* (Mazhar et al., 2023) and developed under the name of BAN by Novozymes. Another amylase, Stainzyme, has also been derived from a *Bacillus* species (Sarmiento et al., 2015). An extracellular α-amylase has been reportedly obtained from *Bacillus cereus* GA6 (Kuddus and Saima, 2013) and *Bacillus cohnii* US147 (Ghorbel et al., 2009) species.

3.4.3 Pharmaceutically Important Enzymes

Arginine deiminase, isolated from bacterial species namely *Pseudomonas* (Galkin et al., 2004; Lu et al., 2006), *Lactobacillus* (De Angelis et al., 2002), and *Lactococcus* (Kim et al., 2007), has found extensive application in the treatment of arginine-auxotrophic tumors like hepatocellular carcinomas and melanomas. Pegylated arginine deiminase (ADI-PEG20) has been developed and is currently in phase I and II clinical trials to assess its efficacy against arginine-auxotrophic carcinomas (Chang et al., 2021).

In 1963, Broome demonstrated the effect of asparaginase in cancer and reported its cancer regression capability (Kenari et al., 2011). The constant depletion of asparagine results in cell death and has been studied since 1950s. Bacteria namely *Escherichia coli* (Kenari et al., 2011) and *Erwinia chrysanthemi* (Kotzia and Labrou, 2007) have been exploited for their potential to produce asparaginase. Asparaginase treatment has shown inhibitory effects on four different cancer cell lines GBM-ES, U87, GL-261, and DAOY (Panosyan et al., 2014). Phase II studies were conducted to evaluate the efficiency of L-asparaginase against NK/T-cell lymphoma (Jaccard et al., 2011).

3.5 Antioxidants

Oxidative stress arises from elevated levels of free intracellular oxygen radicals referred to as Reactive Oxygen Species (ROS). This can result in several problems such as protein denaturation, DNA damage, and eventually carcinogenesis and cell death. Antioxidants help curb the damage caused by these ROS (Wang et al., 2017). Exopolysaccharides, collected from marine bacterial sources namely *Aerococcus uriaeequi* (Wang et al., 2018) and *Microbacterium aurantiacum* FSW-25 (Sran et al., 2019), have gained attention for their antioxidant properties owing to their ability to quench hydroxyl ions. The carcinogenic effect of the synthetic antioxidants, butylated hydroxyanisole (BHA) and butylated hydroxytoluene (BHT), was first reported in 1955. Hence, naturally derived potential antioxidants like carotenoids, polyphenols, vitamins, and sterols constitute an interesting and an important aspect to be studied (Shahidi and Ambigaipalan, 2015).

Violacein, a violet-coloured pigment produced by *Chromobacterium violaceum* (Durán and Menck, 2001), *Janthinobacterium lividum* DSM1522T (Pantanella et al., 2007), and *Pseudomonas* sp. 520P1 and 710P1 (Yada et al., 2008) is made of three structural units: a 5-hydroxyindole, an oxindole, and a 2-pyrrolidone. It exhibits remarkable antioxidant properties against reactive oxygen species (namely hydroxyl and superoxide) and nitrogen reactive species (such as nitric oxide and peroxynitrite). In Konzen et al. (2006) has reported the myeloperoxidase (MPO) inhibition activity of violacein. *Streptomyces glaucescens* strain NEAE-H, when cultivated in peptone-yeast extract iron agar, produced a black pigment which was later identified as melanin. The melanin extracted by acid treatment showed promising antioxidant properties. The extracted melanin was capable of scavenging nearly half of the reactive radicals yielding comparable results to the activity of standard ascorbic acid. The pigment also demonstrated cytotoxic properties against 3 cancer cell lines with the most potent activity against the skin cancer cell line (HFB4) (El-Naggar and El-Ewasy, 2017). Melanin sourced from the bacterial strain *Actinoalloteichus* sp. MA-32 has also demonstrated similar antioxidant properties (Manivasagan et al., 2013). *Pseudomonas aeruginosa*, a causative agent of various nocosomal infections, has been reported to secrete a compound known as pyocyanin (DeBritto et al., 2020). Pyocyanin belongs to a group of compounds known as phenazines, responsible for the virulence of the organism (Hall et al., 2016). Nevertheless, pyocyanin also exhibits potent antioxidant properties due to its radical scavenging ability. Consequently, this pigment has been reported to exhibit significant potential as a natural antioxidant following rigorous toxicological evaluation (Sengupta and Bhowal, 2021).

3.6 Carotenoids

Canthaxanthin (4,4′-diketo-β-carotene) is a bicyclic carotenoid and a polyene pigment. Canthaxanthin has been isolated from *Bradyrhizobium* sp. strain ORS278, accounting for 85% of the total carotenoid content in the bacterial species, the highest compared to other photosynthetic bacteria (Hannibal et al., 2000). It has also been sourced from *Micrococcus roseus* (Tizabi and Hill, 2023) and other extremely halophilic bacteria. Canthaxanthin is a potent antioxidant species with the capacity to inhibit lipid peroxidation (Cantrell et al., 2003). Furthermore, it is reported to enhance alkaline phosphatase activity, an enzyme crucial for the detoxification of lipopolysaccharides during infectious diseases and chronic inflammatory conditions (Esatbeyoglu and Rimbach, 2017).

Astaxanthin is a red–orange coloured 3,3′-dihydroxy-β,β-carotene-4,4′-dione structured pigment and is responsible for the red–orange colour of salmons and lobsters. Astaxanthin production has been reported in multiple bacterial strains such as *Paracoccus carotinifaciens* strain E-396T (Patel et al., 2019) and *Agrobacterium aurantiacum* (Yokoyama and Miki, 1995). In addition to its ability to scavenge superoxides, this carotenoid can also inhibit NF-κB and Wnt/β-catenin signaling pathways, promoting apoptosis in cancer cells (Kavitha et al., 2013).

Zeaxanthin is responsible for the yellow colour of fruits. It is an oxygenated carotenoid (Sajilata et al., 2008) previously isolated from plants. In the past few years, microbial sources have also been investigated for the production of zeaxanthin. Bacterial species such as *Paracoccus zeaxanthinifaciens* strain ATCC 21588 (Berry et al., 2003; Joshi and Singhal, 2016), *Flavobacterium multivorum* ATCC 55238 (Bhosale and Bernstein, 2004), *Mesoflavibacter zeaxanthinifaciens* (Asker et al., 2007), and *Muricauda lutaonensis* CC-HSB-11T (Hameed et al., 2011) have been reported to produce the yellow coloured pigment. Zeaxanthin is known to accumulate in the retinal region (Li et al., 2017) and has been associated with reduction in age-dependent macular degeneration and cataract (Liu et al., 2014). Bi et al. (2013) has reported the role of zeaxanthin in promoting Bak and Bax, pro-apoptotic protein formation and induced apoptosis in human melanoma cell lines SP6.5 and C918 (Bi et al., 2013).

Staphyloxanthin, a red–orange membrane bound carotenoid, was reportedly isolated from *Staphylococcus aureus* (Pelz et al., 2005). This carotenoid pigment has the ability to scavenge superoxide radicals, hydroxyl radicals, and hydrochloride. Staphyloxanthin has also been reported to prevent lipid peroxidation in the bilayer membranes of *Staphylococcus aureus* (Clauditz et al., 2006).

4 Bioactive Compounds from Cyanobacteria

Cyanobacteria are considered to be amongst the most primitive life forms on earth. They are photosynthetic prokaryotes that have a cell wall structurally similar to a Gram-negative bacterium. Their cell wall is composed of a thin peptidoglycan layer and an outer membrane made of lipopolysaccharide. All cyanobacteria are unicellular yet have various physiological and cellular capacities to support the adaption for colonization in varied environments. They can be found in diverse biotope like marine, terrestrial, and freshwater and in extreme habitats like deserts, polar regions, thermal springs, and hypersaline lakes. Cyanobacteria constitute one of the very few microorganisms which can convert atmospheric nitrogen to organic nitrogen and can act as a natural biofertilizer. They can produce bioactive molecules like lipopeptides and terpenes that include antitumor agents, antimicrobial agents, antioxidants, cytotoxic agents, antiviral agents, and immunosuppressive agents. Cyanobacteria have a wide range of applications in the pharmaceutical and biotechnological fields.

A total of 260 bioactive compound families are attributed to cyanobacteria as their secondary metabolites. These can be classified on the basis of their chemical classes or biological activity. The chemical classes reported include peptides, lipopeptides, terpenes, and lipids. The bioactivity demonstrated by these classes include anti-inflammatory, antioxidant, antimicrobial, and cytotoxic properties. Singh and colleagues reported 110 compound families to exhibit cytotoxic activity against different cell lines whereas, 54 and 43 compound families demonstrated lethality and antibacterial activity respectively (Singh et al., 2005).

4.1 Anti-microbial Properties

4.1.1 Antibacterial Compounds

Mundt and team have reported the fatty acids of *Oscillatoria redekei* HUB 051 to exhibit anti-bacterial activity against hospital-based multidrug resistant *Staphylococcus aureus* strain (Mundt et al., 2003). The unsaturated hydroxylate fatty acids produced by the cyanobacteria have the ability to alter cell membrane permeability, to inhibit useful enzymes, and to form a layer surrounding the cells resulting in the bacteriocidic property (Mundt et al., 2003).

Nostophycin, a cyclic peptide from *Nostoc calcicola*, showed antibacterial activity against *Escherichia coli* and *Staphylococcus aureus*. At lower concentrations, nostophycin has been reported to exhibit bactericidal effects, not giving enough time for the bacteria to replicate in great numbers (Gupta and Vyas, 2021). Also, Noscomin obtained from *Nostoc commune* (Rawat and Bhargava, 2011) and Carbamidocyclophanes obtained from *Nostoc* sp. CAVN 10 (Bhateja et al., 2006) have been reported to demonstrate moderate activity against *Staphylococcus aureus*.

Scytophycins, a macrolide isolated from the cyanobacteria *Scytonema pseudohofmanni*, exhibits weak antibacterial activity. Macrolides in general inhibit bacterial protein synthesis by binding to the large ribosomal subunit (Poehlsgaard and Douthwaite, 2002).

4.1.2 Antifungal Compounds

Fungal infections are most effective against immunosuppressed individuals. Limitations of present antifungal drugs have led to the search for novel bioactive molecules. Vestola et al. (2014) has reported the effect of hassallidins, a cyclic glycopeptide isolated from the cyanobacterial species *Nostoc* sp., *Aphanizomenon gracile*, *Tolypothrix* sp., *Nostocales anobacteria*, and *Cylindrospermopsis raciborskii*, against the fungus *Candida albicans*. The ultrastructure of *Candida albicans* was affected by the action of Hassallidin A. Hassallidin A inhibited the secretory vesicle docking and the membrane fusion, thereby disturbing the integrity of plasma membrane (Neuhof et al., 2006). Another antifungal molecule belonging to the lyngbyabellins family, Hectochlorin, was isolated from *Lyngbya majuscule*. Hectochlorin, reported by Marquez et al. (2002), demonstrated antifungal activity against *Candida albicans*. Lyngbyabellins function by disrupting the microfilament networks and arresting cell cycle at the cytokinesis phase (Han et al., 2005a; Luesch et al., 2000).

4.1.3 Antiviral Compounds

Viral infections are quick to spread. The World Health Organization report estimated approximately 1 million deaths in the year 2017 due to Human Immunodeficiency Virus (HIV), the causative agent of AIDS. AIDS is still an epidemic in the regions of West and Central Africa as reported by UNICEF in 2023. Cyanobacterial bioactive compounds have shown promising activity against HIV. An antiviral agent, Cyanovirin-N, has been isolated from the cyanobacteria *Nostoc ellipsosporum* (Fidor et al., 2019) and *Cyanothece* sp. (Matei et al., 2016). Cyanovirin-N has been reported to bind with the glycoprotein gp120, present on the envelope of HIV, thus interfering with the binding of gp120 and the CD4+ receptor of the target cell and inhibiting the fusion of HIV and host cell thereby effectively terminating viral transmission (Matei et al., 2016). The antiviral potential of this bioactive compound was tested against a range of viral agents. Cyanovirin-N has demonstrated antiviral activity against HIV-1, HIV-2, Simian Immunodeficiency Virus (SIV), Measles Virus, Ebola, and Influenza virus (Dey et al., 2000; Matei et al., 2016).

Calcium spirulan is a sulphated polysaccharide, isolated from *Arthrospira platensis* (previously known as *Spirulina platensis*) (Wollina et al., 2018). Ca-spirulan has been reported to show activity against HIV-1, HSV-1, and other enveloped viruses. Ca-spirulan prevents the attachment of the virus to the cell surface and also inhibits the fusion of infected and uninfected CD4+ cells which greatly lowers the infectivity (Mader et al., 2016). Compared to other compounds, this molecule has a lower anticoagulant property and a longer half-life in blood (Wollina et al., 2018), making it a suitable lead for drug development.

4.2 Anticancer Compounds

The growing resistance of cancer cells against existing drugs has created an urgency to find novel drug molecules. As per the report of the American Cancer Society, cancer has claimed the lives of 7.6 million individuals. Various cyanobacterial compounds have been reported to possess anticancer properties.

Swinholide-type molecule, a macrolide initially isolated from *Theonella swinhoei* (Robles-Bañuelos et al., 2022), was also obtained from other cyanobacterial species like *Symploca* sp., *Geitlerinema* sp. (Andrianasolo et al., 2005), and *Nostoc* sp. (Humisto et al., 2018). This bioactive molecule was tested against cell lines like KB (Robles-Bañuelos et al., 2022), Neuro-2a, and MDAMB-435 (Andrianasolo et al., 2005) to determine its anticancer potential. The macrolide molecule functions by disrupting the stability of actin filaments. One of the dimeric macrolides first binds to two G-actin molecules forming a tertiary molecule that inhibits the formation of filamentous actin (F-actin). In addition to this, they can also break the filamentous actin strands, which is an interesting area for anticancer drug development (Andrianasolo et al., 2005).

Carmaphycin is a proteasome inhibitory compound isolated from the marine cyanobacteria *Symploca* sp. The compound has been tested to be cytotoxic against human lung adenocarcinoma (H-460) and colon cancer (HCT-116) cell lines (Pereira et al., 2012). Proteosome is a multi-enzyme complex that is involved in many cell-cycle regulatory protein degradation pathways, important for controlling cellular mechanisms related to cancer. Proteosomes may affect various other cellular processes like oncogenic transformations, apoptosis, and tumor suppression. It has been established that the effect of proteasome inhibitors are more potent in cancer-forming cells than in normal cells (Adams, 2004).

In 1990, a research study led by Schwartz first reported the isolation of a cyclic peptide named Cryptophycin from *Nostoc* sp. (Polyzois et al., 2020), that demonstrated anti-cancer activity against HCT116 colon carcinoma, MX-1 breast carcinoma, and H125 non-small cell lung carcinoma (Shih and Teicher, 2001). The action mechanism of this compound includes microtubule depolymerization which causes the cell cycle to be arrested at the prometaphase/metaphase (G2/M) interface resulting in apoptosis (Kessel and Luo, 2000). Two chemical analogs, cryptophycin 249 and 309, were reported to be under clinical trials for improved stability and water solubility (Ding et al., 2008; Liang et al., 2005).

Two histone deacetylase (HDACs) inhibitors largazole (Taori et al., 2008) and santacruzamate A (Pavlik et al., 2013) were isolated from the cyanobacteria *Symploca* sp. HDACs are significant inhibitors of gene expression and also play a role in the interaction of DNA-histone and binding of transcription factors (Bolden et al., 2006). Largazole demonstrates inhibitory properties to the class 1 HDACs and ubiquitin-activating enzyme (E1). It also alters the binding affinity of E2F transcription factor 1 arresting the cell cycle at G1 phase on the human lung cancer cell line (Wu et al., 2013). Santacruzamate A is a structural analog of suberoylanilide hydroxamic acid (SAHA) which is a clinically approved HDAC inhibitor and is used to treat refractory cutaneous T-cell lymphoma. Santacruzamate A functions as an inhibitor of HDAC2a and is comparatively less effective in the inhibition of HDAC4 or HDAC6. It exhibits anti-cancer activity against HCT-116 colon carcinoma cells, HuT-78 cutaneous T-cell lymphoma cells, and human dermal fibroblast (hDF) cells (Pavlik et al., 2013).

4.3 Carotenoids

Cyanobacteria are photosynthetic microorganisms and can produce three groups of pigments i.e., Chlorophylls, Phycobiliproteins, and Carotenoids (Masojídek et al., 2013). Carotenoids are widely known for their antioxidant, anti-inflammatory, and anti-tumor properties. These molecules can be used as colouring agents and in cosmetic industries owing to their antioxidant and anti-aging characteristics (Pagels et al., 2020). Carotenoids protect the photosynthetic machinery of the cell by absorbing light in the wavelength region of 400–500 nm. Carotenoids can inhibit photosensitized oxidation and function as a quencher of singlet oxygen. The quenching mechanism works by electron exchange between singlet oxygen and singlet carotenoid state forming triplet oxygen and triplet carotenoid state and eventually resulting in energy transfer (Wada et al., 2013).

A total of 14 carotenoids were extracted from the cyanobacteria *Aphanothece microscopica Nageli*. The major carotenoids reported from the cyanobacteria were β-carotene and echinenone. The carotenoid extract exhibited peroxyl radical scavenging properties—7.3 times more potent than α-tocopherol (Patias et al., 2017). Carotenoid extract from *Arthrospira platensis* showed scavenging activities against hydrophobic Reactive Oxygen species (ROS), DPPH, and ABTS radical. Commercially available spirulina powder is made from dried bodies of *Arthrospira platensis* and is used as a source of these pigments in food and cosmetic industries (Park et al., 2018). In 1989, Schwartz and Shklar reported the antitumor properties of the *Arthrospira platensis* extract (Pagels et al., 2021). This cyanobacterial extract was tested against DMBA-induced oral carcinoma (HCPC-1 cell line) cells. The cells were implanted into a hamster cheek pouch and treated. The treated hamsters showed tumor regression along with inhibition of cytokine tumor necrosis factor (TNFα) (Pagels et al., 2021).

Some other cyanobacterium species namely *Cephalothrix lacustris*, *Leptolyngbya boryana*, *Nodosilinea nodulosa*, and *Leptothoe* sp. have been studied for their carotenoid production and anti-inflammatory properties. Extracts rich in β-carotene and zeaxanthin exhibited the ability to scavenge superoxide radical and nitrite oxide radical as well as exhibited anti-inflammatory properties against Lipopolysaccharide (LPS) induced inflammation in RAW264.7 macrophage cells (Rodrigues et al., 2024). Bai et al. (2005) has also demonstrated that β-carotene can suppresses NF-κβ activation and NOS expression in LPS-induced RAW264.7 cells.

4.4 Anti-inflammatory Compounds

In recent times, the onset of chronic inflammatory diseases has initiated extensive research on anti-inflammatory agents. These inflammatory diseases include inflammatory bowel disease, rheumatoid arthritis, multiple sclerosis, and psoriasis. Anti-inflammatory agents, in general, block molecules that may cause inflammation in the body.

The anti-inflammatory agent aeruginosin was isolated from the cyanobacterium species *Microcystis aeruginosa* (Lifshits and Carmeli, 2012), *Nostoc* sp. (Kapuścik et al., 2013) and *Microcystis viridis* (Anas et al., 2017). Aeruginosin can down-regulate the synthesis of proinflammatory mediators like Interleukin-8 (IL-8) and inter-cellular adhesion molecule 1 (ICAM-1). Aeruginosin is also capable of inhibiting the translocation of NF-κβ to the nucleus (Kapuścik et al., 2013). Ishida and group have reported the inhibitory activity of aeruginosin against trypsin, thrombin, and plasmin (Demay et al., 2019).

A phycobiliprotein named Phycocyanin, common in cyanobacterial species, displayed antioxidant, anti-inflammatory, neuroprotective, and hepatoprotective properties as reviewed by Romay et al. (2003). Phycocyanin has been reported to inhibit edema and reduce prostaglandin and leukotrienes levels, reduce tumor necrosis factor-alpha (TNFα), and the levels of nitrite in the serum of phycocyanin-treated mice (Romay et al., 2001). The anti-inflammatory activity of phycocyanin can be credited to their ROS scavenging property, their ability to inhibit lipid peroxidation (Bhat & Madyastha, 2000) and cyclooxygenase (COX) (Reddy et al., 2000).Scytonemin has been reported from the species *Anabaena variabilis*, *Cylindrospernum* sp. (Mushir and Fatma, 2012), *Nostoc commune* (Venckus et al., 2018), and *Scytonema javanicum* (Asencio and Hoffmann, 2013). The pigment plays a role in the protection of the cyanobacterial species from UV radiations. The enzyme polo-like kinase 1 is inhibited due to the action of Scytonemin. Polo-like kinase 1 is an enzyme that aids in the G2-M phase transition involving phosphorylation and activation of cdc25C, a cell cycle regulatory protein. The inhibition of polo-like kinase 1 enzyme inhibits cell proliferation thereby demonstrating the anti-inflammatory and anti-proliferative properties of Scytonemin (Roshak et al., 2000; Stevenson et al., 2002).

5 Bioactive Compounds from Fungi

The diverse world of fungi is no stranger when it comes to bioprospecting for valuable metabolites. Amongst almost 23,000 secondary metabolites obtained from microorganisms, 42% account to be of fungal origin (de Castro et al., 2020). These metabolites include pigments, antitumor agents, immunosuppressants, antioxidants and, most importantly, antibiotics, all of which have extensive applications in human health and research.

5.1 Antimicrobial Properties

5.1.1 Antibacterial Compounds

Humans can benefit from the innate defensive mechanisms of fungi that produce antibiotics to combat infection from microorganisms since both fungi and humans (animals) share common microbial antagonists like *Pseudomonas aeruginosa*, *Staphylococcus aureus*, and *Escherichia coli*. Consequently, fungi have been reported to produce important antibacterials (Ranadive et al., 2013).

Penicillium sp. LS54, a fungal strain associated with sponges, was reported by Liu et al. (2019b). The fungal strain produced penicillilactone A, a 7-membered lactone with a furan core, exhibiting activity against *Vibrio harveyi*. Three bioactive molecules, Communol A, F, and G, found in the extracts of the fungal strain *Penicillium commune* 518, was reported to have anti-bacterial activity against *Escherichia coli* and *Enterobacter aerogenes* (Wang et al., 2012). The bacterial species of MRSA (methicillin-resistant *Staphylococcus aureus*) and *Escherichia coli* were inhibited by pyrrospirones C–F and I obtained from *Penicillium* sp. ZZ380 (Song et al., 2018a). Antimicrobial activity against *Staphylococcus aureus* and *Escherichia coli* was also demonstrated by penicipyrrodiether A, derived from the marine fungal strain *Penicillium* sp. ZZ380 (Song et al., 2018b). Aspergicin, an alkaloid obtained from the co-culture of the fungal strains *Aspergillus* sp. FSY-01 and *Aspergillus* sp. FSW-02 isolated from *Avicennia marina*, exhibited antibacterial activity against *Staphylococcus aureus*, *Escherichia coli*, *Bacillus dysenteriae*, *Bacillus subtilis*, *Bacillus proteus*, and *Staphylococcus epidermidis* (Zhu et al., 2011).

The endophytic fungus *Xylaria* sp., associated with a Chinese medicinal plant *Sophora tonkinensis*, is the source of two novel cyclopentapeptides namely Xylapeptide A and Xylapeptide B. Xylapeptide A contains L-picolinic acid in its cyclopentapeptide structure and exhibits selective antibacterial characteristics against *Bacillus subtilis* and *Bacillus cereus* (Xu et al., 2017a). Xu et al. (2017a) also reported the structure of Xylapeptide B, consisting of a L-proline residue instead of L-picolinic acid and exhibiting a stronger and broader spectrum of antibacterial activities against *Bacillus megaterium* and *Staphylococcus aureus*. Zainuddin et al. (2010) reported antimicrobial activity exhibited by the fungal species *Fasciatispora nypae* against both bacteria and fungus. This antimicrobial activity was attributed to the production of 2,2,7-trimethyl-2H-chromen-5-ol by the isolated strain. The ethyl acetate extract of *Gelasinospora endodonta*, an endophyte to *Bruguiera sexangular*, exhibited activity against *Escherichia coli* (Zhang et al., 2021). Similarly, the ethyl acetate extract obtained from another endophyte, *Nigrospora* sp. isolated from the host *Ocimum basilicum var. thyrsiflora*, was reported to possess antibacterial activity against *Staphylococcus epidermidis*, *Vibrio cholerae*, *Escherichia coli*, and *Bacillus cereus* (Atiphasaworn et al., 2017).

5.1.2 Antifungal Compounds

Two antifungal agents, Penicitide A (Gao et al., 2011a) and Penicisteroid A (Gao et al., 2011b) were obtained from the strain *Penicillium chrysogenum* QEN-24S. Gao et al. (2011a) reported the activity of Penicitide A against *Alternaria brassicae*. Penicisteroid A, a tetrahydroxy and

C-16-acetoxy steroid, have demonstrated activity against *Aspergillus niger* and *Alternaria brassicae* (Gao et al., 2011b). An antifungal agent, Insuetolide A—obtained from the marine fungal species *Aspergillus insuetus* OY-207, was reported to show activity against the filamentous fungus *Neurospora crassa* (Cohen et al., 2011).

Rogozhin et al. (2018) reported an alkalophilic fungal species, *Emericellopsis alkaline*, that produces emericellipsin A. Emericellipsin A is a short linear peptide containing seven amino acid residues, two out of which are conventional (alanine and isoleucine), the other five amino acid residues being 3-methylproline (3MP), 2-Amino-4-methyl-6-hydroxy-8-oxo decanoic acid (AHMOD), 2-aminoisobutyrate (AIB), isovaline, and β-alanine. Emericellipsin A displayed broad spectrum antifungal activity wherein it inhibited all fungal growth of *Candida* species like *Candida tropicales* 1402, *Candidans albicans* 1582 and *Aspergillus* species like *Aspergillus niger* 219, *Aspergillus fumigatus* 163, *Aspergillus flavus* 905 (Rogozhin et al., 2018).

5.1.3 Antiviral Compounds

Rapid mutations, antigenic shifts, and drifts render the development of antiviral drugs difficult. Although fungi are an established source of bioactive compounds with potential antiviral properties, development of such drugs is still at its initial stages.

Three sesquiterpenoid antiviral agents which are xanthone derivatives namely grisephenone A, stachybogrisephenone B, and 3,6,8-trihydroxy-1-methylxanthone were reportedly obtained from the marine fungus *Stachybotry* sp. HH1 ZDDS1F1-2. The molecules when tested against *Enterovirus*-71 demonstrated their significant antiviral potential (Qin et al., 2015). Two novel peptides isolated from the marine fungus *Scytalidium* sp. were reported to have potential antiviral properties (Rowley et al., 2003). Halovir A, produced by the fungal species, demonstrated inhibitory effects against HSV 1 by interacting with the viral envelope. Similarly, lactone based bioactive molecules sourced from *Aspergillus terreus* SCSGAF0162 namely 11a-dehydroxyisoterreulactone A, Isobutyrolactone II, Aspernolide A, and Arisugacin A also exhibited activity against HSV (Nong et al., 2014). Azaphilone, (+)-sclerotiorin, a fungal polyketide reported in the extracts of the strain *Penicillium sclerotiorum* PSU-A13—an endophyte in the leaves of *Garcinia atroviridis*, showed antiviral activity against HIV. The activity was exhibited specifically against the integrase and protease enzymes of the virus (Arunpanichlert et al., 2010; Gao et al., 2013). Another endophytic fungal species, *Phoma* sp., isolated from the roots of *Aconitum vilmorinianum*, was reported for its antiviral properties credited to the production of 14-nordrimane, a lactone, acting against the viral agent influenza A (A/Puerto Rico/8/34, H1N1) (Liu et al., 2019a).

5.2 Antioxidants

Reactive Oxygen Species form inside a cell due to various endogenous and exogenous influences and may lead to complications such as DNA/RNA mutations, lipid peroxidation, and protein peroxidation, which in turn can lead to altered cellular responses through a cascade of events. Hence, antioxidants are supplemented through diet for the maintenance of health and prevention of diseases. Synthetic antioxidants are carcinogenic and toxic, a matter of fact which has necessitated the search for natural alternatives (Yang et al., 2018).

Species belonging to the fungal genera *Aspergillus*, *Penicillium*, *Emericella*, *Fusarium*, *Paecilomyces*, *Trichoderma*, *Phoma*, *Drechslera*, *Microsporum*, *Curvularia*, and *Alternaria* synthesize anthraquinones that have antibacterial, herbicidal, insecticidal, and fungicidal properties (Afroz Toma et al., 2023; Huang et al., 2017; Vidal et al., 2023). Huang et al. (2017) reported the production of two anthraquinones, Aspergilol A and Aspergilol B, from the marine fungus *Aspergillus versicolor* SCSIO 41502. Aspergilol A and Aspergilol B were reported to demonstrate antioxidant activity. An anthraquinone found in the broth of a marine fungal species *Chromolaenicola* sp. (HL-114-33-R04) was reported by Vidal et al. (2023). The anthraquinone [1,8-dihydroxy-9,10-anthraquinone (danthron)] possessed antioxidant, anti-angiogenic, antimetastatic, and anticancer properties. In both tumor and endothelial cells, antioxidant properties of danthron were demonstrated by a decrease in intracellular ROS production and an increase in intracellular sulfhydryl groups.

A dihydroxyanthraquinone compound called physcion, obtained from the endophytic fungal species *Aspergillus versicolor* SB5, demonstrated antioxidant potential against ABTS, DPPH, O^2, and NO radicals, and also played an important role in the inhibition of lipid peroxidation (Elawady et al., 2023). An endophytic fungus, *Fusarium oxysporum*, was isolated from the leaves of a medicinal tree from the Colombian rainforest, *Otoba gracilipes* (Caicedo et al., 2019). The crude exopolysaccharide extract obtained from the fungal spores of *Fusarium oxysporum* demonstrated antioxidant property through free radical scavenging using 2,2-diphenyl-1-picrylhydrazyl (DPPH).

5.3 Anti-cancer Compounds

Cancer is a multi-factorial disease characterized by uncontrolled proliferation of cells. Treatment of cancer through chemotherapy introduces harmful chemical compounds into the body that leave severe detriments, such as a compromised immune system, thereby exposing the patient to the risk of contracting other diseases. Decrease or deceleration of oncoprotein activity is therefore one of the most commonly employed cancer treatment approaches. Natural products and their derivatives have been developed as anticancer drugs and have been identified as a promising alternative to oncogene protein modulators (Demain and Vaishnav, 2011).

The compounds, Acaromycin A and (+)-Cryptosporin, obtained from the marine fungal species *Acaromyces ingoldii* FS121, showed a reduction in cell growth in the cancer cell lines NCI-H460, MCF-7, HepG-2, and SF-268 (Gao et al., 2016). Asperolide E, a tetranorlabdane diterpenoid obtained from the marine fungus *Aspergillus wentii* SD-310, demonstrated cytotoxic activity against the cancer cell lines MCF-7, NCI-H446, and HeLa (Li et al., 2016). A neutral water-soluble polysaccharide (W-NTRP) isolated from the fruiting bodies of *Trametes robiniophila* (Huaier) was reported by Sun et al. (2013). Huaier is a sandy beige mushroom belonging to the class Hymenomycetes. W-NTRP demonstrated anti-cancer properties against three human cholangiocarcinoma cell line (QBC939, Sk-ChA-1 and MZChA-1). Also, W-NTRP exhibited no cytotoxicity to L-929 normal cells. Additionally, W-NTRP was shown to stimulate macrophages through increased pinocytic ability and induced nitric oxide production due to the upregulation of iNOS activity hinting at the potential of W-NTRP as a viable lead in developing drugs for human cholangiocarcinoma. Indole derivative compounds, Varioloid A and Varioloid B, sourced from the endophytic fungus *Paecilomyces variotii* EN-291, demonstrated anti-cancer activity against the cell line HepG2, HCT116, and A549 (Zhang et al., 2016). Wentilactone A obtained from *Aspergillus dimorphicus* SD317 was reported to induce cell apoptosis. Lv et al. (2013) reported that Wentilactone A inhibit cell cycle at the G2/M phase in human lung carcinoma.

5.4 Carotenoids

Microbial pigments are derived from a range of organisms, such as bacteria, yeast, fungi, and microalgae. They are used in food, medicine, colorants, and imaging (Chatragadda and Dufossé, 2021; Rather et al., 2023). Fungi, among many other microorganisms, are interesting due to their rapid growth patterns and their ability to be genetically altered to stimulate the production of more pigments (Meruvu and dos Santos, 2021). Advances in fermenters and downstream processing have made it possible to use low-cost, easily controlled fungal systems for pigment extraction and purification on a large scale (Bhatnagar and Aoyagi, 2022; Singh et al., 2021).

5.4.1 β-Carotene

The yellowish carotenoid pigment β-carotene, also referred to as pro-vitamin A, possesses significant disease-fighting potential and act as an antioxidant. According to reports, the different fungal species producing beta-carotene include *Mucor circinelloides*, *Phycomyces blakesleeanus* (Avalos et al., 2017), *Blakeslea trispora* (Afroz Toma et al., 2023), *Sclerotium rolfsii*, *Cercospora nicotianae*, *Penicillium* sp., *Aspergillus giganteus*, and *Sclerotinia sclerotiorum* (Han et al., 2005b). The major focus has been on yeasts like *Saccharomyces cerevisiae* and *Yarrowia lipolytica* because of their robustness for large-scale fermentation and developed advanced genetic manipulation tools (Guo et al., 2024).

5.4.2 Lycopene

Lycopene is insoluble in water and is highly susceptible to heat and oxidation. Also referred to as psi-carotene, lycopene is an unsaturated carotenoid molecule. *Fusarium sporotrichioides* (Afroz Toma et al., 2023) and *Talaromyces amestolkiae* (de Oliveira et al., 2021) are the examples of lycopene producing fungal species.

5.4.3 Torulene

Torulene, a 3,4-didehydro-β,γ-carotene, is orange or orange-red in colour (Rapoport et al., 2021). The antioxidative potential of this carotenoid molecule was reported to be lesser than β-carotene (Dimitrova et al., 2013). However, experiments on rat and mouse revealed anti-cancerous properties of the pigment (Kot et al., 2018). Fungal species reported for torulene production include *Sporidiobolus*, *Rhodotorula*, *Dioszegia*, *Rhodosporidium*, *Neurospora*, *Cystophilobasidium*, and *Sporobolomyces* (Kot et al., 2018).

5.4.4 Torularhodin

Torularhodin or 3,4-didehydro-β,γ-caroten-16-oic acid also exhibits antimicrobial and antioxidative properties and has applications in cosmetics, food, and feed industries (Kot et al., 2018). Torularhodin, isolated from the fungus *Rhodotorula glutinis*, was reported to scavenge peroxyl radicals and inhibit singlet oxygen decomposition better than carotene (Sakaki et al., 2002). Torularhodin is reportedly produced by fungal species belonging to the genera *Sporobolomyces, Cystofilobasidium*, *Sporidiobolus*, *Rhodotorula*, and *Rhodosporidium* (Kot et al., 2018).

6 Bioactive Compounds from Microalgae

6.1 Anti-microbial Property

Microalgae can synthesize various peptides, fatty acids, and sulfated compounds that have anti-microbial activity (Prakash et al., 2011). Therefore, the extracted metabolites from the algal biomass are studied and utilized for their anti-microbial properties. The microalgal long-chain fatty acids are reported to have properties against human pathogens such as *Pseudomonas aeruginosa, Staphylococcus epidermidis*, *Staphylococcus aureus*, and *Escherichia coli*. The common long chain fatty acids produced by microalgae are hexadecatrienoic acid, γ-linolenic acid, lauric acid, arachidonic acid, oleic acid, palmitoleic acid, lactic acid, docosahexaenoic acid, and eicosapentaenoic acid (Amaro et al., 2011; Smith et al., 2010).

Although less studied, the action mechanism of long-chain fatty acids is understood to affect the cell membrane. The damaged cell membrane leads to loss of intracellular components whereas the entry of foreign molecules restricts nutrient absorption and cellular respiration. Fatty acid chain length and unsaturation determine the damage incurred. Molecules with more than 10 carbon chain length may induce lysis to the bacterial protoplast (Amaro et al., 2011). The sulfated extracellular polysaccharides, produced and secreted by microalgae, also have acidic properties which render them as potential therapeutic agents. The most common sugar moieties in the polysaccharides include galactose, xylose, and glucose. The sulfate group(s) present determine the characteristics of the polysaccharide. Polysaccharides with high sulfate content possess higher antiviral activity and are associated with positive cell surface charge that may inhibit the penetration of a virus into the host cell (Arad and Levy-Ontman, 2010; Amaro et al., 2011; de Jesus Raposo et al., 2014).

6.1.1 Anti-bacterial Compounds

In the recent past, research attention of the scientific community has been significantly directed towards the rising concern of antibacterial-resistant bacteria. Primarily, research conducted in this field, deals with the study of molecular mechanisms related to resistance and development or discovery of a novel antibacterial agent to counter pathogenic antibiotic-resistant strains. The first antibacterial activity from the microalga *Chlorella* was reported by Pratt and group in 1944 (Pradhan et al., 2014). A mix of fatty acids named Chlorelin, obtained from the microalga *Chlorella*, demonstrated inhibition against Gram-positive and Gram-negative bacteria. The multiple layers of a Gram-negative bacterial cell wall hinder the penetration of the bioactive antibiotic molecules thereby rendering the Gram-negative bacteria less susceptible to anti-bacterial agents compared to the Gram-positive species (Ghasemi et al., 2004, 2007; Ördög et al., 2004).

The diatom, *Phaeodactylum tricornutum*, can produce a long-chain polyunsaturated fatty acid called Eicosapentaenoic Acid (20:5) (EPA). EPA, obtained from the cellular lysate of the diatom, contributed to anti-bacterial activity against Gram positive and negative bacterial species like *Staphylococcus epidermidis*, *Bacillus weihenstephanensis*, and *Bacillus cereus* and also to Methicillin-resistant *Staphylococcus aureus* (MRSA) strains 252 and 16a (Ward and Singh, 2005). EPA is also a precursor for aldehydes and is found in the cell membrane of microalgae, acting as a defense barrier, owing to its toxic effects on grazers and copepods (d'Ippolito et al., 2004; Jüttner, 2001). Other long-chain bioactive fatty acids from *Phaeodactylum tricornutum* like hexadecatrienoic acid and palmitoleic acid were also reported to demonstrate antibacterial activities against *Staphylococcus aureus* and several non-marine Gram-positive pathogenic bacterial strains at micromolar concentrations (Smith et al., 2010). Short-chain fatty acid extracts from the microalgae *Haematococcus pluvialis*, containing methyl lactic acids and butanoic acids, were found to be associated with anti-bacterial activity against *Escherichia coli* and *Staphylococcus aureus* (Santoyo et al., 2009). The fatty acid extract from *Scenedesmus costatum* demonstrated antibacterial properties against aquaculture bacterial strains. Interestingly the fatty acid chain was reported to be more than 10 carbons long concluding that long-chain fatty acids induce protoplast lysis (Guedes et al., 2011a). Kokou et al. (2012) reported inhibition of the bacterial cultures *Vibrio splendidus*, *Vibrio alginolyticus*, *Vibrio lentus*, *Vibrio anguillarum*, and *Vibrio parahaemolyticus* by the microalgae *Isochrysis sp.*, *Chlorella minutissima*, *Nannochloropsis sp.*, *Tetraselmis chui*, and *Arthrospira platensis*. The microalga *Fischerella ambigua* produces a molecule called parsiguine. Parsiguine possesses anti-fungal and anti-bacterial activity and has been reported to restrict *Candida krusei* ATCC 44507 and *Staphylococcus epidermidis* PTCC 1114 (Ghasemi et al., 2004).

Diatoms like *Skeletonema marinoi*, *Pseudonitzschia delicatissima,* and *Thalassiosira rotula* produce polyunsaturated aldehydes as bioactive molecules (Vidoudez and Pohnert, 2008). An example of such a bioactive biomolecule is decadienal, which exhibits antimicrobial properties against pathogens like *Haemophilus influenza*, Staphylococcus *epidermidis*, *Escherichia coli*, *Pseudomonas aeruginosa*, *Staphylococcus aureus*, and MRSA. Decadienal is also potent against marine Gram-negative bacteria like *Vibrio anguillarum*, *Psychrobacter immobilis*, *Photobacterium phosphoreum*, *Aeromonas hydrophila*, and *Alteromonas haloplankti* and against marine Gram-positive bacteria like *Planococcus citreus* and *Micrococcus luteus* (Smith et al., 2010). Research on the culture extracts of *Euglena viridis* revealed the potential of the microalga against the proliferation of *Escherichia coli*, *Pseudomonas fluorescence*, *Pseudomonas putida*, *Pseudomonas aeruginosa*, *Edwardsiella tarda*, *Vibrio harveyi*, *Vibrio anguillarum*, *Aeromonas hydrophila*, *Vibrio parahaemolyticus*, *Vibrio fluvialis*, and *Vibrio alginolyticus* (Das et al., 2005). The extract of the diatom *Skeletonema costatum* was also reported to have an inhibitory effect on bacteria, restricting the growth of bacterial species like *Listeria monocytogenes*, *Vibrio anguillarum*, *Vibrio sp.* VRP, *Vibrio sp.* S322, and *Vibrio mytili* T (Terekhova et al., 2009). Another study by Ghasemi et al. (2007) demonstrated the inhibitory effect of *Chlamydomonas reinhardtii*, *Chlorella vulgaris*, and *Chroococcus disperses* on *Salmonella typhi*, *Staphylococcus aureus*, *Staphylococcus epidermidis*, *Escherichia coli*, *Bacillus subtilis*, and *Pseudosomonas aeruginosa*.

6.1.2 Anti-fungal Compounds

Since 1970, fungal infections have been on the rise, primarily due to the onset of immunosuppressive viral infections such as AIDS and widespread indiscriminate utilization of anti-bacterial and steroidal drugs that have immunosuppressive consequences increasing invasive *Aspergillus* species and *Candida* species-associated infections (Bajpai et al., 2019). Candidiasis is common in women, and negligence in treatment can lead to infertility and arthritis (Pellati et al., 2008; Yordanov et al., 2004).

The bioactive molecule from the microalga *Fischerella ambigua,* Parsiguine, has exhibited anti-fungal activity against the yeast *Candida krusei* ATCC 44507 (Ghasemi et al., 2004). Freshwater microalgal species namely *Oscillatoria sp.* and *Chlorococcus sp.* have demonstrated anti-fungal activities against *Saccharomyces cerevisiae*, *Candida albicans,* and *Candida tropicalis* (Stirk and van Staden, 2022). Ethanol extracts of *Haematococcus pluvialis* have exhibited antifungal activity against *Candida albicans* and *Aspergillus niger* owing to the presence of methyl lactate and butanoic acid (Santoyo et al., 2009). Anti-fungal potential of *Chlamydomonas reinhardtii* was reported by Ghasemi et al. (2007) against *Aspergillus fumigatus*, *Candida kefyr*, and *Aspergillus niger*. Diethyl ether extract of *Scenedesmus obliquus* was tested for anti-fungal properties against *Aspergillus steinii*, *Aspergillus flavus*, *Aspergillus parasiticus*, *Aspergillus carbonarius*, *Aspergillus westerdijikia*, *Aspergillus ochraceus*, *Fusarium proliferatum*, *Fusarium verticillioides*, and *Penicillium verrucosum*. The diethyl extract exhibited anti-fungal activity against all the fungal species. On the contrary, ethanolic, methanolic, acetone, chloroform, and aqueous extracts of *Scenedesmus obliquus* did not show anti-fungal activities against all the listed species (Marrez et al., 2019). A polyol molecule named Karatungiols A, extracted from *Amphidinium sp.*, could effectively inhibit *Aspergillus niger* (Washida et al., 2006). Interestingly, the anti-fungal properties of microalgal extracts can differ based on the cultivation strategy. *Chlorella vulgaris* and *Chlorella minutissima* extracts possessed strong activity against the fungal species *Aspergillus niger* and *Fusarium oxysporum* when Iroko tree extract water was used for cultivation. The anti-fungal activity decreased when Bold's Basal Media and Istanbul's wastewater were used for cultivation (Vehapi et al., 2018).

6.1.3 Anti-viral Compounds

Mutations against specific drugs are common in viral agents. Hence, the search for novel anti-viral bioactive molecules is a necessity to counter the rising number of viral diseases and mortality. Anti-viral molecules from microalgae constitute a relatively less explored field of study. Since bioactive molecules from microalgae are relatively less exposed to the virus pool, mutations against the anti-viral molecules are yet to occur (Bratchkova and Kroumov, 2020). In 1958, a research team led by Gerber, first documented anti-viral activity from algae (Wei et al., 2022). Extracted polysaccharides from the red alga *Gelidium cartilagenium* demonstrated anti-viral activity against mumps and influenza B virus, thereby protecting embryonated eggs.

Viral infections generally have 3 stages. The initial stage comprises the adsorption of the viral agents to the host cells followed by invasion in the cells. In the second stage, the host cell makes multiple copies of the virus. The third stage includes exit of the viral particles from the hijacked host cell (Amaro et al., 2011). Acyclovir, an anti-viral molecule, inhibits the viral activity of Herpes Simplex Virus in the second stage (Snoeck, 2000). On the contrary, the anti-HSV factor of *Dunaliella spp.* inhibits the viral activity in the initial stage, hindering adsorption and invasion (Bratchkova and Kroumov, 2020). Pheophorbide-like molecules were reported to be present in the *Dunaliella primolecta* extract through NMR and MS analysis with anti-HSV-1 activity (Bratchkova and Kroumov, 2020). Sulfated polysaccharides, produced by the cell wall of the microalga *Porphyridium* sp., have been reported for their anti-viral activity and have been tested against Herpes Simplex Virus 1 and 2 and Varicella Zoster Virus (Huleihel et al., 2001).

Cellular extracts of *Isochrysis galbana var Tiso*, *Chlorella autotrophica*, *Porphyridium cruentum*, *Dunaliella tertiolecta,* and *Ellipsoidon sp.* inhibited stage 2 of Viral Hemorrhagic Septicemia Virus (VHSV). Similarly, the extracts of *Ellipsoidon* sp., *Chlorella autotrophica*, and *Porphyridium cruentum* also showed activity against the African Swine Fever Virus (ASFV) (Bratchkova and Kroumov, 2020). Naviculan, a sulfated polysaccharide obtained from *Navicula directa*, has been reported to demonstrate antiviral activities against HSV 1, HSV 2,

Influenza virus, and HIV. Naviculan restricts the viral replication in the initial stages. Further studies have revealed that the sulfated polysaccharide interferes with the attachment of HIV by restricting the viral glycoprotein 160 and the host CD4 receptor fusion (Lee et al., 2006). A sulfated polysaccharide, p-KG03—isolated from the microalga *Gyrodinium impudicum*, was tested against Encephalomyocarditis Virus (EMCV) and was reported to show anti-viral activity (Yim et al., 2004). Ahirwar et al. (2022) documented that polysaccharides A and B, produced by *Cochlodinium polykrikoides*, exhibited anti-viral activity against influenza virus types A and B, human immunodeficiency virus type 1, and respiratory syncytial virus types A and B.

6.2 Fatty Acids

Fatty acids can be differentiated on the basis of unsaturation or on the basis of the hydrocarbon chain length. Based on the length of the hydrocarbon chain, the different types of fatty acids include Short-chain Fatty Acid (SCFA), Medium-chain Fatty Acid (MCFA), Long-chain Fatty Acid (LCFA), and Very-long-chain Fatty Acid (VLCFA). The unsaturated fatty acids are classified into Monounsaturated Fatty Acid (MUFA) and Polyunsaturated Fatty Acid (PUFA).

6.2.1 Short Chain Fatty Acids and Medium Chain Fatty Acids

The significance of short chain fatty acids lies in their contribution to the gut immunity and as energy sources for the colon cells. Short chain fatty acids can regulate innate immune cells such as macrophages, neutrophils, and dendritic cells. Also, SCFAs aid in the regulation of B and T cell function and differentiation (Yao et al., 2022). Presence of a short chain fatty acid (C6:0) in the microalga *Nannochloropsis* sp. was reported by Melanie and Fithriani (2020). The content of the Hexanoic acid (C6:0) however was only 3.93% of the total fatty acid content at pH 3 and was further reduced to 1.48% of the total fatty acid content at pH 5. Caprylic Acid (C8:0) was accumulated in *Galderia sulphuraria* cells when grown under autotrophic conditions (Barone et al., 2020).

Medium Chain Fatty Acids are a source of energy and play significant roles in the metabolic system including regulation of glucose and lipid metabolism and hormone secretion (Huang et al., 2021). The microalga *Prymnesium parvum* was reported to produce and accumulate a medium chain fatty acid C14:0 (Maltsev and Maltseva, 2021). In the algal species *Emiliania huxleyi*, C14:0 is reported to be one of the major fatty acids present (Maltsev and Maltseva, 2021). Lang et al. (2011) reported C14:0 as one of the fatty acid constituents in the microalgae *Chlamydomonas asymmetrica* and *Chlorella sp.* Similarly, the production of C14:0 medium chain fatty acid was also reported in *Rhizoclonium riparium* by Osuna-Ruiz et al. (2019) and in *Dunaliella salina* by Azachi et al. (2002). However, the increase in salt concentration of the cultivation medium positively influenced the accumulation of fatty acids C14:0 and C14:2, thereby resulting in an increase of their concentration (Azachi et al., 2002).

6.2.2 Long Chain Fatty Acids

Long Chain Fatty Acids like palmitic (C16:0) acid, palmitoleic (C16:1) acid, stearic (C18:0) acid, oleic (C18:1) acid, linoleic (C18:2) acid, and linolenic (C18:3) acid are most commonly found in microalgae (Liu and Liu, 2017). Long chain fatty acids play a part as metabolic intermediates, energy source, membrane constituents, and in immune system as signaling molecules (Hidalgo et al., 2021; Nakamura et al., 2014). The content of palmitic acid in different *Chlamydomonas* strains have been reported by Lang et al. (2011). In *Chlamydomonas applanate*, the content of palmitic acid was found to be 88.57%. In *Chlamydomonas media* palmitic acid was reported to be 88.12%. Studies on other *Chlamydomonas* strains like *Chlamydomonas callosa*, *Chlamydomonas monadina*, *Chlamydomonas maxima*, and *Chlamydomonas proteus* revealed the palmitic acid content to be 54.0%, 57.36%, 57.49%, and 59.36% respectively. The microalga *Nannochloropsis* sp. was also reported to contain 61.06% palmitic acid (Melanie and Fithriani, 2020). High concentrations of Stearic Acid (C18:0) was found to be present in microalgae like *Parachlorella kessleri* and *Tetracystis texensis*: 71.50% and 66.79% respectively (Lang et al., 2011).

Monounsaturated Fatty Acids (MUFAs) have been associated with playing preventive roles in obesity, diabetes, and cardiovascular diseases (Schwingshackl and Hoffmann, 2012). Microalga like *Nannochloropsis* sp. can produce palmitoleic acid (C16:1) amounting to about 23.51% of the total fatty acid content at a slightly acidic pH of 5 (Melanie and Fithriani, 2020). Another monounsaturated fatty acid, oleic acid is also found in microalgae. 21.91% of oleic acid was reportedly accumulated in *Isochrysis galbana* (Nielsen et al., 2019). In *Chlorella vulgaris* SDEC-3 M, the oleic acid constituent was reported to be 20.02% (Qi et al., 2019). The fatty acid C18:1 was reported to be 73.10% in *Tribonema aequale* (Maltsev and Maltseva, 2021).

Very long chain fatty acids are rarely observed in microalgae. Lignoceric Acid (C24:0) has been reported to be found in the microalgae *Lagerheimia hindakii* (Lang et al., 2011), *Rhodomonas baltica* (Patil et al., 2007), and *Zygnema stellinum* (Ghazala and Shameel, 2005). Fatty acid chain of C28:8 has been reported in the microalgae *Scrippsiella* sp., *Prorocentrum mexicanum*, *Gymnodinium* sp., and

Fragilidium sp. The presence of another C28 fatty acid, C28:7 has been reported in *Symbiodinium microadriaticum* and *Prorocentrum micans* (Maltsev and Maltseva, 2021).

6.2.3 Polyunsaturated Fatty Acids

Omega-6 fatty acids include fatty acids like Dihomo-Gamma-Linolenic Acid (DGLA) and Linoleic Acid (LA) whereas omega-3 fatty acids include Docosapentaenoic Acid (DPA), Docosahexaenoic Acid (DHA), and Eicosapentaenoic Acid (EPA). These fatty acids are also referred to as essential fatty acids.

The linolenic acid content in the microalgae *Heterococcus endolithicus* and *Trebouxia simplex* was reported to be 53.37% and 45.90% of the total fatty acid content respectively (Lang et al., 2011). The presence of linolenic acid was also confirmed in other microalgal species such as *Bracteacoccus bullatus* strain MZ–Ch11 (Mamaeva et al., 2018), *Bracteacoccus bullatus* strain MZ–Ch32 (Maltsev et al., 2020), and *Chlorella sorokiniana* (Maltsev et al., 2020) constituting 13.90%, 23.80%, and 36.0% linolenic acid respectively. Gamma-linolenic acid, one of the essential fatty acids belonging to the omega-6 fatty acid family, has been reported to be rarely present in microalgae. The microalgae, *Chlorococcum* sp. (strain 2076), *Chlamydomonas zebra* (strain 25.86), and *Cryptomonas* sp. (strain 20.88) were reported to accumulate 28.7%, 18.3%, and 17.7% GLA respectively (Lang et al., 2011).

Arachidonic acid (ARA) is an omega-6 fatty acid. Different studies have reported that ARA constitutes 34.30%, 41.3%, and 58.90% of the total fatty acid content in *Khawkinea quartana* (strain 1204-9) (Lang et al., 2011), *Rhabdomonas incurve* (Lang et al., 2011), and *Parietochloris incisa* (Khozin-Goldberg et al., 2002) respectively.

Alpha-linolenic acid or α-linolenic acid (ALA) is an essential fatty acid belonging to the omega-3 fatty acid family. *Chlamydomonas reinhardtii* and commercially available *Chlorella* were reported to contain 42.4% and 23.4% ALA of the total fatty acid content respectively (Darwish et al., 2020). *Dunaliella tertiolecta* (Nielsen et al., 2019), *Chaetopeltis orbicularis* (Lang et al., 2011), *Carteria* (Lang et al., 2011), *Scenedesmus obliquus* (Oliveira et al., 2020) also have high ALA content accounting upto 60.2%, 57.5%, 54.6% and 41.17% of their total fatty acid content respectively. High ALA content was also reported in the fatty acid composition of microalgae like *Poterioochromonas malhamensis* (strain 933-1d)—39.50% (Lang et al., 2011), *Nannochloropsis* sp.—35.70% (Melanie and Fithriani, 2020), *Ellipsoidion parvum* (strain 40.86)—42.00% (Lang et al., 2011), *Chilomonas* sp. (strain 977-2b)—28.90% (Lang et al., 2011), *Rhodomonas salina*—21.14% (Nielsen et al., 2019), *Tribonema aequale* (strain 880-1)—25.30% (Lang et al., 2011), and *Rhodomonas lens*—16.00% (Maltsev and Maltseva, 2021).

Stearidonic acid (SDA), an omega-3 fatty acid and a derivative of ALA is also reported in various microalgal studies, especially in the species *Tetraselmis suecica*—11.75% (Nielsen et al., 2019). *Ankistrodesmus*, *Chlamydocapsa*, and *Chlamydomonas* strains can also accumulate SDA upto 21.4%, 27.2%, and 52.3% of the total fatty acid content (Lang et al., 2011) respectively. *Amphidinium carterae* (strain Amp) (Lang et al., 2011) and *Scrippsiella* sp. (Maltsev and Maltseva, 2021) have also been reported for SDA amounts comprising 33.00% and 10.60% of the total fatty acid content respectively.

Microalgae also serve as a source of poly unsaturated fatty acids like Eicosapentaenoic Acid (EPA) and Docosahexaenoic Acid (DHA). *Chlamydomonas allensworthii* (Lang et al., 2011) and *Tetraselmis suecica* (Nielsen et al., 2019) have been reported to contain 24.00% EPA and 19.47% EPA respectively. Similarly, EPA content as reported in various microalgal cultures include *Phaeodactylum tricornutum* (strain 1090-1b)—23.80% (Lang et al., 2011), *Biddulphia aurica*—26.0%, *Hemiselmis brunescens*—11.00%, *Rhodomonas lens*—13.00%, *Rhodomonas salina*—15.22%, *Nannochloropsis salina*—44.20%, and *Monodus subterraneus*—37.10% (Maltsev and Maltseva, 2021).

Significant production and accumulation of DHA has been reported for the algal species *Crypthecodinium cohnii*, *Crypthecodinium* sp., and *Schizochytrium* sp. (Maltsev and Maltseva, 2021). Apart from these strains, various other microalgal species have also been reported to produce and accumulate DHA such as *Ceratium horridum* (strain Cer) that accumulates 29.30% of the total fatty acids as DHA (Lang et al., 2011), *Gymnodinium sanguineum* that accumulates 24.20% of the total fatty acids as DHA, and *Isochrysis galbana* that accumulates 28.37% of the total fatty acids as DHA (Nielsen et al., 2019). Besides, *Chlorella* sp. has been reported to accumulate upto 26.13% DHA (Sivaramakrishnan and Incharoensakdi, 2020) and *Pavlova* sp. (NIVA-4/92) has been reported to accumulate upto 13.20% DHA (Patil et al., 2007). On the contrary, DPA is found in very few microalgal species such as *Pyrocystis lunula* (SAG 2014)—41.08% and *Trachelomonas volvocina* (SAG 1283-4)—23.66% (Lang et al., 2011).

6.3 Carotenoids

Carotenoids are made up of a long hydrocarbon chain. This hydrocarbon backbone consists of isoprene units (Gong and Bassi, 2016). Carotenoids are typically divided into two major groups: xanthophylls and carotenes. Carotenes consist only of hydrocarbons whereas the carotenoids consisting of an oxygenated functional group are referred to as xanthophylls (Guedes et al., 2011b). The presence of hydroxyl and keto groups make the xanthophylls hydropilic (Gong and Bassi, 2016). Therefore, xanthophylls are also associated with membranes and protein binding (Guedes et al., 2011b). Carotenoids are further divided into primary or secondary. Primary carotenoids like lutein are involved structurally and functionally with the cellular photosynthetic system (Jin et al., 2003b). On the other hand, secondary carotenoids such as asthaxanthin accumulate in the cells post-exposure to environmental stress conditions (Guedes et al., 2011b). Apart from being involved in photosynthesis, carotenoids can also function as anti-oxidants and are sensitive to oxygen, light, and heat (Gong and Bassi, 2016). Carotenoids play a major protective role in human health, mainly attributed to their antioxidant functionality. The scavenging potential of carotenoids can reduce the free radicals such as ROS (Reactive Oxygen Species) thereby providing protection against immunocompromisation, cardiovascular diseases, arthritis, diabetes, neurodegeneration, and cataract (Dufossé et al., 2005; Varela et al., 2015).

6.3.1 Lutein

The primary carotenoid, lutein, is present in microalgae. Lutein is a xanthophyll and constitutes one hydroxyl group at each end of the hydrocarbon chain (Alves-Rodrigues and Shao, 2004). In plants, lutein acts as an antioxidant protecting the plants from light-induced free radical generation (Demmig-Adams et al., 1996). Lutein screens out blue light, which has the highest energy, thereby protecting the plants from the reactive oxygen species generation (Alves-Rodrigues and Shao, 2004). In humans, lutein is present in high concentrations in the eyes and provides protection against the high energy blue light (Alves-Rodrigues and Shao, 2004). Light induced oxidative stress can damage the eyes and oxidation and precipitation of proteins, present in the eyes, result in the formation of cataract.

Commercially, besides being used as a food colourant, lutein is also used to intensify the yellow colour of egg yolk (Lin et al., 2015). Lutein has also found its application in pharmaceutical, cosmetic, and aquaculture feed industries (Cezare-Gomes et al., 2019). The microalga *Chlorella protothecoides* can produce high levels of lutein with an yield upto 200 mg L^{-1} under heterotrophic conditions (Wei et al., 2008). *Dunaliella salina*, *Scenedesmus almeriensis*, and *Galdieria sulphuraria* have also been reported to produce lutein at concentrations of about 6.6 ± 0.9 mg g^{-1}, 5.3 mg g^{-1}, and 0.4 ± 0.1 mg g^{-1} respectively (Sun et al., 2016).

6.3.2 β-Carotene

Beta carotene is a yellow-orange coloured primary carotenoid pigment which serves as a precursor of Vitamin A (Rajesh et al., 2017). Microalga such as *Dunaliella bardawil* demonstrate immunity against high energy light like UV-A and blue light induced stress owing to the presence of the beta carotene pigment (Lorenz and Cysewski, 2000). Apart from being a major Vitamin A precursor, beta carotene also possesses antioxidant properties to quench ROS (Hughes, 2001). Studies have also reported the role of beta carotene in the immune system. It helps to increase the T and B cell proliferation, increases natural killer cell cytotoxicity, and restricts mutagenesis (Hughes, 2001). *Dunaliella salina* has been reported to produce 1.25–2.45 mg L^{-1} of beta carotene depending on the cultivation parameters employed (Hejazi et al., 2004; Pruvost et al., 2008). Applications of this carotenoid is widespread as food colour in food and feed industries, as retinol in cosmetic industries, and as an anticancer agent in pharmaceutical industries (Chiu et al., 2017; Saini and Keum, 2018). A comparative adjunctive study on oral cancer treatment using *Dunaliella salina* and conventional treatment demonstrated the anti-inflammatory and anti-proliferative effects of beta carotene on the cancer cells when treatment was conducted with *Dunaliella salina* biomass (Chiu et al., 2017).

6.3.3 Zeaxanthin

Zeaxanthin is a yellow-orange coloured xanthophyll. Similar to lutein, it is associated with cellular protection against high energy light (Cezare-Gomes et al., 2019). Zeaxanthin is also used in the food industry as a colourant and in the cosmetic industry as a potent antioxidant (Bhosale and Bernstein, 2005). When cultivated under high light intensity, *Dunaliella salina* can potentially serve as a sustainable source for zeaxanthin production (2.62 mg g^{-1}); while a mutant strain of the same organism can produce upto 4.18 mg g^{-1} of zeaxanthin under high intensity light (Jin et al., 2003a). Another microalga, *Porphyridium cruentum* can also produce zeaxanthin upto 97.4% of the total carotenoid (Schubert et al., 2006). Together with lutein, the presence of zeaxanthin also plays a protective role in the macular and lens region (Alves-Rodrigues and Shao, 2004).

6.3.4 Canthaxanthin

Canthaxanthin is a dark red secondary carotenoid derived from beta carotene and is a more potent antioxidant than its precursor molecule (Cezare-Gomes et al., 2019). Canthaxanthin is extensively used as a food colourant and in fish and poultry culture feed (Cezare-Gomes et al., 2019). Studies have reported the positive effect of the carotenoid, when in association with other biomolecules, in deterrence of light induced carcinogenesis in mice and aid towards the maintenance of the immune system (Suhonen and Plosila, 2009). Microalgae like *Chlorella emersonii* (Chekanov, 2023), *Scenedesmus komarekii* (Přibyl et al., 2015), and *Chlorella zofingiensis* (Pelah et al., 2004), when cultivated under high light intensity and nitrogen limitation, produce canthaxanthin.

6.3.5 Astaxanthin

Astaxanthin is a red coloured esterified secondary keto-carotenoid with antioxidant properties (Cezare-Gomes et al., 2019). Astaxanthin is generally produced when the microalgal culture is in a stress environment such as under high light intensity, in high saline conditions, or in a limited nutrient environment. Similar to the formation of canthaxanthin, beta carotene acts as a precursor molecule for astaxanthin and the enzymes, β-carotene ketolase and hydroxylase, catalyze the formation of astaxanthin from beta carotene (Rajesh et al., 2017).

The production and accumulation of astaxanthin has been reported in many microalgal species like *Chlamydomonas nivalis*, *Chlorococcum* sp., *Botryococcus braunii*, *Chlorella zofingiensis*, *Neochloris wimmeri*, *Chloromonas nivalis*, *Scenedesmus* sp., *Protosiphon botryoides*, *Scotiellopsis oocystiformis*, *Trachelomonas volvocina*, and *Haematococcus pluvialis* (Orosa et al., 2000; Qin et al., 2008; Rajesh et al., 2017). The antioxidant activity of astaxanthin is considered to be greater than Vitamin E and beta carotene. Astaxanthin also has health benefitting anti-inflammatory effects and neuroprotective properties. Inhibition of carcinogenesis, liver fibrosis, and cardiovascular disease onset has also been attributed to this red pigment (Guerin et al., 2003; Monroy-Ruiz et al., 2011).

7 Conclusion

The chapter emphasizes on how diverse microorganisms can produce secondary metabolites and how important they are as novel sources of naturally occurring bioactive molecules. Bioactive substances are essential to human health and is part of a rapidly growing market. In order to meet the escalating demand, the conventional methods of synthesizing these compounds through chemical synthesis or extraction from tissues are not sustainable or environmentally friendly. On the other hand, microorganisms are comparatively easy to cultivate and are potentially more efficient for production of bioactive molecules. Though microorganisms are diverse, only a handful have been cultured and their ability to produce secondary metabolites for pharmaceutical and nutraceutical applications have been reported. Thus, exploration of microorganisms with the ability to accumulate significant bioactive molecules, the roles of such high-value metabolites, and their respective modes of action requires extensive research and development.

Acknowledgements Devargya Ganguly is thankful to the Science and Engineering Research Board, India (File No. SRG/2021/001256) for the financial support. Devargya Ganguly is grateful to his mentor, Dr. Gargi Goswami, for her constant guidance and support. Sneha Sengupta, Soumajit Ghosh, and Subhojit Sengupta are thankful to Dr. Gargi Goswami for entrusting them with the responsibility and opportunity to write this book chapter.

References

Adams, J. (2004). The development of proteasome inhibitors as anticancer drugs. *Cancer Cell, 5*, 417–421. https://doi.org/10.1016/S1535-6108(04)00120-5.

Afroz Toma, M., Rahman, M. H., Rahman, M. S., et al. (2023). Fungal pigments: Carotenoids, riboflavin, and polyketides with diverse applications. *Journal of Fungi, 9*, 454. https://doi.org/10.3390/jof9040454.

Ahirwar, A., Kesharwani, K., Deka, R., et al. (2022). Microalgal drugs: A promising therapeutic reserve for the future. *Journal of Biotechnology, 349*, 32–46. https://doi.org/10.1016/j.jbiotec.2022.03.012.

Alvarez-Macarie, E., & Baratti, J. (2000). Short chain flavour ester synthesis by a new esterase from *Bacillus licheniformis*. *Journal of Molecular Catalysis B: Enzymatic, 10*, 377–383. https://doi.org/10.1016/S1381-1177(99)00109-5.

Alves-Rodrigues, A., & Shao, A. (2004). The science behind lutein. *Toxicology Letters, 150*, 57–83. https://doi.org/10.1016/j.toxlet.2003.10.031.

Amaning Danquah, C., Minkah, P. A. B., Osei Duah Junior, I., et al. (2022). Antimicrobial compounds from microorganisms. *Antibiotics, 11*, 285. https://doi.org/10.3390/antibiotics11030285.

Amaro, H. M., Guedes, A. C., & Malcata, F. X. (2011). Antimicrobial activities of microalgae: An invited review. In A. Méndez-Vilas (Ed.), *Science against microbial pathogens: communicating current research and technological advances* (Vol.2, pp. 1272–1284). Formatex Microbiology Book Series.

Anas, A. R. J., Mori, A., Tone, M., et al. (2017). FVIIa-sTF and Thrombin inhibitory activities of compounds isolated from *Microcystis aeruginosa* K-139. *Marine Drugs, 15*, 275. https://doi.org/10.3390/md15090275.

Andrianasolo, E. H., Gross, H., Goeger, D., et al. (2005). Isolation of swinholide A and related glycosylated derivatives from two field collections of marine cyanobacteria. *Organic Letters, 7*, 1375–1378. https://doi.org/10.1021/ol050188x.

Arad, S. (Malis), & Levy-Ontman, O. (2010). Red microalgal cell-wall polysaccharides: Biotechnological aspects. *Current Opinion in Biotechnology, 21*, 358–364. https://doi.org/10.1016/j.copbio.2010.02.008.

Arunpanichlert, J., Rukachaisirikul, V., Sukpondma, Y., et al. (2010). Azaphilone and isocoumarin derivatives from the endophytic fungus *Penicillium sclerotiorum* PSU-A13. *Chemical and Pharmaceutical Bulletin, 58*, 1033–1036. https://doi.org/10.1248/cpb.58.1033.

Asencio, A. D., & Hoffmann, L. (2013). Chemosystematic evaluation of the genus *Scytonema* (Cyanobacteria) based on occurrence of phycobiliproteins, scytonemin, carotenoids and mycosporine-like amino acid compounds. *European Journal of Phycology, 48*, 331–344. https://doi.org/10.1080/09670262.2013.836682.

Asker, D., Beppu, T., & Ueda, K. (2007). *Mesoflavibacter zeaxanthinifaciens* gen. nov., sp. nov., a novel zeaxanthin-producing marine bacterium of the family Flavobacteriaceae. *Systematic and Applied Microbiology, 30*, 291–296. https://doi.org/10.1016/j.syapm.2006.12.003.

Atiphasaworn, P., Monggoot, S., Gentekaki, E., et al. (2017). Antibacterial and antioxidant constituents of extracts of endophytic fungi isolated from *Ocimum basilicum var. thyrsiflora* leaves. *Current Microbiology, 74*, 1185–1193. https://doi.org/10.1007/s00284-017-1303-1.

Avalos, J., Nordzieke, S., Parra, O., et al. (2017). *Carotenoid production by filamentous fungi and yeasts*. Springer. https://doi.org/10.1007/978-3-319-58829-2_8.

Azachi, M., Sadka, A., Fisher, M., et al. (2002). Salt induction of fatty acid elongase and membrane lipid modifications in the extreme halotolerant alga *Dunaliella salina*. *Plant Physiology, 129*, 1320–1329. https://doi.org/10.1104/pp.001909.

Bai, S. K., Lee, S. J., Na, H. J., et al. (2005). β-Carotene inhibits inflammatory gene expression in lipopolysaccharide-stimulated macrophages by suppressing redox-based NF-κB activation. *Experimental & Molecular Medicine, 37*, 323–334. https://doi.org/10.1038/emm.2005.42.

Baindara, P., & Mandal, S. M. (2020). Bacteria and bacterial anticancer agents as a promising alternative for cancer therapeutics. *Biochimie, 177*, 164–189. https://doi.org/10.1016/j.biochi.2020.07.020.

Baindara, P., Singh, N., Ranjan, M., et al. (2016). Laterosporulin10: A novel defensin like Class IId bacteriocin from *Brevibacillus* sp. strain SKDU10 with inhibitory activity against microbial pathogens. *Microbiology, 162*, 1286–1299. https://doi.org/10.1099/mic.0.000316.

Bajaj, B. K., & Manhas, K. (2012). Production and characterization of xylanase from *Bacillus licheniformis* P11(C) with potential for fruit juice and bakery industry. *Biocatalysis and Agricultural Biotechnology, 1*, 330–337. https://doi.org/10.1016/j.bcab.2012.07.003.

Bajpai, V. K., Khan, I., Shukla, S., et al. (2019). Invasive fungal infections and their epidemiology: Measures in the clinical scenario. *Biotechnology and Bioprocess Engineering, 24*, 436–444. https://doi.org/10.1007/s12257-018-0477-0.

Bansal, M., Poonia, A., Kolluri, S.R.P., & Vasundhara. (2023). Introduction on bioactive compounds, sources and their potential applications. In M. Thakur, & T. Belwal (Eds.), *Bioactive components: A sustainable system for good health and well-being* (pp. 3–26). Singapore: Springer Nature. https://doi.org/10.1007/978-981-19-2366-1_1.

Barone, R., De Napoli, L., Mayol, L., et al. (2020). Autotrophic and heterotrophic growth conditions modify biomolecole production in the microalga *Galdieria sulphuraria* (Cyanidiophyceae, Rhodophyta). *Marine Drugs, 18*, 169. https://doi.org/10.3390/md18030169.

Běhal, V. (2006). Mode of action of microbial bioactive metabolites. *Folia Microbiologica, 51*, 359–369. https://doi.org/10.1007/BF02931577.

Bérdy, J. (2005). Bioactive microbial metabolites. *The Journal of Antibiotics, 58*, 1–26. https://doi.org/10.1038/ja.2005.1.

Berry, A., Janssens, D., Hümbelin, M., et al. (2003). *Paracoccus zeaxanthinifaciens* sp. nov., a zeaxanthin-producing bacterium. *International Journal of Systematic and Evolutionary Microbiology, 53*, 231–238. https://doi.org/10.1099/ijs.0.02368-0.

Bhat, V. B., & Madyastha, K. M. (2000) C-phycocyanin: A potent peroxyl radical scavenger in vivo and in vitro.*Biochemical and Biophysical Research Communications, 275*(1), 20–25. https://doi.org/10.1006/bbrc.2000.3270

Bhateja, P., Mathur, T., Pandya, M., et al. (2006). Activity of blue green microalgae extracts against *in vitro* generated *Staphylococcus aureus* with reduced susceptibility to vancomycin. *Fitoterapia, 77*, 233–235. https://doi.org/10.1016/j.fitote.2006.01.009.

Bhatnagar, S., & Aoyagi, H. (2022). Thermal and UV degradation kinetics of water-soluble extracellular pigment produced by *Talaromyces purpurogenus*. *Food and Bioprocess Technology, 15*, 606–619. https://doi.org/10.1007/s11947-021-02733-9.

Bhatnagar, I., & Kim, S. K. (2010). Immense essence of excellence: Marine microbial bioactive compounds. *Marine Drugs, 8*, 2673–2701. https://doi.org/10.3390/md8102673.

Bhosale, P., & Bernstein, P. S. (2004). β-Carotene production by *Flavobacterium multivorum* in the presence of inorganic salts and urea. *Journal of Industrial Microbiology and Biotechnology, 31*, 565–571. https://doi.org/10.1007/s10295-004-0187-9.

Bhosale, P., & Bernstein, P. S. (2005). Microbial xanthophylls. *Applied Microbiology and Biotechnology, 68*, 445–455. https://doi.org/10.1007/s00253-005-0032-8.

Bi, M. C., Rosen, R., Zha, R. Y., et al. (2013). Zeaxanthin induces apoptosis in human uveal melanoma cells through Bcl-2 family proteins and intrinsic apoptosis pathway. *Evidence-Based Complementary and Alternative Medicine, 2013*, 205082. https://doi.org/10.1155/2013/205082.

Birch-Machin, M. A., & Bowman, A. (2016). Oxidative stress and ageing. *British Journal of Dermatology, 175*, 26–29. https://doi.org/10.1111/bjd.14906.

Bolden, J. E., Peart, M. J., & Johnstone, R. W. (2006). Anticancer activities of histone deacetylase inhibitors. *Nature Reviews Drug Discovery, 5*, 769–784. https://doi.org/10.1038/nrd2133.

Bratchkova, A., & Kroumov, A. D. (2020). Microalgae as producers of biologically active compounds with antibacterial, antiviral, antifungal, antialgal, antiprotozoal, antiparasitic and anticancer activity. *Acta Microbiologica Bulgarica, 36*, 3.

Caicedo, N. H., Davalos, A. F., Puente, P. A., et al. (2019). Antioxidant activity of exo-metabolites produced by *Fusarium oxysporum*: An endophytic fungus isolated from leaves of *Otoba gracilipes*. *Microbiology Open, 8*, e903. https://doi.org/10.1002/mbo3.903.

Calder, P. C. (2015). Functional roles of fatty acids and their effects on human health. *Journal of Parenteral and Enteral Nutrition, 39*, 18S-32S. https://doi.org/10.1177/0148607115595980.

Cañeque, T., Gomes, F., Mai, T. T., et al. (2015). Synthesis of marmycin A and investigation into its cellular activity. *Nature Chemistry, 7*, 744–751. https://doi.org/10.1038/nchem.2302.

Cantrell, A., McGarvey, D. J., George Truscott, T., et al. (2003). Singlet oxygen quenching by dietary carotenoids in a model membrane environment. *Archives of Biochemistry and Biophysics, 412*, 47–54. https://doi.org/10.1016/S0003-9861(03)00014-6.

Cezare-Gomes, E. A., Mejia-da-Silva, L. D. C., Pérez-Mora, L. S., et al. (2019). Potential of microalgae carotenoids for industrial application. *Applied Biochemistry and Biotechnology, 188*, 602–634. https://doi.org/10.1007/s12010-018-02945-4.

Chaisuwan, W., Phimolsiripol, Y., Chaiyaso, T., et al. (2021). The antiviral activity of bacterial, fungal, and algal polysaccharides as bioactive ingredients: Potential uses for enhancing immune systems and preventing viruses. *Frontiers in Nutrition*, 8. https://doi.org/10.3389/fnut.2021.772033.

Chandra, P., Sharma, R. K., & Arora, D. S. (2020). Antioxidant compounds from microbial sources: A review. *Food Research International, 129*, 108849. https://doi.org/10.1016/j.foodres.2019.108849.

Chang, K. Y., Chiang, N. J., Wu, S. Y., et al. (2021). Phase 1b study of pegylated arginine deiminase (ADI-PEG 20) plus Pembrolizumab in advanced solid cancers. *OncoImmunology, 10*, 1943253. https://doi.org/10.1080/2162402X.2021.1943253.

Chatragadda, R., & Dufossé, L. (2021). Ecological and biotechnological aspects of pigmented microbes: A way forward in development of food and pharmaceutical grade pigments. *Microorganisms, 9*, 637. https://doi.org/10.3390/microorganisms9030637.

Chaturvedi, A. D., & Tiwari, K. L. (2013). Effect of household detergents (Surfactants) degraded through aquatic fungi. *Recent Research in Science and Technology, 5*, 12–16.

Chauhan, D., Catley, L., Li, G., et al. (2005). A novel orally active proteasome inhibitor induces apoptosis in multiple myeloma cells with mechanisms distinct from Bortezomib. *Cancer Cell, 8*, 407–419. https://doi.org/10.1016/j.ccr.2005.10.013.

Chekanov, K. (2023). Diversity and distribution of carotenogenic algae in Europe: A review. *Marine Drugs, 21*, 108. https://doi.org/10.3390/md21020108.

Chiu, H. F., Liao, J. Y., Lu, Y. Y., et al. (2017). Anti-proliferative, anti-inflammatory and pro-apoptotic effects of *Dunaliella salina* on human KB oral carcinoma cells. *Journal of Food Biochemistry, 41*, e12349. https://doi.org/10.1111/jfbc.12349.

Chng, C., Lum, A. M., Vroom, J. A., et al. (2008). A key developmental regulator controls the synthesis of the antibiotic erythromycin in *Saccharopolyspora erythraea*. *Biological Sciences, 105*, 11346–11351. https://doi.org/10.1073/pnas.0803622105.

Clauditz, A., Resch, A., Wieland, K. P., et al. (2006). Staphyloxanthin plays a role in the fitness of *Staphylococcus aureus* and its ability to cope with oxidative stress. *Infection and Immunity, 74*, 4950–4953. https://doi.org/10.1128/iai.00204-06.

Cohen, E., Koch, L., Thu, K. M., et al. (2011). Novel terpenoids of the fungus *Aspergillus insuetus* isolated from the Mediterranean sponge *Psammocinia* sp. collected along the coast of Israel. *Bioorganic & Medicinal Chemistry, 19*, 6587–6593. https://doi.org/10.1016/j.bmc.2011.05.045.

Correia, R. T., Borges, K. C., Medeiros, M. F., & Genovese, M. I. (2012). Bioactive compounds and phenolic-linked functionality of powdered tropical fruit residues. *Food Science and Technology International, 18*, 539–547. https://doi.org/10.1177/1082013211433077.

Cruz, R., & Wuest, W. M. (2023). Beyond ergosterol: Strategies for combatting antifungal resistance in *Aspergillus fumigatus* and *Candida auris*. *Tetrahedron, 133*, 133268. https://doi.org/10.1016/j.tet.2023.133268.

Cruz, J. C. S., Maffioli, S. I., Bernasconi, A., et al. (2017). Allocyclinones, hyperchlorinated angucyclinones from *Actinoallomurus*. *The Journal of Antibiotics, 70*, 73–78. https://doi.org/10.1038/ja.2016.62.

Cundliffe, E., Bate, N., Butler, A., et al. (2001). The tylosin-biosynthetic genes of *Streptomyces fradiae*. *Antonie Van Leeuwenhoek, 79*, 229–234. https://doi.org/10.1023/A:1012065300116.

d'Ippolito, G., Tucci, S., Cutignano, A., et al. (2004). The role of complex lipids in the synthesis of bioactive aldehydes of the marine diatom *Skeletonema costatum*. *Biochimica et Biophysica Acta (BBA)-Molecular and Cell Biology of Lipids, 1686*, 100–107. https://doi.org/10.1016/j.bbalip.2004.09.002.

Darwish, R., Gedi, M. A., Akepach, P., et al. (2020). *Chlamydomonas reinhardtii* is a potential food supplement with the capacity to outperform chlorella and spirulina. *Applied Sciences, 10*, 6736. https://doi.org/10.3390/app10196736.

Das, B. K., Pradhan, J., Pattnaik, P., et al. (2005). Production of antibacterials from the freshwater alga *Euglena viridis* (Ehren). *World Journal of Microbiology and Biotechnology, 21*, 45–50. https://doi.org/10.1007/s11274-004-1555-3.

de Angelis, M., Mariotti, L., Rossi, J., et al. (2002). Arginine catabolism by sourdough lactic acid bacteria: Purification and characterization of the arginine deiminase pathway enzymes from *Lactobacillus sanfranciscensis* CB1. *Applied and Environmental Microbiology, 68*, 6193–6201. https://doi.org/10.1128/AEM.68.12.6193-6201.2002.

de Castro, I., Mendo, S., & Caetano, T. (2020). Antibiotics from haloarchaea: What can we learn from comparative genomics? *Marine Biotechnology, 22*, 308–316. https://doi.org/10.1007/s10126-020-09952-9.

de Oliveira, C. Y. B., Viegas, T. L., Lopes, R. G., et al. (2020). A comparison of harvesting and drying methodologies on fatty acids composition of the green microalga *Scenedesmus obliquus*. *Biomass and Bioenergy, 132*, 105437. https://doi.org/10.1016/j.biombioe.2019.105437.

de Oliveira, F., Hirai, P. R., Teixeira, M. F. S., et al. (2021). *Talaromyces amestolkiae* cell disruption and colorant extraction using imidazolium-based ionic liquids. *Separation and Purification Technology, 257*, 117759. https://doi.org/10.1016/j.seppur.2020.117759.

de Jesus Raposo, M. F., Morais A. M. M. B., & de Morais, R. M. S. C. (2014). Influence of sulphate on the composition and antibacterial and antiviral properties of the exopolysaccharide from *Porphyridium cruentum*. Life Sciences 101:56–63. https://doi.org/10.1016/j.lfs.2014.02.013.

DeBritto, S., Gajbar, T. D., Satapute, P., et al. (2020). Isolation and characterization of nutrient dependent pyocyanin from *Pseudomonas aeruginosa* and its dye and agrochemical properties. *Scientific Reports, 10*, 1542. https://doi.org/10.1038/s41598-020-58335-6.

Demain, A. L. (2014). Importance of microbial natural products and the need to revitalize their discovery. *Journal of Industrial Microbiology and Biotechnology, 41*, 185–201. https://doi.org/10.1007/s10295-013-1325-z.

Demain, A. L., & Vaishnav, P. (2011). Natural products for cancer chemotherapy. *Microbial Biotechnology, 4*, 687–699. https://doi.org/10.1111/j.1751-7915.2010.00221.x.

Demay, J., Bernard, C., Reinhardt, A., & Marie, B. (2019). Natural products from cyanobacteria: Focus on beneficial activities. *Marine Drugs, 17*, 320. https://doi.org/10.3390/md17060320.

Demmig-Adams, B., Gilmore, A. M., & Iii, W. W. A. (1996). *In vivo* functions of carotenoids in higher plants. *The FASEB Journal, 10*, 403–412. https://doi.org/10.1096/fasebj.10.4.8647339.

Dey, B., Lerner, D. L., Lusso, P., et al. (2000). Multiple antiviral activities of cyanovirin-N: Blocking of Human Immunodeficiency Virus Type 1 gp120 interaction with CD4 and coreceptor and inhibition of diverse enveloped viruses. *Journal of Virology, 74*, 4562–4569. https://doi.org/10.1128/jvi.74.10.4562-4569.2000.

Dimitrova, S., Pavlova, K., Lukanov, L., et al. (2013). Production of metabolites with antioxidant and emulsifying properties by Antarctic strain *Sporobolomyces salmonicolor* AL1. *Applied Biochemistry and Biotechnology, 169*, 301–311. https://doi.org/10.1007/s12010-012-9983-2.

Ding, Y., Seufert, W. H., Beck, Z. Q., & Sherman, D. H. (2008). Analysis of the Cryptophycin P450 epoxidase reveals substrate tolerance and cooperativity. *Journal of the American Chemical Society, 130*, 5492–5498. https://doi.org/10.1021/ja710520q.

Długosz, M., & Trylska, J. (2009). Aminoglycoside association pathways with the 30S Ribosomal subunit. *The Journal of Physical Chemistry B, 113*, 7322–7330. https://doi.org/10.1021/jp8112914.

Dufossé, L., Galaup, P., Yaron, A., et al. (2005). Microorganisms and microalgae as sources of pigments for food use: A scientific oddity or an industrial reality? *Trends in Food Science & Technology, 16*, 389–406. https://doi.org/10.1016/j.tifs.2005.02.006.

Durán, N., & Menck, C. F. M. (2001). *Chromobacterium violaceum*: A review of pharmacological and industrial perspectives. *Critical Reviews in Microbiology, 27*, 201–222. https://doi.org/10.1080/20014091096747.

Elawady, M. E., Hamed, A. A., Alsallami, W. M., et al. (2023). Bioactive metabolite from endophytic *Aspergillus versicolor* SB5 with anti-acetylcholinesterase, anti-inflammatory and antioxidant activities: *In vitro* and *in silico* studies. *Microorganisms, 11*, 1062. https://doi.org/10.3390/microorganisms11041062.

El-Naggar, N. E. A., & El-Ewasy, S. M. (2017). Bioproduction, characterization, anticancer and antioxidant activities of extracellular melanin pigment produced by newly isolated microbial cell factories *Streptomyces glaucescens* NEAE-H. *Scientific Reports, 7*, 42129. https://doi.org/10.1038/srep42129.

Esatbeyoglu, T., & Rimbach, G. (2017). Canthaxanthin: From molecule to function. *Molecular Nutrition & Food Research, 61*, 1600469. https://doi.org/10.1002/mnfr.201600469.

Feling, R. H., Buchanan, G. O., Mincer, T. J., et al. (2003). Salinosporamide A: A highly cytotoxic proteasome inhibitor from a novel microbial source, a marine bacterium of the new genus *Salinospora*. *Angewandte Chemie International Edition, 42*, 355–357. https://doi.org/10.1002/anie.200390115.

Fidor, A., Konkel, R., & Mazur-Marzec, H. (2019). Bioactive peptides produced by cyanobacteria of the genus *Nostoc*: A review. *Marine Drugs, 17*, 561. https://doi.org/10.3390/md17100561.

Frankel, E. N., & Meyer, A. S. (2000). The problems of using one-dimensional methods to evaluate multifunctional food and biological antioxidants. *Journal of the Science of Food and Agriculture, 80*, 1925–1941. https://doi.org/10.1002/1097-0010(200010)80:13%3c1925::AID-JSFA714%3e3.0.CO;2-4.

Frediansyah, A., Sofyantoro, F., Alhumaid, S., et al. (2022). Microbial natural products with antiviral activities, including anti-SARS-CoV-2: A review. *Molecules, 27*, 4305. https://doi.org/10.3390/molecules27134305.

Galasso, C., Corinaldesi, C., & Sansone, C. (2017). Carotenoids from marine organisms: Biological functions and industrial applications. *Antioxidants, 6*, 96. https://doi.org/10.3390/antiox6040096.

Galkin, A., Kulakova, L., Sarikaya, E., et al. (2004). Structural insight into arginine degradation by arginine deiminase, an antibacterial and parasite drug target. *Journal of Biological Chemistry, 279*, 14001–14008. https://doi.org/10.1074/jbc.M313410200.

Gao, S. S., Li, X. M., Du, F. Y., et al. (2011a). Secondary metabolites from a marine-derived endophytic fungus *Penicillium chrysogenum* QEN-24S. *Marine Drugs, 9*, 59–70. https://doi.org/10.3390/md9010059.

Gao, S. S., Li, X. M., Li, C. S., et al. (2011b). Penicisteroids A and B, antifungal and cytotoxic polyoxygenated steroids from the marine alga-derived endophytic fungus *Penicillium chrysogenum* QEN-24S. *Bioorganic & Medicinal Chemistry Letters, 21*, 2894–2897. https://doi.org/10.1016/j.bmcl.2011.03.076.

Gao, J. M., Yang, S. X., & Qin, J. C. (2013). Azaphilones: Chemistry and biology. *Chemical Reviews, 113*, 4755–4811. https://doi.org/10.1021/cr300402y.

Gao, X. W., Liu, H. X., Sun, Z. H., et al. (2016). Secondary metabolites from the deep-sea derived fungus *Acaromyces ingoldii* FS121. *Molecules, 21*, 371. https://doi.org/10.3390/molecules21040371.

Ghasemi, Y., Yazdi, M. T., Shafiee, A., et al. (2004). Parsiguine, a novel antimicrobial substance from *Fischerella ambigua*. *Pharmaceutical Biology, 42*, 318–322. https://doi.org/10.1080/13880200490511918.

Ghasemi, Y., Moradian, A., Mohagheghzadeh, A., et al. (2007). Antifungal and antibacterial activity of the microalgae collected from paddy fields of Iran: Characterization of antimicrobial activity of *Chroococcus dispersus*. *Journal of Biological Sciences, 7*, 904–910. https://doi.org/10.3923/jbs.2007.904.910.

Ghazala, B., & Shameel, M. (2005). Phytochemistry and bioactivity of some freshwater green algae from Pakistan. *Pharmaceutical Biology, 43*, 358–369. https://doi.org/10.1080/13880200590951838.

Ghorbel, R. E., Maktouf, S., Massoud, E. B., et al. (2009). New thermostable amylase from *Bacillus cohnii* US147 with a broad pH applicability. *Applied Biochemistry and Biotechnology, 157*, 50–60. https://doi.org/10.1007/s12010-008-8278-0.

Gong, M., & Bassi, A. (2016). Carotenoids from microalgae: A review of recent developments. *Biotechnology Advances, 34*, 1396–1412. https://doi.org/10.1016/j.biotechadv.2016.10.005.

Guedes, A. C., Amaro, H. M., & Malcata, F. X. (2011a). Microalgae as sources of high added-value compounds—a brief review of recent work. *Biotechnology Progress, 27*, 597–613. https://doi.org/10.1002/btpr.575.

Guedes, A. C., Amaro, H. M., & Malcata, F. X. (2011b). Microalgae as sources of carotenoids. *Marine Drugs, 9*, 625–644. https://doi.org/10.3390/md9040625.

Guerin, M., Huntley, M. E., & Olaizola, M. (2003). *Haematococcus* astaxanthin: Applications for human health and nutrition. *Trends in Biotechnology, 21*, 210–216. https://doi.org/10.1016/S0167-7799(03)00078-7.

Gulder, T. A. M., & Moore, B. S. (2010). Salinosporamide natural products: Potent 20 S proteasome inhibitors as promising cancer chemotherapeutics. *Angewandte Chemie International Edition, 49*, 9346–9367. https://doi.org/10.1002/anie.201000728.

Guo, Q., Peng, Q. Q., Li, Y. W., et al. (2024). Advances in the metabolic engineering of *Saccharomyces cerevisiae* and *Yarrowia lipolytica* for the production of β-carotene. *Critical Reviews in Biotechnology, 44*, 337–351. https://doi.org/10.1080/07388551.2023.2166809.

Gupta, V., & Vyas, D. (2021). Antimicrobial effect of a cyclic peptide Nostophycin isolated from wastewater cyanobacteria, *Nostoc calcicola*. *Current Botany*, 94–101. https://doi.org/10.25081/cb.2021.v12.6612.

Hall, S., McDermott, C., Anoopkumar-Dukie, S., et al. (2016). Cellular effects of pyocyanin, a secreted virulence factor of *Pseudomonas aeruginosa*. *Toxins, 8*, 236. https://doi.org/10.3390/toxins8080236.

Ham, S. S., Kim, S. H., Moon, S. Y., et al. (2009). Antimutagenic effects of subfractions of Chaga mushroom (*Inonotus obliquus*) extract. *Mutation Research/genetic Toxicology and Environmental Mutagenesis, 672*, 55–59. https://doi.org/10.1016/j.mrgentox.2008.10.002.

Hameed, A., Arun, A. B., Ho, H. P., et al. (2011). Supercritical carbon dioxide micronization of zeaxanthin from moderately thermophilic bacteria *Muricauda lutaonensis* CC-HSB-11T. *Journal of Agricultural and Food Chemistry, 59*, 4119–4124. https://doi.org/10.1021/jf200444q.

Han, B., McPhail, K. L., Gross, H., et al. (2005a). Isolation and structure of five lyngbyabellin derivatives from a Papua New Guinea collection of the marine cyanobacterium *Lyngbya majuscula*. *Tetrahedron, 61*, 11723–11729. https://doi.org/10.1016/j.tet.2005.09.036.

Han, J. R., Zhao, W. J., Gao, Y. Y., & Yuan, J. M. (2005b). Effect of oxidative stress and exogenous β-carotene on sclerotial differentiation and carotenoid yield of *Penicillium* sp. PT95. *Letters in Applied Microbiology, 40*, 412–417. https://doi.org/10.1111/j.1472-765X.2005.01697.x.

Hannibal, L., Lorquin, J., D'Ortoli, N. A., et al. (2000). Isolation and characterization of canthaxanthin biosynthesis genes from the photosynthetic bacterium *Bradyrhizobium* sp. strain ORS278. *Journal of Bacteriology*, 142. https://doi.org/10.1128/jb.182.13.3850-3853.2000.

Hejazi, M. A., Holwerda, E., & Wijffels, R. H. (2004). Milking microalga *Dunaliella salina* for β-carotene production in two-phase bioreactors. *Biotechnology and Bioengineering, 85*, 475–481. https://doi.org/10.1002/bit.10914.

Hetz, C., Bono, M. R., Barros, L. F., & Lagos, R. (2002). Microcin E492, a channel-forming bacteriocin from *Klebsiella pneumoniae*, induces apoptosis in some human cell lines. *Proceedings of the National Academy of Sciences, 99*, 2696–2701. https://doi.org/10.1073/pnas.052709699.

Hidalgo, M. A., Carretta, M. D., & Burgos, R. A. (2021). Long chain fatty acids as modulators of immune cells function: Contribution of FFA1 and FFA4 receptors. *Frontiers in Physiology*, 12. https://doi.org/10.3389/fphys.2021.668330.

Huang, Z., Nong, X., Ren, Z., et al. (2017). Anti-HSV-1, antioxidant and antifouling phenolic compounds from the deep-sea-derived fungus *Aspergillus versicolor* SCSIO 41502. *Bioorganic & Medicinal Chemistry Letters, 27*, 787–791. https://doi.org/10.1016/j.bmcl.2017.01.032.

Huang, L., Gao, L., & Chen, C. (2021). Role of medium-chain fatty acids in healthy metabolism: A clinical perspective. *Trends in Endocrinology & Metabolism, 32*, 351–366. https://doi.org/10.1016/j.tem.2021.03.002.

Hughes, D. A. (2001). Dietary carotenoids and human immune function. *Nutrition, 17*, 823–827. https://doi.org/10.1016/S0899-9007(01)00638-4.

Huleihel, M., Ishanu, V., & Tal, J. (2001). Antiviral effect of red microalgal polysaccharides on Herpes simplex and Varicella zoster viruses. *Journal of Applied Phycology, 13*, 127–134.

Humisto, A., Jokela, J., Liu, L., et al. (2018). The swinholide biosynthesis gene cluster from a terrestrial cyanobacterium, *Nostoc* sp. strain UHCC 0450. *Applied and Environmental Microbiology, 84*, e02321-e2417. https://doi.org/10.1128/AEM.02321-17.

Inès, M., & Dhouha, G. (2015). Glycolipid biosurfactants: Potential related biomedical and biotechnological applications. *Carbohydrate Research, 416*, 59–69. https://doi.org/10.1016/j.carres.2015.07.016.

Iwai, R., Uchida, S., Yamaguchi, S., et al. (2023). Effects of LPS from *Rhodobacter sphaeroides*, a purple non-sulfur bacterium (PNSB), on the gene expression of rice root. *Microorganisms, 11*, 1676. https://doi.org/10.3390/microorganisms11071676.

Jaccard, A., Gachard, N., Marin, B., et al. (2011). Efficacy of L-asparaginase with methotrexate and dexamethasone (AspaMet-Dex regimen) in patients with refractory or relapsing extranodal NK/T-cell lymphoma, a phase 2 study. *Blood, 117*, 1834–1839. https://doi.org/10.1182/blood-2010-09-307454.

Janek, T., Łukaszewicz, M., Rezanka, T., & Krasowska, A. (2010). Isolation and characterization of two new lipopeptide biosurfactants produced by *Pseudomonas fluorescens* BD5 isolated from water from the Arctic Archipelago of Svalbard. *Bioresource Technology, 101*, 6118–6123. https://doi.org/10.1016/j.biortech.2010.02.109.

Janek, T., Krasowska, A., Radwańska, A., & Łukaszewicz, M. (2013). Lipopeptide biosurfactant pseudofactin II induced apoptosis of melanoma A 375 cells by specific interaction with the plasma membrane. *PLoS ONE, 8*, e57991. https://doi.org/10.1371/journal.pone.0057991.

Jang, K. H., Nam, S. J., Locke, J. B., et al. (2013). Anthracimycin, a potent anthrax antibiotic from a marine-derived actinomycete. *Angewandte Chemie*, 52. https://doi.org/10.1002/anie.201302749. https://doi.org/10.1002/anie.201302749.

Javed, S., Azeem, F., Hussain, S., et al. (2018). Bacterial lipases: A review on purification and characterization. *Progress in Biophysics and Molecular Biology, 132*, 23–34. https://doi.org/10.1016/j.pbiomolbio.2017.07.014.

Jin, E., Feth, B., & Melis, A. (2003a). A mutant of the green alga *Dunaliella salina* constitutively accumulates zeaxanthin under all growth conditions. *Biotechnology and Bioengineering, 81*, 115–124. https://doi.org/10.1002/bit.10459.

Jin, E.-S., Polle, J. E. W., Lee, H.-K., et al. (2003b). Xanthophylls in Microalgae: From biosynthesis to biotechnological mass production and application. *Journal of Microbiology and Biotechnology, 13*, 165–174.

Joshi, C., & Singhal, R. S. (2016). Modelling and optimization of zeaxanthin production by *Paracoccus zeaxanthinifaciens* ATCC 21588 using hybrid genetic algorithm techniques. *Biocatalysis and Agricultural Biotechnology, 8*, 228–235. https://doi.org/10.1016/j.bcab.2016.10.004.

Jüttner, F. (2001). Liberation of 5,8,11,14,17-eicosapentaenoic acid and other polyunsaturated fatty acids from lipids as a grazer defense reaction in epilithic diatom biofilms. *Journal of Phycology, 37*, 744–755. https://doi.org/10.1046/j.1529-8817.2001.00130.x.

Kapuścik, A., Hrouzek, P., Kuzma, M., et al. (2013). Novel aeruginosin-865 from *Nostoc* sp. as a potent anti-inflammatory agent. *ChemBioChem, 14*, 2329–2337. https://doi.org/10.1002/cbic.201300246.

Kardani, K., & Bolhassani, A. (2021). Antimicrobial/anticancer peptides: Bioactive molecules and therapeutic agents. *Immunotherapy, 13*, 669–684. https://doi.org/10.2217/imt-2020-0312.

Karray, F., Darbon, E., Oestreicher, N., et al. (2007). Organization of the biosynthetic gene cluster for the macrolide antibiotic spiramycin in *Streptomyces ambofaciens*. *Microbiology, 153*, 4111–4122. https://doi.org/10.1099/mic.0.2007/009746-0.

Kavitha, K., Kowshik, J., Kishore, T. K. K., et al. (2013). Astaxanthin inhibits NF-κB and Wnt/β-catenin signaling pathways via inactivation of Erk/MAPK and PI3K/Akt to induce intrinsic apoptosis in a hamster model of oral cancer. *Biochimica et Biophysica Acta (BBA)-General Subjects, 1830*, 4433–4444. https://doi.org/10.1016/j.bbagen.2013.05.032.

Kenari, S. L. D., Alemzadeh, I., & Maghsodi, V. (2011). Production of l-asparaginase from *Escherichia coli* ATCC 11303: Optimization by response surface methodology. *Food and Bioproducts Processing, 89*, 315–321. https://doi.org/10.1016/j.fbp.2010.11.002.

Kessel, D., & Luo, Y. (2000). Cells in cryptophycin-induced cell-cycle arrest are susceptible to apoptosis. *Cancer Letters, 151*, 25–29. https://doi.org/10.1016/S0304-3835(99)00409-7.

Khozin-Goldberg, I., Bigogno, C., Shrestha, P., & Cohen, Z. (2002). Nitrogen starvation induces the accumulation of arachidonic acid in the freshwater green alga *Parietochloris Incisa* (trebuxiophyceae)[1]. *Journal of Phycology, 38*, 991–994. https://doi.org/10.1046/j.1529-8817.2002.01160.x.

Kim, J. E., Jeong, D. W., & Lee, H. J. (2007). Expression, purification, and characterization of arginine deiminase from *Lactococcus lactis* ssp. *lactis* ATCC 7962 in *Escherichia coli* BL21. *Protein Expression and Purification, 53*, 9–15. https://doi.org/10.1016/j.pep.2006.12.002.

Kirti, K., Amita, S., Priti, S., et al. (2014). Colorful world of microbes: Carotenoids and their applications. *Advances in Biology, 2014*, e837891. https://doi.org/10.1155/2014/837891.

Kokou, F., Makridis, P., Kentouri, M., & Divanach, P. (2012). Antibacterial activity in microalgae cultures. *Aquaculture Research, 43*, 1520–1527. https://doi.org/10.1111/j.1365-2109.2011.02955.x.

Konzen, M., De Marco, D., Cordova, C. A. S., et al. (2006). Antioxidant properties of violacein: Possible relation on its biological function. *Bioorganic & Medicinal Chemistry, 14*, 8307–8313. https://doi.org/10.1016/j.bmc.2006.09.013.

Kot, A. M., Błażejak, S., Gientka, I., et al. (2018). Torulene and torularhodin: "New" fungal carotenoids for industry? *Microbial Cell Factory, 17*, 49. https://doi.org/10.1186/s12934-018-0893-z.

Kotzia, G. A., & Labrou, N. E. (2007). L-Asparaginase from *Erwinia Chrysanthemi* 3937: Cloning, expression and characterization. *Journal of Biotechnology, 127*, 657–669. https://doi.org/10.1016/j.jbiotec.2006.07.037.

Kuddus, M., & Saima, R. (2013). Cold-active detergent-stable extracellular α-amylase from *Bacillus cereus* GA6: Biochemical characteristics and its perspectives in laundry detergent formulation. *Journal of Biochemical Technology, 4*, 636–644.

Lang, I., Hodac, L., Friedl, T., & Feussner, I. (2011). Fatty acid profiles and their distribution patterns in microalgae: A comprehensive analysis of more than 2000 strains from the SAG culture collection.

BMC Plant Biology, 11, 124. https://doi.org/10.1186/1471-2229-11-124.

Laws, M., Shaaban, A., & Rahman, K. M. (2019). Antibiotic resistance breakers: Current approaches and future directions. *FEMS Microbiology Reviews, 43*, 490–516. https://doi.org/10.1093/femsre/fuz014.

Lee, J. B., Hayashi, K., Hirata, M., et al. (2006). Antiviral Sulfated Polysaccharide from *Navicula directa*, a diatom collected from deep-sea water in Toyama Bay. *Biological and Pharmaceutical Bulletin, 29*, 2135–2139. https://doi.org/10.1248/bpb.29.2135.

Lee, Y., Puumala, E., Robbins, N., et al. (2020). Antifungal drug resistance: Molecular mechanisms in *Candida albicans* and beyond. *Chemical Reviews, 121*, 3390–3411. https://doi.org/10.1021/acs.chemrev.0c00199.

Li, X. L., Zhang, W. H., Wang, Y. D., et al. (2014). A high-detergent-performance, cold-adapted lipase from *Pseudomonas stutzeri* PS59 suitable for detergent formulation. *Journal of Molecular Catalysis B: Enzymatic, 102*, 16–24. https://doi.org/10.1016/j.molcatb.2014.01.006.

Li, X. D., Li, X., Li, X. M., et al. (2016). Tetranorlabdane diterpenoids from the deep sea sediment-derived fungus *Aspergillus wentii* SD-310. *Planta Medica, 82*, 877–881. https://doi.org/10.1055/s-0042-102965.

Li, B., Vachali, P. P., Shen, Z., et al. (2017). Retinal accumulation of zeaxanthin, lutein, and β-carotene in mice deficient in carotenoid cleavage enzymes. *Experimental Eye Research, 159*, 123–131. https://doi.org/10.1016/j.exer.2017.02.016.

Liang, J., Moore, R. E., Moher, E. D., et al. (2005). Cryptophycins-309, 249 and other cryptophycin analogs: Preclinical efficacy studies with mouse and human tumors. *Investigational New Drugs, 23*, 213–224. https://doi.org/10.1007/s10637-005-6729-9.

Lifshits, M., & Carmeli, S. (2012). Metabolites of *Microcystis aeruginosa* bloom material from Lake Kinneret, Israel. *Journal of Natural Products, 75*, 209–219. https://doi.org/10.1021/np200909x.

Lin, J. H., Lee, D. J., & Chang, J. S. (2015). Lutein production from biomass: Marigold flowers versus microalgae. *Bioresource Technology, 184*, 421–428. https://doi.org/10.1016/j.biortech.2014.09.099.

Liu, H., & Liu, W. (2017). Concentration and distributions of fatty acids in algae, submerged plants and terrestrial plants from the northeastern Tibetan Plateau. *Organic Geochemistry, 113*, 17–26. https://doi.org/10.1016/j.orggeochem.2017.08.008.

Liu, X. H., Yu, R. B., Liu, R., et al. (2014). Association between lutein and zeaxanthin status and the risk of cataract: A meta-analysis. *Nutrients, 6*, 452–465. https://doi.org/10.3390/nu6010452.

Liu, S. S., Jiang, J. X., Huang, R., et al. (2019a). A new antiviral 14-nordrimane sesquiterpenoid from an endophytic fungus *Phoma* sp. *Phytochemistry Letters, 29*, 75–78. https://doi.org/10.1016/j.phytol.2018.11.005.

Liu, Y., Ding, L., Fang, F., & He, S. (2019b). Penicillilactone A, a novel antibacterial 7-membered lactone derivative from the sponge-associated fungus *Penicillium* sp. LS54. *Natural Product Research, 33*, 2466–2470. https://doi.org/10.1080/14786419.2018.1452012.

Lorenz, R. T., & Cysewski, G. R. (2000). Commercial potential for *Haematococcus* microalgae as a natural source of astaxanthin. *Trends in Biotechnology, 18*, 160–167. https://doi.org/10.1016/S0167-7799(00)01433-5.

Lu, X., Li, L., Wu, R., et al. (2006). Kinetic analysis of *Pseudomonas aeruginosa* arginine deiminase mutants and alternate substrates provides insight into structural determinants of function. *Biochemistry, 45*, 1162–1172. https://doi.org/10.1021/bi051591e.

Luesch, H., Yoshida, W. Y., Moore, R. E., et al. (2000). Isolation, structure determination, and biological activity of lyngbyabellin A from the marine cyanobacterium *Lyngbya majuscula*. *Journal of Natural Products, 63*, 611–615. https://doi.org/10.1021/np990543q.

Lv, C., Hong, Y., Miao, L., et al. (2013). Wentilactone A as a novel potential antitumor agent induces apoptosis and G2/M arrest of human lung carcinoma cells, and is mediated by HRas-GTP accumulation to excessively activate the Ras/Raf/ERK/p53-p21 pathway. *Cell Death & Disease, 4*, e952–e952. https://doi.org/10.1038/cddis.2013.484.

Mader, J., Gallo, A., Schommartz, T., et al. (2016). Calcium spirulan derived from *Spirulina platensis* inhibits herpes simplex virus 1 attachment to human keratinocytes and protects against herpes labialis. *Journal of Allergy and Clinical Immunology, 137*, 197-203. e3. https://doi.org/10.1016/j.jaci.2015.07.027.

Maltsev, Y., & Maltseva, K. (2021). Fatty acids of microalgae: Diversity and applications. *Reviews in Environmental Science and Biotechnology, 20*, 515–547. https://doi.org/10.1007/s11157-021-09571-3.

Maltsev, Y. I., Maltseva, I. A., Maltseva, S. Y., & Kulikovskiy, M. S. (2020). Biotechnological potential of a new strain of *Bracteacoccus bullatus* (Sphaeropleales, Chlorophyta) as a promising producer of omega-6 polyunsaturated fatty acids. *Russian Journal of Plant Physiology, 67*, 185–193. https://doi.org/10.1134/S1021443720010124.

Mamaeva, A., Namsaraev, Z., Maltsev, Y., et al. (2018). Simultaneous increase in cellular content and volumetric concentration of lipids in *Bracteacoccus bullatus* cultivated at reduced nitrogen and phosphorus concentrations. *Journal of Applied Phycology, 30*, 2237–2246. https://doi.org/10.1007/s10811-018-1471-9.

Manivasagan, P., Venkatesan, J., Senthilkumar, K., et al. (2013). Isolation and characterization of biologically active melanin from *Actinoalloteichus* sp. MA-32. *International Journal of Biological Macromolecules, 58*, 263–274. https://doi.org/10.1016/j.ijbiomac.2013.04.041.

Marquez, B. L., Watts, K. S., Yokochi, A., et al. (2002). Structure and absolute stereochemistry of hectochlorin, a potent stimulator of actin assembly. *Journal of Natural Products, 65*, 866–871. https://doi.org/10.1021/np0106283.

Marrez, D. A., Naguib, M. M., Sultan, Y. Y., & Higazy, A. M. (2019). Antimicrobial and anticancer activities of *Scenedesmus obliquus* metabolites. *Heliyon, 5*, e01404. https://doi.org/10.1016/j.heliyon.2019.e01404.

Martin, G. D. A., Tan, L. T., Jensen, P. R., et al. (2007). Marmycins A and B, cytotoxic pentacyclic C-glycosides from a marine sediment-derived actinomycete related to the genus *Streptomyces*. *Journal of Natural Products, 70*, 1406–1409. https://doi.org/10.1021/np060621r.

Masojídek, J., Torzillo, G., & Koblížek, M. (2013). Photosynthesis in microalgae. In *Handbook of microalgal culture* (pp. 21–36). Wiley & Sons Ltd. https://doi.org/10.1002/9781118567166.ch2.

Matei, E., Basu, R., Furey, W., et al. (2016). Structure and glycan binding of a new cyanovirin-N homolog. *Journal of Biological Chemistry, 291*, 18967–18976. https://doi.org/10.1074/jbc.M116.740415.

Maurer, K. H. (2004). Detergent proteases. *Current Opinion in Biotechnology, 15*, 330–334. https://doi.org/10.1016/j.copbio.2004.06.005.

Mazhar, H., Ullah, I., Ali, U., et al. (2023). Optimization of low-cost solid-state fermentation media for the production of thermostable lipases using agro-industrial residues as substrate in culture of *Bacillus amyloliquefaciens*. *Biocatalysis and Agricultural Biotechnology, 47*, 102559. https://doi.org/10.1016/j.bcab.2022.102559.

McAuliffe, O., Ross, R. P., & Hill, C. (2001). Lantibiotics: Structure, biosynthesis and mode of action. *FEMS Microbiology Reviews, 25*, 285–308. https://doi.org/10.1111/j.1574-6976.2001.tb00579.x.

Melanie, S., & Fithriani, D. (2020). Combination of cell disruption method and pH variation as pre-treatment for lipid extraction of *Nannochloropsis* sp. *IOP Conference Series: Earth and Environmental Science, 404*, 012022. https://doi.org/10.1088/1755-1315/404/1/012022.

Meruvu, H., & dos Santos, J. C. (2021). Colors of life: A review on fungal pigments. *Critical Reviews in Biotechnology, 41*, 1153–1177. https://doi.org/10.1080/07388551.2021.1901647.

Monciardini, P., Iorio, M., Maffioli, S., et al. (2014). Discovering new bioactive molecules from microbial sources. *Microbial Biotechnology, 7*, 209–220. https://doi.org/10.1111/1751-7915.12123.

Monroy-Ruiz, J., Sevilla, M. Á., Carrón, R., & Montero, M. J. (2011). Astaxanthin-enriched-diet reduces blood pressure and improves cardiovascular parameters in spontaneously hypertensive rats. *Pharmacological Research, 63*, 44–50. https://doi.org/10.1016/j.phrs.2010.09.003.

Mundt, S., Kreitlow, S., & Jansen, R. (2003). Fatty acids with antibacterial activity from the cyanobacterium *Oscillatoria redekei* HUB 051. *Journal of Applied Phycology, 15*, 263–267. https://doi.org/10.1023/A:1023889813697.

Mushir, S., & Fatma, T. (2012). Monitoring stress responses in cyanobacterial scytonemin–screening and characterization. *Environmental Technology, 33*, 153–157. https://doi.org/10.1080/09593330.2011.553842.

Nakamura, M. T., Yudell, B. E., & Loor, J. J. (2014). Regulation of energy metabolism by long-chain fatty acids. *Progress in Lipid Research, 53*, 124–144. https://doi.org/10.1016/j.plipres.2013.12.001.

Nakayama, M., Nakajima-Kambe, T., Katayama, H., et al. (2008). High catalase production by *Rhizobium radiobacter* strain 2–1. *Journal of Bioscience and Bioengineering, 106*, 554–558. https://doi.org/10.1263/jbb.106.554.

Neuhof, T., Seibold, M., Thewes, S., et al. (2006). Comparison of susceptibility and transcription profile of the new antifungal hassallidin A with caspofungin. *Biochemical and Biophysical Research Communications, 349*, 740–749. https://doi.org/10.1016/j.bbrc.2006.08.110.

Nielsen, B. L. H., Gøtterup, L., & Jørgensen, T. S. et al. (2019). n-3 PUFA biosynthesis by the copepod Apocyclops royi documented using fatty acid profile analysis and gene expression analysis. *Biology Open 8*, bio038331.https://doi.org/10.1242/bio.038331

Nong, X.-H., Wang, Y.-F., Zhang, X.-Y., et al. (2014). Territrem and butyrolactone derivatives from a marine-derived fungus *Aspergillus Terreus*. *Marine Drugs, 12*, 6113–6124. https://doi.org/10.3390/md12126113.

Ördög, V., Stirk, W. A., Lenobel, R., et al. (2004). Screening microalgae for some potentially useful agricultural and pharmaceutical secondary metabolites. *Journal of Applied Phycology, 16*, 309–314. https://doi.org/10.1023/B:JAPH.0000047789.34883.aa.

Orosa, M., Torres, E., Fidalgo, P., & Abalde, J. (2000). Production and analysis of secondary carotenoids in green algae. *Journal of Applied Phycology, 12*, 553–556. https://doi.org/10.1023/A:1008173807143.

Osuna-Ruiz, I., Nieves-Soto, M., Manzano-Sarabia, M. M., et al. (2019). Gross chemical composition, fatty acids, sterols, and pigments in tropical seaweed species off Sinaloa, Mexico. *Ciencias Marinas, 45*, 101–120. https://doi.org/10.7773/cm.v45i3.2974.

Oswald, E., Nougayrède, J. P., Taieb, F., et al. (2005). Bacterial toxins that modulate host cell-cycle progression. *Current Opinion in Microbiology, 8*, 83–91. https://doi.org/10.1016/j.mib.2004.12.011.

Pagels, F., Salvaterra, D., Amaro, H. M., et al. (2020). Bioactive potential of *Cyanobium* sp. pigment-rich extracts. *Journal of Applied Phycology, 32*, 3031–3040. https://doi.org/10.1007/s10811-020-02213-1.

Pagels, F., Vasconcelos, V., & Guedes, A. C. (2021). Carotenoids from Cyanobacteria: Biotechnological potential and optimization strategies. *Biomolecules, 11*, 735. https://doi.org/10.3390/biom11050735.

Paiva, A. D., de Oliveira, M. D., de Paula, S. O., et al. (2012). Toxicity of bovicin HC5 against mammalian cell lines and the role of cholesterol in bacteriocin activity. *Microbiology, 158*, 2851–2858. https://doi.org/10.1099/mic.0.062190-0.

Panosyan, E. H., Wang, Y., Xia, P., et al. (2014). Asparagine depletion potentiates the cytotoxic effect of chemotherapy against brain tumors. *Molecular Cancer Research, 12*, 694–702. https://doi.org/10.1158/1541-7786.MCR-13-0576.

Pantanella, F., Berlutti, F., Passariello, C., et al. (2007). Violacein and biofilm production in *Janthinobacterium lividum*. *Journal of Applied Microbiology, 102*, 992–999. https://doi.org/10.1111/j.1365-2672.2006.03155.x.

Park, W. S., Kim, H. J., Li, M., et al. (2018). Two classes of pigments, carotenoids and c-phycocyanin, in spirulina powder and their antioxidant activities. *Molecules, 23*, 2065. https://doi.org/10.3390/molecules23082065.

Parvathy, K. S., Negi, P. S., & Srinivas, P. (2009). Antioxidant, antimutagenic and antibacterial activities of curcumin-β-diglucoside. *Food Chemistry, 115*, 265–271. https://doi.org/10.1016/j.foodchem.2008.12.036.

Patel, K. A., Dhandhukia, P. C., & Thakker, J. N. (2019). Study of biosynthesis and statistical optimization of medium components for β-carotene production using marine *Paracoccus* sp. OC1 isolated from Gulf of Khambhat. *Journal of Pure and Applied Sciences, 27*, 26–40.

Patias, L. D., Fernandes, A. S., Petry, F. C., et al. (2017). Carotenoid profile of three microalgae/cyanobacteria species with peroxyl radical scavenger capacity. *Food Research International, 100*, 260–266. https://doi.org/10.1016/j.foodres.2017.06.069.

Patil, V., Källqvist, T., Olsen, E., et al. (2007). Fatty acid composition of 12 microalgae for possible use in aquaculture feed. *Aquaculture International, 15*, 1–9. https://doi.org/10.1007/s10499-006-9060-3.

Pavlik, C. M., Wong, C. Y. B., Ononye, S., et al. (2013). Santacruzamate A, a potent and selective histone deacetylase inhibitor from the panamanian marine cyanobacterium *Symploca* sp. *Journal of Natural Products, 76*, 2026–2033. https://doi.org/10.1021/np400198r.

Pelah, D., Sintov, A., & Cohen, E. (2004). The effect of salt stress on the production of canthaxanthin and astaxanthin by *Chlorella zofingiensis* grown under limited light intensity. *World Journal of Microbiology and Biotechnology, 20*, 483–486. https://doi.org/10.1023/B:WIBI.0000040398.93103.21.

Pellati, D., Mylonakis, I., Bertoloni, G., et al. (2008). Genital tract infections and infertility. *European Journal of Obstetrics & Gynecology and Reproductive Biology, 140*, 3–11. https://doi.org/10.1016/j.ejogrb.2008.03.009.

Pelz, A., Wieland, K. P., Putzbach, K., et al. (2005). Structure and biosynthesis of staphyloxanthin from *Staphylococcus aureus*. *Journal of Biological Chemistry, 280*, 32493–32498. https://doi.org/10.1074/jbc.M505070200.

Pereira, A. R., Kale, A. J., Fenley, A. T., et al. (2012). The carmaphycins: New proteasome inhibitors exhibiting an α, β-Epoxyketone warhead from a marine cyanobacterium. *ChemBioChem, 13*, 810–817. https://doi.org/10.1002/cbic.201200007.

Pham, J. V., Yilma, M. A., Feliz, A., et al. (2019). A review of the microbial production of bioactive natural products and biologics. *Frontiers in Microbiology*, 10. https://doi.org/10.3389/fmicb.2019.01404.

Poehlsgaard, J., & Douthwaite, S. (2002). The macrolide binding site on the bacterial ribosome. *Current Drug Targets-Infectious Disorders, 2*, 67–78. https://doi.org/10.2174/1568005024605927.

Polyzois, A., Kirilovsky, D., Dufat, T. H., et al. (2020). Effects of modification of light parameters on the production of cryptophycin, cyanotoxin with potent anticancer activity, in *Nostoc* sp. *Toxins, 12*, 809. https://doi.org/10.3390/toxins12120809.

Pradhan, J., Das, S., & Das, B. K. (2014). Antibacterial activity of freshwater microalgae: A review. *African Journal of Pharmacy and Pharmacology, 8*, 809–818. https://doi.org/10.5897/AJPP2013.0002.

Prakash, J. W., Antonisamy, J. M., & Jeeva, S. (2011). Antimicrobial activity of certain fresh water microalgae from Thamirabarani River, Tamil Nadu, South India. *Asian Pacific Journal of Tropical Biomedicine, 1*, S170–S173. https://doi.org/10.1016/S2221-1691(11)60149-4.

Přibyl, P., Cepák, V., Kaštánek, P., & Zachleder, V. (2015). Elevated production of carotenoids by a new isolate of *Scenedesmus* sp. *Algal Research, 11*, 22–27. https://doi.org/10.1016/j.algal.2015.05.020.

Pruvost, J., Cornet, J.-F., & Legrand, J. (2008). Hydrodynamics influence on light conversion in photobioreactors: An energetically consistent analysis. *Chemical Engineering Science, 63*, 3679–3694. https://doi.org/10.1016/j.ces.2008.04.026.

Qi, F., Pei, H., Mu, R., et al. (2019). Characterization and optimization of endogenous lipid accumulation in *Chlorella vulgaris* SDEC-3M ability to rapidly accumulate lipid for reversing nightly lipid loss. *Biotechnology for Biofuels, 12*, 151. https://doi.org/10.1186/s13068-019-1493-9.

Qin, S., Liu, G. X., & Hu, Z. Y. (2008). The accumulation and metabolism of astaxanthin in *Scenedesmus obliquus* (Chlorophyceae). *Process Biochemistry, 43*, 795–802. https://doi.org/10.1016/j.procbio.2008.03.010.

Qin, C., Lin, X., Lu, X., et al. (2015). Sesquiterpenoids and xanthones derivatives produced by sponge-derived fungus *Stachybotry* sp. HH1 ZSDS1F1-2. *The Journal of Antibiotics, 68*, 121–125. https://doi.org/10.1038/ja.2014.97.

Rajesh, K., Rohit, M. V., & Venkata Mohan, S. (2017). Microalgae-based carotenoids production. In R. P. Rastogi, D. Madamwar, & A. Pandey (Eds.), *Algal green chemistry* (pp. 139–147). Amsterdam: Elsevier. https://doi.org/10.1016/B978-0-444-63784-0.00007-2.

Ranadive, K., Belsare, M. H., Deokule, S., et al. (2013). Glimpses of antimicrobial activity of fungi from world. *Journal on New Biological Reports, 2*(2), 142–162.

Rani, A., Saini, K. C., Bast, F., et al. (2021a). A review on microbial products and their perspective application as antimicrobial agents. *Biomolecules, 11*, 1860. https://doi.org/10.3390/biom11121860.

Rani, A., Saini, K. C., Bast, F., et al. (2021b). Microorganisms: A potential source of bioactive molecules for antioxidant applications. *Molecules, 26*, 1142. https://doi.org/10.3390/molecules26041142.

Rapoport, A., Guzhova, I., Bernetti, L., et al. (2021). Carotenoids and some other pigments from fungi and yeasts. *Metabolites, 11*, 92. https://doi.org/10.3390/metabo11020092.

Rather, L. J., Mir, S. S., Ganie, S. A., et al. (2023). Research progress, challenges, and perspectives in microbial pigment production for industrial applications-a review. *Dyes and Pigments, 210*, 110989. https://doi.org/10.1016/j.dyepig.2022.110989.

Rawat, D., & Bhargava, S. (2011). Bioactive compounds from *Nostoc* species. *Current Research in Pharmaceutical Sciences, 1*, 48–54.

Raymond Oluoch, K., Welander, U., Raymond Oluoch, K., et al. (2006). Hydrogen peroxide degradation by immobilized cells of alkaliphilic *Bacillus halodurans*. *Biocatalysis and Biotransformation, 24*, 215–222. https://doi.org/10.1080/10242420600662669.

Reddy, C. M., Bhat, V. B., Kiranmai, G., et al. (2000). Selective inhibition of cyclooxygenase-2 by C-phycocyanin, a biliprotein from *Spirulina platensis*. *Biochemical and Biophysical Research Communications, 277*, 599–603. https://doi.org/10.1006/bbrc.2000.3725.

Rios, N. S., Pinheiro, B. B., Pinheiro, M. P., et al. (2018). Biotechnological potential of lipases from *Pseudomonas*: Sources, properties and applications. *Process Biochemistry, 75*, 99–120. https://doi.org/10.1016/j.procbio.2018.09.003.

Robertsen, H. L., & Musiol-Kroll, E. M. (2019). Actinomycete-derived polyketides as a source of antibiotics and lead structures for the development of new antimicrobial drugs. *Antibiotics, 8*, 157. https://doi.org/10.3390/antibiotics8040157.

Robles-Bañuelos, B., Durán-Riveroll, L. M., Rangel-López, E., Pérez-López, H. I., & González-Maya, L. (2022). Marine cyanobacteria as sources of lead anticancer compounds: A review of families of metabolites with cytotoxic, antiproliferative, and antineoplastic effects. *Molecules, 27*, 4814. https://doi.org/10.3390/molecules27154814.

Rodrigues, L., Morone, J., Hentschke, G. S., et al. (2024). Anti-Inflammatory activity of cyanobacteria pigment extracts: Physiological free radical scavenging and modulation of iNOS and LOX activity. *Marine Drugs, 22*, 131. https://doi.org/10.3390/md22030131.

Rogozhin, E. A., Sadykova, V. S., Baranova, A. A., et al. (2018). A novel lipopeptaibol emericellipsin A with antimicrobial and antitumor activity produced by the extremophilic fungus *Emericellopsis alkalina*. *Molecules, 23*, 2785. https://doi.org/10.3390/molecules23112785.

Romay, C., Delgado, R., Remirez, D., et al. (2001). Effects of phycocyanin extract on tumor necrosis factor-α and nitrite levels in serum of mice treated with endotoxin. *Arzneimittel-Forschung, 51*, 733–736. https://doi.org/10.1055/s-0031-1300107.

Romay, C., Gonzalez, R., Ledon, N., et al. (2003). C-Phycocyanin: A biliprotein with antioxidant, anti-inflammatory and neuroprotective effects. *Current Protein and Peptide Science, 4*, 207–216. https://doi.org/10.2174/1389203033487216.

Roshak, A. K., Capper, E. A., Imburgia, C., et al. (2000). The human *polo*-like kinase, PLK, regulates cdc2/cyclin B through phosphorylation and activation of the cdc25C phosphatase. *Cellular Signalling, 12*, 405–411. https://doi.org/10.1016/S0898-6568(00)00080-2.

Rowley, D. C., Kelly, S., Kauffman, C. A., et al. (2003). Halovirs A-E, new antiviral agents from a marine-derived fungus of the genus *Scytalidium*. *Bioorganic & Medicinal Chemistry, 11*, 4263–4274. https://doi.org/10.1016/S0968-0896(03)00395-X.

Ryeol Park, S., Won Park, J., Hee Ban, Y., et al. (2013). 2-Deoxystreptamine-containing aminoglycoside antibiotics: Recent advances in the characterization and manipulation of their biosynthetic pathways. *Natural Product Reports, 30*, 11–20. https://doi.org/10.1039/C2NP20092A.

Saini, R. K., & Keum, Y.-S. (2018). Carotenoid extraction methods: A review of recent developments. *Food Chemistry, 240*, 90–103. https://doi.org/10.1016/j.foodchem.2017.07.099.

Sajilata, M. G., Singhal, R. S., & Kamat, M. Y. (2008). The carotenoid pigment zeaxanthin—A review. *Comprehensive Reviews in Food Science and Food Safety, 7*, 29–49. https://doi.org/10.1111/j.1541-4337.2007.00028.x.

Sakaki, H., Nochide, H., Komemushi, S., & Miki, W. (2002). Effect of active oxygen species on the productivity of torularhodin by *Rhodotorula glutinis* No. 21. *Journal of Bioscience and Bioengineering, 93*, 338–340. https://doi.org/10.1016/S1389-1723(02)80040-8.

Santoyo, S., Rodríguez-Meizoso, I., Cifuentes, A., et al. (2009). Green processes based on the extraction with pressurized fluids to obtain potent antimicrobials from *Haematococcus pluvialis* microalgae. *LWT-Food Science and Technology, 42*, 1213–1218. https://doi.org/10.1016/j.lwt.2009.01.012.

Sarmiento, F., Peralta, R., & Blamey, J. M. (2015). Cold and hot extremozymes: Industrial relevance and current trends. *Frontiers in*

Bioengineering and Biotechnology, 3. https://doi.org/10.3389/fbioe.2015.00148.

Schubert, N., García-Mendoza, E., & Pacheco-Ruiz, I. (2006). Carotenoid composition of marine red algae[1]. *Journal of Phycology, 42*, 1208–1216. https://doi.org/10.1111/j.1529-8817.2006.00274.x.

Schwingshackl, L., & Hoffmann, G. (2012). Monounsaturated fatty acids and risk of cardiovascular disease: Synopsis of the evidence available from systematic reviews and meta-analyses. *Nutrients, 4*, 1989–2007. https://doi.org/10.3390/nu4121989.

Sengupta, S., & Bhowal, J. (2021). Study on the antioxidant and cytotoxic properties of pyocyanin extracted from *Pseudomonas aeruginosa*. In D. Ramkrishna, S. Sengupta, S. Dey Bandyopadhyay, & A. Ghosh (Eds.), *Advances in bioprocess engineering and technology* (pp. 133–141). Singapore: Springer. https://doi.org/10.1007/978-981-15-7409-2_13.

Shahidi, F., & Ambigaipalan, P. (2015). Phenolics and polyphenolics in foods, beverages and spices: Antioxidant activity and health effects–a review. *Journal of Functional Foods, 18*, 820–897. https://doi.org/10.1016/j.jff.2015.06.018.

Shih, C., & Teicher, B. A. (2001). Cryptophycins: A novel class of potent antimitotic antitumor depsipeptides. *Current Pharmaceutical Design, 7*, 1259–1276. https://doi.org/10.2174/1381612013397474.

Singh, S., & Bajaj, B. K. (2017). Potential application spectrum of microbial proteases for clean and green industrial production. *Energy, Ecology and Environment, 2*, 370–386. https://doi.org/10.1007/s40974-017-0076-5.

Singh, S., Kate, B. N., & Banerjee, U. C. (2005). Bioactive compounds from cyanobacteria and microalgae: An overview. *Critical Reviews in Biotechnology, 25*, 73–95. https://doi.org/10.1080/07388550500248498.

Singh, S., Nimse, S. B., Mathew, D. E., et al. (2021). Microbial melanin: Recent advances in biosynthesis, extraction, characterization, and applications. *Biotechnology Advances, 53*, 107773. https://doi.org/10.1016/j.biotechadv.2021.107773.

Sivaramakrishnan, R., & Incharoensakdi, A. (2020). Plant hormone induced enrichment of *Chlorella* sp. omega-3 fatty acids. *Biotechnology for Biofuels, 13*, 7. https://doi.org/10.1186/s13068-019-1647-9.

Smith, V. J., Desbois, A. P., & Dyrynda, E. A. (2010). Conventional and unconventional antimicrobials from fish, marine invertebrates and micro-algae. *Marine Drugs, 8*, 1213–1262. https://doi.org/10.3390/md8041213.

Snoeck, R. (2000). Antiviral therapy of herpes simplex. *International Journal of Antimicrobial Agents, 16*, 157–159. https://doi.org/10.1016/S0924-8579(00)00233-8.

Song, T., Chen, M., Chai, W., et al. (2018a). New bioactive pyrrospirones C–I from a marine-derived fungus *Penicillium* sp. ZZ380. *Tetrahedron, 74*, 884–891. https://doi.org/10.1016/j.tet.2018.01.015.

Song, T., Chen, M., Ge, Z. W., et al. (2018b). Bioactive penicipyrrodiether A, an adduct of GKK1032 analogue and phenol a derivative, from a marine-sourced fungus *Penicillium* sp. ZZ380. *The Journal of Organic Chemistry, 83*, 13395–13401. https://doi.org/10.1021/acs.joc.8b02172.

Sran, K. S., Bisht, B., Mayilraj, S., & Roy Choudhury, A. (2019). Structural characterization and antioxidant potential of a novel anionic exopolysaccharide produced by marine *Microbacterium aurantiacum* FSW-25. *International Journal of Biological Macromolecules, 131*, 343–352. https://doi.org/10.1016/j.ijbiomac.2019.03.016.

Srinivasan, R., Kannappan, A., Shi, C., & Lin, X. (2021). Marine bacterial secondary metabolites: A treasure house for structurally unique and effective antimicrobial compounds. *Marine Drugs, 19*, 530. https://doi.org/10.3390/md19100530.

Stevenson, C. S., Capper, E. A., Roshak, A. K., et al. (2002). The identification and characterization of the marine natural product Scytonemin as a novel antiproliferative pharmacophore. *Journal of Pharmacology and Experimental Therapeutics, 303*, 858–866. https://doi.org/10.1124/jpet.102.036350.

Stincone, P., & Brandelli, A. (2020). Marine bacteria as source of antimicrobial compounds. *Critical Reviews in Biotechnology, 40*, 306–319. https://doi.org/10.1080/07388551.2019.1710457.

Stirk, W. A., & van Staden, J. (2022). Bioprospecting for bioactive compounds in microalgae: Antimicrobial compounds. *Biotechnology Advances, 59*, 107977. https://doi.org/10.1016/j.biotechadv.2022.107977.

Suhonen, R., & Plosila, M. (2009). The effect of beta-carotene in combination with canthaxanthin, Ro 8–8427 (Phenoro®), in treatment of polymorphous light eruptions. *Dermatologica, 163*, 172–176. https://doi.org/10.1159/000250156.

Sun, Y., Sun, T., Wang, F., et al. (2013). A polysaccharide from the fungi of Huaier exhibits anti-tumor potential and immunomodulatory effects. *Carbohydrate Polymers, 92*, 577–582. https://doi.org/10.1016/j.carbpol.2012.09.006.

Sun, G. Y., Simonyi, A., Fritsche, K. L., et al. (2018). Docosahexaenoic acid (DHA): An essential nutrient and a nutraceutical for brain health and diseases. *Prostaglandins, Leukotrienes and Essential Fatty Acids, 136*, 3–13. https://doi.org/10.1016/j.plefa.2017.03.006.

Sun, Z., Li, T., Zhou, Z., & Jiang, Y. (2016). Microalgae as a source of lutein: Chemistry, biosynthesis, and carotenogenesis. In C. Posten, & S. Feng Chen (Eds.), *Microalgae biotechnology* (pp. 37–58). Cham: Springer International Publishing. https://doi.org/10.1007/10_2015_331.

Taori, K., Paul, V. J., & Luesch, H. (2008). Structure and activity of largazole, a potent antiproliferative agent from the floridian marine cyanobacterium *Symploca* sp. *Journal of the American Chemical Society, 130*, 1806–1807. https://doi.org/10.1021/ja7110064.

Tareq, F. S., Lee, H.-S., Lee, Y.-J., et al. (2015). Ieodoglucomide C and Ieodoglycolipid, new glycolipids from a marine-derived bacterium *Bacillus licheniformis* 09IDYM23. *Lipids, 50*, 513–519. https://doi.org/10.1007/s11745-015-4014-z.

Terekhova, V. E., Aizdaicher, N. A., Buzoleva, L. S., & Somov, G. P. (2009). Influence of extrametabolites of marine microalgae on the reproduction of the bacterium *Listeria monocytogenes*. *Russian Journal of Marine Biology, 35*, 355–358. https://doi.org/10.1134/S1063074009040129.

Tizabi, D., & Hill, R. T. (2023). *Micrococcus* spp. as a promising source for drug discovery: A review. *Journal of Industrial Microbiology and Biotechnology, 50*, kuad017. https://doi.org/10.1093/jimb/kuad017.

Valentão, P., Fernandes, E., Carvalho, F., et al. (2002). Antioxidant activity of *Hypericum androsaemum* infusion: Scavenging activity against superoxide radical, hydroxyl radical and hypochlorous acid. *Biological and Pharmaceutical Bulletin, 25*, 1320–1323. https://doi.org/10.1248/bpb.25.1320.

van der Maarel, M. J. E. C., van der Veen, B., Uitdehaag, J. C. M., et al. (2002). Properties and applications of starch-converting enzymes of the α-amylase family. *Journal of Biotechnology, 94*, 137–155. https://doi.org/10.1016/S0168-1656(01)00407-2.

Varela, J. C., Pereira, H., Vila, M., & León, R. (2015). Production of carotenoids by microalgae: Achievements and challenges. *Photosynthesis Research, 125*, 423–436. https://doi.org/10.1007/s11120-015-0149-2.

Vehapi, M., Yilmaz, A., & Özçimen, D. (2018). Antifungal activities of *Chlorella vulgaris* and *Chlorella minutissima* microalgae cultivated in bold basal medium, wastewater and tree extract water against *Aspergillus niger* and *Fusarium oxysporum*. *Romanian Biotechnological Letters*.

Venckus, P., Paliulis, S., Kostkevičiene, J., et al. (2018). CARS microscopy of scytonemin in cyanobacteria *Nostoc commune*. *Journal of Raman Spectroscopy, 49*, 1333–1338. https://doi.org/10.1002/jrs.5388.

Vestola, J., Shishido, T. K., Jokela, J., et al. (2014). Hassallidins, antifungal glycolipopeptides, are widespread among cyanobacteria and are the end-product of a nonribosomal pathway. *Proceedings of the National Academy of Sciences, 111*, E1909–E1917. https://doi.org/10.1073/pnas.1320913111.

Vidal, I., Torres-Vargas, J. A., Sánchez, J. M., et al. (2023). Danthron, an anthraquinone isolated from a marine fungus, is a new inhibitor of angiogenesis exhibiting interesting antitumor and antioxidant properties. *Antioxidants, 12*, 1101. https://doi.org/10.3390/antiox12051101.

Vidoudez, C., & Pohnert, G. (2008). Growth phase-specific release of polyunsaturated aldehydes by the diatom *Skeletonema marinoi*. *Journal of Plankton Research, 30*, 1305–1313. https://doi.org/10.1093/plankt/fbn085.

Wada, N., Sakamoto, T., & Matsugo, S. (2013). Multiple roles of photosynthetic and sunscreen pigments in cyanobacteria focusing on the oxidative stress. *Metabolites, 3*, 463–483. https://doi.org/10.3390/metabo3020463.

Wang, J., Liu, P., Wang, Y., et al. (2012). Antimicrobial aromatic polyketides from gorgonian-associated fungus, *Penicillium commune* 518. *Chinese Journal of Chemistry, 30*, 1236–1242. https://doi.org/10.1002/cjoc.201100640.

Wang, Y., Wu, Y., Wang, Y., et al. (2017). Antioxidant Properties of probiotic bacteria. *Nutrients, 9*, 521. https://doi.org/10.3390/nu9050521.

Wang, C., Fan, Q., Zhang, X., et al. (2018). Isolation, characterization, and pharmaceutical applications of an exopolysaccharide from *Aerococcus Uriaeequi*. *Marine Drugs, 16*, 337. https://doi.org/10.3390/md16090337.

Wang, L., Lu, H., & Jiang, Y. (2023). Natural polyketides act as promising antifungal agents. *Biomolecules, 13*, 1572. https://doi.org/10.3390/biom13111572.

Ward, O. P., & Singh, A. (2005). Omega-3/6 fatty acids: Alternative sources of production. *Process Biochemistry, 40*, 3627–3652. https://doi.org/10.1016/j.procbio.2005.02.020.

Washida, K., Koyama, T., Yamada, K., et al. (2006). Karatungiols A and B, two novel antimicrobial polyol compounds, from the symbiotic marine dinoflagellate *Amphidinium* sp. *Tetrahedron Letters, 47*, 2521–2525. https://doi.org/10.1016/j.tetlet.2006.02.045.

Wei, D., Chen, F., Chen, G., et al. (2008). Enhanced production of lutein in heterotrophic *Chlorella protothecoides* by oxidative stress. *Science in China Series C-Life Sciences, 51*, 1088–1093. https://doi.org/10.1007/s11427-008-0145-2.

Wei, Q., Fu, G., Wang, K., et al. (2022). Advances in research on antiviral activities of sulfated polysaccharides from seaweeds. *Pharmaceuticals, 15*, 581. https://doi.org/10.3390/ph15050581.

Wollina, U., Voicu, C., Gianfaldoni, S., et al. (2018). *Arthrospira platensis*–potential in dermatology and beyond. *Open Access Macedonian Journal of Medical Sciences, 6*, 176–180. https://doi.org/10.3889/oamjms.2018.033.

Wright, G. D. (2014). Something old, something new: Revisiting natural products in antibiotic drug discovery. *Canadian Journal of Microbiology, 60*, 147–154. https://doi.org/10.1139/cjm-2014-0063.

Wu, S. J., Fotso, S., Li, F., et al. (2007). Amorphane sesquiterpenes from a marine *Streptomyces* sp. *Journal of Natural Products, 70*, 304–306. https://doi.org/10.1021/np050358e.

Wu, L. C., Wen, Z. S., Qiu, Y. T., et al. (2013). Largazole arrests cell cycle at G1 Phase and triggers proteasomal degradation of E2F1 in lung cancer cells. *ACS Medical Chemistry Letters, 4*, 921–926. https://doi.org/10.1021/ml400093y.

Xu, W. -F., Hou, X. -M., Yao, F. -H., et al. (2017a). Xylapeptide A, an antibacterial cyclopentapeptide with an uncommon l-pipecolinic acid moiety from the associated fungus *Xylaria* sp. (GDG-102). *Scientific Reports, 7*, 6937. https://doi.org/10.1038/s41598-017-07331-4.

Xu, Z., He, H., Zhang, S., et al. (2017b). Characterization of feruloyl esterases produced by the four *Lactobacillus* Species: *L. amylovorus*, *L. acidophilus*, *L. farciminis* and *L. fermentum*, isolated from ensiled corn stover. *Frontiers in Microbiology*, 8. https://doi.org/10.3389/fmicb.2017.00941.

Xue, Y., & Sherman, D. H. (2001). Biosynthesis and combinatorial biosynthesis of pikromycin-related macrolides in *Streptomyces venezuelae*. *Metabolic Engineering, 3*, 15–26. https://doi.org/10.1006/mben.2000.0167.

Yada, S., Wang, Y., Zou, Y., et al. (2008). Isolation and characterization of two groups of novel marine bacteria producing violacein. *Marine Biotechnology, 10*, 128–132. https://doi.org/10.1007/s10126-007-9046-9.

Yang, X., Sun, Z., Wang, W., et al. (2018). Developmental toxicity of synthetic phenolic antioxidants to the early life stage of zebrafish. *Science of the Total Environment, 643*, 559–568. https://doi.org/10.1016/j.scitotenv.2018.06.213.

Yao, Y., Cai, X., Fei, W., et al. (2022). The role of short-chain fatty acids in immunity, inflammation and metabolism. *Critical Reviews in Food Science and Nutrition, 62*, 1–12. https://doi.org/10.1080/10408398.2020.1854675.

Yim, J. H., Kim, S. J., Ahn, S. H., et al. (2004). Antiviral effects of sulfated exopolysaccharide from the marine microalga *Gyrodinium impudicum* Strain KG03. *Marine Biotechnology, 1*, 17–25. https://doi.org/10.1007/s10126-003-0002-z.

Yokoyama, A., & Miki, W. (1995). Composition and presumed biosynthetic pathway of carotenoids in the astaxanthin-producing bacterium *Agrobacterium aurantiacum*. *FEMS Microbiology Letters, 128*, 139–144. https://doi.org/10.1111/j.1574-6968.1995.tb07513.x.

Yordanov, M., Danova, S., & Ivanovska, N. (2004). Inflammation induced by inoculation of the joint with *Candida albicans*. *Inflammation, 28*, 127–132. https://doi.org/10.1023/B:IFLA.0000039558.03872.52.

Zainuddin, N., Alias, S. A., Lee, C. W., et al. (2010). *Antimicrobial Activities of Marine Fungi from Malaysia, 53*, 507–513. https://doi.org/10.1515/bot.2010.067.

Zhang, P., Li, X.-M., Mao, X.-X., et al. (2016). Varioloid A, a new indolyl-6,10b-dihydro-5aH-[1]benzofuro[2,3-b]indole derivative from the marine alga-derived endophytic fungus *Paecilomyces variotii* EN-291. *Beilstein Journal of Organic Chemistry, 12*, 2012–2018. https://doi.org/10.3762/bjoc.12.188.

Zhang, J., Li, B., Qin, Y., et al. (2020). A new abyssomicin polyketide with anti-influenza A virus activity from a marine-derived *Verrucosispora* sp. MS100137. *Applied Microbiology and Biotechnology, 104*, 1533–1543. https://doi.org/10.1007/s00253-019-10217-2.

Zhang, X., Xu, Z., Ma, J., et al. (2021). Phylogenetic diversity, antimicrobial and antioxidant potential and identification of bioactive compounds from culturable endophytic fungi associated with mangrove *Bruguiera sexangula* (Lour.) Poir. *Current Microbiology, 78*, 479–489. https://doi.org/10.1007/s00284-020-02314-7.

Zhu, F., Chen, G., Chen, X., et al. (2011). Aspergicin, a new antibacterial alkaloid produced by mixed fermentation of two marine-derived mangrove epiphytic fungi. *Chemistry of Natural Compounds, 47*, 767–769. https://doi.org/10.1007/s10600-011-0053-8.

Role of Biotechnology in Improvement and Regulation of Biosynthesis of Bioactive Compounds

Sandeep Kour, Tania Sagar, Nipunta, Shajaat Hussain, Nisha Kapoor, and Ritu Mahajan

Abstract

Bioactive compounds are produced by plants in huge quantities and are a valuable source for modern pharmacy, food additives, agrochemicals and cosmetics. For thousands of years, plants have been utilized in traditional medicines in different parts of the world. Research on plants has been substantially aided by the recent advancements in biotechnology in the areas of molecular biology, structural and functional genomics, highly efficient analytical chemical techniques and tissue culture. The present chapter is focused on the use of new advances in understanding the production of bioactive compounds in plants such as phenolics, alkaloids, flavonoids, terpenoids, tannins, glycosides, volatile oils etc. Although, these plant based products may have a similar upstream pathway that leads to a small number of common precursors but they differ in specific alterations that results in incredibly diverse structures. Understanding of the different pathways involved in biosynthesizes of these valuable compounds is essential for their large scale and cost-effective production, which in turn will not only improve human well-being but also conserve plant resources.

Keywords

In vitro cultures • Bioactive compounds • Omics • Secondary metabolites • Genome editing

S. Kour · T. Sagar · Nipunta · S. Hussain · N. Kapoor · R. Mahajan (✉)
Plant Molecular Biology Lab, School of Biotechnology, University of Jammu, Jammu (J&K), India
e-mail: ritufeb@gmail.com

1 Introduction

Plants and microbes are cell factories for the synthesis of many natural products, referred as primary and secondary metabolites respectively (Bruce, 2022). The primary metabolites are requisite for the survival of the organisms, whereas the secondary metabolites are not crucial for their survival but have adaptive roles, for instance, defense, adaptation to the environmental stresses, intra/inter species competition/communication and signaling. Therefore, in comparison to primary metabolism, secondary metabolism more closely mimics plant environment (Thirumurugan et al., 2018). The shikimate pathway is a key central pathway that draws primary metabolites into the secondary metabolism to synthesize a broad-spectrum of secondary metabolites. This pathway is present only in the plants and microbes and absent in animals (Santos-Sanchez et al., 2019).

Plant synthesizes four principal classes of secondary metabolites, namely alkaloids, terpenoids, glucosinolates and phenolic compounds (Reshi et al., 2023). Terpenoids are key metabolites essential for the plant interaction with their surroundings. The volatile nature of terpenoids provides a way for interaction with other organisms such as herbivores, neighboring plants and predators. Therefore, potential use of volatile terpenoids is a best alternative approach for pest control strategies (Boncan et al., 2020). There are many key terpenoids that have potent application in agriculture and pharmaceutical industries, namely menthol, eucalyptol, camphor, artemisinin, taxadiene etc. (Barut et al., 2022).

Alkaloids are specialized bioactive compounds having nitrogen as a distinctive atom present in their molecular structures. Most of the alkaloids are poisonous by nature and therefore, function as a defense against herbivorous attacks and microbial infections (Uzor, 2020). Many alkaloids exhibit a broad range of pharmacological properties, for instance, morphine (analgesic), quinine (antimalarial) and homoharringtonine (anticancer) (Dixit et al., 2021).

J. Devi (ed.), *Biotechnological Intervention in Production of Bioactive Compounds*, Sustainable Landscape Planning and Natural Resources Management, https://doi.org/10.1007/978-3-031-76859-0_3

Phenolics being the largest group of plant secondary metabolites possess a common chemical structure consisting of an aromatic ring with one or more hydroxyl moieties and are further divided into distinct groups such as flavonoids, stilbenes, tannins, phenolic acids and lignans (Zhang et al., 2022). Phenolics perform numerous functions like symbiotic nitrogen fixation, physiological regulators, producing yellow, red or blue pigmentation in flowers and fruits, inhibitors of cell cycles and chemical messengers (Kabera et al., 2014).

As per World Health Organization report, approximately 80% of population in developing nations and around 60% of the global population rely on phytomedicine for both traditional medicine practices and primary healthcare requirements (Laloo et al., 2021). About 25% of currently recommended drugs have at least one plant-based compound or ingredient (Singh et al., 2018).

With the advancement in biotechnological approaches, bioactive compounds are isolated in order to address the urgent need for the development of potent medications (Guo et al., 2024). Biotechnological techniques encompass engineering of metabolic pathways, genome editing, in vitro strategies (Selwal et al., 2023) (Fig. 1). Due to increase in market demand of phytochemicals and low production of bioactive substances, researchers have focused on the use of various biotechnological tools. These approaches may be considered as an important and potent substitute for the synthesis of secondary metabolites from plants to enhance their industrial production and lessen the over use of natural resources (Isah et al., 2018). Additionally, bioactive compounds have varied pharmacological properties and significant role against anti-diabetic, antipyretic, anti-cancerous, anti-diuretic, anti-atherosclerotic activities.

2 In Vitro Culture: An Alternative for Secondary Metabolite Production

Plant based biotechnological techniques has provided alternative means to obtain and examine the chemical richness through various in vitro propagation methods in order to generate natural products for the pharmaceutical sector (Lautie et al., 2020). The cell culture approach serves as a foundation for the synthesis of valuable secondary metabolites (Efferth, 2019). Diverse biotechnology methods offer a useful substitute for secondary metabolite production under strictly regulated conditions (Alamgir, 2017). Hence, in vitro culture methods like organ culture of plant, imparts plant tissue as a supplier of natural and organic products (Mahajan

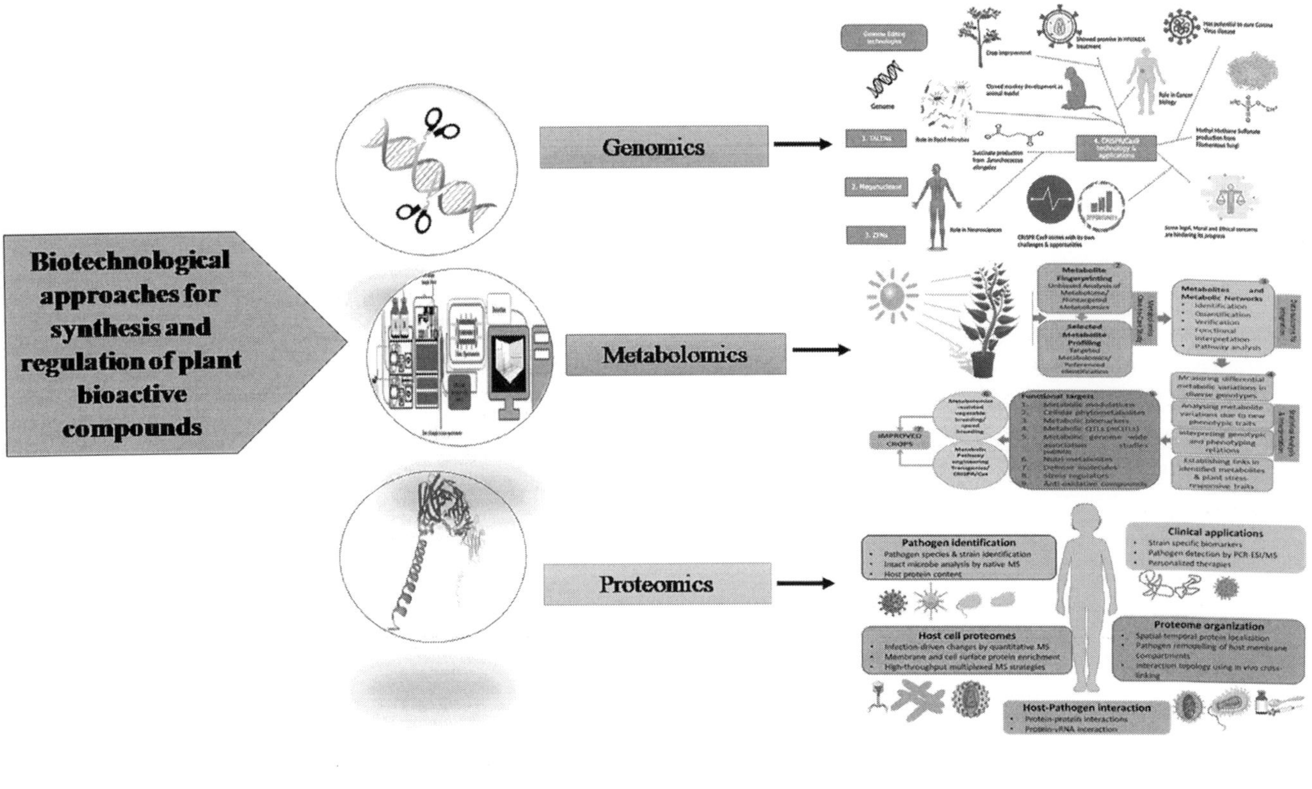

Fig. 1 Integration of omics approaches to elicit the synthesis of bioactive compounds

et al., 2016). Several approaches leveraging cell culture processes have been extensively studied with the aim of enhancing the production and regulating the flow of targeted secondary metabolite biosynthesis (Newman & Cragg, 2016).

Plant cell as well as tissue culture provides a mean for phytochemical production in addition to plant propagation (Hussain et al., 2023; Sharma et al., 2021). Under sterile conditions, any plant part including stems, leaves, meristems or roots can be employed for indefinite plant multiplication and bioactive compounds production. Owing to its numerous benefits, in vitro culture approaches have proven to be an effective strategy for secondary metabolite production (Mohaddab et al., 2022).

Large scale production using micropropagation for various medicinal plants have been effectively accomplished either by organogenesis (Arumugam et al., 2020) or by somatic embryogenesis (Bertero et al., 2020). Organogenesis is a method of micropropagation in which organs are grown from tissues or cells. In particular, organogenesis based plant regeneration entails the shoot induction and growth from an explant that is subsequently transferred to another medium for the induction of roots (Devrnja et al., 2021). However, somatic cells can generate somatic embryos by a method known as somatic embryogenesis, which in turn can grow into saplings in a suitable medium (Soliman et al., 2018). This approach has made it possible to carry out physiological, morphological and genetic alterations (Isah et al., 2018). Therefore, in vitro propagation offers a sustainable choice to the massive production of economically valuable plant species. This can confer a way to source plant material that is likely to yield phytochemicals during the entire year without being limited by seasonal factors (Espinosa-Leal et al., 2018).

2.1 Cell and Suspension Cultures in Secondary Metabolite Production

New avenues for the commercial use of plants and their bioactive compounds should be made possible by the advancement in biotechnology, particularly in the discipline of plant cell culture approaches. Production of in vitro cultures leads to bioactive compounds along with enhancement in cell biomass (Nowak et al., 2021). In laboratory conditions, plant cells that promote the formation of callus by means of a high auxin concentration or by coordinating the auxin and cytokinin ratio are commonly employed for suspension cultures that can be developed from callus for the synthesis of bioactive compounds (Efferth, 2019). Furthermore, immobilizing the hairy root plant's cell system is an effective method for producing pertinent bioactive substances (Kawka et al., 2021).

Bioactive compounds are produced by plant cells as defense measures. Considering this, a strategy called elicitation is used in which substances called elicitors are employed that set off the defense mechanism to enhance the secondary metabolites production. Hence, various researchers have demonstrated how to use elicitors to increase the synthesis of bioactive compounds (Khan et al., 2019). Similarly, it is well studied that plant growth hormones can regulate the secondary metabolite production (Jamwal et al., 2018). Bernabe-Antonio et al. (2021) reported a significant accumulation of biomass in cell suspension culture containing kinetin and α-napthalene acetic acid from *Eysenhardtia platycarpa*. Also, they observed the suspension's dichloromethane extracts containing flavonoids and phenolic compounds with significant antifungal properties. Similarly, Perez-Mendoza et al. (2020) recorded that callus can amalgamate the similar bioactive compounds as *Rosmarinus officinalis* produces in its leaves. Furthermore, Haida et al. (2020) in *Clinacanthus nutans* leaves in vitro culture remarked the improvement in phenolic compounds and antioxidant activity.

2.2 Bioreactors for Eminent Production

The potential of in vitro cultures to synthesize secondary metabolites give rise to bioreactors which are used to produce natural compounds on a large scale (Eibl et al., 2018). In addition, bioreactors are self-sustained systems with sterile surroundings and consistent culture conditions regarding temperature, pH, aeration and agitation along with air as well as liquid outlet and inlet pathways for the significant proliferation of tissues and somatic embryos (Pilarek et al., 2018). Various products like ginsenosides which are utilized as bleaching agents and additives, shikonin used in lipsticks, paclitaxel used together with the anti-cancer medication are produced by means of bioreactor culture (Ritala et al., 2022).

Mehring et al. (2020) observed maximum yield of triterpenoids from cambial meristematic cells of *Ocimum basilicum* using wave-mixed portable bioreactors that resulted in 1.75 fold increase in yield. Kokotkiewicz et al. (2021) reported the substantial biomass yield with elevated essential oil levels in *Verbena officinalis* in vitro shoot cultures developed in temporary immersion bioreactors supplemented with indole-3-butyric acid. Pramita et al. (2018) recorded highest flavonoid production from *Gynura procumbens* by treating the MS medium with indole-3-acetic acid and benzyl adenine in the temporary immersion bioreactors. Thus, bioreactors may enable an exciting new era industrial production by increasing the process efficiency for more lucrative products derived from plants.

2.3 Elicitation

The application of agents that cause plants and cells to go into in vitro defense mode is regarded as a highly effective biotechnological technique for producing secondary metabolites. Elicitor is delineated as an element or factor that when added or modified in in vitro cultures, enhances the secondary metabolites biosynthetic potential. Generally, elicitors can be classified as either biotic or abiotic. The quality as well as quantity of synthesized metabolites can be significantly determined by a number of factors like type of elicitor, its degree of concentration and duration of exposure (Chamkhi et al., 2021).

Abiotic stimulants have a substantial influence on secondary metabolites production. In *Silybum marianum*, methyl jasmonate addition as well as co-treatment of silver nanoparticles and methyl jasmonate in a hydroponic medium showed the most promising accretion of flavonoids and phenolics during extended exposure intervals of 120–144 h (Mubeen et al., 2023). Chavan et al. (2021) reported application of jasmonic acid in *Salacia chinensis* callus culture increased the mangiferin, total flavonoid and phenolic content, thus indicating their highest potential for antioxidants. Furthermore, under salt stress, the growth of *Carum copticum* resulted in an increase in total phenolic content and antioxidant properties (Razavizadeh et al., 2020). Similarly, in callus culture of *Fagonia indica*, elicitation with nanoparticles increased the secondary metabolite production (Khan et al., 2021).

Biological substances like microbial compounds and polysaccharides can be utilized as biotic elicitors. Kanthaliya et al. (2023) conducted a study to analyze the results of different biotic elicitors like pectin, yeast extract and alginate on growth and accumulation of metabolites in in vitro grown shoot cultures of *Pueraria tuberosa* and observed a significant increase in biomass and metabolite content in comparison to untreated control. Bakhtiari and Golkar (2021) reported chitosan application in *Lepidium sativum* callus cultures resulted in an increase in total phenolic and lepidin content by 19.87 times compared to the control. Moreover, in cell suspension of *Silybum marianum*, elicitation by chitosan enhanced the silymarin production and showed strong anti-inflammatory and antioxidant properties (Shah et al., 2021). The inclusion of a fungal elicitor in the *Corylus avellana* cell suspension culture improved the production of paclitaxel (Farhadi et al., 2020).

Studies on eliciting hairy root cultures revealed the possibility of producing secondary metabolites in higher amount (Malarz et al., 2022). In hairy root cultures of *Arachis hypogaea* reported that co-treatment of methyl jasmonate, cyclodextrin, magnesium chloride and hydrogen peroxide resulted in enhanced the prenylated stilbenoids production (Sharma et al., 2022). Qiu et al. (2021) revealed that chitosan application in *Psammosilene tunicoides* hairy root culture increased the accumulation of gypsogenin and quillaic acid. In contrast, Kobtrakul et al. (2023) observed that elicitation by salicylic acid in *Cannabis sativa* hairy root cultures resulted in 1.4 and 1.9 fold enhancement in the production of epifriedeland and friedelin respectively, which may eventually result in the acquisition of phytopharmaceutical benefits.

3 Genomics and Transcriptomics

Advances in genetic engineering have produced large-scale systems for the ample extraction of valuable natural resources. One helpful method for detecting and identifying the genes involved in metabolic pathways is genome-wide expression profile analysis (Kaur et al., 2019). The application of "omics" technology has enhanced the research of transcriptome and metabolite profiles in many tissues under various physiological and pathological conditions (Table 1). New understandings of the biological mechanisms and genes involved in secondary biosynthetic pathways is via the use of metagenomic and genome/transcriptome research techniques (Tian et al., 2021).

Omics techniques have been used to investigate the synthesis of proteins (proteomics), metabolites (metabolomics), genes (genomics), and mRNA (transcriptomics) in particular tissues under specific environmental circumstances (Kaur et al., 2020). Tandem mass tags (TMT), which were created in conjunction with transcriptome technology to accurately quantify proteins, are frequently used to assess the relative abundance of proteins and genes across treatment and control groups, particularly for those experiencing different environmental stresses (Zhang et al., 2018). To study the mutualistic interaction between the two counterparts, genes implicated in such a beneficial relationship must be identified and characterized. It is possible to define functional features and molecular interactions, which makes it possible to forecast endophytes ability to produce new, bioactive secondary metabolites and antibacterial medication candidates that might address the widespread problem of antibiotic resistance (Kaul et al., 2016). The application of bioinformatics and functional genomic techniques, which screen and identify enzyme-coding genes on specific secondary biosynthesis routes from many of the original medicinal plant species, would significantly expedite the research of the secondary metabolites biosynthesis pathways.

One prominent use of genome sequencing is for de novo genome and transcriptome sequences, which have proven to be extremely useful across many species for understanding plant metabolism. In recent years, many plant species have

Table 1 Multi-omics methods for exploring the secondary metabolite synthesis

S. No.	Name of species	Omics techniques	Metabolite	References
01	*Perilla frutescens*	Transcriptomics, Proteomics and Metabolomics	Terpenoids and flavonoids	(Jiang et al., 2020)
02	*Angelica sinensis*	Transcriptomics, Metabolomics	Phenylpropanoid	(Yang et al., 2021)
03	*Triticum aestivum* L	Transcriptomics, Metabolomics	Anthocyanin	(Wang et al., 2021)
04	*Trigonella graecum* L	Transcriptomics, Proteomics	Trigonelline	(Naika et al., 2022)
05	*Aspergillus oryzae*	Transcriptomics, Proteomics and Metabolomics	Ergosterol, Squalene	(Jia et al., 2024)
06	*Nicotiana tabacum* L	Proteomic, Metabolomics	Aroma precursors	(Liu et al., 2022)
07	*Agrimonia pilosa*	Transcriptomics, Proteomics and Metabolomics	Pilosanidin	(Kim et al., 2019)
08	*Papaver somniferum*	Proteomics and Metabolomics	Benzylisoquinoline alkaloids	(Martínez-Esteso et al., 2015)
09	*Corydalis yanhusuo*	Transcriptomics, Metabolomics	Benzylisoquinoline alkaloid	(Xu et al., 2021)
10	*Triticum aestivum*	Transcriptomics, Metabolomics	Phenolics	(Ma et al., 2022a, b)

genome sequences that may be utilized to determine secondary metabolite pathways because of advancements in annotation software, assembly methods, robust computational resources, and genomic technologies (Liu et al., 2021). Using a highly homozygous genotype, Ma et al. (2021) reported an enhanced genome assembly of Danshen and discovered a tandem gene array for cytochrome P450 using gene expansion analysis.

Nowadays, transcriptome sequencing is a common experimental method for finding new genes, determining the expression level of genes, and finding new transcripts. Transcriptome sequencing has revolutionized studies on the genetics and functional genomics of plants, particularly for non-model species whose genomes have not been sequenced. Transcriptome analysis has been used on a number of plant species which revealed the functional potential of genes and transcripts linked to the generation of bioactive compounds in certain metabolic pathways (Zameer et al., 2022). RNA-Seq is an essential technique for finding gene predictions and mining functional analysis, which enhances gene ontology understanding mechanisms of biological processes, molecular activities, and cellular components.

Transcriptomics studies on wide range of plant species have resulted in improved breeding, selection and cultivation practices by plant breeders better and understanding their regulatory processes and functional genes (Tyagi et al., 2022). The bioactive chemicals found in leaves, stems, and roots can be examined and their potential genes can be identified using transcriptome sequencing that are involved in the creation of tissues and manufacture of active compounds. A comparative transcriptome study identified 103 differentially expressed genes involved in the biosynthesis of polysaccharides, including genes linked to fructose and mannose, glycosyltransferase and the production of flavonoids. There is a chance that some of the candidate genes involved in the control of photoperiod, metabolism of starch and sucrose, and hormone-induced activation of signaling pathways are also involved in stem development (Zhou et al., 2020).

Using the Illumina sequencing technology, Singh et al. (2017) sequenced the transcriptome of *Trillium govanianum*, obtaining 69,174 transcripts and identifying several genes involved in the synthesis of steroidal saponins and the biosynthesis of other secondary metabolites. The discovery of physiologically active metabolites and the creation of molecular markers functionally linked to these metabolites are two important applications of genetic modification (Loke et al., 2017). Although variety of genomic sequences available for medicinal plants is restricted, next generation sequencing has been more popular in order to provide a complete transcriptome resource for these plants. Databases and other resources produced by this research are expected to be very important in the near future for the investigation and use of metabolites from medicinal plants (Tripathi et al., 2016). Numerous large-scale medical plant genomes have been successfully sequenced, and as more and more genomes of medicinal plants have been revealed, the sequencing of medicinal plants has entered the development stage. The high level of genome repeat in plants makes it challenging to precisely assemble their genomes using NGS technology, which is another challenge (Cheng et al., 2021).

4 Proteomics

Proteomics has become a promising technique for identifying important proteins or enzymes involved in the production of natural products. The primary fields of study in proteomics are protein–protein interactions, protein localization, identification, and protein translational changes. Moreover, protein–protein interactions, post-translational and transcriptional modifications, and protein cellular site which influence how proteins functions can be revealed (Mahajan & Gupta, 2010, Kosova et al., 2018). Proteome profiling is carried out using a variety of techniques, including MALDI-TOF, 2D- gel electrophoresis, and mass spectrometry, which measures the charge and mass of protein fragments to determine whether there have been any modification to the peptide sequences (Chaudhary et al., 2019). During cellular metabolism, systemic changes enable the proteomic techniques employed in the analysis of bioactive molecule synthesis and provide health-promoting properties (Kim et al., 2016). The bioactive compounds such as oridonin, curdioneand and geniposide were investigated from the intricate biological pathways using proteomics (Hashiguchi et al., 2017).

Song et al. (2015) studied proteins associated to flavonoid and anthocyanin production during ripening of strawberry and peptides were separated using multiple reaction monitoring (MRM) technique which is based on isoelectric point. MRM based targeted proteomics is a potent proteomic technique that has not been applied much in the study of plant proteomics, particularly secondary metabolites proteomics (Martinez-Esteso et al., 2015). Naik et al. (2023) revealed proteomics technique increases the proteins which are associated with protein production or amino acid degradation in cherry tomatoes. By examining proteome datasets of *Dendrobium huoshanense*, cotynylated proteins were identified in the metabolic pathways of photosynthesis, the Calvin cycle, and the synthesis of alkaloids and polysaccharides (Wu et al., 2022a). Similarly, two-dimensional electrophoretic and MALDI-TOF/TOF mass spectrometry studies identified twenty differently expressed proteins of *Lepidium draba* linked to photosynthesis, energy metabolism, and other activities to water stress (Jamshidi Goharrizi et al., 2020). Ma et al. (2022b) examined *Sophora alopecuroides* and performed tandem mass tag based proteomic profiling to confirmed salt stress transporters proteins were linked to secondary metabolites production. *Artemisia annua* with high and low artemisinin contents were analysed quantitatively using targeted sub-proteomic analysis, and the high artemisinin concentration was found to be due to increased DBR2 expression (Chen et al., 2020).

While, Liu et al. (2019) reported proteomic changes in rhizomes of *Paris polyphylla* varieties and observed higher efficiency of sucrose consumption in the sugar metabolism which ultimately linked with elevated protein. Nawrot et al. (2017) compared proteins of different developmental stages of *Chelidonium majus* and reported that fruits in the green phase had larger levels of stress and defense-related proteins, while transcription-related proteins were more prevalent in the flowers but active molecular transport and protein folding showed up regulation. This brings new insights for synthetic biology and potential medical applications. Studies on saponins for the glycosylation modifications which convert tri-terpenoid carboxylic acid enzyme into saponin glucosyl transferases in *Centella asiatica* using proteomic isolation technique were also reported (De Costa et al., 2017). Proteomic approach was used to evaluate the production of carotenoids in *Zingiber zerumbet* and discovered that epigenetic modifications increased the effectiveness of carotenoids (Mahadevan et al., 2015). Proteomic approaches have been significant in the field of medicinal plant science not only in clarifying physiological changes at the systemic level but also in isolating critical enzymes involved in the synthesis of valuable compounds (Gray et al., 2017).

5 Genetic Engineering Approaches to Elicit Biosynthesis of Plant Bioactive Compounds

Medicinal and aromatic plants are cell factories for the biosynthesis of numerous natural compounds that are either directly employed as a phytomedicine or indirectly utilized in the drug discovery (Nielsen et al., 2019). These bioactive compounds are species specific and are produced only under adversarial environmental biotic and abiotic stresses for instance metal, UV, salinity, drought and pathogen attack (Sharma et al., 2019).

However, plants naturally synthesize low concentrations of these natural products and over- utilization of these medicinal and aromatic plants for development of pharmaceutical drugs would have deteriorating impact on the environment and ecology. Therefore, the meager concentration of these bioactive compounds in aromatic and medicinal plants in addition to the ever-increasing demand of these natural products for pharmaceutical and industrial usage have prompted research and development into biotechnological synthesis of bioactive compounds (Guo et al., 2024). Therefore, metabolic engineering is a propitious tool to surmount the scarcity of coveted plant compounds by using many approaches such as over-expression of enzymes, increased flux of precursors, by creating sink compartments that store the desired metabolite, suppression of competitive pathway, by increasing the activity of rate-limiting enzymes; introducing new metabolic pathways, and the over expression of regulatory genes or transcription factors (Sharma et al., 2019).

Genetic engineering employs many laboratory-based techniques to modify the genetic material of species, including alteration, addition and deletion of the segment of DNA (Bhattaraia et al., 2023). The functional genomics is a promising approach to study the key genes responsible for the biosynthesis and regulation of bioactive compounds. Therefore, the functional genomics is a imperative for augmenting bioactive compound production or engineering genes located in homologous or heterologous systems. With the dawn of high throughput sequencing technologies, profiling thousands of genes simultaneously is feasible now (Barut et al., 2022). Zhao et al. (2020) explored the over-expression of the precursor approach to enhance the synthesis of tropane alkaloids in the hairy roots of *Atropa belladonna*. Wu et al. (2022b) employed the over-expression of R2R3-MYB transcription factor (*AtMYB12*), for metabolic engineering of the key bioactive flavonoids in hairy roots of *Glycyrrhiza inflate*. Licochalcone A and Echinatin are major chalcones present in rhizomes and roots of *Glycyrrhiza inflate*. Overexpression of transcription factor *AtMYB12* in the hairy roots significantly increased the production of licochalcone by five-fold and echinatin by two-fold.

5.1 CRISPR/Cas9: A Promising Genome Editing Tool

The rise of genome editing tools act as a boon for industrial biotechnology. The traditional genome modification approaches include, engineered endonuclease, RNA interference (RNAi) and gene targeting (Xu et al., 2020). However, their extensive application is restricted by laborious transformation procedures, dearth of vector system and selectable markers and a limited number of screening techniques to detect successful clones. Furthermore, the use of these approaches is limited to only a handful of industrially significant strains and model organisms (Mans et al., 2015). There are primarily three major categories of genome editing technologies; the CRISPR/Cas9 (the clustered regularly interspaced short palindromic repeats/CRISPR-associated protein 9), TALENs (the transcription activator-like effector nucleases) and ZFNs (zinc finger nucleases) (Mohanta et al., 2017).

Both the TALENs and ZFNs are the first and second generation engineered nucleases tools for genome editing respectively, consisting of a DNA binding domain and a *Fok*I cleavage domain that through DNA–protein interactions, creates a double strand cut in the target sequence, thereby, preparing the target site for modifications (Alagoz et al., 2016). Owing to the substandard precision, laborious, intricated and the time consuming assembly and fabrication of DNA-binding module, reduced target cutting efficiency and expensive in nature, ZFNs and TALENs are easily outperformed by the third generation genome editing system, known as the CRISPR/Cas9 technology (Bhuyan et al., 2023).

Conventionally, this CRISPR/Cas9 system has been referred as the adaptive immune system of Archaea and bacteria. It is an antiviral mechanism that helps them to identify and degrade invasive foreign DNA from plasmids and bacteriophages (Bhagwat et al., 2022). This system has two key elements: the gRNA (guide RNA) and Cas9 endonuclease (Son & Park, 2022). The gRNA is a small guiding RNA that comprehends approximately 20 nucleotides at its 5′ end which are complementary to target DNA sequences. The gRNA is essential for guiding Cas9 endonuclease to the target site. Cas9 protein is an RNA dependent DNA endonuclease. The 3′ end of the target sequence contains a short nucleotide repeats referred as the PAM (protospacer adjacent motif) sequence, serves as a Cas9 specific binding site. The Cas9 endonuclease from the *Streptococcus pyogenes* (SpCas9) recognises 5′-NGG-3′sequence as the PAM sequence. This specific binding activates the Cas9 endonuclease to produce a double-strand cut in the target sequence (Zhang et al., 2017). This editing tool can also be employed to introduce simultaneous multiplex editing by designing multiple gRNAs for distinct target sequences (Xie et al., 2015). Researchers mostly used this tool to develop gene knockouts, execute transgene insertion and bring epigenetic changes in the target sequences (Jinek et al., 2012). Therefore, CRISPR/Cas genome editing system is a promising tool for metabolite engineering in medicinal plants to elicit the synthesis of desirable bioactive compound (Das et al., 2024).

The red pigment lycopene is a carotenoid and it was observed that the lycopene content decreases while ripening of the fruits because of the conversion of β-carotene into α-carotene. Li et al. (2018) applied this approach to knockout the expression of beta-lycopene cyclase, lycopene ε-cyclase, lycopene β-cyclase 1 and lycopene β-cyclase 2 genes respectively, responsible for conversion of lycopene to carotene in tomato fruits. This CRISPR/Cas9 editing increased the lycopene content of the tomato fruit by five-fold. Alagoz et al. (2016) utilized this editing tool to knockout the expression of *4′OMT2* gene that performs the regulatory role in the synthesis of benzylisoquinoline alkaloids) in *Papaver somniferum*. Red sage is a chinese medicinal plant that synthesizes rosmarinic acid by phenylpropanoid pathway. Shi et al. (2021) used the CRISPR/Cas9 genome editing approach to identify the key genes involved in the biosynthesis of rosmarinic acid while accomplishing site specific mutagenesis to enhance the effectiveness of this medicinal herb. Wen et al. (2022) utilized CRISPR/CAS for targeted editing of the *FtMYB45* gene in *Fagopyrum tataricum*. The mutagenesis resulted in the elevated flavonoids content in hairy roots of buckwheat, will be potent source for the production of flavonoids. Furthermore, they elucidated the negative regulatory function *FtMYB45* gene in the flavonoid metabolic pathway.

6 Metabolomics

Since phytochemicals have many chemical properties and are complicated, it is not possible to fully study the phytochemical profile of a substance using a single analytical approach. Based on the nature of the targeted phytochemical, plant metabolomics has been thoroughly investigated utilizing a variety of analytical techniques (Sharma & Yadav, 2022). The field of metabolomics has appeared to be one of the major contributors in the identification of bioactive compounds and discover unique phytochemicals for the development of novel drugs (Aslam et al., 2020).

Metabolomics represents the final biochemical characteristics connected to the physiological, developmental, and functional aspects of biological systems. The plant metabolomic method is a useful tool to supplement transcriptome and proteome monitoring investigations. At the structural genomic level, metabolites can be functional entities and their gene clusters are involved in the production of natural products (Guo et al., 2024). Several metabolites that include sugars, phenolics, flavonoids, alkaloids, and amino acids are commonly identified and quantified using liquid chromatography-mass spectrometry (LC–MS), gas chromatography-mass spectrometry (GC–MS), capillary electrophoresis-mass spectrometry, high performance liquid chromatography and nuclear magnetic resonance spectroscopy (NMR) (Marr et al., 2021). The usage of GC–MS and LC–MS for metabolomics analysis is growing due to their high sensitivity (Kaluarachchi et al., 2018). Using mass spectrometry, bioactive compounds are separated and extracted that is limited to the chemicals of volatile and polar characteristics, based upon charge by mass ratio. The m/z value and the chromatographic retention time can also have data acquisition tolerances for non-targeted metabolites, and these consequences are also common to measurements performed with alignment software.

In addition, molecular characterization based on fragmentation is made possible by capillary electrophoresis-mass spectrometry that provides precise mass information and functions in a uniform separation environment for nucleotides and highly charged metabolites (Ren et al., 2018). Furthermore, NMR is a highly useful method for determining the structures of new and unexpected molecules, such as those having different iso-topomer distributions or identical masses (Salem et al., 2020). The application of metabolomics endeavours the nutritional value plant-based foods to improve food processing methods and also aids in understanding the mechanism of action to confirm the effectiveness of conventional treatments.

Three sea buckthorn species namely *Hippopahe gyantsensis*, *H. Neurocarpa*, and *H. tibetana* examined under 1H-NMR combination with DNA barcoding technology showed distinct genetic diversity (Liu et al., 2016). Hu et al. (2019) studied complementary multi-element and metabolomic data which indicated the geographical origin of *Astragalus membranaceus* and *Paeonia albiflora* using 1H NMR and inductively coupled plasma atomic emission spectroscopy. Gonulalan et al. (2020) compared the metabolomic and proteomics profiles of *Passiflora incarnata*, *Valeriana officinalis*, *Hypericum perforatum* and *Melissa officinalis* extracts to treat sleep problems and sedative-anxiolytic diseases.

A concise description of the main metabolomics approaches is given for the analysis of metabolites.

6.1 Liquid Chromatography-Mass Spectrometry (LC–MS)

LC–MS has emerged as the most comprehensive method for measuring a broad variety of different metabolites (Aslam et al., 2022). Metabolite separation is achieved by selecting columns that include reversed phase, hydrophobic interaction and ion exchange and all are based on various chemical properties (Salem et al. 2020). Currently, the most popular reversed phase columns used for LC–MS gradient are C18 and C8. A comparative study was performed between the wild plant and callus extract of *Saraca asoca,* their phytochemical constituents were analysed using LC–MS technique and it was revealed that callus extract contains more bioactive compounds than wild plants, and it could be possible to investigate these substances for the development of potent drugs without harming the wild plant (Vignesh et al., 2022). Ali et al. (2021) reported LC–MS profile of ten Australian herbs and observed the significant presence of phenolic compounds. Similarly, Singh et al. (2022) investigated phytochemicals from herbal formulation (cumin and fenugreek seeds) using HR LC–MS and identified 51 compounds including 9 small peptides and 42 other phytochemicals. Most of the phytochemicals observed in this were belong to flavonoids class.

6.2 Capillary Electrophoresis-Mass Spectrometry (CE-MS)

Capillary electrophoresis is one of the techniques for separating, identifying, and quantifying different bioactive components, especially in food products, clinical diagnostics, environmental monitoring and pharmaceutical analysis (Gackowski et al., 2021). When it is coupled with mass spectrometry, it provides exact information about phytochemicals based on charge-to-mass ratio. In CE-MS, the separation is due to electric field and electro-osmotic flow of the buffer and potential difference causes liquid flow in the capillary. But the ability of CE to detect cations and anions simultaneously, is a crucial aspect that influences the decision to use it. Furthermore, CE has amazing separation efficiency, low solvent consumption, minimal sample size, reduced operational expenses, repeatability, and adaptability of different separation modes and all contributed to its increased popularity in recent years. Przybylska et al. (2021) used CE-MS technique to isolate and identify the anthocyanins present in *Hibiscus sabdariffa* using 200 mML^{-1} ammonia and boric acid as solvents.

6.3 Gas Chromatography-Mass Spectrometry (GC–MS)

GC–MS is one of the most commonly used metabolomic method for determining the concentrations of volatile and semi-volatile organic molecules in a range of samples. Plant metabolites isolated from GC–MS were derivitized and become volatile. These derivaization includes trimethylsilylation, silylation, acylation, alkylation, and methoximation (Salem et al., 2020). GC–MS also provides quantitative data in addition to the linked mass spectral database, which is extremely valuable for establishing a linkage between bioactive chemicals and their pharmacological applications (Kumar et al., 2019).

Casuga et al. (2016) investigated the extracts of *Broussonetia luzonica* using GC–MS technique and identified major three compounds namely lupeol, tritriacontane and *γ-sitosterol* from methanol extract whereas, ethyl acetate extract contained monoacetate, phytol and squalene. Konappa et al. (2020) identified the important functional groups and phytochemicals from *Amomum nilgiricum* leaf and rhizome extracts using fourier transform infrared spectrometer (FTIR) and GC–MS. From leaf and rhizome extracts, the functional characteristic peaks of alkyl halides, carboxylic acids, ketones, phenols, and aromatic amines were present in the FTIR spectra with twenty-five phytochemical components in the extracts.

7 Conclusion

Medicinal plants being valuable source of metabolites are of significant importance for human healthcare. The concurrent occurrence of global natural disasters has threatened the growth of medicinally important plants that has resulted in low yield of active metabolites. Also over- exploitation of plants from the wild resources have resulted in great loss of diversity and habitat destruction. Therefore, biosynthesis of these natural bio-products using biotechnological approaches such as tissue culture, genetic engineering, fermentation and synthetic biology have opened several new avenues for proficient large-scale production of economically important bioactive compounds. Also, the by-products obtained during extraction can contribute to health and food industry, along with sustainable use of natural resources. Hence, these biotechnological approaches must be continuously improved to extract variety of compounds which can assure maximum potential to meet the demands of pharmaceutical and cosmetic industries.

Acknowledgements The authors thankfully acknowledge various funding agencies (Govt. of India) such as Department of Science and Technology (DST), the University Grants Commission (UGC) and the National Medicinal Plants Board (NMPB) for research grants in the form of major research projects. In addition, the authors are thankful to Rashtriya Uchchattar Shiksha Abhiyan (RUSA), the University Grants Commission Special Assistance Programme (UGC-SAP), Promotion of University Research and Scientific Excellence (PURSE) grants, Central facility, Seed grant by University of Jammu and Department of Biotechnology (DBT) funded Bioinformatics center at School of Biotechnology, University of Jammu for their support.

References

Alagoz, Y., Gurkok, T., Zhang, B., & Unver, T. (2016). Manipulating the biosynthesis of bioactive compound alkaloids for next-generation metabolic engineering in opium poppy using CRISPR-Cas 9 genome editing technology. *Science and Reports, 6*(1), 30910.

Alamgir, A. N. (2017). *Therapeutic use of medicinal plants and their extracts: volume 1*. Springer International Publishing AG.

Ali, A., Bashmil, Y. M., Cottrell, J. J., Suleria, H. A., & Dunshea, F. R. (2021). LC-MS/MS-QTOF screening and identification of phenolic compounds from Australian grown herbs and their antioxidant potential. *Antioxidants, 10*(11), 1770. ·

Arumugam, G., Sinniah, U. R., Swamy, M. K., & Lynch, P. T. (2020). Micropropagation and essential oil characterization of *Plectranthus amboinicus* (Lour.) Sprengel, an aromatic medicinal plant. *In Vitro Cell Development Biological Plant, 56*, 491–503.

Aslam, L., Kaur, R., Kapoor, N., & Ritu Mahajan, R. (2020). Phytochemical composition and antioxidant activities of leaf extracts of Viola odorata from Kishtwar, Jammu and Kashmir. *Journal of Herbs, Spices & Medicinal Plants, 26*, 77–88.

Aslam, L., Kaur, R., Hussain, S., Kapoor, N., Ritu Mahajan, R. 2022. LC-MS/MS identification and structural characterization of isolated cyclotides from precursor sequences of *Viola odorata* L. petiole tissue using computational approach. J Biosci. 47, 50.

Bakhtiari, M.A., & Golkar, P. (2021). The effects of callus elicitation on lepidine, phenolic content, and antioxidant activity of *Lepidium sativum* L.: Chitosan and gibberellic acid. *Journal of Plant Growth Regulation, 41*(3), 1148–1160.

Barut, M., Nadeem, M. A., Akgur, O., Tansi, L. S., Aasim, M., Altaf, M. T., et al. (2022). Medicinal and aromatic plants in the omics era: Application of plant breeding and biotechnology for plant secondary metabolite production. *Turk. J. Agric., 46*(2), 182–203.

Bernabé-Antonio, A., Sánchez-Sánchez, A., Romero-Estrada, A., Meza-Contreras, J. C., Silva-Guzmán, J. A., Fuentes-Talavera, F. J., et al. (2021). Establishment of a cell suspension culture of *Eysenhardtia platycarpa*: Phytochemical screening of extracts and evaluation of antifungal activity. *Plants, 10*(2), 414.

Bertero, V. G., Beznec, A., Faccio, P., Auteri, M., Arteaga, M., Bonafede, M., et al. (2020). High-efficiency direct somatic embryogenesis and plant regeneration from leaf base explants of "peperina" (*Minthostachys verticillata*). *In Vitro Cell Development Biological Plant, 56*, 915–919.

Bhagwat, A. C., Patil, A. M., & Saroj, S. D. (2022). CRISPR/Cas 9-based editing in the production of bioactive molecules. *Molecular Biotechnology, 64*, 245–251.

Bhattaraia, S., Bhandaria, S., Yadava, G., & Sapkotab, R. P. (2023). Genetic engineering: An emerging method in plant breeding. *Plant Physiol. Soil Chem., 3*(1), 31–34.

Bhuyan, S. J., Kumar, M., Ramrao-Devde, P., Rai, A. C., Mishra, A. K., Singh, P. K., et al. (2023). Progress in gene editing tools, implications and success in plants: A review. *Front. Genome Ed., 5*, 1272678.

Boncan, D. A. T., Tsang, S. S., Li, C., Lee, I. H., Lam, H. M., Chan, T. F., et al. (2020). Terpenes and terpenoids in plants: Interactions with environment and insects. *International Journal of Molecular Sciences, 21*(19), 7382.

Bruce, S. O. (2022). Secondary metabolites from natural products. *Secondary Metabolites: Trends and Reviews*, 51.

Casuga, F. P., Castillo, A. L., & Corpuz, M. J. A. T. (2016). GC–MS analysis of bioactive compounds present in different extracts of an endemic plant *Broussonetia luzonica* (Blanco) (Moraceae) leaves. *Asian Pacific Journal of Tropical Biomedicine, 6*(11), 957–961.

Chamkhi, I., Benali, T., Aanniz, T., El Menyiy, N., Guaouguaou, F. E., El Omari, N., et al. (2021). Plant-microbial interaction: The mechanism and the application of microbial elicitor induced secondary metabolites biosynthesis in medicinal plants. *Plant Physiology and Biochemistry, 167*, 269–295.

Chaudhary, J., Khatri, P., Singla, P., Kumawat, S., Kumari, A., Vikram, A., et al. (2019). Advances in omics approaches for abiotic stress tolerance in tomato. *Biology, 8*(4), 90.

Chavan, J. J., Kshirsagar, P. R., Jadhav, S. G., Nalavade, V. M., Gurme, S. T., & Pai, S. R. (2021). Elicitor-mediated enhancement of biomass, polyphenols, mangiferin production and antioxidant activities in callus cultures of *Salacia chinensis* L. *3 Biotech, 11*(6), 285.

Chen, M., Yan, T., Ji, L., Dong, Y., Sidoli, S., Yuan, Z., et al. (2020). Comprehensive map of the *Artemisia annua* proteome and quantification of differential protein expression in chemotypes producing high versus low content of artemisinin. *Proteomics, 20*(10), 900310.

Cheng, Q. Q., Ouyang, Y., Tang, Z. Y., Lao, C. C., Zhang, Y. Y., Cheng, C. S., et al. (2021). Review on the development and applications of medicinal plant genomes. *Frontiers in Plant Science, 12*, 791219.

Das, S., Kwon, M., & Kim, J. Y. (2024). Enhancement of specialized metabolites using CRISPR/Cas gene editing technology in medicinal plants. *Frontiers in Plant Science, 15*, 1279738.

De Costa, F., Barber, C. J., Kim, Y. B., Reed, D. W., Zhang, H., Fett-Neto, A. G., et al. (2017). Molecular cloning of an ester-forming triterpenoid: UDP-glucose 28-O-glucosyl-transferase involved in saponin biosynthesis from the medicinal plant *Centella asiatica*. *Plant Science, 262*, 9–17.

Devrnja, N., Krstić-Milošević, D., Janošević, D., Tešević, V., Vinterhalter, B., Savić, J., et al. (2021). *In vitro* cultivation of tansy (*Tanacetum vulgare* L.): A tool for the production of potent pharmaceutical agents. *Protoplasma, 258*, 587–599.

Dixit, S., Shukla, A., Singh, V., & Upadhyay, S. K. (2021). Bioprospecting of natural compounds for industrial and medical applications: Current scenario and bottleneck. *Bioprospecting of Plant Biodiversity for Industrial Molecules*, 53–71.

Efferth, T. (2019). Biotechnology applications of plant callus cultures. *Eng., 5*(1), 50–59.

Eibl, R., Meier, P., Stutz, I., Schildberger, D., Hühn, T., & Eibl, D. (2018). Plant cell culture technology in the cosmetics and food industries: Current state and future trends. *Applied Microbiology and Biotechnology, 102*, 8661–8675.

Espinosa-Leal, C. A., Puente-Garza, C. A., & García-Lara, S. (2018). *In vitro* plant tissue culture: Means for production of biological active compounds. *Planta, 248*, 1–18.

Farhadi, S., Moieni, A., Safaie, N., Sabet, M. S., & Salehi, M. (2020). Fungal cell wall and methyl-β–cyclodextrin synergistically enhance paclitaxel biosynthesis and secretion in *Corylus avellana* cell suspension culture. *Science and Reports, 10*(1), 5427.

Gackowski, M., Przybylska, A., Kruszewski, S., Koba, M., Mądra--Gackowska, K., & Bogacz, A. (2021). Recent applications of capillary electrophoresis in the determination of active compounds in medicinal plants and pharmaceutical formulations. *Molecules, 26* (14), 4141.

Gonulalan, E. M., Nemutlu, E., Bayazeid, O., Koçak, E., Yalçın, F. N., & Demirezer, L. O. (2020). Metabolomics and proteomics profiles of some medicinal plants and correlation with BDNF activity. *Phytomedicine, 74*, 152920.

Gray, N. E., Zweig, J. A., Matthews, D. G., Caruso, M., Quinn, J. F., & Soumyanath, A. (2017). *Centella asiatica* attenuates mitochondrial dysfunction and oxidative stress in a β-exposed hippocampal neurons. *Oxidative Medicine and Cellular Longevity, 1*, 7023091.

Guo, M., Lv, H., Chen, H., Dong, S., Zhang, J., Lu, W., et al. (2024). Strategies on biosynthesis and production of bioactive compounds in medicinal plants. *Chinese Herbal Medicines, 16*(1), 13.

Haida, Z., Nakasha, J. J., & Hakiman, M. (2020). *In vitro* responses of plant growth factors on growth, yield, phenolics content and antioxidant activities of *Clinacanthus nutans* (Sabah Snake Grass). *Plants, 9*(8), 1030.

Hashiguchi, A., Tian, J., & Komatsu, S. (2017). Proteomic contributions to medicinal plant research: From plant metabolism to pharmacological action. *Proteomics, 5*(4), 35.

Hu, S., Yu, X., Li, R., Wu, X., Yang, X., Han, Z., et al. (2019). Analysis of the differential compounds in *Astragalus membranaceus* from different geographical origins by UFLC-IT-TOF/MS. *Journal of Chinese Pharmaceutical Sciences, 50*(1), 59–65.

Hussain, S., Kapoor, N., & Mahajan, R. (2023). High-frequency shoot multiplication and in vitro plantlet regeneration from shoot bud explants in *Curcuma zedoaria* Roscoe. *Journal of Herbs, Spices & Medicinal Plants, 29*(3), 262–273.

Isah, T., Umar, S., Mujib, A., Sharma, M. P., Rajasekharan, P. E., Zafar, N., et al. (2018). Secondary metabolism of pharmaceuticals in the plant *in vitro* cultures: Strategies, approaches, and limitations to achieving higher yield. *Plant Cell, Tissue and Organ Culture, 132*, 239–265.

Jamshidi-Goharrizi, K., Fatehi, F., Nazari, M., Salehi, F., & Maleki, M. (2020). Assessment of changes in the content of sulforaphane and expression levels of CYP79F1 and myrosinase genes and proteomic profile of *Lepidium draba* plant under water-deficit stress induced by polyethylene glycol. *Acta Physiologiae Plantarum, 42*, 1–18.

Jamwal, K., Bhattacharya, S., & Puri, S. (2018). Plant growth regulator mediated consequences of secondary metabolites in medicinal plants. *Journal of Applied Research on Medicinal and Aromatic Plants, 9*, 26–38.

Jia, X., Song, J., Wu, Y., Feng, S., Sun, Z., Hu, Y., et al. (2024). Strategies for the enhancement of secondary metabolite production via biosynthesis gene cluster regulation in *Aspergillus oryzae*. *Journal of Fungus, 10*(5), 312.

Jiang, T., Guo, K., Liu, L., Tian, W., Xie, X., Wen, S., et al. (2020). Integrated transcriptomic and metabolomic data reveal the flavonoid biosynthesis metabolic pathway in *Perilla frutescens* (L.) leaves. *Scientific Reports, 10*(1), 16207.

Jinek, M., Chylinski, K., Fonfara, I., Hauer, M., Doudna, J. A., & Charpentier, E. (2012). A programmable dual-RNA–guided DNA endonuclease in adaptive bacterial immunity. *Sci., 337*(6096), 816–821.

Kabera, J. N., Semana, E., Mussa, A. R., & He, X. (2014). Plant secondary metabolites: Biosynthesis, classification, function and pharmacological properties. *Journal of Pharmacy and Pharmacology, 2*(7), 377–392.

Kaluarachchi, M., Boulangé, C. L., Karaman, I., Lindon, J. C., Ebbels, T. M., Elliott, P., Olson, N. C., et al. (2018). A comparison of human serum and plasma metabolites using untargeted 1 H NMR spectroscopy and UPLC-MS. *Metabolomics, 14*, 1–12.

Kanthaliya, B., Joshi, A., Arora, J., Alqahtani, M. D., & Abd Allah, E. F. (2023). Effect of biotic elicitors on the growth, antioxidant activity and metabolites accumulation in *in vitro* propagated shoots of *Pueraria tuberosa*. *Plants, 12*(6), 1300.

Kaul, S., Sharma, T., & Dhar, M. (2016). Omics tools for better understanding the plant–endophyte interactions. *Frontiers in Plant Science, 7*, 955.

Kaur, R., Aslam, L., Kapoor, N., & Mahajan, R. (2020). Identification and comparative expression analysis of Chalcone Synthase, Flavanone 3-Hydroxylase and Dihydroflavonol 4-reductase genes in wild pomegranate (*Punica granatum* L.) organs. *Braz. J. Bot., 4*, 883–896.

Kaur, R., Kapoor, N., Aslam, L., & Mahajan, R. (2019). Molecular characterization of PgUFGT gene and R2R3-PgMYB transcription factor involved in flavonoid biosynthesis in four tissues of wild pomegranate (*Punica granatum* L.). *Journal of Genetics, 98*, 94.

Kawka, M., Bubko, I., Koronkiewicz, M., Gruber-Bzura, B., Graikou, K., Chinou, I., et al. (2021). Polyurethane foam rafts supported *in vitro* cultures of *Rindera graeca* roots for enhanced production of rinderol, potent proapoptotic naphthoquinone compound. *International Journal of Molecular Sciences, 23*(1), 56.

Khan, T., Khan, T., Hano, C., & Abbasi, B. H. (2019). Effects of chitosan and salicylic acid on the production of pharmacologically attractive secondary metabolites in callus cultures of *Fagonia indica*. *Industrial Crops and Products, 129*, 525–535.

Khan, A. U., Khan, T., Khan, M. A., Nadhman, A., Aasim, M., Khan, N. Z., et al. (2021). Iron-doped zinc oxide nanoparticles-triggered elicitation of important phenolic compounds in cell cultures of *Fagonia indica*. *Plant Cell, Tissue and Organ Culture, 147*(2), 287–296.

Kim, S. W., Gupta, R., Lee, S. H., Min, C. W., Agrawal, G. K., Rakwal, R., et al. (2016). An integrated biochemical, proteomics, and metabolomics approach for supporting medicinal value of *Panax ginseng* fruits. *Frontiers in Plant Science, 7*, 994.

Kim, H. W., Choi, S. Y., Jang, H. S., Ryu, B., Sung, S. H., & Yang, H. (2019). Exploring novel secondary metabolites from natural products using pre-processed mass spectral data. *Science and Reports, 9* (1), 17430.

Kobtrakul, K., Rani, D., Binalee, A., Udomlarp, P., Srichai, T., De-Eknamkul, W., et al. (2023). Elicitation enhances the production of friedelin and epifriedelanol in hairy root cultures of *Cannabis sativa* L. *Frontiers in Plant Science, 14*, 1242584.

Kokotkiewicz, A., Zabiegala, B., Kubica, P., Szopa, A., Bucinski, A., Ekiert, H., et al. (2021). Accumulation of volatile constituents in agar and bioreactor shoot cultures of *Verbena officinalis* L. *Plant Cell, Tissue and Organ Culture, 144*, 671–679.

Konappa, N., Udayashankar, A. C., Krishnamurthy, S., Pradeep, C. K., Chowdappa, S., & Jogaiah, S. (2020). GC–MS analysis of phytoconstituents from *Amomum nilgiricum* and molecular docking interactions of bioactive serverogenin acetate with target proteins. *Science and Reports, 10*(1), 16438.

Kosová, K., Vítámvás, P., Urban, M. O., Prášil, I. T., & Renaut, J. (2018). Plant abiotic stress proteomics: The major factors determining alterations in cellular proteome. *Frontiers in Plant Science, 9*, 325823.

Kumar, D., Singh, L., Antil, R., Kumari, S., & Dahiya, P. (2019). GC-MS analysis and phytochemical screening of methanolic fruit extract of *Citrullus colocynthis* (L.) Schrad. *Journal of Pharmacognosy and Phytochemistry, 8*(3), 3360–3363.

Laloo, D., Kalita, J.M., & Prasad, S. K. (2021). Molecular docking studies of plant-derived bioactive compounds: A comprehensive *in silico* standardization approach. In *Evidence based validation of traditional medicines: A comprehensive approach* (pp. 371–404).

Lautie, E., Russo, O., Ducrot, P., & Boutin, J. A. (2020). Unraveling plant natural chemical diversity for drug discovery purposes. *Frontiers in Pharmacology, 11*, 397.

Li, X., Wang, Y., Chen, S., Tian, H., Fu, D., Zhu, B., et al. (2018). Lycopene is enriched in tomato fruit by CRISPR/Cas9-mediated multiplex genome editing. *Frontiers in Plant Science, 9*, 559.

Liu, Y., Liu, C., Tan, E., Fan, G., Xiang, L., Li, X. D., et al. (2016). Genetic and chemical discrimination of traditional Tibetan medicine sea buckthorn based on DNA barcode and ^{1}H-NMR metabolic method. *China Journal of Chinese Materia Medica, 41*(4), 578–585.

Liu, F., Meng, Y., He, K., Song, F., Cheng, J., Wang, H., et al. (2019). Comparative analysis of proteomic and metabolomic profiles of different species of Paris. *J. Proteom., 200*, 11–27.

Liu, Y., Wang, B., Shu, S., Li, Z., Song, C., Liu, D., et al. (2021). Analysis of the *Coptis chinensis* genome reveals the diversification of protoberberine-type alkaloids. *Nature Communications, 12*(1), 3276.

Liu, A., Yuan, K., Xu, H., Zhang, Y., Tian, J., Li, Q., et al. (2022). Proteomic and metabolomic revealed differences in the distribution and synthesis mechanism of aroma precursors in yunyan 87 tobacco leaf, stem, and root at the seedling stage. *ACS Omega, 7*(37), 33295–33306.

Loke, K. K., Rahnamaie-Tajadod, R., Yeoh, C. C., Goh, H. H., Mohamed-Hussein, Z. A., Zainal, Z., et al. (2017). Transcriptome analysis of *Polygonum minus* reveals candidate genes involved in important secondary metabolic pathways of phenylpropanoids and flavonoids. *Peer J, 5*, e2938.

Ma, Y., Cui, G., Chen, T., Ma, X., Wang, R., Jin, B., et al. (2021). Expansion within the CYP71D subfamily drives the heterocyclization of tanshinones synthesis in *Salvia miltiorrhiza*. *Nature Communications, 12*(1), 685.

Ma, D., Xu, B., Feng, J., Hu, H., Tang, J., Yin, G., et al. (2022a). Dynamic metabolomics and transcriptomics analyses for characterization of phenolic compounds and their biosynthetic characteristics in wheat grain. *Frontiers in Nutrition, 9*, 844337.

Ma, T. L., Li, W. J., Hong, Y. S., Zhou, Y. M., Tian, L., Zhang, X. G., et al. (2022b). TMT based proteomic profiling of *Sophora alopecuroides* leaves reveal flavonoid biosynthesis processes in response to salt stress. *J. Proteom., 253*, 104457.

Mahadevan, C., Jaleel, A., Deb, L., Thomas, G., & Sakuntala, M. (2015). Development of an efficient virus induced gene silencing strategy in the non-model wild ginger-*Zingiber zerumbet* and investigation of associated proteome changes. *PLoS ONE, 10*(4), 0124518.

Mahajan, R., & Gupta, P. (2010). Proteomics: Where genomics leaves off. *Czech Journal of Genetics and Plant Breeding, 46*, 47–53.

Mahajan, R., Kapoor, N., & Billowria, P. (2016). Callus proliferation and *in vitro* organogenesis of *Gloriosa superba*: An endangered medicinal plant. *Annals of Plant Sciences, 5*, 1466–1471.

Malarz, J., Michalska, K., Yudina, Y. V., & Stojakowska, A. (2022). Hairy root cultures as a source of polyphenolic antioxidants: Flavonoids, stilbenoids and hydrolyzable tannins. *Plants, 11*(15), 1950.

Mans, R., Van Rossum, H. M., Wijsman, M., Backx, A., Kuijpers, N. G., Van den Broek, M., et al. (2015). CRISPR/Cas9: a molecular Swiss army knife for simultaneous introduction of multiple genetic modifications in *Saccharomyces cerevisiae*. *FEMS Yeast Research, 15*(2), fov004.

Marr, S., Hageman, J. A., Wehrens, R., van Dam, N. M., Bruelheide, H., & Neumann, S. (2021). LC-MS based plant metabolic profiles of thirteen grassland species grown in diverse neighbourhoods. *Scientific Data, 8*(1), 52.

Martínez-Esteso, M. J., Martínez-Márquez, A., Sellés-Marchart, S., Morante-Carriel, J. A., & Bru-Martínez, R. (2015). The role of proteomics in progressing insights into plant secondary metabolism. *Frontiers in Plant Science, 6*, 504.

Mehring, A., Haffelder, J., Chodorski, J., Stiefelmaier, J., Strieth, D., & Ulber, R. (2020). Establishment and triterpenoid production of *Ocimum basilicum* cambial meristematic cells. *Plant Cell, Tissue and Organ Culture, 143*, 573–581.

Mohaddab, M., El Goumi, Y., Gallo, M., Montesano, D., Zengin, G., Bouyahya, A., et al. (2022). Biotechnology and *in vitro* culture as an alternative system for secondary metabolite production. *Molecules, 27*(22), 8093.

Mohanta, T. K., Bashir, T., Hashem, A., Abd Allah, E.F., & Bae, H. (2017). Genome editing tools in plants. *Genes, 8*(12), 399.

Mubeen, B., Hasnain, A., Jie, W., Zheng, H., Peijnenburg, W.J., Rozali, S.E., et al. 2023. Enhanced production of active photosynthetic and biochemical molecules in *Silybum marianum* L. using biotic and abiotic elicitors in hydroponic culture. *Molecules, 28*(4), 1716.

Naik, B., Kumar, V., Rizwanuddin, S., Chauhan, M., Choudhary, M., Gupta, A. K., et al. (2023). Genomics, proteomics, and metabolomics approaches to improve abiotic stress tolerance in tomato plant. *International Journal of Molecular Sciences, 24*(3), 3025.

Naika, M.B., Sathyanarayanan, N., Sajeevan, R.S., Bhattacharyya, T., Ghosh, P., Iyer, M. S., et al. (2022). Exploring the medicinally important secondary metabolites landscape through the lens of transcriptome data in fenugreek (*Trigonella foenum graecum* L.). *Scientific Reports, 12*(1), 13534.

Nawrot, R., Lippmann, R., Matros, A., Musidlak, O., Nowicki, G., & Mock, H. P. (2017). Proteomic comparison of *Chelidonium majus* L. latex in different phases of plant development. *Plant Physiology and Biochemistry, 112*, 312–325.

Newman, D. J., & Cragg, G. M. (2016). Natural products as sources of new drugs from 1981 to 2014. *Journal of Natural Products, 79*(3), 629–661.

Nielsen, E., Temporiti, M. E. E., & Cella, R. (2019). Improvement of phytochemical production by plant cells and organ culture and by genetic engineering. *Plant Cell Reports, 38*, 1199–1215.

Nowak, B., Kawka, M., Wierzchowski, K., Sykłowska-Baranek, K., & Pilarek, M. (2021). MTMS-based aerogel constructs for immobilization of plant hairy roots: Effects on proliferation of *Rindera graeca* biomass and extracellular secretion of naphthoquinones. *Journal of Functional Biomaterials, 12*(1), 19.

Pérez-Mendoza, M. B., Llorens-Escobar, L., Vanegas-Espinoza, P. E., Cifuentes, A., Ibáñez, E., & Villar-Martínez, A. A. D. (2020). Chemical characterization of leaves and calli extracts of *Rosmarinus officinalis* by UHPLC-MS. *Electrophor, 41*(20), 1776–1783.

Pilarek, M., Sobieszuk, P., Wierzchowski, K., & Dąbkowska, K. (2018). Impact of operating parameters on values of a volumetric mass transfer coefficient in a single-use bioreactor with wave-induced agitation. *Chemical Engineering Research and Design, 136*, 1–10.

Pramita, A. D., Kristanti, A. N., Utami, E. S. W., Manuhara, Y. S. W. (2018). Production of biomass and flavonoid of *Gynura procumbens* (Lour.) Merr shoots culture in temporary immersion system. *Journal of Genetic Engineering and Biotechnology, 16*(2), 639–643.

Przybylska, A., Gackowski, M., & Koba, M. (2021). Application of capillary electrophoresis to the analysis of bioactive compounds in herbal raw materials. *Molecules, 26*(8), 2135.

Qiu, H., Su, L., Wang, H., & Zhang, Z. (2021). Chitosan elicitation of saponin accumulation in *Psammosilene tunicoides* hairy roots by modulating antioxidant activity, nitric oxide production and differential gene expression. *Plant Physiology and Biochemistry, 166*, 115–127.

Razavizadeh, R., Adabavazeh, F., & Komatsu, S. (2020). Chitosan effects on the elevation of essential oils and antioxidant activity of *Carum copticum* L. seedlings and callus cultures under *in vitro* salt stress. *Journal of Plant Biochemistry and Biotechnology, 29*, 473–483.

Ren, J. L., Zhang, A. H., Kong, L., & Wang, X. J. (2018). Advances in mass spectrometry-based metabolomics for investigation of metabolites. *RSC Advances, 8*(40), 22335–22350.

Reshi, Z. A., Ahmad, W., Lukatkin, A. S., & Javed, S. B. (2023). From Nature to lab: A review of secondary metabolite biosynthetic pathways, environmental influences, and *in vitro* approaches. *Metabolites, 13*(8), 895.

Ritala, A., Heiniö, R. L., Häkkinen, S. T., Lille, M., Hyytiäinen-Pabst, T., & Rischer, H. (2022). Tailoring sensory properties of plant cell cultures for food use. *Food Research International, 157*, 111440.

Salem, M. A., Perez de Souza, L., Serag, A., Fernie, A. R., Farag, M. A., Ezzat, S. M., et al. (2020). Metabolomics in the context of plant natural products research: From sample preparation to metabolite analysis. *Metabolites, 10*(1), 37.

Santos-Sanchez, N. F., Salas-Coronado, R., Hernandez-Carlos, B., & Villanueva-Canongo, C. (2019). Shikimic acid pathway in biosynthesis of phenolic compounds. *Plant Physiological Aspects of Phenolic Compounds, 1*, 1–15.

Selwal, N., Rahayu, F., Herwati, A., Latifah, E., Suhara, C., Suastika, I. B. K., et al. (2023). Enhancing secondary metabolite production in plants: Exploring traditional and modern strategies. *Journal of Agricultural Research, 14*, 100702.

Shah, M., Jan, H., Drouet, S., Tungmunnithum, D., Shirazi, J. H., Hano, C., & Abbasi, B. H. (2021). Chitosan elicitation impacts flavonolignan biosynthesis in *Silybum marianum* (L.) Gaertn cell suspension and enhances antioxidant and anti-inflammatory activities of cell extracts. *Molecules, 26*(4), 791.

Sharma, B., & Yadav, D. K. (2022). Metabolomics and network pharmacology in the exploration of the multi-targeted therapeutic approach of traditional medicinal plants. *Plants, 11*(23), 3243.

Sharma, M., Koul, A., Sharma, D., Kaul, S., Swamy, M. K., & Dhar, M. K. (2019). Metabolic engineering strategies for enhancing the production of bio-active compounds from medicinal plants. *Natural Bio-Active Compounds: Biotechnology, Bioengineering, and Molecular Approaches, 3*, 287–316.

Sharma, K., Mishra, K., Senapati, K.K., & Danciu, C. (2021). *Bioactive compounds in nutraceutical and functional food for good human health*. Books on Demand.

Sharma, A. R., Gajurel, G., Ahmed, I., Roedel, K., Medina-Bolivar, F. (2022). Induction of the prenylated stilbenoids arachidin-1 and arachidin-3 and their semi-preparative separation and purification from hairy root cultures of peanut (*Arachis hypogaea* l.). *Molecules, 27*(18), 6118.

Shi, M., Du, Z., Hua, Q., & Kai, G. (2021). CRISPR/Cas9-mediated targeted mutagenesis of bZIP2 in *Salvia miltiorrhiza* leads to promoted phenolic acid biosynthesis. *Industrial Crops and Products, 167*, 113560.

Singh, P., Singh, G., Bhandawat, A., Singh, G., Parmar, R., Seth, R., et al. (2017). Spatial transcriptome analysis provides insights of key gene (s) involved in steroidal saponin biosynthesis in medicinally important herb *Trillium govanianum*. *Science and Reports, 7*(1), 45295.

Singh, A., Rani, R., & Sharma, M. (2018). Medicinal herbs of Punjab (India). *Biological Forum, 10*, 10–27.

Singh, P. K., Singh, J., Medhi, T., & Kumar, A. (2022). Phytochemical screening, quantification, FT-IR analysis, and *in silico* characterization of potential bio-active compounds identified in HR-LC/MS analysis of the polyherbal formulation from Northeast India. *ACS Omega, 7*(37), 33067–33078.

Soliman, H. I., Abo-El-Hasan, F. M., Ayman, S., & Mabrouk, Y. M. (2018). Influence of plant growth regulators on somatic embryogenesis induction in *Seriphidium Herba*-Album. *International Journal of Environment, Agriculture and Biotechnology, 3*(4), 264401.

Son, S., & Park, S. R. (2022). Challenges facing CRISPR/Cas9-based genome editing in plants. *Frontiers in Plant Science, 13*, 902413.

Song, J., Du, L., Li, L., Kalt, W., Palmer, L. C., Fillmore, S., et al. (2015). Quantitative changes in proteins responsible for flavonoid and anthocyanin biosynthesis in strawberry fruit at different ripening stages: A targeted quantitative proteomic investigation employing multiple reaction monitoring. *Journal of Proteomics, 122*, 1–10.

Thirumurugan, D., Cholarajan, A., Raja, S. S., & Vijayakumar, R. (2018). An introductory chapter: Secondary metabolites. *Secondary Metabolites-Sources and Applications, 1*, 13.

Tian, Y. Z., Wang, Z. F., Liu, Y. D., Zhang, G. Z., & Li, G. (2021). The whole-genome sequencing and analysis of a *Ganoderma lucidum* strain provide insights into the genetic basis of its high triterpene content. *Genomics, 113*(1), 840–849.

Tripathi, S., Jadaun, J. S., Chandra, M., & Sangwan, N. S. (2016). Medicinal plant transcriptomes: The new gateways for accelerated understanding of plant secondary metabolism. *Plant Genetic Resources, 14*(4), 256–269.

Tyagi, P., Singh, D., Mathur, S., Singh, A., & Ranjan, R. (2022). Upcoming progress of transcriptomics studies on plants: An overview. *Frontiers in Plant Science, 13*, 1030890.

Uzor, P. F. (2020). Alkaloids from plants with antimalarial activity: A review of recent studies. *Evidence-Based Complementary and Alternative Medicine, 1*, 8749083.

Vignesh, A., Selvakumar, S., & Vasanth, K. (2022). Comparative LC-MS analysis of bioactive compounds, antioxidants and antibacterial activity from leaf and callus extracts of *Saraca asoca*. *Phytomedicine Plus, 2*(1), 100167.

Wang, F., Ji, G., Xu, Z., Feng, B., Zhou, Q., Fan, X., et al. (2021). Metabolomics and transcriptomics provide insights into anthocyanin biosynthesis in the developing grains of purple wheat (*Triticum aestivum* L.). *Journal of Agricultural and Food Chemistry, 69*(38), 11171–11184.

Wen, D., Wu, L., Wang, M., Yang, W., Wang, X., Ma, W., et al. (2022). CRISPR/Cas9-mediated targeted mutagenesis of FtMYB45 promotes flavonoid biosynthesis in tartary buckwheat (*Fagopyrum tataricum*). *Frontiers in Plant Science, 13*, 879390.

Wu, J., Meng, X., Jiang, W., Wang, Z., Zhang, J., Meng, F., et al. (2022a). Qualitative proteome-wide analysis reveals the diverse functions of lysine crotonylation in *Dendrobium huoshanense*. *Frontiers in Plant Science, 13*, 822374.

Wu, Z., Singh, S. K., Lyu, R., Pattanaik, S., Wang, Y., Li, Y., et al. (2022b). Metabolic engineering to enhance the accumulation of bioactive flavonoids licochalcone A and echinatin in *Glycyrrhiza inflata* (Licorice) hairy roots. *Frontiers in Plant Science, 13*, 932594.

Xie, K., Minkenberg, B., & Yang, Y. (2015). Boosting CRISPR/Cas9 multiplex editing capability with the endogenous tRNA-processing system. *PNAS, 112*(11), 3570–3575.

Xu, X., Yuan, Y., Feng, B., & Deng, W. (2020). CRISPR/Cas9-mediated gene-editing technology in fruit quality improvement. *Food Quality and Safety, 4*(4), 159–166.

Xu, D., Lin, H., Tang, Y., Huang, L., Xu, J., Nian, S., et al. (2021). Integration of full-length transcriptomics and targeted metabolomics to identify benzylisoquinoline alkaloid biosynthetic genes in *Corydalis yanhusuo*. *Hortic. Res., 8*, 16.

Yang, J., Zhang, C., Li, W. H., Zhang, T. E., Fan, G. Z., Guo, B. F., et al. (2021). Comprehensive analysis of transcriptomics and metabolomics between the heads and tails of *Angelica Sinensis*: Genes related to phenylpropanoid biosynthesis pathway. *CCHTS, 24*(9), 1417–1427.

Zameer, R., Tariq, S., Noreen, S., Sadaqat, M., & Azeem, F. (2022). Role of transcriptomics and artificial intelligence approaches for the selection of bioactive compounds. *Drug design Using Machine Learning*, 283–317.

Zhang, D., Zhang, H., Li, T., Chen, K., Qiu, J. L., & Gao, C. (2017). Perfectly matched 20-nucleotide guide RNA sequences enable robust genome editing using high-fidelity SpCas9 nucleases. *Genome Biology, 18*, 1–7.

Zhang, C., Yao, X., Ren, H., Chang, J., & Wang, K. (2018). RNA-Seq reveals flavonoid biosynthesis-related genes in pecan (*Carya illinoinensis*) kernels. *Journal of Agriculture and Food Chemistry, 67*(1), 148–158.

Zhang, Y., Cai, P., Cheng, G., & Zhang, Y. (2022). A brief review of phenolic compounds identified from plants: Their extraction, analysis, and biological activity. *Natural Product Communications, 17*(1).

Zhao, T., Li, S., Wang, J., Zhou, Q., Yang, C., Bai, F., et al. (2020). Engineering tropane alkaloid production based on metabolic characterization of ornithine decarboxylase in *Atropa belladonna*. *ACS Synthetic Biology, 9*, 437–448.

Zhou, P., Pu, T., Gui, C., Zhang, X., & Gong, L. (2020). Transcriptome analysis reveals biosynthesis of important bioactive constituents and mechanism of stem formation of *Dendrobium huoshanense*. *Science and Reports, 10*(1), 2857.

Bioactive Compounds for Food, Agricultural and Health Sector: Opportunities and Challenges

Janvika Varma, Urjitsinh Zala, Pranay Punj Pankaj, Vijay Jagdish Upadhye, and Anupama Shrivastav

Abstract

Secondary metabolites that have been isolated from bacteria, fungi, plants, or animals are referred to as bioactive chemicals. Bioactive chemicals are helpful in a variety of applications due to their intrinsic qualities, which include antibacterial, anti-inflammatory, and antioxidant effects. A wide range of applications, from fictionalized biomaterials and cosmetics to bioremediation and alternative fuels, have been made possible by the discovery of unique features of these molecules. Nevertheless, obstacles including extraction techniques, formulation stability, regulatory concerns, and consumer acceptability must be overcome in order to fully explore the potential of bioactive chemicals. Advances in characterization and optimization strategies, coupled with non-conventional extraction techniques such as pressurization, enzymes, pulse electric fields, ultrasound, or microwaves, have made extraction more commercially viable, particularly for agrowastes and organic residues. This has contributed to the development of a sustainable circular economy. Bioactive chemicals have encouraging prospects for sustainable development and better human health by overcoming these obstacles. This chapter highlights the necessity for coordinated efforts to realize the full potential of bioactive chemicals by offering insights into the opportunities and constraints connected with their utilization across many sectors.

Keywords

Multidisciplinary collaboration • Functional foods • Nutraceuticals • Bioactive substances • Food industry • Agriculture • Health sector • Opportunities • Obstacles

1 Introduction

Bioactive compounds are substances found naturally in microorganisms, plants, and animals that have biological impacts on humans. These substances, classified as secondary metabolites in plants, provide bioactive qualities without contributing additional nutritional benefits. Their wide range of applications and potential health advantages have garnered significant attention in the food, agricultural, and health sectors. According to Câmara et al. (2020), these compounds can have pharmacological or toxicological effects on both humans and other animals (Câmara et al., 2020). The production of these compounds is closely linked to plant responses to environmental stresses over time, making them of utmost importance. Plants contain large quantities of various chemicals in materials such as roots, leaves, fruits, flowers, and crops (Botella et al., 2021, p. 534). Additionally, recent research has highlighted that plant wastes are a valuable source of bioactive compounds with a broad range of potential applications (Shirahigue & Ceccato-Antonini, 2020; Chiocchio et al., 2021). Certainly! Here's a revised version of your text with grammatical corrections and improved sentence structure. Changes are highlighted for clarity.

J. Varma · U. Zala
Department of Medical Laboratory Technology, Faculty of Life Health and Allied Sciences, ITM Vocational University, Ravaal, Waghodia, Vadodara, 391760, Gujarat, India

P. P. Pankaj
Department of Zoology, Nagaland University, Lumami, Zunheboto, 798627, Nagaland, India

V. J. Upadhye
Research and Development Cell, Parul Institute of Applied Sciences, Parul University, Limda, Waghodia, Vadodara, 391760, Gujarat, India

A. Shrivastav (✉)
Department of Microbiology, Faculty of Life Health and Allied Sciences, ITM Vocational University, Ravaal, Waghodia, Vadodara, 391760, Gujarat, India
e-mail: anupamashrivastav@gmail.com; anupamashrivastav@yahoo.co.in; anupamam@itmvu.in

J. Devi (ed.), *Biotechnological Intervention in Production of Bioactive Compounds*,
Sustainable Landscape Planning and Natural Resources Management, https://doi.org/10.1007/978-3-031-76859-0_4

These substances accumulate and can be extracted in high concentrations from plant matrices, satisfying human needs and enhancing the value of the raw materials (Sruthi et al., 2023). This capability favors the functional application of these components in a variety of academic subjects. Additionally, the major functional features and physicochemical characteristics of the molecules present are closely linked to this capability. Polyphenols, triterpenes, phytosterols, terpenoids, polysaccharides, capsaicinoids, carotenoids, tocopherols, alkaloids, saponins, glucosinolates, and other substances represent the different types of natural bioactive compounds (Sharma et al., 2023). Moreover, they are extracted from bacteria, fungi, and animals (Câmara et al., 2020). They are also isolated from agro-industrial waste products such as grape peels, avocado peels, and mango seeds. Bioactive substances are associated with multiple health benefits and the prevention of various diseases and metabolic disorders, as evidenced by numerous pharmacological investigations. These elements are extensively researched for their numerous advantages to human health and the environment, demonstrating great potential for use. Phytochemicals have been studied for their natural actions, including antioxidant, anti-inflammatory, anti-cancer, and antidiabetic properties, and are recommended as supplements to the body's natural needs.

Bioactive compounds, including essential oils, flavonoids, tannins, phenolic acids, carotenoids, organosulfur compounds, phytosterols, and tocopherols, are utilized in the food and fermentation industries for processing vegetable oils, meat and seafood products, bakery goods, dairy products, and other items. It has been discovered that bioactive chemicals provide anti-aging qualities, making them appealing to the cosmetic industry. Several studies have indicated that bioactive compounds are used as essential components in biotechnology-oriented companies, particularly those focused solely on agriculture and food-related purposes. As a result, studies have shown how to utilize bioactive chemicals as active ingredients to improve and preserve food quality, boost food nutrition, and support plant defense, among other purposes. Over the years, many extraction strategies have been employed, considering characteristics such as the source matrix's nature, the relative solubility of the bioactive compound and the solvents, as well as the impact of temperature, pressure, pH, and other variables on the extraction's time, yield, and selectivity. Traditional extraction techniques include decoction and maceration, which are extensively used in household settings to extract essential oils and other bioactive compounds, as well as the laboratory-scale Soxhlet extraction method, which employs moderately higher temperatures to circulate solvent throughout a device and facilitate the purification of substances from samples contained in a thimble (Muala et al., 2021). But when relative yields and the specificity of the recovered compounds are taken into account, the drawbacks of these techniques become apparent. They also involve large solvent usage and lengthy extraction durations. For instance, maceration takes two to seven days to produce a suitable extraction, and in the case of decoction (hot water extraction), the solvent-to-crude-extract ratio might range from 4:1 to even about 50:1 (Li et al., 2020).

In addition to the creation of novel, "green," less toxic, more specific, and economical solvent types, numerous strategies have been devised to overcome the aforementioned constraints (Pal & Jadeja, 2022). For example, ionic liquids, deep eutectic solvents, and aqueous two-phase systems have been used to accelerate and improve extraction yields while also helping to extract chemicals that proved too challenging to extract using traditional solvents. A number of characteristics, including the Hansen parameters and solvation parameter models, can be used to compare and optimize solvent-bioactive substance interactions (Lefebvre et al., 2021).

The fundamental idea behind non-conventional extraction techniques is to enhance extraction through the use of ultrasound, microwaves, pressurized liquids, and pulsed electric fields. The primary objective of these techniques is to rupture cell walls, which improves mass transfer and makes effective mixing easier by increasing solvent exposure to the cytoplasmic contents.

Several non-conventional techniques that adhere to the aforementioned principles include pulse electric field-assisted extraction (PEFAE), supercritical fluid extraction (SCFE), pressurized liquid extraction (PLE), microwave-assisted extraction (MAE), and ultrasound-assisted extraction (UAE). Enzyme-driven cell wall breakdown, in conjunction with enzyme-assisted extraction (EAE), has been explored as a potential method for extracting compounds related to the cell wall, as opposed to the cytoplasm. Due to its renewable nature and favorable thermodynamic properties, supercritical fluid extraction is one of the most commonly utilized extraction procedures, with supercritical CO_2 typically used in this process (Gan et al., 2020).

The efficacy and performance of any method are additionally affected by solvent selection; the effects of temperature and pH, which may alter cell wall organization, specificity; the chemical makeup of the substance to be extracted; and the overall viability of the process. An appropriate set of parameters is determined by using algorithmic optimization techniques specific to the process.

Prior evaluations have concentrated on various aspects of the bioactive chemical extraction process. Sources, applications, and extraction techniques have all been covered in distinct documentation. Our goal is to present a thorough analysis of bioactive chemicals used in the food, chemical, and pharmaceutical industries, including their sources, extraction techniques, limitations, characterization, uses, and potential future applications.

However, in light of the acceleration of population growth and the steady rise in food consumption, a number of difficulties have arisen in the creation and maintenance of these components during the matrix exploration process. One of the primary challenges is the widespread use of healthy food, which has recently become a high priority for the populace. The study of plant chemicals with potential antiviral properties has been prompted by the rise in viral diseases such as coronavirus and the prevalence of cardiovascular diseases (Remali & Aizat, 2021). Promising alternatives for the production of biocompounds are being explored, as evidenced by the adoption of technologies that enhance the production potential of bioactive metabolites, metabolic pathway strategies as precursors in the synthesis of phytocompounds, and the combination of various abiotic agents to model the morphological and, consequently, biochemical performance of plants (Kumar et al., 2021; Yoon et al., 2021).

Furthermore, the growth of plant-based diets has been facilitated by the increasing popularity of vegetarianism and veganism, and the rise in their consumption indicates the need for creative solutions to address evolving needs (Rubio et al., 2020; Alcorta et al., 2021). Proteins, carotenoids, phenolic compounds, and other components with significant health-promoting and disease-prevention potential can be obtained from plant bioactive substances. However, there is a growing demand for technological advancements to stimulate consumer interest and large-scale manufacturing. Enzyme-based extraction, subcritical and supercritical fluids, and ultrasound equipment are all utilized in many of these techniques (Samtiya et al., 2021; Hassim et al., 2021). These technological advancements promote higher concentrations of bioactive compounds compared to conventional methods like Soxhlet extraction, maceration, and reflux methods, due to the action of high pressures and temperatures, which enhance solvent penetration into the plant matrix and improve compound extraction.

Additionally, a thorough examination of the quantitative performance of research primarily devoted to the application of plant compounds is positive and confirms the growing significance and attention paid to the topic in recent years. Plant bioactive substances are extensively examined as a cutting-edge and environmentally friendly alternative for sectors connected to agriculture and food. The goal of this review was to investigate the primary sources of bioactive compounds for various application sectors and to conduct a thorough and methodical analysis of the significance of these elements. Additionally, the review discusses the primary uses, advanced methods, and strategies for enhancing the value of bioactive materials, along with potential future developments.

2 Sources of Bioactive Compounds

Food quality, agricultural production, and health are significantly enhanced by bioactive chemicals, which can be found in many different natural sources. These substances are derived from multiple primary sources and include flavonoids, polyphenols, antioxidants, and other phytochemicals.

The secondary metabolites known as "bioactive compounds" are commonly found in plant tissues and include proteins, carbohydrates, alkaloids, vitamins, saponins, polyphenols, gluconates, and total phenolic compounds. Most of these substances are typically thermolabile, and their presence gives fruit and vegetable extracts their antioxidant properties. Anthocyanins, flavonols, flavones, and flavanones are among the primary categories of flavonoids; tannins, stilbenes, and phenolic acids are examples of non-flavonoid chemicals (Montoro et al., 2020; Peanparkdee & Iwamoto, 2020).

Furthermore, the range of uses for bioactive chemicals is extensive, spanning pharmaceutical, medical, and agricultural applications, including weed and pest control. Antioxidant chemicals are also categorized into primary and secondary types. Primary antioxidants can halt the propagation of self-oxidation, delay or even suppress the initiation process, and act as free radical scavengers. Secondary antioxidants, also known as preventatives, provide antioxidant activity by reducing the rate of oxidation processes through various mechanisms. Unlike primary antioxidants, secondary antioxidants do not convert free radicals into stable molecules but instead enhance the antioxidant activity of primary antioxidants.

In addition to supplying H^+ to primary antioxidants, breaking down hydroperoxides into non-radical species, deactivating singlet oxygen, and absorbing UV light, secondary antioxidants also function as pro-oxidant metal ion chelators or catalysts. To prevent oxidative stress, natural antioxidants generally block the production of reactive oxygen species (ROS), thereby shielding cells from oxidative damage.

Fruits, vegetables, and their by-products contain a variety of phenolic chemicals, some of which have biological characteristics. However, it is thought that the bioactive effect results from a combination of several components that together account for the advantageous biological features, rather than originating from a single chemical class or substance. Moreover, not all potential uses of bioactive substances derived from various plant sources have been thoroughly documented to date. Fruit and vegetable pulps are frequently used in scientific research (Aguirre Calvo & Santagapita, 2017). Marine sources, such as algae and fish, are also rich in bioactive substances with a variety of health-promoting qualities. Marine algae provide compounds such as fucoxanthin and phlorotannins, while fish and shellfish offer omega-3 fatty acids, bioactive peptides, and other vital nutrients, which promote cardiovascular and cognitive health.

The process of converting waste into valuable products has gained increasing attention in recent years. Research on the extraction of biocompounds from fruit peels and other residues has shown promising results for their potential applications. Some of these residues are being used as growth media for cultivating microorganisms that produce enzymes or other substances of interest (Scaglia et al., 2020; Jalgaonkar et al., 2020).

There are numerous opportunities for innovation in the food, agricultural, and health sectors provided by these diverse sources of bioactive compounds. However, effectively utilizing these substances requires addressing challenges related to their extraction, stability, bioavailability, and product integration, without compromising their efficacy or safety.

The research literature discusses a wide range of extraction techniques, along with the optimal processing parameters and constraints for each, used to extract bioactive compounds from plant matrices. Over the years, numerous studies have focused on conventional methods such as Soxhlet extraction and the use of chemical solvents. However, cutting-edge technologies have emerged as effective strategies to enhance the extraction of biocompounds from plants. The most commonly employed modern methods include emulsification, enzyme-assisted extraction, subcritical and supercritical fluids, and microwave and ultrasound-assisted techniques.

The correct application of extraction technologies and/or technological integration is a crucial step in the compounding process. The pursuit of advanced technologies and improved extraction efficiency considers several factors that enable a feasible cost–benefit analysis while avoiding environmental contamination or health risks to humans. Therefore, when evaluating the extraction process, key concerns include the extensive use of chemical solvents, challenges with equipment handling and/or unskilled labor, high energy consumption, the risk of corrosion or damage to the extraction equipment, and the significant degradation of the matrix being extracted.

Many studies have yielded promising results regarding the feasibility of using plant matrices to extract compounds of interest, which could extend the shelf life of various plant products and secondary products. Due to the wide range of primary and secondary metabolites found in plants and microorganisms, and their numerous applications across different sectors, a variety of extraction techniques tailored to the specific characteristics of the metabolites are required.

Extraction, the process of isolating a component of interest from a raw source, can be broadly divided into two categories: conventional and non-conventional methods. Conventional techniques include Soxhlet extraction, decoction, and maceration. Most solvents used in industrial-scale extraction—especially hexane—are byproducts of the petrochemical industry. The large-scale use of such solvents, combined with high energy consumption, has a negative impact on the environment. Consequently, non-conventional methods focus on enhancing extraction efficiency through various mechanical or enzymatic techniques while using fewer solvents or specific "green" solvents (Anticona et al., 2020). Table 1 and Fig. 1 illustrate the various methods for extracting bioactive compounds.

3 Techniques for Bioactive Compound Extraction

3.1 Conventional Methods

3.1.1 Maceration

Maceration is a traditional extraction process that has gained significant attention due to its affordability and ease of use. This review examines the principles, applications, benefits, and drawbacks of maceration in the extraction of bioactive substances. Maceration involves soaking plant material in a solvent at room temperature to extract bioactive compounds. The process relies on the solvent diffusing into the plant matrix and dissolving the targeted chemicals. Since this method operates at room temperature, it is often used for the extraction of heat-sensitive compounds, preserving the integrity of thermolabile substances (Malini et al., 2023).

Maceration is utilized in various fields, including:

Pharmaceuticals: Extraction of alkaloids, flavonoids, and glycosides used in drug formulations.

Nutraceuticals: Extraction of vitamins, minerals, and antioxidants for use in dietary supplements.

Cosmetics: Extraction of fragrances, essential oils, and other bioactive ingredients used in skincare and cosmetic products.

Recent research has focused on improving maceration efficiency by optimizing key parameters. Adjusting variables such as solvent type, solvent-to-plant ratio, particle size of the plant material, and extraction time can enhance extraction yield. To overcome its limitations, combining maceration with other techniques like ultrasound-assisted or enzyme-assisted extraction is being explored. These hybrid

Table 1 Extraction technologies applied for different raw materials for bioactive compounds extractions

Plant Spp.	Plant organ	Extraction technology	Operation conditions	Objectives	Results
Vernonia cinerea	Leaf	Soxhlet extraction	Extraction time 120 min, feed/solvent 1:20 g mL^{-1} and ethanol concentration 60% v/v	Phenolic compounds determination and antioxidant activity	Extract yield 10.01 ± 0.85% w/w, phenolic compounds yield 53.96 ± 1.45 mg GAE g^{-1} and flavonoids yield 30.09 ± 0.44 mg QE g^{-1}
Linnophila aromatica	Aerial Parts	Solvent-based extraction	50% Acetone concentration	Antioxidant activity	Phenolic compounds 40.5 mg galic acid and flavonoids 31.1 mg quercetin
Hibiscus cannabinus L.	Seed	Soxhlet and nanoemulsion	Hexane 1 L, extraction time 180 h, temperature 60 °C	Stability of bioactive compounds and antioxidant potential	Conservation 4 °C provided 65% retention of phytosterols
Berberis jaeschkeana	Root	Microwave-assisted extraction	Microwave power 598 W and 2 min of irradiation time	Alkaloids and phenolic compounds determination	Alkaloids up to 46.38 mg g^{-1}
Elaeocarpus serratus L.	Leaf	Ultrasound-assisted extraction	95% Ethanol, solvent ratio 1/10 g mL^{-1}, ultrasonic power under 40 kHz/300 W	Phenolic compounds, flavonoids, myricitrin and antioxidant capacity	Extract yield 26.63%
Orostachys japonicus A. Berger	Various extracts	Subcritical fluid extraction	Water, extraction time 15 min, temperature 220 °C	Phenolic compounds, antioxidant activity and flavonoids determination	Phenolics content 39.9 ± 4.1 mg g^{-1}, flavonoids content 11.4 ± 0.6 mg g^{-1} and antioxidants up to 662.4 ± 17.2 mg g^{-1}

Source Dos Santos et al. (2022)

methods may increase efficiency while retaining the advantages of traditional maceration.

3.1.2 Soxhlet Extraction

Soxhlet extraction is another traditional and widely used technique for removing bioactive compounds from solid matrices, primarily plant materials. Developed by Franz von Soxhlet in 1879, this method has become a standard in chemistry, pharmacognosy, and food science due to its effectiveness in extracting both polar and non-polar compounds.

The Soxhlet extraction process involves several essential components and steps: The Soxhlet extractor setup consists of a condenser, an extraction chamber (thimble), and a distillation flask. A porous cellulose thimble, typically disposable, is used to hold the sample within the extraction chamber. Once the thimble is filled with plant material, it is placed into the extractor. The correct solvent must be selected based on the target compounds. Common solvents include ethanol, methanol, acetone, hexane, and chloroform, which are added to the distillation flask.

The extraction process begins by heating the solvent to reflux. The vapors rise to the condenser, where they are cooled and recondensed into liquid form. The liquid then seeps into the extraction chamber, where it penetrates the plant material and dissolves the bioactive compounds. Once the extraction chamber is filled to a certain level, the solvent, now containing the dissolved compounds, returns to the distillation flask.

Repetition and Concentration: To ensure complete extraction of the compounds, this cycle is repeated several times. The concentrated extract remains when the solvent evaporates during the process. However, the use of this method is limited by its long extraction times (12–24 h), high energy consumption, and issues with selectivity and efficiency (Weggler et al., 2020; Gilabadi et al., 2023).

Soxhlet extraction has wide-ranging applications across multiple industries.

Pharmaceuticals: Used to extract terpenes, alkaloids, flavonoids, and other therapeutic compounds from plant materials.

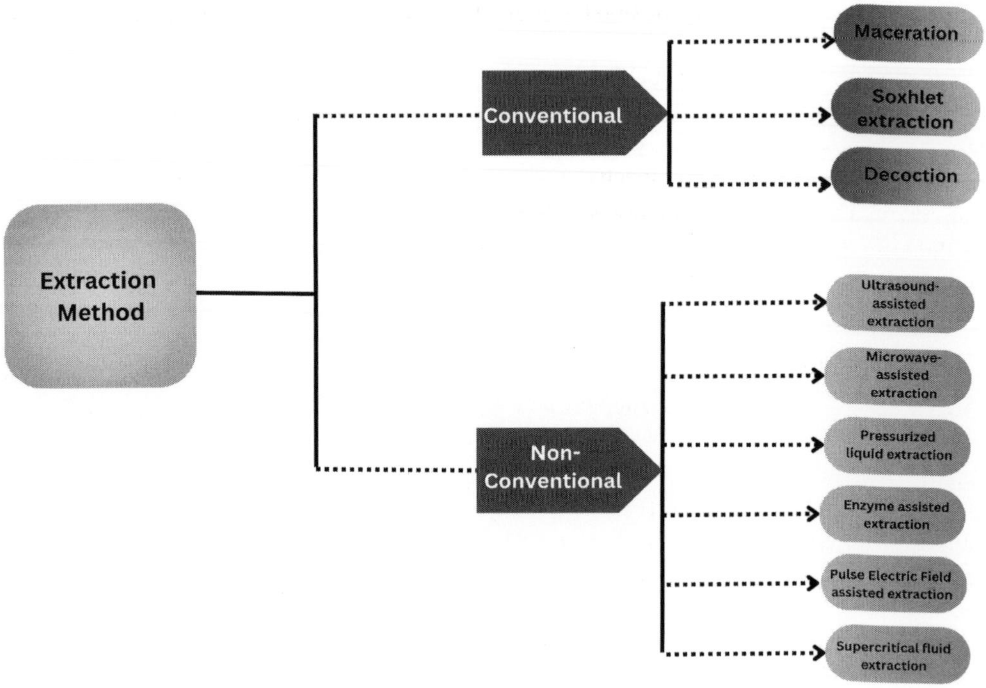

Fig. 1 Extraction methods of bioactive compounds (Dos Santos et al., 2022)

Food and Beverages: Utilized in extracting essential oils and flavors from herbs and spices for seasoning and fragrance.

Environmental Analysis: Employed to extract pollutants from soil and sediments.

Cosmetics: Natural extracts obtained through Soxhlet extraction are used in skincare, haircare, and cosmetic products.

Soxhlet extraction remains a powerful and reliable method for extracting bioactive compounds from plant sources. Its effectiveness and ability to extract a broad range of compounds make it indispensable across various industries, despite some disadvantages such as high solvent consumption and lengthy extraction times. Advances in solvent recovery and the development of more eco-friendly solvents are improving the sustainability of this traditional method.

3.1.3 Decoction

Decoction is one of the earliest and simplest techniques for extracting bioactive compounds from plant sources. This traditional method involves boiling plant materials, such as roots, bark, seeds, or stems, in water to extract water-soluble compounds. Decoction remains widely used in ancient medical systems like Ayurveda, Traditional Chinese Medicine (TCM), and herbalism, as it is an effective method for extracting a variety of bioactive substances.

The extraction times for decoction are relatively short, typically ranging from 5 to 10 min. However, this can be a limitation if the bioactive compounds to be extracted are not highly water-soluble. The process begins with the careful selection of plant material, usually focusing on sturdy, woody parts such as bark and roots. These components are chosen because their structure, once broken down, releases a large amount of bioactive substances. After selection, the plant material is often dried and chopped into small pieces to increase the surface area for extraction and improve efficiency.

The prepared plant material is then added to water and brought to a boil. Once boiling, the mixture is simmered for an extended period. Depending on the toughness of the plant material and the specific compounds being extracted, this process can take anywhere from 30 min to several hours. This prolonged simmering allows the bioactive compounds to break down and be fully released into the water. The mixture is then filtered to remove any remaining solids, producing a concentrated liquid extract rich in the desired bioactive components. This process ensures efficient extraction of valuable compounds with a wide range of applications (Sultana, 2024).

Decoction is used in various fields for extracting bioactive compounds:

Traditional Medicine: Widely used in Western herbalism, TCM, and Ayurveda to prepare therapeutic teas and tonics.

Pharmaceuticals: Employed to extract medicinal plant compounds, especially in the early stages of drug development.

Nutraceuticals: Used in the production of functional foods and health supplements.

Cosmetics: Used to obtain organic extracts for hair and skincare products.

Decoction remains a valuable technique for bioactive compound extraction, particularly where affordability and simplicity are important. Its effectiveness in extracting a wide range of water-soluble bioactive compounds ensures its continued use in both traditional and modern applications, despite its limitations, such as the potential for thermal degradation and the exclusion of non-water-soluble compounds.

3.2 Revolutionary Techniques for Extraction

3.2.1 Pressurized Liquid Extraction (PLE)

Pressurized liquid extraction (PLE), also known as accelerated solvent extraction (ASE), is a proven and highly efficient technique for extracting bioactive compounds from plant materials. By utilizing high pressures and temperatures, PLE enhances the extraction efficiency of bioactive compounds. Due to its speed, reduced solvent consumption, and ability to extract a wide range of chemicals, PLE is valuable for both research and industrial applications. This method involves using solvents at temperatures above, but still below, their critical points to maintain them in a liquid state. PLE leverages the mass transfer properties of solvents under high pressures and temperatures (Zakaria et al., 2020).

The PLE process involves several key steps. Two primary setups are used: static and dynamic, or a combination of both. In the dynamic system, solvent aliquots are continuously pumped at a rate of 0.5–2.5 ml per minute, while in the static method, the extracted solvent is collected every 5–10 min (Soriano et al., 2024). The selection of plant material —leaves, roots, seeds, or flowers—depends on the desired bioactive components. To improve extraction efficiency, the plant material is typically dried and finely ground to increase surface area (Álvarez Alarcon et al., 2023). The prepared plant material is packed into an extraction cell to ensure optimal solvent contact.

During the extraction process, the cell is placed in the PLE apparatus, where it is exposed to a solvent at high temperatures and pressures, typically between 50 and 200 °C and 10–15 MPa. The combination of pressure and heat enhances the solvent's ability to penetrate the plant matrix and dissolve the bioactive compounds. After a predetermined extraction time, the solvent containing the dissolved compounds is collected. The solvent then evaporates, leaving behind a concentrated extract rich in bioactive substances. The sample is moistened with solvent during the process, and the target components desorb from the plant matrix into the extraction solvent.

Temperature, a key PLE parameter, can be adjusted to modify the solvent's physicochemical properties. PLE is often used to analyze complex matrices like food, especially for detecting persistent organic pollutants.

PLE is applied in various fields to extract bioactive compounds:

Pharmaceuticals: For isolating therapeutic compounds such as alkaloids, flavonoids, and terpenes from medicinal plants.

Food and Beverages: To extract nutrients, flavors, and aromas from herbs, spices, and other plant materials.

Cosmetics: For obtaining natural extracts used in makeup, skincare, and haircare products.

Environmental Analysis: To extract pollutants from soil, sediments, and other environmental samples.

Pressurized liquid extraction is a highly effective and versatile method for extracting bioactive compounds from plant materials. Its advantages include improved extraction efficiency, reduced solvent consumption, and adaptability, making it a preferred method for many applications. However, the complex procedure and expensive equipment can limit its widespread use in some settings. Despite these challenges, technological advancements and the growing demand for sustainable extraction methods continue to drive the development and optimization of PLE (López-Bascón et al., 2020).

3.2.2 Ultrasound-Assisted Extraction (UAE)

Ultrasound-Assisted Extraction (UAE) is a sophisticated and widely used technique for extracting bioactive substances from plant materials. Compared to conventional methods, this technology uses ultrasonic waves to enhance extraction efficiency, reduce solvent consumption, and shorten extraction times. Ultrasonic waves create cavitation, which causes bubbles in the medium to implode. UAE is well-regarded for efficiently extracting a wide variety of bioactive compounds, making it valuable in industries such as pharmaceuticals, nutraceuticals, and cosmetics. UAE is used on both commercial and scientific scales. Cavitation disrupts solid particles, causing them to collide and experience macroturbulence. This process forms new pores and enlarges existing ones, increasing the mass transfer rate and allowing solvents to penetrate the biomass (González et al., 2020).

UAE can be configured in various ways depending on the requirements of the extraction process. One method involves directly submerging the mass into the transducer. Although this improves process efficiency, it also increases the risk of contamination (Linares & Rojas, 2022).

There are several key steps in the UAE process. First, selecting the appropriate plant material is critical, depending on the desired bioactive components. Leaves, roots, seeds, and flowers can all be used depending on the intended extract. Next, the plant material is typically crushed and dried to maximize surface area and improve extraction efficiency. The choice of solvent is also crucial, as it depends on the polarity of the target compounds. Common solvents include water, ethanol, methanol, and their combinations.

Once the plant material is prepared, it is mixed with the chosen solvent and exposed to ultrasonic waves, typically generated at frequencies between 20 and 100 kHz. These ultrasonic waves help release bioactive compounds into the solvent and rupture the plant cell walls by breaking down cavitation bubbles, generating localized high temperatures and pressures. After ultrasonic treatment, the mixture is filtered to remove solid residues, and the solvent is evaporated, yielding a concentrated extract rich in bioactive compounds. This method ensures efficient extraction while preserving the integrity of heat-sensitive compounds (Šimat et al., 2024).

The benefits of UAE—such as improved extraction efficiency, reduced solvent use, and versatility—make it a preferred choice in many scenarios. However, its application may be limited in some cases due to the high cost of equipment and scalability challenges. Ongoing research and technological advancements are expected to further optimize UAE for industrial applications (Rastogi, 2023).

3.2.3 Microwave-Assisted Extraction (MAE)

Microwave-Assisted Extraction (MAE) is a sophisticated extraction technique that utilizes microwave energy to heat solvents and plant materials, significantly enhancing the extraction of bioactive compounds. Compared to conventional methods, MAE is recognized for its efficiency, shorter extraction times, and reduced solvent usage. This process has gained increasing popularity across various industries, including food, pharmaceutical, and nutraceutical sectors, due to its ability to extract a wide range of bioactive substances. The method involves irradiating the sample with microwaves (frequency range: 300 MHz–300 GHz) to aid in the extraction process (Rehman et al., 2020). The energy from these microwaves is converted into thermal energy through two mechanisms: ionic conduction and dipole rotation. As the moisture within the cells begins to evaporate, pressure builds up, ultimately causing cell rupture and releasing bioactive compounds (Pinzon et al., 2020).

The MAE process includes several key steps:

1. Selection of Plant Material: The selection of plant material is crucial and depends on the desired bioactive components. Leaves, roots, seeds, and flowers can all be used based on the type of extract needed.
2. Sample Preparation: The plant material is typically dried and finely powdered to maximize surface area and improve extraction efficiency.
3. Solvent Selection: The polarity of the target compounds determines the solvent choice, which is a critical step. Common solvents include water, ethanol, methanol, and their mixtures.
4. Microwave Treatment: The prepared plant material is placed in a microwave extraction system with the selected solvent. Microwave energy, usually between 300 MHz and 300 GHz, rapidly heats the mixture. The oscillation of polar molecules in both the solvent and plant material generates heat, which breaks down the plant cell walls, releasing bioactive compounds into the solvent.
5. Filtration and Concentration: After microwave treatment, the mixture is filtered to remove any solid residues. The solvent is then evaporated to yield a concentrated extract rich in bioactive substances.

MAE is widely used for extracting bioactive compounds across various fields:

Pharmaceuticals: For extracting therapeutic compounds like alkaloids, flavonoids, and phenolics from medicinal plants.

Food and Beverages: For extracting nutrients, flavors, and fragrances from herbs, spices, and other plant materials.

Cosmetics: To obtain natural extracts for use in makeup, skincare, and haircare products.

Nutraceuticals: For producing functional foods and dietary supplements rich in bioactive compounds.

The ability of MAE to use minimal amounts of solvents and achieve fast extraction times has garnered considerable attention. It also allows for better control over process parameters like pressure and temperature, as well as enhanced reproducibility (Milani et al., 2020). The underlying principles of this technique promote uniform temperature distribution, which enhances mass and heat transfer from the sample to the solvent, resulting in improved yields. While the advantages of MAE in terms of extraction efficiency, reduced solvent use, and versatility make it the preferred option in many cases, its use may be limited by the high cost of equipment and scalability challenges. Nonetheless, ongoing research and technological advancements are expected to further optimize MAE and expand its industrial applications.

3.2.4 Enzyme-Assisted Extraction (EAE)

Enzyme-Assisted Extraction (EAE) is a novel and effective technique for extracting bioactive compounds from plant sources. This method enhances the release of desired compounds by breaking down plant cell walls with specific enzymes. EAE is considered efficient and eco-friendly, offering several advantages over conventional extraction methods. Due to its ability to increase the yield and selectivity of bioactive compounds, it has garnered significant interest from industries such as food, pharmaceuticals, and nutraceuticals. EAE is particularly useful for extracting phytochemicals associated with the cell wall, which are difficult to access using other extraction techniques due to the high content of cellulose, hemicellulose, and lignin (Sridhar, 2024). EAE addresses this issue by utilizing enzymes involved in cell wall degradation, such as cellulase, pectinase, and alpha-amylase.

This process offers several benefits, including reduced environmental impact, lower energy consumption, and fewer equipment requirements compared to other methods. It is also highly effective at extracting volatile and heat-sensitive compounds used in fragrances, colorants, pigments, and other products. Numerous bioactive compounds with significant commercial and medicinal value have been successfully isolated using EAE, demonstrating its enormous potential. However, the high cost of enzymes limits its application in large-scale material extraction. Another challenge lies in developing efficient strategies for scaling up these processes (Sridhar, 2024; Martí-Quijal et al., 2021).

Enzyme-Assisted Extraction is a powerful and efficient technique for extracting bioactive substances from plant materials. Its environmental benefits, mild extraction conditions, and higher yield make it a preferred option in many applications. However, in some cases, its use may be restricted due to the high cost of enzymes and the need for precise optimization. With further research and technological advancements, the industrial application of EAE is expected to expand (Abdelmohsen et al., 2022).

3.2.5 Pulse Electric Feld-Assisted Extraction (PEFAE)

Pulse Electric Field-Augmented Extraction (PEFAE) is a novel technique that uses short bursts of high-voltage electric fields to enhance the extraction of bioactive compounds from plant materials. This method is recognized for its efficiency, reduced extraction times, and decreased solvent usage compared to conventional procedures. PEFAE has gained prominence in industries such as food, pharmaceuticals, and nutraceuticals due to its ability to extract a wide range of bioactive substances without compromising their integrity (Aguilar-Machado et al., 2022).

The operation of PEFAE is based on the concept of electroporation. When cells are exposed to intense electric fields, their membranes develop pores once the transmembrane potential exceeds approximately 1 V. This occurs due to enhanced repulsive interactions between membrane constituents, leading to the formation of pores (Gorte et al., 2020).

The PEFAE process involves several key steps to ensure the effective extraction of bioactive compounds:

Selection of Plant Material: Choosing the appropriate plant material, which may include leaves, roots, seeds, or flowers, is the first step.

Sample Preparation: After selection, the plant material is typically dried and finely powdered to increase surface area and improve extraction efficiency.

Preparation of Suspension: The processed plant material is mixed with a solvent to create a suspension. The choice of solvent, based on the polarity of the target components, includes options such as ethanol, water, methanol, and their combinations.

Pulse Electric Field Treatment: The suspension is placed between two electrodes, and brief bursts of high-voltage electric fields, typically ranging from 1 to 50 kV/cm, are applied. This electroporation process creates pores in the cell membranes, disrupting the cell walls and facilitating the release of bioactive compounds into the solvent.

Extraction and Purification: Following the electric field treatment, the suspension is subjected to standard extraction techniques, such as solvent extraction, to remove bioactive substances from the plant matrix. The extract is then filtered to eliminate solid residues, and the solvent is evaporated to obtain a concentrated extract rich in bioactive compounds.

PEFAE maintains the integrity of bioactive compounds while ensuring a successful and efficient extraction process. Typical equipment for static bench-scale operations includes a high-voltage power generator, a digital oscilloscope (for measuring voltage, current, frequency, etc.), and a treatment chamber where the sample is placed (Sukardi et al., 2021).

This method enhances permeability and extraction rates without generating cellular debris, improving the product's purity and making it a viable alternative to other techniques (Canabarro, 2021). Additionally, PEFAE is well-suited for extracting chemicals from wet biomass, as it does not require drying. It also preserves thermally unstable substances by allowing exposure to electric pulses without causing a sharp increase in temperature (Kokkali et al., 2020). However, process parameters such as field strength and pulse number are specifically influenced by the composition of the medium and its ability to generate a potential in response to electric field pulses. The high cost of equipment limits the widespread use of this method.

3.2.6 Supercritical Fluid Extraction (SFE)

Supercritical Fluid Extraction (SFE) is an advanced green technology increasingly used to extract valuable compounds from a variety of sources, including seaweeds, plant matrices, yeasts, and natural pigments. Its many benefits include reduced solvent consumption, shorter extraction times, and environmental friendliness (Banafi et al., 2023; Larocca et al., 2023; Abou Elmaaty et al., 2022). SFE employs fluids such as CO_2, which offer better extraction rates and selectivity compared to traditional solvents. This makes SFE particularly useful for extracting labile chemicals. The method has shown potential in a range of sectors, including food, textiles, cosmetics, and pharmaceuticals. Its applications include extracting antimicrobial pigments for health-promoting uses and recovering phenolic compounds, fatty acids, and carotenes from seaweeds (Abou Elmaaty et al., 2022; Dos Santos et al., 2023). Overall, SFE provides an effective and sustainable way to produce high-quality products, comply with environmental regulations, and promote green practices across various industries (Díaz-Reinoso et al., 2023).

SFE is the method of choice in many industries due to its numerous advantages over traditional extraction techniques. It is renowned for its ability to preserve bioactive, thermo-labile components in essential oils, maintaining the natural composition of heat-sensitive chemicals (Manzoor et al., 2023, p. 685). Additionally, SFE is an environmentally friendly method with excellent selectivity, effectively extracting heat-sensitive and readily oxidized chemicals while using less organic solvent and avoiding hazardous residues in the final product (Li et al., 2022). Furthermore, SFE efficiently removes solvent residues while preserving the unique qualities of natural antioxidants, resulting in higher yields, greater purity, and enhanced bioactivity compared to conventional methods. With these advantages, SFE stands out as a viable technique for extracting bioactive substances from a range of natural sources, offering an effective and environmentally friendly alternative to chemical solvents in separation processes (Vafaei et al., 2022; Irom & Hulle, 2022).

4 Application of Bioactive Compounds

Bioactive chemicals derived from natural sources such as plants, animals, and microorganisms possess a variety of pharmacological and therapeutic effects. These compounds find applications across diverse fields, including medicine, agriculture, and the food industry. Figure 2 illustrates the different applications of bioactive compounds.

4.1 Bioactive Compound Application in Medicine and Health Care

Polyphenols, saponins, terpenes, and alkaloids are examples of bioactive compounds derived from plants and natural sources that have been extensively studied for their potent antioxidant and anti-inflammatory properties. They have also been shown to have a significant positive impact on a variety of diseases and offer substantial health benefits (Kussmann et al., 2023; Dasgupta, 2023).

Biotechnology has significantly advanced the production and application of these bioactive compounds. An important

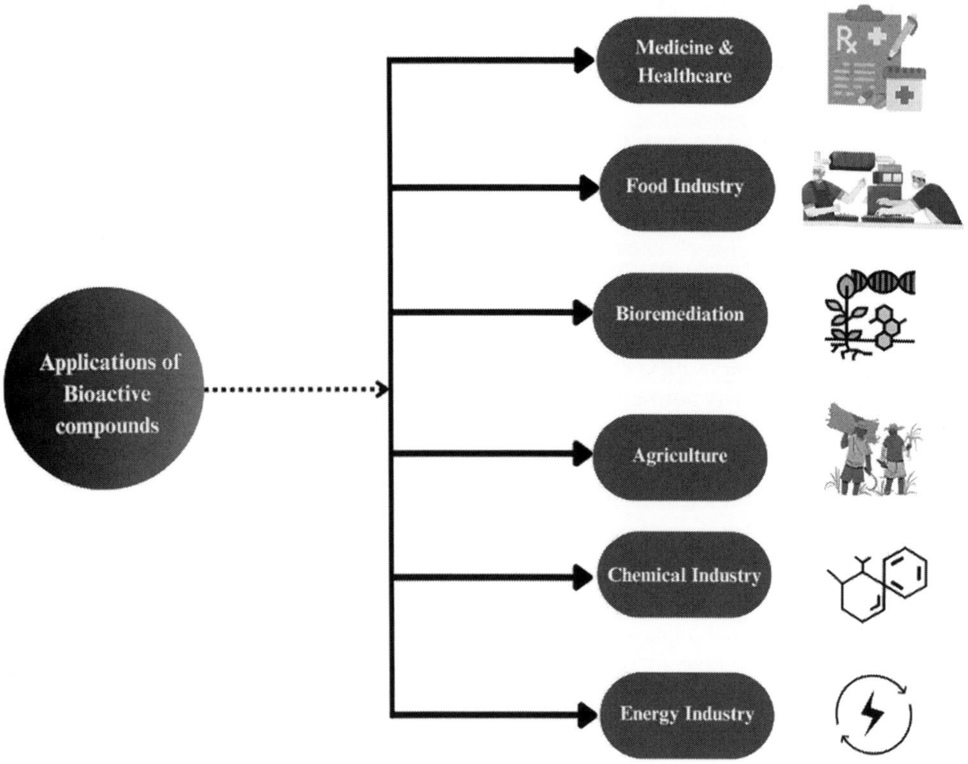

Fig. 2 Applications of bioactive compounds (Dos Santos et al., 2022)

function of biotechnological approaches is to enhance the yield and efficacy of bioactive substances from medicinal plants and crops, which plays a crucial role in preventing diseases such as cancer and cardiovascular disease (Siddiqui et al., 2023). This is especially important given that chronic illnesses are the primary causes of morbidity and mortality globally.

Traditional medical systems, such as Ayurveda and Traditional Chinese Medicine, have long recognized and utilized the health benefits of bioactive substances. These systems have a rich history of using natural remedies to treat and prevent a wide range of illnesses, underscoring the enduring importance of bioactive substances in promoting health (Dasgupta, 2023). The principles and methods from these ancient systems continue to influence contemporary healthcare, providing a foundation for integrating natural remedies with modern medical practices.

The development of bioactive-based treatments has advanced further with the integration of precision nutrition, medicine, and nutrition, particularly in addressing drug-resistant disorders like cancer. By tailoring dietary interventions to an individual's genetic profile, precision nutrition enhances the effectiveness of bioactive substances in disease prevention and treatment (Rekha & Thiruvengadam, 2022). This multidisciplinary approach highlights how natural substances can offer complementary and effective therapies for complex, chronic illnesses.

4.2 Bioactive Compound Application in Food Industry

Bioactive compounds are essential to the food industry because they enhance the nutritional value of food products and offer numerous health benefits. These compounds, including flavonoids, vitamins, and phenolic compounds, are widely present in fruits, vegetables, and agro-industrial by-products such as potato peels. Their anti-inflammatory, antioxidant, and even anticancer properties are well established (Martins et al., 2021).

Recent advancements in encapsulation techniques based on microfluidics and nanofluidics have been developed to improve the stability, bioavailability, and controlled release of bioactives during food preparation. These innovations ensure that bioactive substances effectively reach their intended sites in the body (Siddiqui et al., 2023). The food industry utilizes bioactive compounds from natural sources to enhance the nutritional profile of food items while also promoting sustainable and circular economy practices (Bansal et al., 2022). Further research is necessary to fully explore the potential of these bioactive compounds in creating functional foods with preservative and protective qualities. Addressing the growing consumer demand for natural and nutrient-dense products requires such research (Bansal et al., 2022). By advancing research and innovation, the food industry can leverage these ingredients to develop healthier food options, thereby improving public health and environmental sustainability.

4.3 Bioactive Compound Application in Bioremediation

The role of bioactive compounds in bioremediation is crucial for the removal of pollutants from the environment. Integrating bioactive compounds into bioremediation processes represents a promising approach to addressing environmental contamination. With the diverse range of bioactive chemicals sourced from plants, fungi, bacteria, and mammals (Chaudhary & Garg, 2023)—including polyphenols, vitamins, and organic acids—there is significant potential for their application in remediation technologies. Notably, biosurfactants, as natural bioactive compounds, have demonstrated remarkable properties for enhancing bioremediation processes by aiding in the solubilization and degradation of contaminants (De Oliveira et al., 2023). Biosurfactants are amphiphilic compounds secreted by microorganisms that offer advantages such as environmental compatibility, biocompatibility, and high solubility, making them effective in the removal of toxic pollutants (Chafale & Kapley, 2022). These substances help reduce environmental contamination caused by the improper disposal of oil waste by recovering oil from sewage and sludge through microbially enhanced oil recovery (MEOR) (Chafale & Kapley, 2022). Furthermore, bioaugmentation techniques involve the incorporation of bioactive microorganisms to enhance the breakdown of hydrophobic contaminants, highlighting the significance of biological substances in bioremediation efforts. Additionally, biogas, due to its unique properties, is a viable option for various industries, including biology, as it can recover pollution from human activities (Molina et al., 2022). These compounds have demonstrated surfactant characteristics in various industrial settings, including mining, petroleum, and bioremediation. Utilizing biologically active substances, particularly bioactive compounds, in bioremediation techniques can explore novel approaches to pollution issues and promote long-term environmental reform initiatives.

4.4 Bioactive Compound Application in Agriculture

In contemporary agriculture, bioactive chemicals are essential for providing long-term solutions for crop protection and improvement. These substances, derived from a variety of sources including lipids, essential oils, microbial metabolites, and plant extracts, are encapsulated using cutting-edge methods such as spray- and freeze-drying. Encapsulation enhances their stability and effectiveness, making them attractive candidates for use as biofungicides, biopesticides, and biofertilizers.

Bio-insecticides made from plant extracts, phytohormones, and rhizobacteria that promote plant growth are proving effective in protecting crops from insect pests with minimal environmental impact (Anguiano-Cabello et al., 2023). These bio-insecticides specifically target pests, reducing the need for broad-spectrum conventional insecticides that can harm pollinators and beneficial organisms.

Moreover, aquaculture is discovering new uses for agro-industrial waste, which is rich in bioactive substances such as phenolic compounds and terpenes. These substances help aquatic creatures become more resistant to illness and maintain better health. By repurposing waste materials, this technique reduces environmental degradation associated with waste disposal while adding value to industrial processes.

The intersection of medicine and nutrition in agriculture is driving innovation in precision farming. Bioactive chemicals are at the forefront of this movement, fostering the development of sustainable farming methods designed to feed the world's growing population while preserving biodiversity and ecosystem health (Parmar & Parmar, 2023). By utilizing these substances, farmers can improve soil health, maximize crop yields, and reduce their reliance on artificial inputs.

The use of bioactive compounds in agriculture represents a significant shift toward environmentally friendly and sustainable practices. These compounds are paving the way for a more resilient agricultural sector that balances productivity with ecological stewardship by leveraging novel encapsulation techniques and innovative uses of natural resources (Kussmann et al., 2023). Continued research on novel bioactive compounds and their applications holds the potential to revolutionize environmental sustainability and global food security.

4.5 Bioactive Compound Application in Chemical Industry

Bioactive compounds such as polyphenols, carotenoids, vitamins, omega-3 fatty acids, and phytosterols have diverse applications across several sectors, including the chemical industry. Derived from plants, fungi, bacteria, or mammals, these compounds play a crucial role in the development of nutraceuticals, functional foods, and food additives. Their well-researched pharmacological effects, including potential antidiabetic properties, make them valuable in both traditional medical systems and contemporary drug development (Dubey et al., 2022).

Furthermore, bioactive components such as peptides, phytosterols, and fatty acids have shown potential for modulating metabolic processes, gene expression, and enzyme function. This makes them critical elements in the food sector for creating natural, healthy, and cost-effective products (Bansal et al., 2022). The extraction and use of these compounds have substantial economic and environmental impacts. Encapsulation techniques aimed at improving their stability and bioavailability represent a promising approach for their application.

4.6 Bioactive Compound Application in Energy Industry

Bioactive chemicals derived from plants, fungi, and animals are gaining popularity for their dual benefits in health and energy applications. These molecules, which include polyphenols, carotenoids, vitamins, and omega-3 fatty acids, are not only essential for human health but also possess qualities that enhance sustainability in the energy sector.

In recent years, the extraction of bioactive chemicals from energy crops has emerged as a crucial technique for improving economic efficiency and sustainability. Industries can add value to crops by extracting these molecules before utilizing them for energy purposes such as biofuels, bioethanol, or biogas generation. This approach aligns with the principles of the bioeconomy, which emphasizes maximizing the use of biomass resources (Pai et al., 2022).

Bioactive substances extracted from organic wastes and agricultural residues have become commercially viable thanks to the use of green solvents and innovative extraction methods. These techniques not only reduce environmental impact but also advance a circular economy within the energy sector. By repurposing organic residues and agro-wastes to extract bioactive chemicals, industries can lower waste output and generate additional revenue.

Furthermore, incorporating bioactive compounds into industrial processes enhances the value of biomass that would otherwise remain unused. This valorization makes energy crop production more economically feasible and fosters innovation across various fields. The energy sector has experienced growth and technological advancement due to the use of these molecules, with other industries, including food, pharmaceuticals, and chemicals, also benefiting.

In summary, bioactive substances from diverse biological sources are vital for enhancing the sustainability and economic viability of the energy sector. Their integration into energy production processes fosters innovation, improves resource efficiency, and supports the transition to a more sustainable energy future. By effectively harnessing these substances, industries can simultaneously address health needs, environmental challenges, and economic imperatives in a synergistic manner.

5 Limitations of Bioactive Compounds

Bioactive substances have a lot of potential health advantages, but there are a lot of obstacles in the way of their full therapeutic potential and use in preserving human health.

5.1 Bioavailability

The low bioavailability of bioactive chemicals is one of their main drawbacks. Bioavailability refers to the percentage of a substance that enters the bloodstream and is able to affect biological processes. Numerous bioactive substances, such as resveratrol from grapes and curcumin from turmeric, have low bioavailability due to factors such as rapid metabolism, poor water solubility, and instability in the gastrointestinal tract. These factors lead to restricted absorption and rapid excretion from the body, which diminishes their efficacy in achieving therapeutic effects.

For example, curcumin's rapid metabolism and limited solubility reduce its absorption and bioavailability, thereby diminishing its therapeutic benefits. Similarly, resveratrol suffers from rapid metabolism and excretion, making it difficult for the body to achieve therapeutic levels.

5.2 Potential Toxicity

Though typically regarded as benign, bioactive chemicals can be harmful, particularly when taken in excessive concentrations or by vulnerable individuals. For instance, excessive intake of catechins from green tea may harm the liver. Furthermore, drug interactions can occur, negatively impacting the efficacy of medications. These potential hazards underscore the importance of carefully assessing and monitoring dosage, especially when used therapeutically (Liu et al., 2022).

5.3 Variability

Factors such as plant genetics, soil properties, temperature fluctuations, and harvesting techniques can significantly impact the composition of bioactive chemicals found in natural sources. This variability in bioactive chemicals poses a challenge to standardizing research and clinical applications. Variations in composition can affect potency and effectiveness, complicating the development of reliable dosage recommendations and consistent therapeutic outcomes (Himani et al., 2022).

5.4 Individual Variations

A notable constraint is the variation in individual responses to bioactive substances. Differences in metabolism, genetic makeup, and underlying health conditions can all influence how people absorb, process, and react to bioactive compounds. This variability limits the broad applicability of bioactive substances for general health maintenance, as what works for one person may not be as effective for another (Robberecht et al., 2021).

In conclusion, while bioactive compounds have the potential to benefit health, their limitations—such as bioavailability, potential toxicity, compositional variability, and individual responses—highlight the need for further research, improved delivery methods, and personalized approaches to optimize their safety and efficacy. Addressing these issues is essential for maximizing the benefits of bioactive compounds in promoting human health.

6 Prospective Challenges and Scope of Bioactive Compounds

The field of bioactive compounds presents a diverse array of opportunities and challenges across various sectors. These compounds, derived from plants, fungi, and microbes, have long been recognized for their potential health benefits and industrial applications. Historically, extracting bioactive compounds was a time-consuming and solvent-intensive process. However, recent advancements in extraction techniques, particularly the use of green solvents and non-traditional methods, are enhancing the commercial viability of these compounds while supporting sustainability.

One of the most promising applications of bioactive compounds is in the treatment of diabetes mellitus. These molecules have the potential to be more effective than traditional drugs, with fewer adverse effects. This represents a significant advancement in diabetes management, offering patients more options and potentially improving their quality of life. Additionally, research into bioactive compounds produced by endophytic fungi has unveiled new possibilities for industrial applications. Despite their potential, the production of these compounds is still in its early stages, necessitating further technological advancements. CRISPR-Cas9 techniques could significantly enhance the synthesis and effectiveness of bioactive compounds derived from these fungal sources (Dos Santos et al., 2022).

Plant-derived bioactive compounds, in particular, have garnered substantial interest due to their therapeutic properties. They are especially noted for their potential in treating neurodegenerative diseases. Their antioxidant, anti-inflammatory, and neuroprotective properties make them promising candidates for developing new treatments for disorders such as Alzheimer's and Parkinson's (Anand et al., 2023). The evolving landscape of bioactive compounds underscores the need for ongoing research and innovation. Overcoming existing challenges, such as improving extraction efficiency and increasing yield, is crucial for fully realizing the potential of bioactive compounds across various applications.

7 Conclusion

The exploration and utilization of bioactive compounds in the food, agricultural, and health sectors present promising opportunities for enhancing human health and promoting sustainable practices. The inherent properties of these compounds, such as antibacterial, anti-inflammatory, and antioxidant effects, make them valuable across various applications, from functional foods and nutraceuticals to cosmetics and bioremediation. However, the full potential of bioactive compounds can only be realized by addressing the challenges associated with their extraction, formulation, stability, and regulatory concerns. Advances in extraction techniques, such as the use of non-conventional methods like pressurization, enzymes, pulse electric fields, ultrasound, and microwaves, have made the commercial extraction of these compounds more viable, particularly from agro-wastes and organic residues. To fully harness the benefits of bioactive compounds, it is essential to foster multidisciplinary collaboration and innovative research. This includes developing sustainable and efficient extraction methods, improving consumer acceptability, and navigating regulatory landscapes. By overcoming these obstacles, bioactive compounds can contribute significantly to sustainable development and improved human health. The coordinated efforts highlighted in this chapter emphasize the importance of ongoing research and innovation in realizing the full potential of bioactive compounds across diverse sectors, ultimately leading to a more sustainable and healthier future.

References

Abdelmohsen, U. R., Sayed, A. M., & Elmaidomy, A. H. (2022). Natural products' extraction and isolation-between conventional and modern techniques. *Frontiers in Natural Products*, 1. https://doi.org/10.3389/fntpr.2022.873808.

Abou Elmaaty, T., Sayed-Ahmed, K., Elsisi, H., & Magdi, M. (2022). Optimization of extraction of natural antimicrobial pigments using supercritical fluids: A review. *Processes, 10*(10), 2111. https://doi.org/10.3390/pr10102111.

Aguilar-Machado, D., Montañez, J., Raso, J., & Martínez, J. M. (2022). Other applications of pulsed electric fields technology for the food industry. *Pulsed Electric Fields Technology for the Food Industry*, 439–466. https://doi.org/10.1007/978-3-030-70586-2_15.

Aguirre Calvo, T. R., & Santagapita, P. R. (2017). Encapsulation of a free-solvent extract of lycopene in alginate-Ca(II) beads containing sugars and biopolymers. *Chemical and Biological Technologies in Agriculture, 4*(1). https://doi.org/10.1186/s40538-017-0099-3.

Alcorta, A., Porta, A., Tárrega, A., Alvarez, M. D., & Vaquero, M. P. (2021). Foods for plant-based diets: Challenges and innovations. *Foods, 10*(2), 293. https://doi.org/10.3390/foods10020293.

Álvarez Alarcon, N., Ortiz-Ramírez, J. S., Ortega-Barbosa, J. P., Ballesteros-Vivas, D., Rodríguez-Varela, L. I., & Parada Alfonso, F. (2023). Pressurized liquid extraction (PLE). *Green Extraction Techniques in Food Analysis*, 357–395. https://doi.org/10.2174/9789815049459123030011.

Anand, K., Kumar, V., Prasher, I. B., Sethi, M., Raj, H., Ranjan, H., Chand, S., & Pandey, G. K. (2023). Bioactive molecules from fungal endophytes and their applications in pharmaceutical industries: Challenges and future scope. *Journal of Basic Microbiology, 63*(7), 690–708. https://doi.org/10.1002/jobm.202200696.

Anguiano-Cabello, J. C., Laredo-Alcalá, E. I., Barrera-Martínez, C., Hernández-Castillo, F. D., Ilyina, A., Segura-Ceniceros, P., et al. (2023). Plants and microorganism bio-compounds on agriculture applications by nanotechnology. *Microbial Biodiversity, Biotechnology and Ecosystem Sustainability*, 137–159. https://doi.org/10.1007/978-981-19-4336-2_8.

Anticona, M., Blesa, J., Frigola, A., & Esteve, M. J. (2020). High biological value compounds extraction from citrus waste with non-conventional methods. *Foods, 9*(6), 811. https://doi.org/10.3390/foods9060811.

Banafi, A., Wee, S. K., Tiong, A. N. T., Kong, Z. Y., Saptoro, A., & Sunarso, J. (2023). Modeling of supercritical fluid extraction bed: A critical review. *Chemical Engineering Research and Design, 193*, 685–712. https://doi.org/10.1016/j.cherd.2023.04.001.

Bansal, M., Poonia, A., Kolluri, S. R. P., & Vasundhara. (2022). Introduction on bioactive compounds, sources and their potential applications. *Bioactive Components*, 3–26. https://doi.org/10.1007/978-981-19-2366-1_1.

Botella, M. Á., Hernández, V., Mestre, T., Hellín, P., García-Legaz, M. F., Rivero, R. M., et al. (2021). Bioactive compounds of tomato fruit in response to salinity, heat and their combination. *Agriculture, 11*(6), 534. https://doi.org/10.3390/agriculture11060534.

Câmara, J. S., Albuquerque, B. R., Aguiar, J., Corrêa, R. C. G., Gonçalves, J. L., Granato, D., et al. (2020). Food bioactive compounds and emerging techniques for their extraction: Polyphenols as a case study. *Foods, 10*(1), 37. https://doi.org/10.3390/foods10010037.

Canabarro, N. (2021). *Review for "Pulsed electric fields- and ultrasound-assisted green extraction of valuable compounds from Origanum vulgare L. and Thymus serpyllum L"*. https://doi.org/10.1111/ijfs.15159/v1/review3.

Chafale, A., & Kapley, A. (2022). Biosurfactants as microbial bioactive compounds in microbial enhanced oil recovery. *Journal of Biotechnology, 352*, 1–15. https://doi.org/10.1016/j.jbiotec.2022.05.003.

Chaudhary, M., & Garg, A. P. (2023). Bioactive compounds from plants and animals: A review. *International Journal of Advanced Biochemistry Research, 7*(1), 105–117. https://doi.org/10.33545/26174693.2023.v7.i1b.175.

Chiocchio, I., Mandrone, M., Tomasi, P., Marincich, L., & Poli, F. (2021). Plant secondary metabolites: An opportunity for circular economy. *Molecules, 26*(2), 495. https://doi.org/10.3390/molecules26020495.

Dasgupta, S. C. (2023). Bioactive compounds from medicinal plants and its therapeutic uses in the traditional healthcare system. *Sustainable Development and Biodiversity*, 525–537. https://doi.org/10.1007/978-981-19-9936-9_19.

De Oliveira, M. A., Silva, F. F., Tarabal, V. S., Livio, D. F., Meira, H. G. R., Gonçalves, P. F. R., et al. (2023). Recent patents on the industrial application of biosurfactants. *Recent Patents on Biotechnology, 17*(4), 376–394. https://doi.org/10.2174/1872208317666230117143535.

Díaz-Reinoso, B., Balboa, E. M., Moure, A., & Domínguez, H. (2023). Supercritical extraction of marine antioxidants. *Marine Antioxidants*, 73–87. https://doi.org/10.1016/b978-0-323-95086-2.00026-6.

Dos Santos, L. C., Torres, T. M. S., Campos, D. F., Ghiglieno, F. G., & Martínez, J. (2023). Supercritical fluid extraction applied to food wastewater processing. *Advanced Technologies in Wastewater Treatment*, 179–215. https://doi.org/10.1016/b978-0-323-88510-2.00007-5.

Dos Santos, M. S., Wancura, J. H., Oro, C. E., Dallago, R. M., & Tres, M. V. (2022). Opportunities and challenges of plant bioactive compounds for food and agricultural-related areas. *Phyton, 91*(6), 1105–1127. https://doi.org/10.32604/phyton.2022.020913.

Dubey, R. S., Naidu, M., Shukla, A. K., Shukla, A. K., Kumar, M., Verma, S., & Mourya, P. K. (2022). Application of bioactive molecules in the treatment and management of type-1 diabetic disease. *The Scientific Temper, 13*(2), 265–281. https://doi.org/10.58414/scientifictemper.13.2.2022.265-281.

Gan, Y., Xu, D., Zhang, J., Wang, Z., Wang, S., Guo, H., et al. (2020). Rana chensinensis ovum oil based on CO_2 supercritical fluid extraction: Response surface methodology optimization and unsaturated fatty acid ingredient analysis. *Molecules, 25*(18), 4170. https://doi.org/10.3390/molecules25184170.

Gilabadi, S., Stanyon, H., DeCeita, D., Pendry, B. A., & Galante, E. (2023). Simple and effective method for the extraction of silymarin from Silybum marianum (L.) gaertner seeds. *Journal of Herbal Medicine, 37*, 100619. https://doi.org/10.1016/j.hermed.2022.100619.

González, M., Barrios, S., Budelli, E., Pérez, N., Lema, P., & Heinzen, H. (2020). Ultrasound assisted extraction of bioactive compounds in fresh and freeze-dried Vitis vinifera cv Tannat grape pomace. *Food and Bioproducts Processing, 124*, 378–386. https://doi.org/10.1016/j.fbp.2020.09.012.

Gorte, O., Nazarova, N., Papachristou, I., Wüstner, R., Leber, K., Syldatk, C., et al. (2020). Pulsed electric field treatment promotes lipid extraction on fresh oleaginous yeast saitozyma podzolica DSM 27192. *Frontiers in Bioengineering and Biotechnology*, 8. https://doi.org/10.3389/fbioe.2020.575379.

Hassim, N., Markom, M., Rosli, M. I., & Harun, S. (2021). Publisher correction: Scale-up approach for supercritical fluid extraction with ethanol–water modified carbon dioxide on Phyllanthus niruri for safe enriched herbal extracts. *Scientific Reports, 11*(1). https://doi.org/10.1038/s41598-021-96669.

Himani, Upadhyay, P., Kumar Mahawer, S., Kumar, R., & Prakash, O. (2022). Chapter 2 plant protection through agrochemicals and its consequences. *Plant Protection*, 25–44. https://doi.org/10.1515/9783110771558-002.

Irom, B. S., & Hulle, N. R. S. (2022). Recent advancements and challenges in supercritical fluid extraction methods and their applications in food industries. *Recent Advances in Food Biotechnology*, 371–385. https://doi.org/10.1007/978-981-16-8125-7_18.

Jalgaonkar, K., Mahawar, M. K., Bibwe, B., & Kannaujia, P. (2020). Postharvest profile, processing and waste utilization of dragon fruit (Hylocereus Spp.): A review. *Food Reviews International, 38*(4), 733–759. https://doi.org/10.1080/87559129.2020.1742152.

Kokkali, M., Martí-Quijal, F. J., Taroncher, M., Ruiz, M.-J., Kousoulaki, K., & Barba, F. J. (2020). Improved extraction efficiency of antioxidant bioactive compounds from tetraselmis chuii and phaeodoactylum tricornutum using pulsed electric fields. *Molecules, 25* (17), 3921. https://doi.org/10.3390/molecules25173921.

Kumar, K., Srivastav, S., & Sharanagat, V. S. (2021). Ultrasound assisted extraction (UAE) of bioactive compounds from fruit and vegetable processing by-products: A review. *Ultrasonics Sonochemistry, 70*, 105325. https://doi.org/10.1016/j.ultsonch.2020.105325.

Kussmann, M., Abe Cunha, D. H., & Berciano, S. (2023). Bioactive compounds for human and planetary health. *Frontiers in Nutrition*, 10. https://doi.org/10.3389/fnut.2023.1193848.

Larocca, V., Martino, M., Trupo, M., Magarelli, R. A., Spagnoletta, A., & Ambrico, A. (2023). Evaluation of carbon dioxide supercritical fluid extraction (CO2-SFE) on carotenoids recovery from red yeast cells. *Biomass Conversion and Biorefinery*. https://doi.org/10.1007/s13399-023-04434-z.

Lefebvre, T., Destandau, E., & Lesellier, E. (2021). Selective extraction of bioactive compounds from plants using recent extraction techniques: A review. *Journal of Chromatography A, 1635*, 461770. https://doi.org/10.1016/j.chroma.2020.461770.

Li, J., Niu, D., Zhang, Y., & Zeng, X.-A. (2020). Physicochemical properties, antioxidant and antiproliferative activities of polysaccharides from Morinda citrifolia L. (Noni) based on different extraction methods. *International Journal of Biological Macromolecules, 150*, 114–121. https://doi.org/10.1016/j.ijbiomac.2019.12.157.

Li, L., Chen, M., Zeng, Y., & Liu, G. (2022). Application and perspectives of supercritical fluid technology in the nutraceutical industry. *Advanced Sustainable Systems, 6*(7). https://doi.org/10.1002/adsu.202200055.

Linares, G., & Rojas, M. L. (2022). Ultrasound-assisted extraction of natural pigments from food processing by-products: A review. *Frontiers in Nutrition*, 9. https://doi.org/10.3389/fnut.2022.891462.

Liu, B., Chen, Z., Zhang, Y., Liu, E., Han, S., Gong, Z., & Xiao, W. (2022). Epigallocatechin-3-gallate + L-theanine/β-cyclodextrin inclusion complexes enhance epigallocatechin-3-gallate bioavailability and its lipid-lowering and weight loss effects. *Journal of Functional Foods, 90*, 104998. https://doi.org/10.1016/j.jff.2022.104998.

López-Bascón, M., & Luque de Castro, M. (2020). Soxhlet Extraction. *Liquid-Phase Extraction*, 327–354. https://doi.org/10.1016/b978-0-12-816911-7.00011-6.

Malini, G., Saranya, B., & Parameswari, A. (2023). Review on extraction techniques for medicinal and aromatic plants. *International Journal of Pharmaceutical Chemistry and Analysis, 10*(4), 253–256. https://doi.org/10.18231/j.ijpca.2023.042.

Manzoor, S., Rashid, R., Ahmad, M., Masoodi, F., Junaid, P. M., & Parvez, S. (2023). Advantages of essential oil extraction using supercritical fluid. *Essential Oils*, 685–705. https://doi.org/10.1002/9781119829614.ch31.

Martí-Quijal, F. J., Barba, F. J., & Ramon-Mascarell, F. (2021). Effect of pulsed electric fields on the recovery of antioxidant protein extracts from fish side streams. *Foods*, 2021. https://doi.org/10.3390/foods2021-10962.

Milani, G., Curci, F., Cavalluzzi, M. M., Crupi, P., Pisano, I., Lentini, G., et al. (2020). Optimization of microwave-assisted extraction of antioxidants from bamboo shoots of phyllostachys pubescens. *Molecules, 25*(1), 215. https://doi.org/10.3390/molecules25010215.

Molina, G., Usmani, Z., Sharma, M., Yasri, A., & Gupta, V. K. (2022). Microbes in agri-forestry. *Biotechnology*. https://doi.org/10.1201/9781003110477.

Montoro, P., Serreli, G., Gil, K. A., D'Urso, G., Kowalczyk, A., & Tuberoso, C. I. G. (2020). Evaluation of bioactive compounds and antioxidant capacity of edible feijoa (Acca sellowiana (O. Berg) Burret) flower extracts. *Journal of Food Science and Technology, 57*(6), 2051–2060. https://doi.org/10.1007/s13197-020-04239-2.

Muala, W. C. B., Desobgo, Z. S. C., & Jong, N. E. (2021). Optimization of extraction conditions of phenolic compounds from Cymbopogon citratus and evaluation of phenolics and aroma profiles of extract. *Heliyon, 7*(4). https://doi.org/10.1016/j.heliyon.2021.e06744.

Pai, S., Hebbar, A., & Selvaraj, S. (2022). A critical look at challenges and future scopes of bioactive compounds and their incorporations in the food, energy, and pharmaceutical sector. *Environmental Science and Pollution Research, 29*(24), 35518–35541. https://doi.org/10.1007/s11356-022-19423-4.

Pal, C. B. T., & Jadeja, G. C. (2022). Optimization and kinetics of polyphenol recovery from raw mango (Mangifera indica L.) peel using a glycerol-sodium acetate deep eutectic solvent system. *Biomass Conversion and Biorefinery, 14*(2), 2453–2465. https://doi.org/10.1007/s13399-022-02550-w.

Parmar, D. S., & Parmar, A. (2023). Bioactive compounds in Argo-industrial waste: A review on the prospects for the application in aquaculture. *Journal of Entomology and Zoology Studies, 11*(1), 118–125. https://doi.org/10.22271/j.ento.2023.v11.i1b.9133.

Peanparkdee, M., & Iwamoto, S. (2020). Encapsulation for improving in vitro gastrointestinal digestion of plant polyphenols and their applications in food products. *Food Reviews International, 38*(4), 335–353. https://doi.org/10.1080/87559129.2020.1733595.

Pinzon, I. A. P., Razal, R. A., Mendoza, R. C., Carpio, R. B., & Eusebio, R. C. P. (2020). Parametric study on microwave-assisted extraction of runo (Miscanthus sinensis Andersson) dye and its application to paper and cotton fabric. *Biotechnology Reports*, 28. https://doi.org/10.1016/j.btre.2020.e00556.

Rastogi, N. K. (2023). Developments in osmotic dehydration of foods. *Drying Technology in Food Processing*, 241–304. https://doi.org/10.1016/b978-0-12-819895-7.00015-8.

Rehman, M. U., Abdullah, Khan, F., & Niaz, K. (2020). Introduction to natural products analysis. *Recent Advances in Natural Products Analysis*, 3–15. https://doi.org/10.1016/b978-0-12-816455-6.00001-9.

Rekha, K., & Thiruvengadam, M. (2022). Production and biomedical applications of bioactive compounds. *Processes, 10*(9), 1830. https://doi.org/10.3390/pr10091830.

Remali, J., & Aizat, W. M. (2021). A review on plant bioactive compounds and their modes of action against coronavirus infection. *Frontiers in Pharmacology*, 11. https://doi.org/10.3389/fphar.2020.589044.

Robberecht, H., Bruyne, T. D., & Hermans, N. (2021). Association between selenium status and metabolic syndrome and its biomarkers. *Trace Elements and Electrolytes, 38*(1), 13–20. https://doi.org/10.5414/tex01646.

Rubio, N. R., Xiang, N., & Kaplan, D. L. (2020). Plant-based and cell-based approaches to meat production. *Nature Communications, 11*(1). https://doi.org/10.1038/s41467-020-20061-y.

Samtiya, M., Aluko, R. E., Dhewa, T., & Moreno-Rojas, J. M. (2021). Potential health benefits of plant food-derived bioactive components: An overview. *Foods, 10*(4), 839. https://doi.org/10.3390/foods10040839.

Scaglia, B., D'Incecco, P., Squillace, P., Dell'Orto, M., De Nisi, P., Pellegrino, L., Botto, A., Cavicchi, C., & Adani, F. (2020). Development of a tomato pomace biorefinery based on a CO_2-supercritical extraction process for the production of a high value lycopene product, bioenergy and digestate. *Journal of Cleaner Production, 243*, 118650. https://doi.org/10.1016/j.jclepro.2019.118650.

Sharma, P., Gangwar, A., Rani, N., & Singh, R. (2023). Nutrition and neurological disorders. *Treatments, Nutraceuticals, Supplements, and Herbal Medicine in Neurological Disorders*, 797–814. https://doi.org/10.1016/b978-0-323-90052-2.00018-4.

Shirahigue, L. D., & Ceccato-Antonini, S. R. (2020). Agro-industrial wastes as sources of bioactive compounds for food and fermentation industries. *Ciência Rural, 50*(4). https://doi.org/10.1590/0103-8478cr20190857.

Siddiqui, S. A., Farooqi, M. Q. U., Bhowmik, S., Zahra, Z., Mahmud, M. C., Assadpour, E., et al. (2023). Application of micro/nano-fluidics for encapsulation of food bioactive compounds-principles, applications, and challenges. *Trends in Food Science & Technology, 136*, 64–75. https://doi.org/10.1016/j.tifs.2023.03.025.

Šimat, V., Frleta, R., Čagalj, M., & Skroza, D. (2024). Technological and analytical aspects of bioactive compounds and nutraceuticals from marine algae. *Bioactive Compounds and Nutraceuticals from Dairy, Marine, and Nonconventional Sources*, 159–196. https://doi.org/10.1201/9781003452768-7.

Soriano, Y., Andreu, V., & Picó, Y. (2024). Pressurized liquid extraction of organic contaminants in environmental and food samples. *TrAC Trends in Analytical Chemistry, 173*, 117624. https://doi.org/10.1016/j.trac.2024.117624.

Sridhar, J. (2024). *Review for "Enzyme-assisted extraction of rapeseed oil with minimum water addition: A proof-of-concept study"*. https://doi.org/10.1111/ijfs.17030/v2/review1.

Sruthi, D., Dhanalakshmi, M., Yashavantha Rao, H., Parthasarathy, R., & Jayabaskaran, C. (2023). Extraction, isolation, and characterization of phytochemicals, the bioactive compounds of plants. *Recent Frontiers of Phytochemicals*, 1–8. https://doi.org/10.1016/b978-0-443-19143-5.00033-5.

Sukardi, Safitri, P., & Setyawan, H. Y. (2021). Comparative study on the extraction and quantification of polyphenols from moringa oleifera leaves using maceration with pulsed electric field (PEF) as pretreatment and microwave-assisted extraction (MAE). In *Proceedings of the International Conference on Innovation and Technology* (*ICIT* 2021). https://doi.org/10.2991/aer.k.211221.019.

Sultana, S. (2024). Plants flavonoids: Chemistry, metabolism and biological activities. *Journal of Population Therapeutics & Clinical Pharmacology*, 2299–2308. https://doi.org/10.53555/jptcp.v31i6.6883.

Vafaei, N., Rempel, C. B., Scanlon, M. G., Jones, P. J. H., & Eskin, M. N. A. (2022). Application of supercritical fluid extraction (SFE) of tocopherols and carotenoids (Hydrophobic Antioxidants) compared to non-SFE methods. *AppliedChem, 2*(2), 68–92. https://doi.org/10.3390/appliedchem2020005.

Weggler, B. A., Gruber, B., Teehan, P., Jaramillo, R., & Dorman, F. L. (2020). Inlets and sampling. *Separation Science and Technology*, 141–203. https://doi.org/10.1016/b978-0-12-813745-1.00005-2.

Yoon, H. I., Kim, H. Y., Kim, J., & Son, J. E. (2021). Quantitative analysis of UV-B radiation interception and bioactive compound contents in Kale by leaf position according to growth progress. *Frontiers in Plant Science*, 12. https://doi.org/10.3389/fpls.2021.667456.

Zakaria, S. M., Mustapa Kamal, S. M., Harun, M. R., Omar, R., & Siajam, S. I. (2020). Extraction of phenolic compounds from Chlorella sp. microalgae using pressurized hot water: Kinetics study. *Biomass Conversion and Biorefinery*, 12(6), 2081–2089. https://doi.org/10.1007/s13399-020-01027-y.

Biotechnological Application of Health-Promising Bioactive Compounds

Akanksha Singh and Pradeep Singh Negi

Abstract

Bioactive compounds are naturally occurring chemicals that exert significant biological effects, including therapeutic and preventive actions against diseases. These compounds play a crucial role in synthesizing essential biomolecules and modulating biological functions, making them invaluable in various fields of health and science. This chapter systematically categorizes bioactive compounds based on their sources, which include plants, agro-industrial by-products, animals, and microbes. Each source contributes unique chemical properties and health benefits, making bioactive compounds highly versatile. Emphasis is placed on the biotechnological production of these compounds, utilizing advanced techniques such as microbial fermentation, plant cell culture, metabolic engineering, and recombinant DNA technology. These methodologies increase the efficiency of bioactive compound production and promote sustainability by reducing the need for natural harvesting, thus supporting biodiversity conservation efforts. The application section delves into the vast uses of bioactive compounds, particularly their role in developing innovative solutions across multiple industries such as pharmaceuticals, the food industry, cosmetics, and environmental management. In pharmaceuticals, bioactive compounds are pivotal in drug discovery and development of antibiotics, enzyme inhibitors, and vaccines, effectively demonstrating their capacity to treat and prevent a range of diseases. In the food industry, these compounds are critical for improving food safety and enhancing nutritional value by serving as natural alternatives to synthetic additives. Industrial applications highlight the role of bioactive compounds in producing biopolymers and biofuels, positioning them as key contributors to sustainable technological advancements. The chapter concludes by underscoring the importance of ongoing research into new sources of bioactive compounds and optimizing their properties for broader applications. Future directions propose enhancing biotechnological processes and expanding their utilization in environmental biotechnology to exploit further their potential in promoting human health and environmental sustainability. The chapter presents bioactive compounds as vital agents in bridging natural biological benefits with technological innovation, driving future advancements in science and industry.

Keywords

Bioactives • Biotechnology • Health • Pharmaceuticals • Therapeutics

A. Singh · P. S. Negi (✉)
Department of Fruit and Vegetable Technology, CSIR-Central Food Technological Research Institute, Mysuru, 570020, India
e-mail: psnegi@cftri.res.in

1 Introduction

Bioactive compounds are naturally occurring chemical constituents that exert biological effects on living organisms. These effects include therapeutic and preventive actions against various diseases, roles in synthesizing biomolecules, modulating biological functions, and preventing microbial contamination. Additionally, they serve as carriers for drugs, enzymes, and nutrients uniquely positioning them at the intersection of nature and technology (Sytar & Smetanska, 2022). The sources of these compounds are diverse. Plants are perhaps the most significant contributors, offering a plethora of phytochemicals, like phenolic acids, flavonoids, and terpenoids, which have been documented for their antioxidant, anti-inflammatory, and anticarcinogenic properties (Twaij & Hasan, 2022). Animal-derived bioactives, such as omega-3 fatty acids from fish oil and conjugated linoleic acid from dairy products, are noted for their cardiovascular and metabolic health benefits (Alagawany et al., 2022). Fungi and microorganisms are not to be overlooked,

J. Devi (ed.), *Biotechnological Intervention in Production of Bioactive Compounds*,
Sustainable Landscape Planning and Natural Resources Management, https://doi.org/10.1007/978-3-031-76859-0_5

as they produce a range of important bioactive metabolites, like penicillin from *Penicillium*, which not only marked the dawn of antibiotics but also represents a bioactive compound with profound therapeutic effects (Pham et al., 2019).

Beyond their intrinsic biological activities, these compounds are pivotal in the development of functional foods, nutraceuticals, and pharmacological agents. In functional foods, bioactive compounds are utilized to fortify and enhance the nutritional profile of everyday consumables. For example, phytosterols derived from plants are incorporated into dairy products to help in reducing cholesterol levels (Mannarino et al., 2009). Similarly, beverages enriched with polyphenols from berries demonstrate enhanced antioxidant properties, appealing to health-conscious consumers (García-Alonso et al., 2006). In the pharmaceutical industry, bioactive compounds underpin the development of novel drugs through both traditional and biotechnological methods. Taxol, a complex diterpene derived from the Pacific yew tree, is a powerful chemotherapeutic agent used in cancer treatment (Cech & Oberlies, 2023). Nutraceuticals represent another significant application of bioactive compounds. These products bridge the gap between food and pharmaceuticals by providing medicinal benefits through dietary intake. Omega-3 fatty acid supplements are marketed for their efficacy in reducing inflammation and supporting heart health based on extensive research into their biological effects (Gupta & Gupta, 2020). Moreover, the role of bioactive compounds extends into the cosmetics industry, where their antioxidant and anti-aging properties are harnessed in skincare products (Hernandez et al., 2021). In environmental applications, bioactive compounds contribute to sustainability and ecological health through bioremediation processes, aiding the breakdown of pollutants and enhancing soil and water quality (Pal et al., 2020). Additionally, the development of biopesticides derived from natural bio-actives offers a less harmful alternative to synthetic pesticides, providing effective pest control while minimizing ecological damage (Kumar et al., 2021). In industrial applications, bioactive compounds have facilitated the production of biopolymers and biofuels, ushering in a new era of green technology (Jiménez-Gómez & Cecilia, 2020).

Integrating bioactive compounds across these diverse sectors highlights their remarkable versatility and emphasizes the ongoing necessity for innovative research to explore new sources and expand their applications. This interdisciplinary approach pushes the boundaries of science and industry and offers sustainable solutions that harness the synergies between biology and technology, paving the way for future advancements and environmental stewardship.

2 Sources of Health-Promising Bioactive Compounds

All living organisms synthesize bioactive compounds as part of their natural defense mechanisms. The rich biodiversity of our planet is evident in the vast array of bioactive compounds produced. Principal sources of bio-actives include plants encompassing fruits-vegetables, agro-industrial by-products, animals, and microbes from terrestrial and marine ecosystems.

2.1 Fruits and Vegetables as the Sources of Bioactive Compounds

Food serves as a source of essential nutrients critical for life, along with various bioactive compounds that promote health and help to prevent diseases. Dietary guidelines advocate for a substantial intake of plant-based foods, recommending the consumption of at least 400 g of fruits and vegetables daily. This consumption is linked to a lower incidence of several health issues, including hypertension, obesity, coronary heart disease, stroke, various cancers, eye conditions, asthma, osteoporosis, neurological diseases, and sleep disorders (Pengpid & Peltzer, 2020; Ramírez-Salazar et al., 2021). Fruits and vegetables are rich in micronutrients and bioactive substances, like vitamins and phytochemicals, including phenolic compounds, carotenoids, essential minerals like potassium, calcium, magnesium, amino acids, and dietary fiber (Rashmi & Negi, 2020). Over 5,000 distinct phytochemicals are identified in these foods, although many still remain unidentified (Liu, 2013). Phenolic acids such as ferulic and caffeic acids found in coffee, berries, and cereals exhibit strong antioxidant and anti-inflammatory properties. These acids have been used in various commercial applications, including skin care products for UV protection and topical formulations to prevent skin aging (Espíndola et al., 2019). Another class of compounds, flavonoids (including flavanols and flavones) are renowned for their potent anti-inflammatory and antioxidant capabilities. Catechins and epigallocatechin gallate from tea and cocoa are widely recognized for their health benefits and have been incorporated into nutraceuticals and UV-protection creams (Bae et al., 2020). Isoflavones, such as daidzein and genistein, primarily found in soy, play critical roles in managing diabetes, cancer, and inflammation. They are incorporated into food supplements to maintain post-menopausal bone density and hormonal balance (Alshehri et al., 2021; Sharifi-Rad et al., 2021). Moreover, tannins like ellagitannins and procyanidins are well known for their cardiovascular and gut health benefits, finding applications in functional foods and health products focused on oral and gut health (Banc et al., 2023; Chen et al., 2022). Carotenoids and amino acids also demonstrate significant health-promoting properties. Zeaxanthin and lutein are pivotal in eye health and are used extensively in functional foods, and dietary supplements to reduce the risk of chronic eye diseases (Bouyahya et al., 2021; Mitra et al., 2021). Essential amino acids, like tryptophan and phenylalanine, not only serve as precursors to vital neurotransmitters but also enhance mood, sleep–wake cycle, and cognitive functions, making them suitable for inclusion in dietary supplements (Friedman, 2018; Rana et al., 2022). Dietary fibers serve as prebiotics and help in maintaining gut homeostasis (Cui et al., 2019). Thus, as sources of bioactive compounds, fruits and vegetables play an indispensable role in enhancing human health and have wide utility in various biotechnological applications. These applications range from food preservation and pharmaceuticals to nutraceuticals and functional foods. Table 1 presents a comprehensive inventory of prevalent bioactive compounds in fruits and vegetables and their associated health benefits and potential biotechnological utilities.

Table 1 Fruits and vegetables as the sources of bioactive compounds

Class	Compound	Sources	Potential health benefits	Biotechnological application	References
Phenolic acid	Caffeic acid	Coffee, berries, olives	Antioxidant, anti-inflammatory	Used in skin care products for UV protection	(Espíndola et al., 2019)
	Ferulic acid	Cereals, coffee, apples, oranges	Antioxidant, photoprotective, anti-inflammatory	Used in topical formulation to prevent skin aging	(Srinivasan et al., 2007)
	Sinapic acid	Citrus fruits, berries	Antioxidant, antimicrobial	Food preservatives, pharmaceuticals application in anti-aging creams	(Chen, 2016)
	Hydroxycinnamic acid	Coffee, blueberries, kiwis	Antioxidants	Functional food development for cardiovascular health	(Teixeira et al., 2013)
	Quinic acid	Coffee, apples, peaches	Antioxidant, supports digestive health	Health supplements for digestive wellness	(Benali et al., 2022)
	Chlorogenic acid	Coffee, sunflower seeds, prunes	Antioxidant, antihypertensive, weight management	Functional beverages and supplements	(Yu et al., 2022)
	Gallic acid	Nuts, tea, grapes, mango, pomegranate, berries	Antioxidant, prevention of lipid peroxidation, Anti-inflammatory, anti-melanogenic, antiviral, neuroprotective	Tanning stabilization of collagen in the leather industry	(Badhani et al., 2015)
	Protocatechuic acid	Tea, mango, berries	Neuroprotective, antitumor, anti-osteoporotic activity	Pharmaceutical application in cancer-prevention	(Song et al., 2020)
	Syringic acid	Olives, pumpkin, grapes, palm	Antiendotoxic, anti-microbial, anti-inflammatory activities	Lactase-based detoxification, ozonation, bioremediation	(Srinivasulu et al., 2018)
	Hydroxy-benzoic acid	Apple, tea, strawberries, grapes	Antimicrobial, antioxidant	As a preservative in cosmetics, flavoring agent	(del Olmo et al., 2017)
	p-Coumaric acid	Tomatoes, carrots, garlic	Antioxidant, anti-melanogenic effect	In cosmetic and UV protection lotion	(Boo, 2019)

(continued)

Table 1 (continued)

Class	Compound	Sources	Potential health benefits	Biotechnological application	References
Stilbenes	Resveratrol	Red grapes, berries	Anti-angiogenic, anti-inflammatory, antioxidant	Cosmetic and dermatological application	(Ratz-Łyko & Arct, 2019; Tian & Liu, 2020)
Lignans	Secoiariciresinol	Whole grain cereals, legumes, seeds, nuts	Anticancer, antioxidant, anti-obesity	Dietary supplement for hormonal balance	(Park, 2014)
Flavanols	Catechin	Tea, cocoa, apples	Antioxidant, anti-allergic, anticancer, heart health, weight management	UV-protection cream, nutraceutical application	(Bae et al., 2020)
	Epigallocatechin gallate	Tea	Antioxidant, anticancer, fat reduction	Nutraceuticals for metabolic and cardiovascular health	(Singh et al., 2011)
Isoflavones	Genistein	Soya	Antidiabetic, anticancer, anti-inflammatory	Hormonal balance, anticancer, osteoporosis prevention	(Sharifi-Rad et al., 2021)
	Daidzein	Soya	Immune regulator, anti-cancerous, anti-allergic, maintains post-menopausal bone density	Food supplements	(Alshehri et al., 2021)
Anthocyanins	Delphinidin	Berries, grapes, teas	Anti-oxidant, anti-inflammation, anti-diabetes, anti-obesity, cardiovascular protection, neuroprotection, and anti-cancer	Colorant in food industry, supplement	(Chen et al., 2019)
	Cyanidin	Tomatoes, grapes	Antioxidant, anti-inflammatory, lipid balance	Fortification in the food industry	(Shozib et al., 2021)
	Malvidin	Grapes	Cardiovascular protection	Wine industry, food supplement	(Merecz-Sadowska et al., 2023)
Flavanones	Hesperidin	Citrus fruits	Antioxidant, anti-inflammatory, vascular health	Citrus-based supplements, functional juices	(Stanisic et al., 2018)
	Naringenin	Grapefruit, tomatoes	Antioxidant, anti-inflammatory, lipid balance	Functional foods and beverages	(Salehi, Fokou et al., 2019a, b)
Flavones	Luteolin	Peppers, celery, parsley	Anti-inflammatory, anticancer, antioxidant	Dietary supplements for anti-inflammatory purposes	(Rakoczy et al., 2023)
	Apigenin	Chamomile, parsley	Anti-anxiety, anticancer, anti-inflammatory, sleep-promoting	Pharmaceutical and nutraceutical applications	(Majma Sanaye et al., 2022; Salehi et al., 2019a, b; Yoon et al., 2023)
Flavonols	Quercetin	Onions, apples, berries	Anti-inflammatory, antihistamine, cardiovascular health	Supplements for respiratory and cardiovascular health	(Papakyriakopoulou et al., 2022; Yang et al., 2020)
	Kaempferol	Kale, beans, tea	Antioxidant, anticancer	Nutraceuticals for cancer prevention	(Ma et al., 2023; Periferakis et al., 2022)
	Myricetin	Berries, nuts, herbs	Antioxidant, anti-inflammatory, anticancer	Dietary supplements and functional foods	(Imran et al., 2021)

(continued)

Table 1 (continued)

Class	Compound	Sources	Potential health benefits	Biotechnological application	References
Tannins	Procyanidins	Apples, cocoa, cinnamon	Antioxidant, cardiovascular, and oral health	Functional foods and oral health products	(Chen et al., 2022)
	Gallotannins	Tea, sumac, witch hazel	Antioxidant, antimicrobial, astringent	Pharmaceutical applications for skin health	(He, 2022)
	Ellagitannins	Pomegranates, berries, nuts	Antioxidant, anticancer, gut health	Nutraceuticals and functional foods for gut health	(Banc et al., 2023)
Amino acids	Tryptophan	Nuts, pomegranate, citrus fruits	Precursor to serotonin and melatonin, mood regulation, sleep	Dietary supplements, functional foods	(Friedman, 2018)
	Histidine	Soya, nuts, berries	Precursor to histamine, immune response, digestive health	Nutritional supplements, medical foods	(Holeček, 2020)
	Phenylalanine	Nuts, seeds, soya	Precursor to tyrosine, dopamine, adrenaline, and norepinephrine	Dietary supplements for mood enhancement	(Rana et al., 2022)
	Tyrosine	Soya, fruits	Precursor to dopamine, norepinephrine, and adrenaline, affects mood and stress response	Nutritional supplements for cognitive and physical performance	
	L- theanine	Tea	Modulates neurotransmitters like GABA and dopamine, promoting relaxation	Dietary supplements for relaxation and sleep improvement	(Türközü & Şanlier, 2017)
Mineral	Magnesium	Nuts, seeds, green leafy vegetables	Plays a role in supporting GABA function, which promotes relaxation and sleep	As supplement	(Papadopol & Nechifor, 2011)
Carotenoid	Beta-Carotene	Carrots, sweet potatoes, spinach	Antioxidant, precursor to Vitamin A, eye health, skin health	Dietary supplements, food colorants	(Ebadi et al., 2022)
	Lycopene	Tomatoes, watermelon, pink grapefruit	Antioxidant, supports heart and prostate health	Food supplements, functional foods, cosmetics	(Khan et al., 2021)
	Lutein	Kale, spinach, corn, egg yolks	Eye health, may reduce risk of chronic eye diseases	Dietary supplements, functional foods	(Mitra et al., 2021)
	Zeaxanthin	Corn, saffron, orange, bell peppers	Eye health, protects against UV and blue light	Dietary supplements, functional foods	(Bouyahya et al., 2021)
	Astaxanthin		Antioxidant, supports skin and joint health	Dietary supplements, aquaculture feed, cosmetics	(Stachowiak & Szulc, 2021)

2.2 Agro-Industrial Byproducts as the Sources of Bioactive Compounds

Despite adequate food production to sustain the global population, the 2017 Global Hunger Index highlighted that millions of the population worldwide face hunger and malnutrition (Klaus et al., 2017). Agricultural practices generate substantial residues, which are a significant source of industrial waste. Food loss varies by region in the world, with figures ranging from around 6% in New Zealand, and Australia to over 20% in parts of Asia. This equates to an estimated 221 million tons of food being lost or wasted annually, from the production stage onwards. India alone produces nearly 1% of its GDP in the form of food wastage and ranks 107th in the global hunger index (FAO, 2021; Food Wastage in India, 2024). Processed food forms such as pastes, juices, and sauces lead to considerable fruit by-products like seeds, skins, and pomace. The term "food by-products" is preferred over "food waste" to emphasize their potential for developing value-added products. Environmental and economic challenges arise from the disposal of organic wastes, which produce greenhouse gases and occupy landfill space. However, valorizing these by-products can reduce pollution and create new markets and income streams. Effective waste management can be cost-effective for achieving food security, and sustainability (Reguengo et al., 2022).

Historically, food by-products have been utilized for livestock feed or compost. More recently, their potential for producing bioenergy has been explored (Dey et al., 2024). Furthermore, they offer a promising source of dietary nutrients and bioactive compounds with antioxidant and antimicrobial properties. The search for natural bioactive compounds for managing and preventing neurodegenerative diseases and sleep disorders is ongoing. These compounds can interact with proteins, DNA, and other molecules, inactivate free radicals, and affect the gut microbiota, supporting health promotion and disease prevention (Rojas-García et al., 2023). Recently kiwi fruit peel has been reported to improve sleep in mice (Kim et al., 2022). Similarly, lemon seed is also reported to promote sleep (Rahman et al., 2022). Orange peel also has been reported to provide protection against Alzheimer's disease (Abd El-Aziz et al., 2023). Other than protecting against diseases, bio-actives from fruit and vegetable by-products are used for various biotechnological applications; a comprehensive list of these by-products is provided in Table 2.

2.3 Animals as the Sources of Bioactive Compounds

Animals are indispensable sources of bioactive compounds that have profound implications for human health. These bioactive molecules, which are essential for their survival, also confer significant benefits to humans. Over the past decades, a variety of natural compounds from animal origins have been systematically extracted, identified, and harnessed as dietary or therapeutic agents to treat, prevent, and alleviate a spectrum of diseases and health conditions (Zhang et al., 2015). Among the proteins and peptides derived from animals, collagen from fish, bovine, and porcine skin is notable for its effects on skin and joint health and is widely used in cosmetics and dietary supplements (Avila Rodríguez et al., 2018). Gelatin, obtained from animal bones, skins, and connective tissues, serves as a food-gelling agent and aids in wound healing, with applications in the food industry and pharmaceutical formulations (Alipal et al., 2021). Lactoferrin, sourced from milk, is recognized for its antimicrobial properties and immune support, finding its way into infant formula and health supplements (H. Ashraf et al., 2024a, 2024b). Fatty acids, such as omega-3 from fish oil and krill oil, are crucial for cardiovascular and cognitive health and are prevalent in dietary supplements and functional foods (Alagawany et al., 2022). Conjugated Linoleic Acid, found in dairy and beef, is employed in weight management and immune support products, particularly in weight loss supplements (Bauman et al., 2020). Vitamins also play a critical role. Vitamin D from fish liver oils and fatty fish is essential for bone health and immune function, and is widely used in fortified foods and dietary supplements (Janoušek et al., 2022). Vitamin B12, present in meat, eggs, and dairy, supports red blood cell formation and neurological function and is a staple in nutritional supplements and fortified foods (Sobczyńska-Malefora et al., 2021). Minerals like calcium and iron from animal sources contribute significantly to human health; calcium enhances bone strength and cardiovascular function and is included in dairy products and fortified non-dairy alternatives, while the iron is vital for hemoglobin formation and muscle function, commonly added to fortified cereals and supplements (Libera et al., 2021; Masotti et al., 2020). The distinctive chemical structures of certain bioactive compounds enable them to act as substrates for the synthesis of vital biomolecules or as modulators that influence the structure and function of

Table 2 Agro-industrial byproducts as the sources of bioactive compounds

Class	Compound	Sources	Potential health benefits	Biotechnological application	References
Flavonoids	Quercetin, catechins	Citrus peels, grape pomace	Antioxidant, anti-inflammatory, cardiovascular support	Nutraceuticals, functional foods	(Almanza-Oliveros et al., 2024; Andrade et al., 2022)
Phenolic acids	Ferulic acid, chlorogenic acid	Coffee grounds, cereal bran	Antioxidant, anti-inflammatory, anti-diabetic	Cosmeceuticals, dietary supplement	(Acevedo et al., 2013; Fărcaş et al., 2022)
Carotenoids	Lycopene, β-Carotene	Tomato pomace, carrot tops, pumpkin seeds	Antioxidant, anti-cancer, supports skin health, vision support, immune system boost	Food coloring, supplements, pharmaceuticals nutraceuticals, food fortification	(Aniceto et al., 2022; Ninčević Grassino et al., 2023)
Essential oils	Limonene, citral	Citrus peels	Aromatherapy, antimicrobial, stress relief	Flavoring, cosmetics, pharmaceuticals	(Andrade et al., 2022)
Dietary fiber	Pectin	Apple pomace, mango peel, citrus peel	Digestive health, cholesterol management	Food industry (thickener, stabilizer), pharmaceuticals	(Sarangi et al., 2023)
	Cellulose	Wheat bran	Digestive health, weight management	Food additives, pharmaceuticals	(Vu et al., 2022)
Proteins	Peptides	Legume waste	Muscle recovery, blood pressure regulation	Functional foods, sports nutrition	(Kamani et al., 2023)
Amino acids	Melatonin	Grape pomace	Regulation of sleep cycles, antioxidant	Supplements, pharmaceuticals	(Almanza-Oliveros et al., 2024)
	Serotonin	Grape pomace	Mood regulation	Pharmaceuticals, dietary supplements	(Monteiro et al., 2021)
	Tryptophan	Banana peel, pumpkin seeds, soybean meal	Precursor to serotonin, sleep aid	Nutraceuticals, dietary supplements	(Soeliono et al., 2023)

organs and tissues. Their applications extend beyond nutritional and therapeutic realms into the food, biotechnological, and agriculture industries, highlighting the extensive influence of animal-derived bio-actives across various sectors. A comprehensive list of animal-derived bioactive is provided in Table 3.

2.4 Microbes as the Sources of Bioactive Compounds

Microorganisms, encompassing a vast array of microscopic life forms such as viruses, bacteria, archaea, fungi, protozoa, and algae, are prolific producers of unique chemical compounds. These organisms contribute significantly to the pool of secondary metabolites, with actinomycetes and fungi each accounting for approximately 42% of known compounds, while eubacteria contribute the remaining 16% (Demain, 2014). These secondary metabolites are diverse, including substances like growth hormones, pigments, antibiotics, and antitumor agents. Although these compounds do not play a role in the primary growth and development of the microorganisms themselves, they hold immense potential for human health applications (Singh et al., 2017). Secondary metabolites of microbes have been proven to be critical in the development of life-saving drugs (Giurini et al., 2024). The production of these metabolites is often triggered by the activation of otherwise dormant gene clusters, which do not express under normal conditions. This makes the expression of these genes crucial for tapping into the chemical diversity of microorganisms. The metabolites they produce have wide-ranging applications, notably in the pharmaceutical, nutraceutical, and cosmetic industries, where they are valued for their biotechnological potential (Zhao et al., 2019). Various studies have highlighted the classification of these compounds based on their chemical structures and explored their roles in various industries. Research has emphasized the antioxidant potential of these microorganisms, suggesting that they could be leveraged in the food, agriculture, nutraceutical, and pharmaceutical sectors. A detailed list of

Table 3 Animals as the sources of bioactive compounds

Class	Compound	Sources	Potential health benefits	Biotechnological application	References
Proteins and peptides	Collagen	Fish, bovine, porcine skin	Skin health, joint health	Cosmetics, dietary supplements	(Avila Rodríguez et al., 2018)
	Gelatin	Bones, skins, and connective tissues of animals	Food gelling agent, wound healing	Food industry, pharmaceutical formulations	(Alipal et al., 2021)
	Lactoferrin	Milk	Antimicrobial, immune support	Infant formula, health supplements	(Ashraf et al., 2024a, 2024b)
Fatty acids	Omega-3	Fish oil, krill oil	Cardiovascular health, cognitive function	Dietary supplements, functional foods	(Alagawany et al., 2022)
	Conjugated Linoleic Acid	Dairy and beef	Weight management, immune support	Weight loss supplements	(Bauman et al., 2020)
Vitamins	Vitamin D	Fish liver oils, fatty fish	Bone health, immune function	Fortified foods, dietary supplements	(Janoušek et al., 2022)
	Vitamin B12	Meat, eggs, dairy	Red blood cell formation, neurological function	Nutritional supplements, fortified foods	(Sobczyńska-Malefora et al., 2021)
Mineral	Calcium	Dairy	Bone strength, cardiovascular function	Dairy products, fortified non-dairy alternatives	(Masotti et al., 2020)
	Iron	Red meat, poultry, fish	Hemoglobin formation, muscle function	Fortified cereals, supplements	(Libera et al., 2021)

various bioactive compounds from microbial sources, their health benefits, and their biotechnological application, is mentioned in Table 4.

3 Biotechnological Production of Bioactive Compounds

Biotechnological production of bioactive compounds is a rapidly expanding field that integrates the principles of biology and technology to harness the biosynthetic capabilities of living organisms, including microorganisms, plants, and animal cells. This sophisticated approach focuses on producing a vast array of substances known for their beneficial health effects, such as antioxidants, enzymes, vitamins, and secondary metabolites with pharmaceutical properties (Gill et al., 2020; Namitha & Negi, 2013). Microbial fermentation and plant cell culture are the most common techniques employed. Microbial fermentation serves as a cornerstone, utilizing microorganisms to produce essential bioactive compounds (Sharma et al., 2020). Plant cell culture technology, meanwhile, has evolved as a pivotal technique in this realm. This enhances production efficiency and supports sustainable practices by reducing the need for natural plant harvesting, which can be detrimental to biodiversity (Ozyigit et al., 2023). Metabolic engineering is also employed to increase the yields of specific flavonoids and polyphenols by modifying the metabolic pathways in host plants or microorganisms. The overexpression of the gene CHS (chalcone synthase), a key enzyme in the flavonoid biosynthesis pathway, significantly boosts specific flavonoid production. This genetic manipulation facilitates greater control over the quantity and quality of the produced compounds (Hwang et al., 2023). Similarly, recombinant DNA technology and CRISPR-Cas9 are used to enhance the production of bioactive compounds by introducing specific genes into the DNA of host cells, thereby producing higher yields of desired compounds. Moreover, CRISPR-Cas9 technology has enabled precise genetic modifications, such as deactivating certain genes that inhibit production or introducing new biosynthetic pathways to enhance compound synthesis. This technology has been used to increase the production of specific flavonoids in yeast, demonstrating the potential to create enhanced strains for industrial-scale production (Dhokane et al., 2023; Liu et al., 2021). The use of bioreactors for plant cell culture also represents a significant advancement in biotechnological production. Cells from plants known for their high polyphenolic content, such as grapes or berries, can be cultured in bioreactors to produce specific compounds like resveratrol or quercetin. This controlled environment ensures a consistent supply of these bioactive compounds and allows for the modulation of growth conditions to optimize the yield and quality of extracts (Hao et al., 2020; Motolinía-Alcántara et al., 2021). Enzymatic biotransformation is crucial in enhancing the

Table 4 Microbes as the sources of bioactive compounds

Class	Compound	Sources	Biological activity	Biotechnological application	References
Antibiotics	Penicillin	*Penicillium chrysogenum*	Antibacterial	Pharmaceutical	(Tan & Tatsumura, 2015)
	Streptomycin	*Streptomyces griseus*	Antibacterial	Pharmaceutical	(Waksman et al., 1946)
	Reuterin	*Lactobacillus reuteri*	Antibacterial	Pharmaceutical	(Gyawali & Ibrahim, 2014)
	Nisin	*Lactococcus lactis*	Antibacterial	Pharmaceutical	(Cobb et al., 2013; Gyawali & Ibrahim, 2014)
	Vancomycin	*Amycolalatopsis orientalis*	Antibacterial	Pharmaceutical	(Dasgupta, 2012)
	Tetracycline	*Streptomyces rimosus*	Antibacterial	Pharmaceutical	(Demain, 2009)
	Erythromycin	*Saccharopolyspora erythraea*	Antibacterial	Pharmaceutical	(Cobb et al., 2013)
	Amphotericin B	*Streptomyces nodosus*	Antifungal	Pharmaceutical	(Tevyashova et al., 2013)
	Leodoglucomide C	*Bacillus licheniformis*	Antifungal	Pharmaceutical	(Giddings & Newman, 2013)
Enzymes	Amylase	*Bacillus* spp.	Starch breakdown	Food processing, textile industry	(Gangadharan et al., 2008)
	Lipase	*Aspergillus niger*	Fat breakdown	Used in detergents, food industry	(Mahadik et al., 2002)
Protein and peptides	Bleomycin	*Streptomyces verticillus*, *Streptoalloteichus hindustanus*	Antitumor and anticancer	Pharmaceutical	(Demain & Vaishnav, 2011)
	Paclitaxel	*Taxomyces andreanae*	Antitumor and anticancer	Pharmaceutical	(Strobel & Daisy, 2003)
	Epothilone	*Sorangium cellulosum*	Antitumor and anticancer	Pharmaceutical	(Katz & Baltz, 2016)
	Depsipeptide	*Chromobacterium violaceum*	Antitumor and anticancer	Pharmaceutical	(Zhou et al., 2015)
	Daunorubicin	*Streptomyces peucetius*	Antitumor and anticancer	Pharmaceutical	(Giddings & Newman, 2013)
	Camptothecin	*Entrophospora infrequens*	Antitumor and anticancer	Pharmaceutical	(Puri et al., 2006)
	Cryptophycins	Cyanobacteria	Antitumor and anticancer	Pharmaceutical	(Bhatnagar & Kim, 2010)
Alkaloid	Spinosad	*Saccharopolyspora spinosa*	Targeting nicotinic acetylcholine receptors	Agricultural industry	(Dayan et al., 2009)
	Okaramine	*Penicillium simplicissimum* AK-40	Insecticide	Pharmaceuticals	(Matsuda, 2018)
Macrocyclic lactone	Abamectin	*Streptomyces avermitilis*	Insecticide, antihelmethic	Agricultural industry	(Dayan et al., 2009)
Mycotoxins	Roquefortine C	*Penicillium roqueforti*	Bactericidal	Pharmaceuticals	(García-Estrada et al., 2011)
	Aflatoxin	*Aspergillus flavus*	Hepatocarcinogen	Pharmaceuticals	(Pitt, 2000)
	Ochratoxin A	*Aspergillus ochraceus*	Key enzyme in gluconeogenesis	Pharmaceuticals	(van der Merwe et al., 1965)
	Zearalenone	*Fusarium graminearum*	Binds on estrogen receptors	Pharmaceuticals	(Pitt, 2000)

(continued)

Table 4 (continued)

Class	Compound	Sources	Biological activity	Biotechnological application	References
Vitamins	Vitamin B12	*Propionibacterium freudenreichii*	Essential for nerve function and blood cell production	Nutritional supplements, fortified foods	(Martens et al., 2002)
Organic acids	Citric acid	*Aspergillus niger*	Antioxidant, preservative	Food and beverage industry, pharmaceuticals	(El-Holi & Al-Delaimy, 2003; Schrickx et al., 1995)
	Lactic acid	*Lactobacillus* spp.	pH regulation, preservative	Food preservation, biodegradable plastics production	(Abedi & Hashemi, 2020; Hujanen et al., 2001)
Polysaccharides	Xanthan gum	*Xanthomonas campestris*	Thickening agent, stabilizer	Food industry, oil drilling, pharmaceuticals	(Krishna Leela & Sharma, 2000)
Biopesticides	Spinosad	*Saccharopolyspora spinosa*	Insecticide	Organic agriculture, pest control	(Guojun et al., 2016)
Amino acids	Glutamic acid	*Corynebacterium glutamicum*	Flavor enhancer (umami)	Food industry (as monosodium glutamate)	(Wendisch et al., 2016)

bioavailability and activity of flavonoids and polyphenols. Enzymes such as glucosyltransferases can add sucrose molecules to these compounds, increasing their solubility and stability, which is essential for their effective use in various applications, including pharmaceuticals and cosmetics. This enzymatic modification can also produce novel derivatives with enhanced or targeted therapeutic properties (Tao et al., 2022). Additionally, the integration of biotechnologically produced bio-actives with nanotechnology has opened new avenues for developing advanced delivery systems that improve the efficacy and targeted delivery of these compounds. Encapsulating polyphenols in biodegradable nanoparticles can enhance their absorption in the human body and provide sustained release, which is beneficial for therapeutic applications (Marcillo-Parra et al., 2021). As the field continues to evolve, the integration of these compounds into biotechnological innovations is poised to drive significant advancements in health and industry, fully harnessing their potential for future applications.

4 Application of Bioactive Compounds

Bioactive compounds, derived from diverse biological sources, play a pivotal role in advancing biotechnological applications across various sectors (Fig. 1). Naturally occurring substances, including antioxidants, peptides, and phytochemicals, are crucial for developing innovative solutions in healthcare, agriculture, and environmental management. These compounds can be extracted, optimized, and utilized through biotechnological processes to enhance disease prevention, crop protection, and pollution mitigation. Integrating bioactive compounds into biotechnology promises enhanced efficacy in traditional applications and opens the door to novel uses in emerging industries. As research progresses, the potential to harness these compounds for sustainable and efficient technologies continues to expand, significantly impacting both the economic and ecological fronts.

4.1 Pharmaceutical Applications

Bioactive compounds derived from various biological sources are increasingly recognized for their potential in pharmaceutical applications. These natural compounds, including alkaloids, terpenoids, polyphenols, amino acids, and fatty acids, are integral to discovering and developing new drugs, antibiotics, enzyme inhibitors, and vaccines. A detailed exploration of their roles across these areas is presented in following sections.

4.1.1 Drug Discovery and Development

From ancient times, traditional medicines were dependent on natural sources to treat many diseases. These natural sources, encompass bioactive compounds, impart biological activity to protect against various diseases or disease-causing agents. To date, various natural bioactive compounds with potent pharmacological properties have been identified. Plants, animals, and microbes play important roles in drug discovery, and more than 85% of the world's total population is dependent on the traditional medicine system to combat disease (Wangchuk, 2018). Morphine, the first plant-derived natural bioactive compound in 1803 from *Papaver*

Fig. 1 Various applications of bioactive compounds

somniferum, marked the inception of the drug discovery era (Krishnamurti & Rao, 2016). Around 70,000 plant species are being used for medicinal applications, and from this, about 20% of these plant species are used for medicinal purposes in India alone (Nasim et al., 2022). Hypericin, shikonin, wogonin, pinocembrin, kaempferitrin and piperine are the plant-derived bioactive molecules used in cancer treatment as immunogenic cell death inducers (Chand et al., 2022; Chelora et al., 2019; Yin et al., 2016). Berberine and trigonelline are used in antidiabetic drugs (Bahmani et al., 2016). Quinine is a well-known antimalarial drug derived from the bark of *Chincona* spp (Mohammadi et al., 2020). Taxol is used for lung, ovarian and breast cancer (Mamadalieva & Mamedov, 2020). Genistein is used in Alzheimer's disease (Viña et al., 2022). Capsaicin is used in pain relievers (Pasierski & Szulczyk, 2022). Valerian, lavender, hop, chamomile, and rosemary are used for sleep-inducing drugs (Guadagna et al., 2020). These uses underscore the critical role of natural bioactive compounds in modern pharmaceutical research and development, highlighting their potential to address a wide array of health challenges.

4.1.2 Antibiotics

The "Golden Age of Antibiotics" commenced with the discovery of penicillin in 1929, marking a transformative chapter in medical history as the first antibiotic to be mass-produced. The pursuit of natural products has been integral to the development of new antibiotic treatments. Notably, significant antibiotics derived from natural sources include non-ribosomal peptides, polyketides, and aminoglycosides (Wright, 2014). Polyketides, synthesized by PKS (polyketide synthases), are chemically diverse and play crucial roles in a variety of applications beyond healthcare, including agriculture and industry. A prominent example is pikromycin, the first polyketide antibiotic identified from *Streptomyces venezuelae* in 1950, which has shown high efficacy against multi-drug resistant respiratory pathogens (Jung et al., 2006). Another essential polyketide, erythromycin A, discovered in 1952 by *Saccharopolyspora erythraea*, is a broad-spectrum antibiotic. It is particularly valuable in treating a range of bacterial infections, such as gastrointestinal, and respiratory ailments, syphilis, whooping cough, and acne, and is especially beneficial for patients with penicillin allergies (Pham et al., 2019). Penicillin, derived from *P. notatum*, specifically targets Gram-positive bacteria and effectively combats diseases, including pneumonia, scarlet fever, gonorrhea, diphtheria, and meningitis (Tan & Tatsumura, 2015). Similarly, vancomycin, classified as a non-ribosomal peptide antibiotic, is synthesized via non-ribosomal peptide synthetases (NRPS). It stands out for its effectiveness against several formidable pathogens, including methicillin-resistant *Staphylococcus aureus* (MRSA). In the realm of aminoglycosides, these antibiotics inhibit protein synthesis by binding to the 30S subunit of the bacterial ribosome (Dasgupta, 2012). Streptomycin discovered in 1944 and the pioneer of this class, was notably effective against pulmonary tuberculosis (Sakula, 1988). The emergence of bacterial resistance prompted the development of semi-synthetic derivatives like amikacin and netilmicin to maintain the clinical utility of aminoglycosides (Park et al., 2013). Furthermore, natural antimicrobials are essential in ensuring food safety, particularly in the food industry. Lactic acid bacteria, for example, produce various antimicrobial compounds, such as nisin and reuterin (Li & Vederas, 2009). Nisin, a bacteriocin from *Lactococcus lactis*, has been approved for use in over 50 countries for food preservation and combats Gram-positive bacteria resistant to conventional antibiotics. Reuterin, produced by *Limosilactobacillus reuteri*, is effective against pathogens and spoilage microorganisms in dairy and meat products (Gyawali & Ibrahim, 2014). Overall, the exploration and application of natural bioactive compounds continue to be paramount in developing new antibiotics, reflecting a critical intersection of biotechnology and pharmacology that addresses both medical needs and contributes to public health safety.

4.1.3 Enzyme Inhibitors

Enzyme inhibitors play a crucial role in managing various diseases by regulating biological pathways that are mediated by specific enzymes. Enzyme inhibitors derived from natural products are pivotal in treating diseases linked to enzyme dysfunction or overexpression. These bioactive compounds hold immense potential for drug development due to their diverse structures and potent bioactivity. One key group of natural enzyme inhibitors includes glutathione S-transferase inhibitors, which are used to combat drug resistance in cancer therapy by blocking pathways that cancer cells use for detoxification. For instance, despite its limitations, such as side effects and lack of isoenzyme specificity, ethacrynic acid showcases these inhibitors' therapeutic potential. Similarly, acetylcholinesterase (AChE) inhibitors such as galanthamine, derived from plants like snowdrops and daffodils, are crucial in managing neurodegenerative disorders like Alzheimer's (Mohamed et al., 2011). By inhibiting the breakdown of acetylcholine, these compounds help enhance cognitive function and mitigate symptoms associated with memory loss. α-glucosidase inhibitors, like acarbose obtained from microorganisms, help manage diabetes mellitus by delaying the absorption of carbohydrates, thereby moderating post-meal blood sugar levels. Other natural compounds, particularly flavonoids, have also demonstrated significant α-glucosidase inhibitory activity (Hossain et al., 2020). In the cosmetic industry, tyrosinase inhibitors such as kojic acid are valuable for controlling melanin production, thus addressing skin hyperpigmentation disorders. Kojic acid inhibits melanin synthesis, making it a popular ingredient for skin-lightening products (Zolghadri et al., 2019). Additionally, cyclooxygenase (COX) inhibitors derived from natural sources like salicylates (e.g., aspirin from willow bark) have been instrumental in developing anti-inflammatory drugs. These inhibitors prevent the enzyme from producing prostaglandins, which are the key mediators of inflammation (Xiao et al., 2023). Overall, the exploration of the critical role of natural products as sources of enzyme inhibitors highlights their extensive applications across various medical and cosmetic fields. Understanding these compounds' biosynthetic pathways and structure–activity relationships is essential for optimizing their therapeutic potential.

4.1.4 Vaccine Development

Natural bioactive compounds such as saponins, beta-glucans, polysaccharides, and glycoproteins are increasingly pivotal in enhancing the efficacy and safety of vaccines through their roles as adjuvants and immunopotentiators. Saponins have been utilized to augment immune responses, facilitating the induction of robust and balanced humoral and cell-mediated immunity, which is crucial for vaccine effectiveness (Wang, 2021). Beta-glucans also contribute significantly by activating key immune cells like natural killer cells, and macrophages critical for a strong immunogenic response necessary for the success of many vaccines. Polysaccharides are invaluable in extending the duration and intensity of immune responses, making them essential for vaccines that require long-lasting protection. They help in sustaining an immune response over time, which is beneficial for vaccines against chronic or recurring infections (Vetvicka et al., 2020). Glycoproteins, due to their unique structural features, play a specific role in modulating the immune system to recognize and react to vaccine antigens more effectively. This is particularly important in designing vaccines that target elusive pathogens where a precise immune response is critical for protection. Chitosan stands out for its application in mucosal vaccines, which enhances local immune responses at the mucosal surfaces, the primary entry point for many pathogens. Its biocompatibility and ability to adhere to mucosal tissues make it an excellent candidate for delivering antigens directly to the sites where an initial immune response is most effective (Margolin et al., 2018). These natural bioactive compounds are integral to developing and optimizing vaccine formulations. Their natural immunomodulatory properties improve the efficacy and safety of vaccines and underscore the potential of natural resources in advancing modern vaccine technology. Through strategic incorporation into vaccine designs, these compounds help in achieving more effective and targeted vaccination strategies, which are crucial for combating a wide array of diseases.

4.2 Food Industry Applications

Natural bioactive compounds are revolutionizing the food industry by providing safer, more sustainable alternatives to synthetic additives. These compounds, derived from a wide range of natural sources, offer multifunctional benefits that enhance food safety, nutritional value, and health-promoting properties. This section explores their impactful applications across various domains within the food industry, such as food preservatives, nutraceuticals, and functional foods, illustrating their pivotal role in advancing food technology and consumer health.

4.2.1 Food Preservatives

Natural bioactive compounds have gained significant attention as viable alternatives to synthetic preservatives due to their potential to inhibit microbial growth and oxidation, thereby extending the shelf life of food products. The essential oils derived from herbs such as thyme and oregano are rich in phenolic compounds like thymol and carvacrol. These compounds have been demonstrated to exhibit strong antimicrobial and antifungal properties, making them effective against a broad spectrum of foodborne pathogens (Pandey et al., 2017). Research has shown that thymol can disrupt the cell membrane of bacteria, leading to leakage of vital cell components and eventualy cell death (Escobar et al., 2020). Similarly, natural antioxidants such as ascorbic acid (vitamin C), tocopherols (vitamin E), and plant extracts rich in polyphenols play a crucial role in preventing the oxidative degradation of fats and oils in foods, which helps in maintaining their quality and nutritional value. Ascorbic acid acts as a potent reducing agent, neutralizing free radicals and thereby preventing oxidative damage (Sheraz et al., 2011). Tocopherols, a class of chemical compounds of which vitamin E is a member, have been found to stabilize lipid membranes and protect against lipid peroxidation, a key factor in extending the shelf life of perishable products (Chu et al., 2023). Furthermore, polyphenols, abundant in various fruits, vegetables, and whole grains, contribute as antioxidants and metal chelators, binding iron and copper ions that catalyze oxidative reactions. This dual functionality enhances their preservative effect, particularly beneficial in complex food matrices (Gokoglu, 2018). These scientific findings underscore the potential of natural bioactive compounds to replace or reduce the use of synthetic additives in food preservation.

4.2.2 Nutraceuticals

Nutraceuticals are defined as specialized products derived from food sources that provide health benefits beyond basic nutritional value. These products utilize natural bioactive compounds known for their health-enhancing properties. For example, omega-3 fatty acids, predominantly sourced from marine organisms, are acclaimed for their cardiovascular benefits. Research has demonstrated that omega-3 s can reduce triglyceride levels, lower blood pressure, and decrease arrhythmia risks, contributing significantly to heart health (Gupta & Gupta, 2020). Curcumin, extracted from turmeric, exhibits potent anti-inflammatory and antioxidant properties. It modulates inflammatory molecules like cytokines and COX-2 enzymes, offering therapeutic benefits for chronic conditions such as arthritis (Kotha & Luthria, 2019). Similarly, resveratrol, found in grapes and berries, is noted for its anti-aging and cardioprotective effects. It enhances vascular function and reduces oxidative stress by activating sirtuin pathways associated with longevity (Prakash et al., 2024). Incorporating these bioactive compounds into nutraceuticals not only aids in managing health conditions but also promotes disease prevention by improving overall body resilience. This proactive approach to health maintenance underscores the growing importance of nutraceuticals in modern dietary practices.

4.2.3 Functional Foods

Functional foods are enhanced with natural bioactive compounds, transcending basic nutrition to improve health and reduce disease risk. Dairy products fortified with probiotics like *Lactobacillus* and *Bifidobacterium* improve gut health and support the immune system, potentially reducing the risk of chronic ailments such as irritable bowel syndrome and certain cancers (Damián et al., 2022). Beverages enriched with plant sterols effectively reduce cholesterol levels. Structurally similar to cholesterol, plant sterols compete for absorption in the digestive system, significantly lowering LDL cholesterol (Decloedt et al., 2017). Additionally, fibers like beta-glucan and inulin increase dietary fiber content, manage blood sugar, and enhance digestive health. Beta-glucan from oats improves insulin resistance and glucose control (Shen et al., 2016), while inulin from chicory root promotes beneficial gut bacteria growth, improving nutrient absorption and bowel regularity (Nwafor et al., 2017). These enhancements in functional foods meet consumer demands for health-supportive foods backed by scientific evidence of their health benefits.

Incorporating natural bioactive compounds into food industry applications responds to the growing consumer demand for natural and health-promoting products, aligning with global trends toward sustainability and wellness. These applications showcase the versatility of natural compounds in enhancing food quality, nutritional value, and health benefits, positioning them as essential elements in advancing the modern food industry.

4.3 Industrial Applications

Natural bioactive compounds have found pivotal roles across various industrial sectors, where their unique properties drive innovation and efficiency. These compounds are at the forefront of industrial sustainability and technological advancement, from biopolymers that offer sustainable alternatives to traditional plastics to biofuels that reduce reliance on fossil fuels and enzyme technology that enhances production processes.

4.3.1 Biopolymers

Natural bioactive compounds are increasingly recognized for their versatility and sustainability, particularly in the production of biopolymers. These organic compounds play a pivotal role in developing environment friendly materials that are biodegradable and exhibit properties comparable to those of synthetic polymers. Polysaccharides such as chitosan and alginate, derived from crustacean shells and seaweed, are used to produce films and coatings that are both robust and compostable (Jiménez-Gómez & Cecilia, 2020). Similarly, proteins like zein, extracted from corn, are utilized in creating edible packaging solutions. These biopolymers degrade naturally, reducing environmental impact and dependency on non-renewable petroleum-based plastics (Bayer, 2021). Research indicates that biodegradable packaging made from these natural compounds can significantly lower carbon footprints due to their renewable origins and energy-efficient decomposition processes. Additionally, using agricultural waste and by-products in biopolymer production not only helps in waste reduction but also promotes a circular economy by valorizing materials that would otherwise contribute to landfills. This approach aligns with sustainable development goals to minimize environmental pollution while enhancing material efficiency.

4.3.2 Biofuels

In the production of biofuels, natural bioactive compounds play a pivotal role in developing sustainable energy solutions. Biofuels like bioethanol and biodiesel are primarily derived from biomass, which includes forest debris, agricultural residues, and other organic waste materials (Wani et al., 2023). These biofuels are produced using enzymatic hydrolysis, where enzymes facilitate the break down of complex carbohydrates in the biomass into simpler sugars that are then fermented to alcohol in the case of bioethanol or trans-esterified into fatty acid esters for biodiesel (Cruz et al., 2020). The use of natural bioactive compounds, particularly enzymes such as cellulases and lipases, enhances the efficiency of these processes. Cellulases help in breaking down cellulose into glucose, which is a crucial step in converting plant materials into bioethanol (Ingle et al., 2017). Meanwhile, lipases are used in biodiesel production to catalyze the transesterification of fats and oils into methyl esters and glycerol, significantly speeding up the reaction compared to traditional chemical catalysts and reducing the need for high temperatures and pressures (Chioke et al., 2018). Studies have reported that integrating these bioactive compounds can reduce the energy input required for bioconversion processes by up to 50%, thereby lowering the overall carbon footprint of biofuel production. Moreover, these natural catalysts are biodegradable and nontoxic, minimizing environmental contamination and enhancing the sustainability of the biofuel sector (Ruangrit et al., 2023). This biological route provides an eco-friendlier alternative to fossil fuels and leverages renewable resources, promoting a circular economy and contributing significantly to global energy sustainability goals.

4.3.3 Enzyme Technology

Enzyme technology is a rapidly evolving field where natural bioactive compounds demonstrate profound impacts across various industries. As biological catalysts, enzymes are integral to numerous industrial processes, ranging from food production to waste management and bio-remediation. Their unique ability to catalyze specific biochemical reactions under mild conditions makes them highly desirable in sectors seeking to reduce energy consumption and chemical use (Bharathiraja et al., 2017). Scientifically, enzymes such as proteases, cellulases, and amylases are harvested from microbial sources or plants. They can be genetically modified to enhance their stability, activity, and specificity under harsh industrial conditions like extreme pH or high temperatures (Adrio & Demain, 2014). The thermostable amylases derived from extremophilic microorganisms are used in starch processing industries to operate efficiently at high temperatures, significantly speeding up the process and reducing the risk of microbial contamination (Ibrahim & Ma, 2017). Furthermore, enzyme technology facilitates the development of environmentally benign processes. In the textile industry, enzymes are used for bio-polishing and bio-bleaching of fabrics, drastically reducing the amount of water, energy, and harsh chemicals typically required. This not only improves the sustainability of the processes but also enhances the quality of the final product. The role of enzymes in waste management is another testament to their versatility. Enzymes such as ligninases and pectinases are employed to break down complex organic molecules in agricultural and municipal waste, turning them into simpler compounds that are easier to manage or repurpose, thus contributing to waste valorization strategies (de Castro Bizerra et al., 2024). The enzyme technology leverages the specific and efficient action of natural bioactive compounds to drive industrial processes that are more sustainable, cost-effective, and environment friendly.

4.4 Cosmetics and Personal Care

In the cosmetics and personal care industry, the growing preference for natural bioactive compounds is increasingly influenced by consumers who favour products that are not only effective but also sustainable and safe for prolonged use (Lavuri et al., 2022). These natural compounds, sourced from a diverse plants, marine organisms, and even microbial species, offer various benefits, from anti-inflammatory and anti-aging properties to enhanced skin hydration and protection against UV radiation (Kusumawati & Indrayanto, 2013). One of the major features of natural bioactive compounds is their ability to interact with the skin in ways that mimic the body's processes. Antioxidants derived from sources like green tea and grape seeds can help combat oxidative stress, a key factor in aging and skin degradation (Jadoon et al., 2015). Similarly, hyaluronic acid provides unmatched hydration by binding water molecules to the skin, promoting a plump and youthful appearance (Makvandi et al., 2020). Moreover, the anti-inflammatory properties of compounds like aloe vera and turmeric are invaluable in treating skin conditions such as acne and eczema, offering relief without the harsh side effects associated with synthetic ingredients (Hęś et al., 2019). This aligns with the holistic approach favored in modern dermatology, where the focus is on supporting skin health at its foundational levels rather than merely masking symptoms.

These bioactive compounds also extend to hair care, where ingredients like keratin-based peptides derived from natural sources strengthen hair structure and improve its overall health (Fernandes et al., 2012). As the demand for clean beauty continues to grow, the cosmetic industry is likely to embrace the potential of these natural compounds further to enhance beauty and foster well-being and sustainability. Natural bioactive compounds are reshaping the landscape of cosmetics and personal care by offering safer, more sustainable, and highly effective alternatives to traditional synthetic ingredients. Their ability to nurture and heal the body aligns perfectly with the evolving consumer ethos of health and environmental consciousness, making them a cornerstone in the future of personal care products.

5 Conclusion

Bioactive compounds from plants, animals, fungi, and microorganisms have proven invaluable in various biotechnological applications. Their diverse chemical properties enable their use in developing functional foods, nutraceuticals, pharmaceuticals, cosmetics, and environmental solutions. Integrating bioactive compounds in these sectors enhances the efficacy of traditional applications and opens new avenues for innovative and sustainable practices. The health-promoting properties of these compounds, such as anti-inflammatory, antioxidant, and antimicrobial activities, highlight their potential in disease prevention and health maintenance. Future research should focus on exploring new sources of bioactive compounds and elucidate their molecular mechanisms of action. Advancements in biotechnology can facilitate the extraction and optimization of these compounds, enhancing their stability and efficacy in various applications. Additionally, interdisciplinary studies combining genomics, metabolomics, and bioinformatics can provide deeper insights into the biosynthetic pathways and regulatory mechanisms of bioactive compounds. Developing sustainable and scalable bioprocesses to produce bioactive compounds will be crucial in meeting the growing demand for natural and health-promoting products. Furthermore, expanding the application of bioactive compounds in environmental biotechnology, such as bioremediation and biosorption, can contribute to environmental sustainability and pollution mitigation. Overall, the continuous exploration and innovative application of bioactive compounds holds significant promise for improving human health and environmental well-being.

References

Abd El-Aziz, N. M., Shehata, M. G., Alsulami, T., Badr, A. N., Elbakatoshy, M. R., Ali, H. S., et al. (2023). Characterization of orange peel extract and its potential protective effect against aluminum chloride-induced Alzheimer's disease. *Pharmaceuticals, 16*(1), Article 1. https://doi.org/10.3390/ph16010012.

Abedi, E., & Hashemi, S. M. B. (2020). Lactic acid production–producing microorganisms and substrates sources-state of art. *Heliyon, 6*(10), e04974. https://doi.org/10.1016/j.heliyon.2020.e04974.

Acevedo, F., Rubilar, M., Scheuermann, E., Cancino, B., Uquiche, E., Garces, M., et al. (2013). Spent coffee grounds as a renewable source of bioactive compounds. *Journal of Biobased Materials and Bioenergy, 7*, 420–428. https://doi.org/10.1166/jbmb.2013.1369.

Adrio, J. L., & Demain, A. L. (2014). Microbial enzymes: Tools for biotechnological processes. *Biomolecules, 4*(1), Article 1. https://doi.org/10.3390/biom4010117.

Alagawany, M., Elnesr, S. S., Farag, M. R., El-Sabrout, K., Alqaisi, O., Dawood, M. A. O., et al. (2022). Nutritional significance and health benefits of omega-3, -6 and -9 fatty acids in animals. *Animal*

Biotechnology, 33(7), 1678–1690. https://doi.org/10.1080/10495398.2020.1869562.

Alipal, J., Mohd Pu'ad, N. A. S., Lee, T. C., Nayan, N. H. M., Sahari, N., Basri, H., et al. (2021). A review of gelatin: Properties, sources, process, applications, and commercialisation. *Materials Today: Proceedings, 42*, 240–250. https://doi.org/10.1016/j.matpr.2020.12.922.

Almanza-Oliveros, A., Bautista-Hernández, I., Castro-López, C., Aguilar-Zárate, P., Meza-Carranco, Z., Rojas, R., et al. (2024). Grape pomace—advances in its bioactivity, health benefits, and food applications. *Foods, 13*(4), 580. https://doi.org/10.3390/foods13040580.

Alshehri, M. M., Sharifi-Rad, J., Herrera-Bravo, J., Jara, E. L., Salazar, L. A., Kregiel, D., et al. (2021). Therapeutic potential of isoflavones with an emphasis on daidzein. *Oxidative Medicine and Cellular Longevity, 2021*, 6331630. https://doi.org/10.1155/2021/6331630.

Andrade, M. A., Barbosa, C. H., Shah, M. A., Ahmad, N., Vilarinho, F., Khwaldia, K., et al. (2022). Citrus by-products: Valuable source of bioactive compounds for food applications. *Antioxidants, 12*(1), 38. https://doi.org/10.3390/antiox12010038.

Aniceto, J. P. S., Rodrigues, V. H., Portugal, I., & Silva, C. M. (2022). Valorization of tomato residues by supercritical fluid extraction. *Processes, 10*(1), Article 1. https://doi.org/10.3390/pr10010028.

Ashraf, M. F., Zubair, D., Bashir, M. N., Alagawany, M., Ahmed, S., Shah, Q. A., et al. (2024b). Nutraceutical and health-promoting potential of lactoferrin, an iron-binding protein in human and animal: Current knowledge. *Biological Trace Element Research, 202*(1), 56–72. https://doi.org/10.1007/s12011-023-03658-4.

Ashraf, H., Iahtisham-Ul-Haq, Butt, M. S., Nayik, G. A., Ramniwas, S., Damto, T., et al. (2024). Phytochemical and antioxidant profile of citrus peel extracts in relation to different extraction parameters. *International Journal of Food Properties, 27*(1), 286–299. https://doi.org/10.1080/10942912.2024.2304274.

Avila Rodríguez, M. I., Rodríguez Barroso, L. G., & Sánchez, M. L. (2018). Collagen: A review on its sources and potential cosmetic applications. *Journal of Cosmetic Dermatology, 17*(1), 20–26. https://doi.org/10.1111/jocd.12450.

Badhani, B., Sharma, N., & Kakkar, R. (2015). Gallic acid: A versatile antioxidant with promising therapeutic and industrial applications. *RSC Advances, 5*(35), 27540–27557. https://doi.org/10.1039/C5RA01911G.

Bae, J., Kim, N., Shin, Y., Kim, S.-Y., & Kim, Y.-J. (2020). Activity of catechins and their applications. *Biomedical Dermatology, 4*, 8. https://doi.org/10.1186/s41702-020-0057-8.

Bahmani, M., Shirzad, H., Mirhosseini, M., Mesripour, A., & Rafieian-Kopaei, M. (2016). A review on ethnobotanical and therapeutic uses of fenugreek (Trigonella foenum-graceum L.). *Journal of Evidence-Based Complementary & Alternative Medicine, 21*(1), 53–62. https://doi.org/10.1177/2156587215583405.

Banc, R., Rusu, M. E., Filip, L., & Popa, D.-S. (2023). The impact of ellagitannins and their metabolites through gut microbiome on the gut health and brain wellness within the gut-brain axis. *Foods, 12*(2), 270. https://doi.org/10.3390/foods12020270.

Bauman, D. E., Lock, A. L., Conboy Stephenson, R., Linehan, K., Ross, R. P., & Stanton, C. (2020). Conjugated linoleic acid: Biosynthesis and nutritional significance. In P. L. H. McSweeney, P. F. Fox, & J. A. O'Mahony (Eds.), *Advanced dairy chemistry, volume 2: Lipids* (pp. 67–106). Springer International Publishing. https://doi.org/10.1007/978-3-030-48686-0_3.

Bayer, I. S. (2021). Zein in food packaging. In *Sustainable food packaging technology* (pp. 199–224). Wiley & Sons, Ltd. https://doi.org/10.1002/9783527820078.ch7.

Benali, T., Bakrim, S., Ghchime, R., Benkhaira, N., El Omari, N., Balahbib, A., et al. (2022). Pharmacological insights into the multifaceted biological properties of quinic acid. *Biotechnology and Genetic Engineering Reviews, 0*(0), 1–30. https://doi.org/10.1080/02648725.2022.2122303.

Bharathiraja, S., Suriya, J., Krishnan, M., Manivasagan, P., & Kim, S. -K. (2017). Chapter six—production of enzymes from agricultural wastes and their potential industrial applications. In S. -K. Kim, & F. Toldrá (Eds.), *Advances in food and nutrition research* (Vol. 80, pp. 125–148). Academic Press. https://doi.org/10.1016/bs.afnr.2016.11.003.

Bhatnagar, I., & Kim, S.-K. (2010). Marine antitumor drugs: Status, shortfalls and strategies. *Marine Drugs, 8*(10), 2702–2720. https://doi.org/10.3390/md8102702.

Boo, Y. C. (2019). p-Coumaric acid as an active ingredient in cosmetics: A review focusing on its antimelanogenic effects. *Antioxidants, 8*(8), Article 8. https://doi.org/10.3390/antiox8080275.

Bouyahya, A., El Omari, N., Hakkur, M., El Hachlafi, N., Charfi, S., Balahbib, A., et al. (2021). Sources, health benefits, and biological properties of zeaxanthin. *Trends in Food Science & Technology, 118*, 519–538. https://doi.org/10.1016/j.tifs.2021.10.017.

Cech, N. B., & Oberlies, N. H. (2023). From plant to cancer drug: Lessons learned from the discovery of taxol. *Natural Product Reports, 40*(7), 1153. https://doi.org/10.1039/d3np00017f.

Chand, J., Panda, S. R., Jain, S., Murty, U. S. N., Das, A. M., Kumar, G. J., et al. (2022). Phytochemistry and polypharmacology of cleome species: A comprehensive Ethnopharmacological review of the medicinal plants. *Journal of Ethnopharmacology, 282*, 114600. https://doi.org/10.1016/j.jep.2021.114600.

Chelora, J., Zhang, J., Wan, Y., Cui, X., Zhao, J., Meng, X.-M., et al. (2019). Plant-derived single-molecule-based nanotheranostics for photoenhanced chemotherapy and ferroptotic-like cancer cell death. *ACS Applied Bio Materials, 2*(6), 2643–2649. https://doi.org/10.1021/acsabm.9b00311.

Chen, C. (2016). Sinapic acid and its derivatives as medicine in oxidative stress-induced diseases and aging. *Oxidative Medicine and Cellular Longevity, 2016*, 3571614. https://doi.org/10.1155/2016/3571614.

Chen, H., Wang, W., Yu, S., Wang, H., Tian, Z., & Zhu, S. (2022). Procyanidins and their therapeutic potential against oral diseases. *Molecules, 27*(9), 2932. https://doi.org/10.3390/molecules27092932.

Chen, Z., Zhang, R., Shi, W., Li, L., Liu, H., Liu, Z., et al. (2019). The multifunctional benefits of naturally occurring delphinidin and its glycosides. *Journal of Agricultural and Food Chemistry*. https://doi.org/10.1021/acs.jafc.9b05079.

Chioke, O. J., Ogbonna, C. N., Onwusi, C., & Ogbonna, J. C. (2018). Lipase in biodiesel production. *African Journal of Biochemistry Research, 12*(8), 73–85. https://doi.org/10.5897/AJBR2018.0999.

Chu, C., Chew, S. C., Liew, W., & Nyam, K. (2023). Review article vitamin E: A multi-functional ingredient for health enhancement and food preservation. *Journal of Food Measurement and Characterization, 17*. https://doi.org/10.1007/s11694-023-02042-z.

Cobb, R. E., Luo, Y., Freestone, T., & Zhao, H. (2013). Chapter 10—drug discovery and development via synthetic biology. In H. Zhao (Ed.), *Synthetic biology* (pp. 183–206). Academic Press. https://doi.org/10.1016/B978-0-12-394430-6.00010-8.

Cruz, A. G., Mtz-Enríquez, A. I., Díaz-Jiménez, L., Ramos-González, R., Valdés, J. A. A., Flores, M. E. C., et al. (2020). Production of fatty acid methyl esters and bioactive compounds from citrus wax. *Waste Management, 102*, 48–55. https://doi.org/10.1016/j.wasman.2019.10.021.

Cui, J., Lian, Y., Zhao, C., Du, H., Han, Y., Gao, W., et al. (2019). Dietary fibers from fruits and vegetables and their health benefits via modulation of gut microbiota. *Comprehensive Reviews in Food Science and Food Safety, 18*(5), 1514–1532. https://doi.org/10.1111/1541-4337.12489.

Damián, M. R., Cortes-Perez, N. G., Quintana, E. T., Ortiz-Moreno, A., Garfias Noguez, C., Cruceño-Casarrubias, C. E., et al. (2022). Functional foods, nutraceuticals and probiotics: A focus on human health. *Microorganisms, 10*(5), 1065. https://doi.org/10.3390/microorganisms10051065.

Dasgupta, A. (2012). Advances in antibiotic measurement. *Advances in Clinical Chemistry, 56*, 75–104. https://doi.org/10.1016/b978-0-12-394317-0.00013-3.

Dayan, F. E., Cantrell, C. L., & Duke, S. O. (2009). Natural products in crop protection. *Bioorganic & Medicinal Chemistry, 17*(12), 4022–4034. https://doi.org/10.1016/j.bmc.2009.01.046.

de Castro Bizerra, V., Sales, M. B., Fernandes Melo, R. L., Andrade do Nascimento, J. G., Junior, J. B., França Silva, M. P., et al. (2024). Opportunities for cleaner leather processing based on protease enzyme: Current evidence from an advanced bibliometric analysis. *Renewable and Sustainable Energy Reviews, 191*, 114162. https://doi.org/10.1016/j.rser.2023.114162.

Decloedt, A. I., Van Landschoot, A., Watson, H., Vanderputten, D., & Vanhaecke, L. (2017). Plant-based beverages as good sources of free and glycosidic plant sterols. *Nutrients, 10*(1), 21. https://doi.org/10.3390/nu10010021.

del Olmo, A., Calzada, J., & Nuñez, M. (2017). Benzoic acid and its derivatives as naturally occurring compounds in foods and as additives: Uses, exposure, and controversy. *Critical Reviews in Food Science and Nutrition, 57*(14), 3084–3103. https://doi.org/10.1080/10408398.2015.1087964.

Demain, A. L. (2009). Antibiotics: Natural products essential to human health. *Medicinal Research Reviews, 29*(6), 821–842. https://doi.org/10.1002/med.20154.

Demain, A. L. (2014). Importance of microbial natural products and the need to revitalize their discovery. *Journal of Industrial Microbiology and Biotechnology, 41*(2), 185–201. https://doi.org/10.1007/s10295-013-1325-z.

Demain, A. L., & Vaishnav, P. (2011). Natural products for cancer chemotherapy. *Microbial Biotechnology, 4*(6), 687–699. https://doi.org/10.1111/j.1751-7915.2010.00221.x.

Dey, D., Kundu, K., & Singh, D. (2024). *Bioenergy production from vegetable and fruit waste* (pp. 225–245). https://doi.org/10.1007/978-981-99-7552-5_11.

Dhokane, D., Shaikh, A., Yadav, A., Giri, N., Bandyopadhyay, A., Dasgupta, S., et al. (2023). CRISPR-based bioengineering in microalgae for production of industrially important biomolecules. *Frontiers in Bioengineering and Biotechnology, 11*. https://doi.org/10.3389/fbioe.2023.1267826.

Ebadi, M., Mohammadi, M., Pezeshki, A., & Jafari, S. M. (2022). health benefits of beta-carotene. In S. M. Jafari, A. Rashidinejad, & J. Simal-Gandara (Eds.), *Handbook of food bioactive ingredients: Properties and applications* (pp. 1–26). Springer International Publishing. https://doi.org/10.1007/978-3-030-81404-5_51-1.

El-Holi, M. A., & Al-Delaimy, S. (2003). Citric acid production from whey with sugars and additives by aspergillus niger. *African Journal of Biotechnology*, *2*(10), Article 10. https://doi.org/10.5897/AJB2003.000-1073.

Escobar, A., Pérez, M., Romanelli, G., & Blustein, G. (2020). Thymol bioactivity: A review focusing on practical applications. *Arabian Journal of Chemistry, 13*(12), 9243–9269. https://doi.org/10.1016/j.arabjc.2020.11.009.

Espíndola, K. M. M., Ferreira, R. G., Narvaez, L. E. M., Silva Rosario, A. C. R., da Silva, A. H. M., Silva, A. G. B., et al. (2019). Chemical and pharmacological aspects of caffeic acid and its activity in hepatocarcinoma. *Frontiers in Oncology, 9*, 541. https://doi.org/10.3389/fonc.2019.00541.

Fao, P. (2021). *Regional overview of food security and nutrition in Latin America and the Caribbean 2020*. https://policycommons.net/artifacts/1422037/regional-overview-of-food-security-and-nutrition-in-latin-america-and-the-caribbean-2020/2036109/.

Fărcaş, A. C., Socaci, S. A., Nemeş, S. A., Pop, O. L., Coldea, T. E., Fogarasi, M., et al. (2022). An update regarding the bioactive compound of cereal by-products: Health benefits and potential applications. *Nutrients, 14*(17), 3470. https://doi.org/10.3390/nu14173470.

Fernandes, M. M., Lima, C. F., Loureiro, A., Gomes, A. C., & Cavaco-Paulo, A. (2012). Keratin-based peptide: Biological evaluation and strengthening properties on relaxed hair. *International Journal of Cosmetic Science, 34*(4), 338–346. https://doi.org/10.1111/j.1468-2494.2012.00727.x.

Food Wastage in India. (2024). *From farm to bin (Hidden truth)*. Retrieved May 29, 2024, from https://timesofagriculture.in/food-wastage-in-india-farm-to-bin/.

Friedman, M. (2018). Analysis, nutrition, and health benefits of tryptophan. *International Journal of Tryptophan Research, 11*, 1178646918802282. https://doi.org/10.1177/1178646918802282.

Gangadharan, D., Sivaramakrishnan, S., Nampoothiri, K. M., Sukumaran, R. K., & Pandey, A. (2008). Response surface methodology for the optimization of alpha amylase production by *Bacillus amyloliquefaciens*. *Bioresource Technology, 99*(11), 4597–4602. https://doi.org/10.1016/j.biortech.2007.07.028.

García-Alonso, J., Ros, G., Vidal-Guevara, M. L., & Periago, M. J. (2006). Acute intake of phenolic-rich juice improves antioxidant status in healthy subjects. *Nutrition Research, 26*(7), 330–339. https://doi.org/10.1016/j.nutres.2006.06.004.

García-Estrada, C., Ullán, R. V., Albillos, S. M., Fernández-Bodega, M. Á., Durek, P., von Döhren, H., et al. (2011). A single cluster of coregulated genes encodes the biosynthesis of the mycotoxins roquefortine C and meleagrin in Penicillium chrysogenum. *Chemistry & Biology, 18*(11), 1499–1512. https://doi.org/10.1016/j.chembiol.2011.08.012.

Giddings, L.-A., & Newman, D. J. (2013). Microbial natural products: Molecular blueprints for antitumor drugs. *Journal of Industrial Microbiology & Biotechnology, 40*(11), 1181–1210. https://doi.org/10.1007/s10295-013-1331-1.

Gill, B. S., Navgeet, & Qiu, F. (2020). Chapter 1—Technologies for extraction and production of bioactive compounds. In M. L. Verma, & A. K. Chandel (Eds.), *Biotechnological Production of Bioactive Compounds* (pp. 1–36). Elsevier. https://doi.org/10.1016/B978-0-444-64323-0.00001-1.

Giurini, E. F., Godla, A., & Gupta, K. H. (2024). Redefining bioactive small molecules from microbial metabolites as revolutionary anticancer agents. *Cancer Gene Therapy, 31*(2), 187–206. https://doi.org/10.1038/s41417-023-00715-x.

Gokoglu, N. (2018). Novel natural food preservatives and applications in seafood preservation: A review. *Journal of the Science of Food and Agriculture*, *99*. https://doi.org/10.1002/jsfa.9416.

Guadagna, S., Barattini, D. F., Rosu, S., & Ferini-Strambi, L. (2020). Plant extracts for sleep disturbances: A systematic review. *Evidence-Based Complementary and Alternative Medicine: eCAM, 2020*, 3792390. https://doi.org/10.1155/2020/3792390.

Guojun, Y., Yuping, H., Yan, J., Kaichun, L., & Haiyang, X. (2016). A new medium for improving spinosad production by saccharopolyspora spinosa. *Jundishapur Journal of Microbiology, 9*(6), e16765. https://doi.org/10.5812/jjm.16765.

Gupta, J., & Gupta, R. (2020). Nutraceutical status and scientific strategies for enhancing production of omega-3 fatty acids from microalgae and their role in healthcare. *Current Pharmaceutical*

Biotechnology, 21(15), 1616–1631. https://doi.org/10.2174/1389201021666200703201014.

Gyawali, R., & Ibrahim, S. A. (2014). Natural products as antimicrobial agents. *Food Control, 46*, 412–429. https://doi.org/10.1016/j.foodcont.2014.05.047.

Hao, Y.-J., Cui, X.-H., Li, J.-R., An, X.-L., Sun, H.-D., Piao, X.-C., et al. (2020). Cell bioreactor culture of *Orostachys cartilaginous* A. Bor. and involvement of nitric oxide in methyl jasmonate-induced flavonoid synthesis. *Acta Physiologiae Plantarum, 42*(1), 9. https://doi.org/10.1007/s11738-019-3008-5.

He, H.-F. (2022). Recognition of gallotannins and the physiological activities: From chemical view. *Frontiers in Nutrition, 9*, 888892. https://doi.org/10.3389/fnut.2022.888892.

Hernandez, D. F., Cervantes, E. L., Luna-Vital, D. A., & Mojica, L. (2020). Food-derived bioactive compounds with anti-aging potential for nutricosmetic and cosmeceutical products. *Critical Reviews in Food Science and Nutrition, 61*(22), 3740–3755. https://doi.org/10.1080/10408398.2020.1805407.

Hęś, M., Dziedzic, K., Górecka, D., Jędrusek-Golińska, A., & Gujska, E. (2019). Aloe vera (L.) Webb.: Natural sources of antioxidants–a review. *Plant Foods for Human Nutrition, 74*(3), 255–265. https://doi.org/10.1007/s11130-019-00747-5.

Holeček, M. (2020). Histidine in health and disease: Metabolism, physiological importance, and use as a supplement. *Nutrients, 12*(3), Article 3. https://doi.org/10.3390/nu12030848.

Hossain, U., Das, A. K., Ghosh, S., & Sil, P. C. (2020). An overview on the role of bioactive α-glucosidase inhibitors in ameliorating diabetic complications. *Food and Chemical Toxicology, 145*, 111738. https://doi.org/10.1016/j.fct.2020.111738.

Hujanen, M., Linko, S., Linko, Y.-Y., & Leisola, M. (2001). Optimisation of media and cultivation conditions for L(+)(S)-lactic acid production by Lactobacillus casei NRRL B-441. *Applied Microbiology and Biotechnology, 56*(1), 126–130. https://doi.org/10.1007/s002530000501.

Hwang, H. G., Milito, A., Yang, J.-S., Jang, S., & Jung, G. Y. (2023). Riboswitch-guided chalcone synthase engineering and metabolic flux optimization for enhanced production of flavonoids. *Metabolic Engineering, 75*, 143–152. https://doi.org/10.1016/j.ymben.2022.12.006.

Ibrahim, N. E., & Ma, K. (2017). Industrial applications of thermostable enzymes from extremophilic microorganisms. *Current Biochemical Engineering, 4*(2), 75–98. https://doi.org/10.2174/2212711904666170405123414.

Imran, M., Saeed, F., Hussain, G., Imran, A., Mehmood, Z., Gondal, T. A., et al. (2021). Myricetin: A comprehensive review on its biological potentials. *Food Science & Nutrition, 9*(10), 5854–5868. https://doi.org/10.1002/fsn3.2513.

Ingle, A. P., Rathod, J., Pandit, R., da Silva, S. S., & Rai, M. (2017). Comparative evaluation of free and immobilized cellulase for enzymatic hydrolysis of lignocellulosic biomass for sustainable bioethanol production. *Cellulose, 24*(12), 5529–5540. https://doi.org/10.1007/s10570-017-1517-1.

Jadoon, S., Karim, S., Asad, M. H. H. B., Akram, M. R., Kalsoom Khan, A., Malik, A., et al. (2015). Anti-aging potential of phytoextract loaded-pharmaceutical creams for human skin cell longetivity. *Oxidative Medicine and Cellular Longevity, 2015*(1), 709628. https://doi.org/10.1155/2015/709628.

Janoušek, J., Pilařová, V., Macáková, K., Nomura, A., Veiga-Matos, J., da Silva, D. D., et al. (2022). Vitamin D: Sources, physiological role, biokinetics, deficiency, therapeutic use, toxicity, and overview of analytical methods for detection of vitamin D and its metabolites. *Critical Reviews in Clinical Laboratory Sciences, 59*(8), 517–554. https://doi.org/10.1080/10408363.2022.2070595.

Jiménez-Gómez, C. P., & Cecilia, J. A. (2020). Chitosan: A natural biopolymer with a wide and varied range of applications. *Molecules, 25*(17), Article 17. https://doi.org/10.3390/molecules25173981.

Jung, W. S., Lee, S. K., Hong, J. S. J., Park, S. R., Jeong, S. J., Han, A. R., et al. (2006). Heterologous expression of tylosin polyketide synthase and production of a hybrid bioactive macrolide in Streptomyces venezuelae. *Applied Microbiology and Biotechnology, 72*(4), 763–769. https://doi.org/10.1007/s00253-006-0318-5.

Kamani, M. H., Neji, C., Fitzsimons, S. M., Fenelon, M. A., & Murphy, E. G. (2023). Unlocking the nutritional and functional potential of legume waste to produce protein ingredients. *Critical Reviews in Food Science and Nutrition, 0*(0), 1–19. https://doi.org/10.1080/10408398.2023.2184322.

Katz, L., & Baltz, R. H. (2016). Natural product discovery: Past, present, and future. *Journal of Industrial Microbiology & Biotechnology, 43*(2–3), 155–176. https://doi.org/10.1007/s10295-015-1723-5.

Khan, U. M., Sevindik, M., Zarrabi, A., Nami, M., Ozdemir, B., Kaplan, D. N., et al. (2021). Lycopene: Food sources, biological activities, and human health benefits. *Oxidative Medicine and Cellular Longevity, 2021*, e2713511. https://doi.org/10.1155/2021/2713511.

Kim, D., Yoon, M., Kim, S., Um, M. Y., & Cho, S. (2022). Effects of green kiwifruit peel extract on sleep-wake profiles in mice: A polysomnographic study based on electroencephalogram and electromyogram recordings. *Nutrients, 14*(22), Article 22. https://doi.org/10.3390/nu14224732.

Klaus, von G., Jill, B., Naomi, H., Tracy, B., Nilam, P., Yisehac, Y., et al. (2017). *2017 Global hunger index: The inequalities of hunger*. Intl Food Policy Res Inst.

Kotha, R. R., & Luthria, D. L. (2019). Curcumin: Biological, pharmaceutical, nutraceutical, and analytical aspects. *Molecules, 24*(16), 2930. https://doi.org/10.3390/molecules24162930.

Krishna Leela, J., & Sharma, G. (2000). Studies on xanthan production from Xanthomonas campestris. *Bioprocess Engineering, 23*(6), 687–689. https://doi.org/10.1007/s004499900054.

Krishnamurti, C., & Rao, S. C. (2016). The isolation of morphine by Serturner. *Indian Journal of Anaesthesia, 60*(11), 861–862. https://doi.org/10.4103/0019-5049.193696.

Kumar, P., Kamle, M., Borah, R., Mahato, D. K., & Sharma, B. (2021). Bacillus thuringiensis as microbial biopesticide: Uses and application for sustainable agriculture. *Egyptian Journal of Biological Pest Control, 31*(1), 95. https://doi.org/10.1186/s41938-021-00440-3.

Kusumawati, I., & Indrayanto, G. (2013). Chapter 15—natural antioxidants in cosmetics. In Atta-ur-Rahman (Ed.), *Studies in natural products chemistry* (Vol. 40, pp. 485–505). Elsevier. https://doi.org/10.1016/B978-0-444-59603-1.00015-1.

Lavuri, R., Chiappetta Jabbour, C. J., Grebinevych, O., & Roubaud, D. (2022). Green factors stimulating the purchase intention of innovative luxury organic beauty products: Implications for sustainable development. *Journal of Environmental Management, 301*, 113899. https://doi.org/10.1016/j.jenvman.2021.113899.

Li, J. W. -H., & Vederas, J. C. (2009). Drug discovery and natural products: End of an era or an endless frontier? *Science (New York, N.Y.), 325*(5937), 161–165. https://doi.org/10.1126/science.1168243.

Libera, J., Iłowiecka, K., & Stasiak, D. (2021). Consumption of processed red meat and its impact on human health: A review. *International Journal of Food Science & Technology, 56*(12), 6115–6123. https://doi.org/10.1111/ijfs.15270.

Liu, R. H. (2013). Dietary bioactive compounds and their health implications. *Journal of Food Science, 78*(Suppl 1), A18-25. https://doi.org/10.1111/1750-3841.12101.

Liu, Q., Yang, F., Zhang, J., Liu, H., Rahman, S., Islam, S., Ma, W., & She, M. (2021). Application of CRISPR/Cas9 in crop quality improvement. *International Journal of Molecular Sciences, 22*(8), 4206. https://doi.org/10.3390/ijms22084206.

Ma, X., Zhang, X., Wang, X., Wang, C., & Ma, Y. (2023). The role of kaempferol in gynaecological malignancies: Progress and perspectives. *Frontiers in Pharmacology, 14*. https://doi.org/10.3389/fphar.2023.1310416.

Mahadik, N. D., Puntambekar, U. S., Bastawde, K. B., Khire, J. M., & Gokhale, D. V. (2002). Production of acidic lipase by *Aspergillus niger* in solid state fermentation. *Process Biochemistry, 38*(5), 715–721. https://doi.org/10.1016/S0032-9592(02)00194-2.

Majma Sanaye, P., Mojaveri, M. R., Ahmadian, R., Sabet Jahromi, M., & Bahramsoltani, R. (2022). Apigenin and its dermatological applications: A comprehensive review. *Phytochemistry, 203*, 113390. https://doi.org/10.1016/j.phytochem.2022.113390.

Makvandi, P., Caccavale, C., Della Sala, F., Zeppetelli, S., Veneziano, R., & Borzacchiello, A. (2020). Natural formulations provide antioxidant complement to hyaluronic acid-based topical applications used in wound healing. *Polymers, 12*(8), 1847. https://doi.org/10.3390/polym12081847.

Mamadalieva, N., & Mamedov, N. (2020). *Taxus brevifolia a high-value medicinal plant, as a source of taxol* (pp. 201–218). https://doi.org/10.1007/978-3-030-44930-8_9.

Mannarino, E., Pirro, M., Cortese, C., Lupattelli, G., Siepi, D., Mezzetti, A., et al. (2009). Effects of a phytosterol-enriched dairy product on lipids, sterols and 8-isoprostane in hypercholesterolemic patients: A multicenter Italian study. *Nutrition, Metabolism, and Cardiovascular Diseases: NMCD, 19*(2). https://doi.org/10.1016/j.numecd.2008.03.012.

Marcillo-Parra, V., Tupuna-Yerovi, D. S., González, Z., & Ruales, J. (2021). Encapsulation of bioactive compounds from fruit and vegetable by-products for food application—A review. *Trends in Food Science & Technology, 116*, 11–23. https://doi.org/10.1016/j.tifs.2021.07.009.

Margolin, E., Chapman, R., Williamson, A.-L., Rybicki, E. P., & Meyers, A. E. (2018). Production of complex viral glycoproteins in plants as vaccine immunogens. *Plant Biotechnology Journal, 16*(9), 1531–1545. https://doi.org/10.1111/pbi.12963.

Martens, J.-H., Barg, H., Warren, M., & Jahn, D. (2002). Microbial production of vitamin B12. *Applied Microbiology and Biotechnology, 58*(3), 275–285. https://doi.org/10.1007/s00253-001-0902-7.

Masotti, F., Cattaneo, S., Stuknytė, M., Pica, V., & De Noni, I. (2020). Analytical advances in the determination of calcium in bovine milk, dairy products and milk-based infant formulas. *Trends in Food Science & Technology, 103*, 348–360. https://doi.org/10.1016/j.tifs.2020.07.013.

Matsuda, K. (2018). Okaramines and other plant fungal products as new insecticide leads. *Current Opinion in Insect Science, 30*, 67–72. https://doi.org/10.1016/j.cois.2018.09.010.

Merecz-Sadowska, A., Sitarek, P., Kowalczyk, T., Zajdel, K., Jęcek, M., Nowak, P., et al. (2023). Food anthocyanins: Malvidin and its glycosides as promising antioxidant and anti-inflammatory agents with potential health benefits. *Nutrients, 15*(13), 3016. https://doi.org/10.3390/nu15133016.

Mitra, S., Rauf, A., Tareq, A. M., Jahan, S., Emran, T. B., Shahriar, T. G., et al. (2021). Potential health benefits of carotenoid lutein: An updated review. *Food and Chemical Toxicology, 154*, 112328. https://doi.org/10.1016/j.fct.2021.112328.

Mohamed, T., Zhao, X., Habib, L. K., Yang, J., & Rao, P. P. N. (2011). Design, synthesis and structure–activity relationship (SAR) studies of 2,4-disubstituted pyrimidine derivatives: Dual activity as cholinesterase and Aβ-aggregation inhibitors. *Bioorganic & Medicinal Chemistry, 19*(7), 2269–2281. https://doi.org/10.1016/j.bmc.2011.02.030.

Mohammadi, S., Jafari, B., Asgharian, P., Martorell, M., & Sharifi-Rad, J. (2020). Medicinal plants used in the treatment of Malaria: A key emphasis to artemisia, cinchona, cryptolepis, and tabebuia genera. *Phytotherapy Research: PTR, 34*(7), 1556–1569. https://doi.org/10.1002/ptr.6628.

Monteiro, G. C., Minatel, I. O., Junior, A. P., Gomez-Gomez, H. A., de Camargo, J. P. C., Diamante, M. S., et al. (2021). Bioactive compounds and antioxidant capacity of grape pomace flours. *LWT, 135*, 110053. https://doi.org/10.1016/j.lwt.2020.110053.

Motolinía-Alcántara, E. A., Castillo-Araiza, C. O., Rodríguez-Monroy, M., Román-Guerrero, A., & Cruz-Sosa, F. (2021). Engineering considerations to produce bioactive compounds from plant cell suspension culture in bioreactors. *Plants, 10*(12), 2762. https://doi.org/10.3390/plants10122762.

Namitha, K. K., & Negi, P. S. (2013). Enhancement of natural antioxidants in plants by biosynthetic pathway modulation. In G. Brahmachari (Ed.), *Chemistry and pharmacology of naturally occurring bioactive compounds*. CRC Press.

Nasim, N., Sandeep, I. S., & Mohanty, S. (2022). Plant-derived natural products for drug discovery: Current approaches and prospects. *The Nucleus, 65*(3), 399–411. https://doi.org/10.1007/s13237-022-00405-3.

Ninčević Grassino, A., Rimac Brnčić, S., Badanjak Sabolović, M., Šic Žlabur, J., Marović, R., & Brnčić, M. (2023). Carotenoid content and profiles of pumpkin products and by-products. *Molecules, 28*(2), 858. https://doi.org/10.3390/molecules28020858.

Nwafor, I. C., Shale, K., & Achilonu, M. C. (2017). Chemical composition and nutritive benefits of chicory (Cichorium intybus) as an ideal complementary and/or alternative livestock feed supplement. *The Scientific World Journal, 2017*, 7343928. https://doi.org/10.1155/2017/7343928.

Ozyigit, I. I., Dogan, I., Hocaoglu-Ozyigit, A., Yalcin, B., Erdogan, A., Yalcin, I. E., et al. (2023). Production of secondary metabolites using tissue culture-based biotechnological applications. *Frontiers in Plant Science, 14*, 1132555. https://doi.org/10.3389/fpls.2023.1132555.

Pal, A. K., Singh, J., Soni, R., Tripathi, P., Kamle, M., Tripathi, V., et al. (2020). 10-The role of microorganism in bioremediation for sustainable environment management. In V. C. Pandey & V. Singh (Eds.), *Bioremediation of pollutants* (pp. 227–249). Elsevier. https://doi.org/10.1016/B978-0-12-819025-8.00010-7.

Pandey, A. K., Kumar, P., Singh, P., Tripathi, N. N., & Bajpai, V. K. (2017). Essential oils: Sources of antimicrobials and food preservatives. *Frontiers in Microbiology, 7*. https://doi.org/10.3389/fmicb.2016.02161.

Papadopol, V., & Nechifor, M. (2011). Magnesium in neuroses and neuroticism. In R. Vink & M. Nechifor (Eds.), *Magnesium in the central nervous system*. University of Adelaide Press. http://www.ncbi.nlm.nih.gov/books/NBK507254/.

Papakyriakopoulou, P., Velidakis, N., Khattab, E., Valsami, G., Korakianitis, I., & Kadoglou, N. P. (2022). Potential pharmaceutical applications of quercetin in cardiovascular diseases. *Pharmaceuticals, 15*(8), 1019. https://doi.org/10.3390/ph15081019.

Park, S. R., Park, J. W., Ban, Y. H., Sohng, J. K., & Yoon, Y. J. (2013). 2-Deoxystreptamine-containing aminoglycoside antibiotics: Recent advances in the characterization and manipulation of their biosynthetic pathways. *Natural Product Reports, 30*(1), 11–20. https://doi.org/10.1039/c2np20092a.

Park, J. B. (2014). Chapter 29-Flaxseed secoisolariciresinol diglucoside and visceral obesity: A closer look at its chemical properties, absorption, metabolism, bioavailability, and effects on visceral fat, lipid profile, systemic inflammation, and hypertension. In R. R. Watson (Ed.), *Nutrition in the prevention and treatment of abdominal obesity* (pp. 317–327). Academic Press. https://doi.org/10.1016/B978-0-12-407869-7.00029-5.

Pasierski, M., & Szulczyk, B. (2022). Beneficial effects of capsaicin in disorders of the central nervous system. *Molecules (Basel, Switzerland), 27*(8), 2484. https://doi.org/10.3390/molecules27082484.

Pengpid, S., & Peltzer, K. (2020). Fruit and vegetable consumption is protective from short sleep and poor sleep quality among university students from 28 countries. *Nature and Science of Sleep, 12*, 627–633. https://doi.org/10.2147/NSS.S263922.

Periferakis, A., Periferakis, K., Badarau, I. A., Petran, E. M., Popa, D. C., Caruntu, A., et al. (2022). Kaempferol: Antimicrobial properties, sources, clinical, and traditional applications. *International Journal of Molecular Sciences, 23*(23), 15054. https://doi.org/10.3390/ijms232315054.

Pham, J. V., Yilma, M. A., Feliz, A., Majid, M. T., Maffetone, N., Walker, J. R., et al. (2019). A review of the microbial production of bioactive natural products and biologics. *Frontiers in Microbiology, 10*. https://doi.org/10.3389/fmicb.2019.01404.

Pitt, J. I. (2000). Toxigenic fungi and mycotoxins. *British Medical Bulletin, 56*(1), 184–192. https://doi.org/10.1258/0007142001902888.

Prakash, V., Bose, C., Sunil kumar, D., Cherian, R. M., Thomas, S. S., & Nair, B. G. (2024). Resveratrol as a promising nutraceutical: implications in gut microbiota modulation, inflammatory disorders, and colorectal cancer. *International Journal of Molecular Sciences, 25*(6), Article 6. https://doi.org/10.3390/ijms25063370.

Puri, S. C., Nazir, A., Chawla, R., Arora, R., Riyaz-Ul-Hasan, S., Amna, T., et al. (2006). The endophytic fungus Trametes hirsuta as a novel alternative source of podophyllotoxin and related aryl tetralin lignans. *Journal of Biotechnology, 122*(4), 494–510. https://doi.org/10.1016/j.jbiotec.2005.10.015.

Rahman, M. M., Islam, F., Parvez, A., Azad, M. A. K., Ashraf, G. M., Ullah, M. F., et al. (2022). Citrus limon L. (lemon) seed extract shows neuro-modulatory activity in an in vivo thiopental-sodium sleep model by reducing the sleep onset and enhancing the sleep duration. *Journal of Integrative Neuroscience, 21*(1), Article 1. https://doi.org/10.31083/j.jin2101042.

Rakoczy, K., Kaczor, J., Sołtyk, A., Szymańska, N., Stecko, J., Sleziak, J., et al. (2023). Application of luteolin in neoplasms and nonneoplastic diseases. *International Journal of Molecular Sciences, 24*(21), 15995. https://doi.org/10.3390/ijms242115995.

Ramírez-Salazar, S. A., Herren, C., McCartney, J., & Ortiz García, J. G. (2021). Dietary insights in neurological diseases. *Current Neurology and Neuroscience Reports, 21*(10), 55. https://doi.org/10.1007/s11910-021-01143-w.

Rana, A., Samtiya, M., Dhewa, T., Mishra, V., & Aluko, R. E. (2022). Health benefits of polyphenols: A concise review. *Journal of Food Biochemistry, 46*(10), e14264. https://doi.org/10.1111/jfbc.14264.

Rashmi, H. B., & Negi, P. S. (2020). Health benefits of bioactive compounds from vegetables. In M. Swamy (Ed.), *Plant-derived bioactives* (pp. 115–166). Springer. https://doi.org/10.1007/978-981-15-1761-7_5.

Ratz-Łyko, A., & Arct, J. (2019). Resveratrol as an active ingredient for cosmetic and dermatological applications: A review. *Journal of Cosmetic and Laser Therapy: Official Publication of the European Society for Laser Dermatology, 21*(2), 84–90. https://doi.org/10.1080/14764172.2018.1469767.

Reguengo, L. M., Salgaço, M. K., Sivieri, K., & Maróstica Júnior, M. R. (2022). Agro-industrial by-products: Valuable sources of bioactive compounds. *Food Research International, 152*, 110871. https://doi.org/10.1016/j.foodres.2021.110871.

Rojas-García, A., Fernández-Ochoa, Á., de la Cádiz-Gurrea, M. L., Arráez-Román, D., & Segura-Carretero, A. (2023). Neuroprotective effects of agri-food by-products rich in phenolic compounds. *Nutrients, 15*(2), Article 2. https://doi.org/10.3390/nu15020449.

Ruangrit, K., Chaipoot, S., Phongphisutthinant, R., Duangjan, K., Phinyo, K., Jeerapan, I., et al. (2023). A successful biorefinery approach of macroalgal biomass as a promising sustainable source to produce bioactive nutraceutical and biodiesel. *Biomass Conversion and Biorefinery, 13*(2), 1089–1099. https://doi.org/10.1007/s13399-021-01310-6.

Sakula, A. (1988). Selman Waksman (1888–1973), discoverer of streptomycin: A centenary review. *British Journal of Diseases of the Chest, 82*(1), 23–31. https://doi.org/10.1016/0007-0971(88)90005-8.

Salehi, B., Fokou, P. V. T., Sharifi-Rad, M., Zucca, P., Pezzani, R., Martins, N., et al. (2019a). The therapeutic potential of naringenin: A review of clinical trials. *Pharmaceuticals, 12*(1), 11. https://doi.org/10.3390/ph12010011.

Salehi, B., Venditti, A., Sharifi-Rad, M., Kręgiel, D., Sharifi-Rad, J., Durazzo, A., et al. (2019b). The therapeutic potential of apigenin. *International Journal of Molecular Sciences, 20*(6), 1305. https://doi.org/10.3390/ijms20061305.

Sarangi, P. K., Mishra, S., Mohanty, P., Singh, P. K., Srivastava, R. K., Pattnaik, R., et al. (2023). Food and fruit waste valorisation for pectin recovery: Recent process technologies and future prospects. *International Journal of Biological Macromolecules, 235*, 123929. https://doi.org/10.1016/j.ijbiomac.2023.123929.

Schrickx, J. M., Raedts, M. J., Stouthamer, A. H., & van Verseveld, H. W. (1995). Organic acid production by Aspergillus niger in recycling culture analyzed by capillary electrophoresis. *Analytical Biochemistry, 231*(1), 175–181. https://doi.org/10.1006/abio.1995.1518.

Sharifi-Rad, J., Quispe, C., Imran, M., Rauf, A., Nadeem, M., Gondal, T. A., et al. (2021). Genistein: An integrative overview of its mode of action, pharmacological properties, and health benefits. *Oxidative Medicine and Cellular Longevity, 2021*, 3268136. https://doi.org/10.1155/2021/3268136.

Sharma, R., Garg, P., Kumar, P., Bhatia, S. K., & Kulshrestha, S. (2020). Microbial fermentation and its role in quality improvement of fermented foods. *Fermentation, 6*(4), 106. https://doi.org/10.3390/fermentation6040106.

Shen, X. L., Zhao, T., Zhou, Y., Shi, X., Zou, Y., & Zhao, G. (2016). Effect of oat β-glucan intake on glycaemic control and insulin sensitivity of diabetic patients: A meta-analysis of randomized controlled trials. *Nutrients, 8*(1), 39. https://doi.org/10.3390/nu8010039.

Sheraz, M. A., Ahmed, S., Ahmad, I., Shaikh, R. H., Vaid, F. H. M., & Iqbal, K. (2011, July 1). *Formulation and stability of ascorbic acid in topical preparations*. Surgical Neurology International, EBSCOhost. https://doi.org/10.4103/0975-8453.86296.

Shozib, H. B., Islam, M. M., Mahmud, S. A. S., Bari, M. N., Akter, N., Jahan, S., et al. (2021). Application of cyanidin-3-glucosides as a functional food ingredient in rice-based bakery products. *Saudi Journal of Biological Sciences, 28*(12), 7472–7480. https://doi.org/10.1016/j.sjbs.2021.08.042.

Singh, B. N., Shankar, S., & Srivastava, R. K. (2011). Green tea catechin, epigallocatechin-3-gallate (EGCG): Mechanisms, perspectives and clinical applications. *Biochemical Pharmacology, 82*(12), 1807–1821. https://doi.org/10.1016/j.bcp.2011.07.093.

Singh, R., Kumar, M., Mittal, A., & Mehta, P. K. (2017). Microbial metabolites in nutrition, healthcare and agriculture. *3 Biotech, 7*(1), 15. https://doi.org/10.1007/s13205-016-0586-4.

Sobczyńska-Malefora, A., Delvin, E., McCaddon, A., Ahmadi, K. R., & Harrington, D. J. (2021). Vitamin B12 status in health and disease: A critical review. Diagnosis of deficiency and insufficiency–clinical and laboratory pitfalls. *Critical Reviews in Clinical Laboratory Sciences, 58*(6), 399–429. https://doi.org/10.1080/10408363.2021.1885339.

Soeliono, I., Puspitasari, I., Gani, A. P., & Hadisoewignyo, L. (2023, May 1). *Antioxidant activity and tryptophan content of banana peel*. EnvironmentAsia, EBSCOhost. https://doi.org/10.14456/ea.2023.24.

Song, J., He, Y., Luo, C., Feng, B., Ran, F., Xu, H., et al. (2020). New progress in the pharmacology of protocatechuic acid: A compound ingested in daily foods and herbs frequently and heavily. *Pharmacological Research, 161*, 105109. https://doi.org/10.1016/j.phrs.2020.105109.

Srinivasan, M., Sudheer, A. R., & Menon, V. P. (2007). Ferulic acid: Therapeutic potential through its antioxidant property. *Journal of Clinical Biochemistry and Nutrition, 40*(2), 92–100. https://doi.org/10.3164/jcbn.40.92.

Srinivasulu, C., Ramgopal, M., Ramanjaneyulu, G., Anuradha, C. M., & Suresh Kumar, C. (2018). Syringic acid (SA)-a review of its occurrence, biosynthesis, pharmacological and industrial importance. *Biomedicine & Pharmacotherapy, 108*, 547–557. https://doi.org/10.1016/j.biopha.2018.09.069.

Stachowiak, B., & Szulc, P. (2021). Astaxanthin for the food industry. *Molecules, 26*(9), Article 9. https://doi.org/10.3390/molecules26092666.

Stanisic, D., Costa, A. F., Cruz, G., Durán, N., & Tasic, L. (2018). Chapter 6-applications of flavonoids, with an emphasis on hesperidin, as anticancer prodrugs: Phytotherapy as an alternative to chemotherapy. In Atta-ur-Rahman (Ed.), *Studies in natural products chemistry* (Vol. 58, pp. 161–212). Elsevier. https://doi.org/10.1016/B978-0-444-64056-7.00006-4.

Strobel, G., & Daisy, B. (2003). Bioprospecting for microbial endophytes and their natural products. *Microbiology and Molecular Biology Reviews, 67*(4), 491–502. https://doi.org/10.1128/mmbr.67.4.491-502.2003.

Sytar, O., & Smetanska, I. (2022). Special issue "Bioactive compounds from natural sources (2020, 2021)." *Molecules*, *27*(6), 1929. https://doi.org/10.3390/molecules27061929.

Tan, S. Y., & Tatsumura, Y. (2015). Alexander Fleming (1881–1955): Discoverer of penicillin. *Singapore Medical Journal*, *56*(7), 366–367. https://doi.org/10.11622/smedj.2015105.

Tao, Y., Sun, P., Cai, R., Li, Y., & Jia, H. (2022). Exploring the strategy of ffusing sucrose synthase to glycosyltransferase UGT76G1 in enzymatic biotransformation. *Applied Sciences, 12*(8), 3911. https://doi.org/10.3390/app12083911.

Teixeira, J., Gaspar, A., Garrido, E. M., Garrido, J., & Borges, F. (2013). Hydroxycinnamic acid antioxidants: An electrochemical overview. *BioMed Research International, 2013*, e251754. https://doi.org/10.1155/2013/251754.

Tevyashova, A. N., Olsufyeva, E. N., Solovieva, S. E., Printsevskaya, S. S., Reznikova, M. I., Trenin, A. S., et al. (2013). Structure-antifungal activity relationships of polyene antibiotics of the amphotericin B group. *Antimicrobial Agents and Chemotherapy, 57*(8), 3815–3822. https://doi.org/10.1128/AAC.00270-13.

Tian, B., & Liu, J. (2020). Resveratrol: A review of plant sources, synthesis, stability, modification and food application. *Journal of the Science of Food and Agriculture, 100*(4), 1392–1404. https://doi.org/10.1002/jsfa.10152.

Türközü, D., & Şanlier, N. (2017). L-Theanine, unique amino acid of tea, and its metabolism, health effects, and safety. *Critical Reviews in Food Science and Nutrition, 57*(8), 1681–1687. https://doi.org/10.1080/10408398.2015.1016141.

Twaij, B. M., & Hasan, M. N. (2022). Bioactive secondary metabolites from plant sources: Types, synthesis, and their therapeutic uses. *International Journal of Plant Biology*, *13*(1), Article 1. https://doi.org/10.3390/ijpb13010003.

van der Merwe, K. J., Steyn, P. S., Fourie, L., Scott, D. B., & Theron, J. J. (1965). Ochratoxin A, a toxic metabolite produced by Aspergillus ochraceus Wilh. *Nature, 205*(976), 1112–1113. https://doi.org/10.1038/2051112a0.

Vetvicka, V., Vannucci, L., & Sima, P. (2020). β-Glucan as a new tool in vaccine development. *Scandinavian Journal of Immunology, 91*(2), e12833. https://doi.org/10.1111/sji.12833.

Viña, J., Escudero, J., Baquero, M., Cebrián, M., Carbonell-Asíns, J. A., Muñoz, J. E., et al. (2022). Genistein effect on cognition in prodromal Alzheimer's disease patients. The GENIAL clinical trial. *Alzheimer's Research & Therapy, 14*, 164. https://doi.org/10.1186/s13195-022-01097-2.

Vu, V., Farkas, C., Riyad, O., Bujna, E., Kilin, A., Sipiczki, G., et al. (2022). Enhancement of the enzymatic hydrolysis efficiency of wheat bran using the *Bacillus* strains and their consortium. *Bioresource Technology, 343*, 126092. https://doi.org/10.1016/j.biortech.2021.126092.

Waksman, S. A., Reilly, H. C., & Johnstone, D. B. (1946). Isolation of streptomycin-producing strains of Streptomyces griseus. *Journal of Bacteriology, 52*, 393–397. https://doi.org/10.1128/JB.52.3.393-397.1946.

Wang, P. (2021). Natural and synthetic saponins as vaccine adjuvants. *Vaccines, 9*(3), 222. https://doi.org/10.3390/vaccines9030222.

Wangchuk, P. (2018). Therapeutic applications of natural products in herbal medicines, biodiscovery programs, and biomedicine. *Journal of Biologically Active Products from Nature, 8*(1), 1–20. https://doi.org/10.1080/22311866.2018.1426495.

Wani, A. K., Rahayu, F., Fauziah, L., & Suhara, C. (2023). Advances in safe processing of sugarcane and bagasse for the generation of biofuels and bioactive compounds. *Journal of Agriculture and Food Research, 12*, 100549. https://doi.org/10.1016/j.jafr.2023.100549.

Wendisch, V. F., Jorge, J. M. P., Pérez-García, F., & Sgobba, E. (2016). Updates on industrial production of amino acids using Corynebacterium glutamicum. *World Journal of Microbiology and Biotechnology, 32*(6), 105. https://doi.org/10.1007/s11274-016-2060-1.

Wright, G. D. (2014). Something old, something new: Revisiting natural products in antibiotic drug discovery. *Canadian Journal of Microbiology, 60*(3), 147–154. https://doi.org/10.1139/cjm-2014-0063.

Xiao, T., Cheng, X., Huang, J., Guo, Z., Tao, L., & Shen, X. (2023). Bioactive substances inhibiting COX-2 and cancer cells isolated from the fibrous roots of Alangium Chinense (Lour.) harms. *RSC Advances, 13*(5), 3346–3363. https://doi.org/10.1039/D2RA06931H.

Yang, D., Wang, T., Long, M., & Li, P. (2020). Quercetin: Its main pharmacological activity and potential application in clinical medicine. *Oxidative Medicine and Cellular Longevity, 2020*, 8825387. https://doi.org/10.1155/2020/8825387.

Yin, S.-Y., Efferth, T., Jian, F.-Y., Chen, Y.-H., Liu, C.-I., Wang, A. H. J., et al. (2016). Immunogenicity of mammary tumor cells can be induced by shikonin via direct binding-interference with hnRNPA1. *Oncotarget, 7*(28), 43629–43653. https://doi.org/10.18632/oncotarget.9660.

Yoon, J. H., Kim, M. -Y., & Cho, J. Y. (2023). Apigenin: A therapeutic agent for treatment of skin inflammatory diseases and cancer. *International Journal of Molecular Sciences*, *24*(2), Article 2. https://doi.org/10.3390/ijms24021498.

Yu, Y., Zhang, Z., & Chang, C. (2022). Chlorogenic acid intake guidance: Sources, health benefits, and safety. *Asia Pacific Journal of Clinical Nutrition, 31*(4), 602–610. https://doi.org/10.6133/apjcn.202212_31(4).0003.

Zhang, X., Chen, F., & Wang, M. (2015). Bioactive substances of animal origin. In P. C. K. Cheung & B. M. Mehta (Eds.), *Handbook of food chemistry* (pp. 1009–1033). Springer. https://doi.org/10.1007/978-3-642-36605-5_14.

Zhao, J., Wang, G., Chu, J., & Zhuang, Y. (2019). Harnessing microbial metabolomics for industrial applications. *World Journal of Microbiology and Biotechnology, 36*(1), 1. https://doi.org/10.1007/s11274-019-2775-x.

Zhou, Z., Wang, X., Zhang, H., Sun, J., Zheng, L., Liu, H., et al. (2015). Chromopeptide A, a highly cytotoxic depsipeptide from the marine sediment-derived bacterium *Chromobacterium* sp. HS-13-94. *Acta Pharmaceutica Sinica B, 5*(1), 62–66. https://doi.org/10.1016/j.apsb.2014.11.001.

Zolghadri, S., Bahrami, A., Hassan Khan, M. T., Munoz-Munoz, J., Garcia-Molina, F., Garcia-Canovas, F., et al. (2019). A comprehensive review on tyrosinase inhibitors. *Journal of Enzyme Inhibition and Medicinal Chemistry, 34*(1), 279–309. https://doi.org/10.1080/14756366.2018.1545767.

Biosynthesis of the Immunomodulatory Bioactive Molecule

S. Tejashwini, C. Dhaarini, and M. Gayathri

Abstract

Immunomodulators are the molecules, drugs that have an ability to trigger the immune system either by stimulating or suppressing. It encompasses numerous stages, enzymes and regulatory mechanisms. The actual biosynthesis of immunomodulatory bioactive molecule involves combined biochemical pathway through which organisms produce compounds, that has an ability to regulate immune responses. This chapter delves into the process by which the molecules are synthesized, accentuate their biological origin and their enzymatic mechanisms that is responsible for their production. Understanding the biosynthesis pathway is crucial for illuminating the molecular structure and also the functions of immunomodulatory bioactive molecules which includes cytokinesis, chemokines, growth factors and antibodies. Understanding the biosynthesis of bioactive molecules is key step for application of biotechnology in diverse forms, which includes the optimum utilization of natural products and development of novel therapeutics which may lead to uncover the new possibilities that enhances the ability to manipulate immune responses for therapeutic responses.

Keywords

Immune system • Immunomodulators • Origin • Biosynthesis • Pathways • Regulatory mechanisms

S. Tejashwini · C. Dhaarini · M. Gayathri (✉)
School of Biosciences and Technology, Vellore Institute of Technology, Vellore Campus, Vellore, Tamil Nadu 632014, India
e-mail: gayathrigopinath@vit.ac.in

1 Introduction

Immune system plays a crucial walk-on part in defending against pathogenic septicity and ensuring internal balance. A well-coordinated immune response to threats is essential for maintaining overall health and equilibrium (Garofalo et al., 2020). However, an exaggerated and uncontrolled immune response can become a self-perpetuating cycle with serious consequences, closely linked to the onset of various diseases (Brás et al., 2020). Some diseases stem from chronic inflammation, while others arise from an unusual immune response targeting specific cells, cause to autoimmune conditions. The progression of inflammation exacerbates disease and worsens patient outcomes. There is a justified need to suppress immune responses externally, supported by mounting evidence indicating that chronic inflammation, characterized by elevated ranks of proincitive cytokines, contributes to the pathological process of numerous diseases. These include asthma, rheumatism, shock liver, cardiovascular disease, and common central nervous system disorders like Alzheimer's disease, epilepsy, depression, and altered perception syndrome. Immunomodulation represents a robust field that continues to advance in pharmacokinetics and pharmacodynamics. However, significant challenges persist, However, toxicity remains a concern with the latest interventions (Szebeni et al., 2018). Nevertheless, immunomodulation has demonstrated successful application in clinical settings (Brás et al., 2020; Heneka et al., 2015), particularly in mitigating the side effects associated with pharmacological treatments (Román-Albasini et al., 2020). Immuno stimulators and immunosuppressants are two categories of immunomodulators widely used in medical practice (Jantan et al., 2019). Immunopharmacology, a relatively modern branch of pharmacology, focuses specifically on these immunomodulators. Administering immune stimulators, such as in cases of AIDS, and employing immunosuppressants to manage overactive immune responses are crucial for restoring

J. Devi (ed.), *Biotechnological Intervention in Production of Bioactive Compounds*,
Sustainable Landscape Planning and Natural Resources Management, https://doi.org/10.1007/978-3-031-76859-0_6

normal immune function and prolonging life expectancy. Immunomodulators, when coupled with antigens, are intended to bolster the immune system, which plays a crucial role in distinguishing between self and non-self. Immunization can occur actively, where an antigen stimulates the body to develop its own immune defenses against potential threats, or passively, involving the administration of pre-formed antibodies to individuals already exposed to or at risk of exposure to an antigen. Numerous plant-based immunostimulants have demonstrated activity in enhancing specific or nonspecific immunity (Mukherjee, 2019). Traditionally used for immune rejuvenation and treatment of chronic diseases in many countries, these plants have been studied extensively (Ramanadham & Nageshwari, 2010). For instance, studies have shown that certain plant extracts can increase haemagglutinating antibody (HA) titers and the formation of plaque-forming cells in mice, indicating their potential role in immune stimulation. These findings underscore the significant contribution of plant-derived compounds to rejuvenation therapies and the treatment of chronic ailments in traditional Indian medicine.

2 Classification of Immunomodulatory Molecules

Immunomodulators are categorized into three subdivisions:

(a) Immuno-adjuvants enhance the effectiveness of vaccines and can function specifically as immunostimulants. They play a crucial role in modulating rapid immune responses by selectively influencing immune-protective and immune-destructive T1 (Th1) and T2 (Th2) cells.
(b) With reference to the immunological properties, immuno-adjuvants improve the strength of vaccines and immunization, and can also be referred to as immunostimulants. They directly are involved in the control of quickly produced first-line immune reactions that mediate immune protection or immune destruction by either activating Th1 and T2 cells Immunostimulants are invariably non-selective as far as immune stimulation is a concern. These compounds stimulate both the host's natural immunity and acquired immunity and have different functions as preventives and stimulants, such as immunostimulants that stimulate the immune response as a means of treatment.
(c) Immunosuppressants are drugs that are formally classified into a wide miscellaneous category which is used firmly in combination therapies with autoimmune diseases and allergies). Inhibition of immune reactions and an increased susceptibility to infections can be the consequences of these drugs' therapeutic effects.

2.1 Immuno-Adjuvants

An adjuvant is defined as any substance which supplements the action of an immunogen in producing and restricting the character of the immune response (Box 1). The process of the creation of new vaccine adjuvants has traditionally ranked among some of the sluggish ones that are observed in medicine. Alum, insoluble aluminum salts were the only adjuvants employed in the licensed vaccines which include hepatitis B, diphtheria, tetanus, pertussis, and human papillomavirus vaccines since the licensure of vaccines in the 1920 and 1930s. Nevertheless, in the late 1990s the oil-in-water emulsion adjuvant MF59 was incorporated in to a licensed product in Europe namely Fluad, a trivalent inactivated whole virus vaccine for seasonal influenza for persons aged more than 65 years. In the following 20 years, four more adjuvants were added to other vaccines: For Shingrix and Mosquirix vaccine adjuvants, these include AS01, while adjuvants used in Fendrix and Cervarix vaccines are AS04, for Pandemrix and Arepanrix vaccines, the adjuvant is AS03 and CpG 1018 for Heplisav-B. During this period, many other adjuvants have demonstrated high potency in the preclinical models; nonetheless, most of those have not been approved for use in humans because of safety or tolerability issues. In addition, current adjuvants such as alum, MF59, and the ASO adjuvants are widely applied, but little is known regarding their molecular mechanism in human beings.

2.2 Immunostimulants

Immunostimulants, or immune stimulators, are agents that trigger the immune systems of animals and humans. Aims of preventing illnesses and bolster the body's innate defences against a range of pathogenic infections. These biologically active substances are either sourced naturally or synthesized with a diverse range of chemical aspects and modes of action. Immunostimulants generally stimulate the production of cytokines, crucial for fighting off infectious diseases. They fall into two primary categories: specific immunostimulants, such as antigens used in vaccines to provoke immune responses, and vague immunostimulants, which lack immunological properties but amplify immune responses to other antigens, like adjuvants and non-specific immune stimulators (Labh & Shakya, 2014). Immunostimulants activate various components of the immune system in both humans and animals. They boost vague immunotherapy and immunological prevention by stimulating critical immune system factors suchas cellular ingestion, the complement factor p, and complement systems, mucosal igA antibodies α-, γ-interferons release, T- and B-leukocytes, the synthesis of specific antibodies andinterleukin family, and the production of alveolar surfactant (Li et al., 2019). Many reasons for using immunostimulants in the management of infectious diseases: addressing bacterial antibiotic resistance, mitigating anaphylaxis to antibiotics, counteracting the immunomodulator effe ts of antibiotics; and overcoming the limited efficacy of antibiotics against viral infections.

2.3 Immunosuppressants

The drugs are made to reduce or suppress the activity of the immune system. Initially therapeutic approach to prevent transplant rejection, these drugs are now also used to treat cancer and self-reactive disorder diseases like rheumatoid arthritis. In autoimmune arthritis, the immune system mistakenly attacks healthy tissue instead of pathogens, leading to severe inflammation, particularly in the joints. Immunosuppressants help manage this condition by inhibiting parts of the immune system, thereby reducing prolonged inflammation and pain and slowing the destruction of joint tissue (Tong et al., 2015). However, this suppression increases the risk of both common and serious infections and may lead to a attenuated immune response to vaccinations, including covid 19.

2.3.1 Broad-Spectrum Immuno-Suppressants

Many commonly prescribed drugs for autoimmune arthritis are broad-spectrum immunosuppressants. These include conventional immunomodulatory drugs for rheumatic disease like methotrexate and leflunomide (Arava). Methotrexate works by inhibiting the production of certain inflammatory proteins, or "cytokines," while leflunomide slows the large growth of immune cells. Though their mechanisms differ, many traditional antirheumatic drugshave a broad impact on the immune system. Other arthritis treatments, such as biologic disease modifying antirheumatic drugs (biologics) and tyrosine kinase 2 inhibitors, selective components of the immune system.

3 Biosynthesis

Living organisms function by converting simple substrates into more complex forms required for cellular and metabolic activities important for life. This process is a part of anabolic metabolism, as it leads to the formation of complex molecules such as vitamins. Biosynthetic pathways generate nearly all the organic compounds required by microorganisms (Nabavi et al., 2020). These pathways utilize chemical energy and enzymes that act as catalysts to produce macromolecules. Key building blocks for biosynthesis include amino acids, purines, pyrimidine's, lipid's, sugar's, and enzyme's cofactors (Pietta et al., 2003). Various mechanisms regulate these pathways to check that cells produce the required amounts of products. Anabolism, or biosynthetic metabolism, involves constructing macromolecules from basic building blocks through multi-step, enzyme-mediated processes.

3.1 Flavonoids

The biosynthesis of flavonoids, essentially is more like a interconnected set of biochemical pathways, where a primary metabolite undergoes pathway that yields a product and followed by the other, this at the result brings in the final target molecule (Tariq et al., 2023). The structure of flavonoid is presented in Fig. 1.

The overall biosynthesis of flavonoid can be grouped as following.

* Shikimic Acid Pathway
* Phenylpropanoid Pathway
* Chalcone and Flavonoid Formation.

3.1.1 Formation of Phenylalanine

Shikimic Acid Pathway:

This pathway is important as it forms the primary proiduct wich forms phenyalanine. The end product of shikimic pathway is chorismate.

- Chorismite then undergoes Chorismate → Prephenate → Phenylpyruvate → Phenylalanine

This pathway is crucial for the synthesis of phenylalanine, which, as explained earlier, serves as a precursor for the phenylpropanoid pathway, leading to the production of important secondary metabolites like flavonoids and lignins (Rehan, 2021).

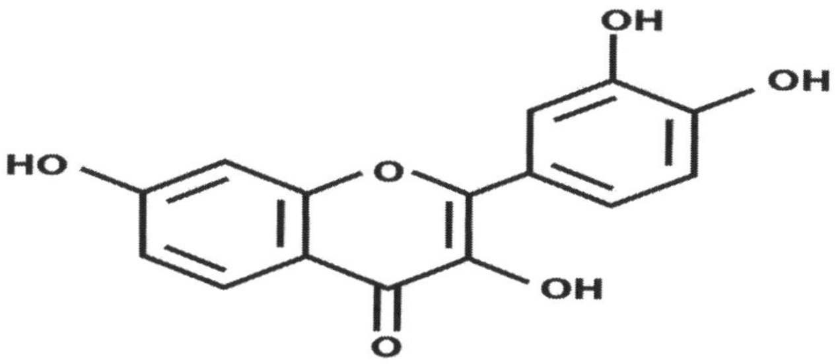

Fig. 1 Structure flavonoid

Conversion of Phenylalanine to Trans-Cinnamic Acid:

- The enzyme Phenylalanine Ammonia Lyase (PAL) catalyzes the first step, where phenylalanine is deaminated to form trans-cinnamic acid.
- This reaction involves the removal of an ammonia group (−NH3) from phenylalanine.

Hydroxylation of Trans-Cinnamic Acid to p-Coumaric Acid:

- The enzyme Cinnamate-4-hydroxylase (C4H) adds a hydroxyl group (−OH) to the 4th position of the aromatic ring of trans-cinnamic acid, converting it to p-coumaric acid.

Activation of p-Coumaric Acid to p-Coumaroyl-CoA:

- The enzyme 4-Coumarate-CoA Ligase activates p-coumaric acid by attaching a CoA (Coenzyme A) group, forming p-coumaroyl-CoA.
- p-Coumaroyl-CoA is a central intermediate in flavonoid biosynthesis (Dias et al., 2021).

Synthesis of Chalcone:

- The enzyme Chalcone Synthase (CHS) catalyzes the condensation of p-coumaroyl-CoA with three molecules of malonyl-CoA to form tetrahydroxychalcone, a precursor to flavonoids (Okoye et al., 2023).
- Chalcones are the first committed step towards flavonoid production and serve as the backbone for several flavonoid subclasses.

Isomerization of Chalcone to Flavanone:

- The enzyme Chalcone-Flavanone Isomerase (CHI) converts chalcone into flavanone, a core flavonoid structure.
- Flavanones are the starting compounds for the formation of various flavonoid subclasses, including flavonols, flavones, anthocyanins, and isoflavones.

Formation of Flavonoid Subclasses:

- Flavonoid-specific enzymes (such as flavonol synthase, flavone synthase, and anthocyanidin synthase) modify flavanones through hydroxylation, methylation, or glycosylation to produce different subclasses of flavonoid (Chávez-González et al., 2020).

3.2 Interleukins

Interleukin production is a naturally constrained process. The messenger RNA responsible for encoding most interleukins is short-lived, leading to brief periods of synthesis. Once synthesized, these molecules are quickly released into circulation. Cellular reactions to interleukins involve both upregulation and downregulation mechanisms, often involving the activation of genes encoding inhibitors that target cytokine receptors. Interleukins (ILs) are produced through a intricate process within immune cells. Here's an overview of how interleukins are synthesized:

Gene Activation: Interleukin synthesis initiates with the activation of specific genes within immune cells like phagocytes (macrophages and dentric cells). Activation is often triggered by stimuli such as antigens, cytokines, or pathogen-associated molecular patterns (PAMPs) (Janeway et al., 2001).

Transcription: Activated genes encoding interleukins undergo transcription, a process where DNA is transcribed into messenger RNA (mRNA) by RNA polymerase enzymes within the cell nucleus.

mRNA Processing: The initial mRNA transcript undergoes processing, including capping, splicing (if necessary), and polyadenylation. These modifications stabilize the mRNA and ready it for translation.

Translation: Mature mRNA exits the nucleus and enters the cytoplasm, binding to ribosomes. Ribosomes then translate the mRNA sequence into a precursor protein form of the interleukin.

Post-Translational Modifications: The precursor protein undergoes various modifications, such as signal peptide cleavage and often glycosylation. These changes are crucial for the interleukin's functional maturation and secretion.

Secretion: Once processed, mature interleukin is either secreted directly into the extracellular space or stored in vesicles for later release. Secretion methods vary based on the specific interleukin and the type of immune cell involved.

Function: Interleukins express their effects by binding to specific receptors on target cells, which intiates signaling pathways that remote immune responses like inflammation, cell proliferation, differentiation, and immune cell activation.

Overall, interleukin synthesis is a tightly regulated process involving multiple cellular steps to ensure effective immune system function in response to diverse stimuli.

3.3 Alkaloids

The enzymatic processes responsible for alkaloid biosynthesis exhibit significant variability based on the type of alkaloid and the organism involved. Below is a generalized overview of the enzymatic reactions commonly associated with alkaloid biosynthesis:

1. Formation of Precursors:

Amino Acid Precursors: Many alkaloids originate from amino acids like tryptophan, tyrosine, phenylalanine, or ornithine. Enzymes such as decarboxylases, transaminases, and other modifying enzymes catalyze the conversion of these amino acids into alkaloid precursors.

2. Condensation Reactions:

Condensation of Intermediates: Enzymes facilitate condensation reactions where intermediates derived from amino acids or other precursors combine to form complex alkaloid structures. These enzymes often include synthases and transferases.

3. Oxidation and Reduction Reactions:

Hydroxylation and Oxidation: Enzymes like cytochrome P450 monooxygenases and other oxidoreductases introduce hydroxyl groups or catalyze oxidation reactions, modifying alkaloid structures.

Reduction: Reductases are enzymes that reduce specific functional groups, altering the chemical properties of alkaloids.

4. Ring Formation:

Cyclization: Enzymes catalyze the formation of ring structures within alkaloids. Cyclases and other ring-forming enzymes play a key role in creating the complex ring systems characteristic of alkaloids.

5. Methylation and Acylation:

Modification Reactions: Enzymes methylate or acylate alkaloid intermediates, adding methyl or acyl groups to specific positions on the alkaloid molecule. This process is crucial for modifying alkaloid bioactivity.

6. Glycosylation:

Attachment of Sugars: Some alkaloids undergo glycosylation, where sugars attach to the alkaloid backbone. Glycosyltransferases catalyze these reactions, often enhancing alkaloid solubility or stability.

7. Transport and Compartmentalization:

Transporter Proteins: Enzymes are involved in transporting alkaloid intermediates or final products across cellular membranes or into specific organelles (e.g., vacuoles), where further modification or storage occurs.

8. Regulatory Enzymes:

Regulation of Pathway: Enzymes control the flow through the alkaloid biosynthetic pathway, responding to environmental or developmental signals. This regulation involves transcription factors, kinases, and feedback inhibition mechanisms.

Alkaloid biosynthesis displays a diverse array of enzymatic reactions and pathways, tailored to the specific alkaloid and the biological context in which it is synthesized. Understanding these processes is essential for unraveling alkaloid biosynthetic pathways and exploring their potential applications in fields such as medicine, agriculture, and biotechnology.

3.4 TNF (Tumor Necrosis Factor)

The synthesis of TNF (Tumor Necrosis Factor) is a crucial process in immune response regulation. It begins with the activation of genes encoding TNF proteins in response to immune stimuli such as infections or inflammatory cytokines like IL-1 and TNF itself, forming an autocrine loop (Zhou et al., 2018). Following gene activation, transcription occurs in the nucleus, where DNA is transcribed into mRNA by RNA polymerase enzymes. The primary messenger ribonucleic acid transcript undergoes processing, including capping, splicing (if necessary), and poly(a) sie recognition to produce mature mRNA. This processed mRNA then moves to the cytoplasm for translation. In the cytoplasm, ribosomes bind to the mature mRNA and translate it into a precursor protein known as pro-TNF or TNF-α. Post-translational modifications are crucial next steps, including cleavage of a signal sequence and addition of a glycosylphosphatidylinositol (GPI) anchor, which prepares pro-TNF for activation.

Pro-TNF is processed into its active form, TNF-α, typically by the protease TNF-α converting enzyme (TACE/ADAM17), either in the Golgi apparatus or on the cell membrane. This proteolytic cleavage releases soluble TNF-α for secretion. Once mature, TNF-α is secreted from the cell in response to specific stimuli or constitutively, utilizing vesicle-mediated secretion processes or direct release mechanisms. Upon secretion, TNF-α acts on target cells via TNFR1 and TNFR2 receptors, initiating signaling cascades that regulate inflammation, immune responses, apoptosis, and other physiological processes. The production of TNF is tightly regulated to maintain immune balance. Dysregulation of TNF synthesis or signaling pathways can lead to chronic inflammation and contribute to the development of autoimmune diseases, inflammatory disorders, and cancer (Ren et al., 2020).

3.5 Sapogenin

Saponins have the ability to trigger immune cell activation and stimulate the release of cytokines by engaging multiple signaling pathways associated with immune function, including TLRs, NF-κB/MAPK, and hippo-YAP pathways (Yendo et al., 2016). This activation initiates innate immune responses and supports the development of antigen-specific immune responses.

The biosynthesis of sapogenins typically involves several steps:

Sapogenins often originate from larger precursor molecules in plants, such as saponins. Saponins are glycosides containing a hydrophobic aglycone portion, which can undergo hydrolysis to yield sapogenins (Warshakoon et al., 2009). Before proceeding further kindly look into the Fig. 5 for the understanding of saponin formation. Saponins are natural glycosides that consist of a sugar moiety attached to a triterpenoid or steroidal aglycone, called sapogenin. The process of synthesizing sapogenins from saponins involves the removal of the sugar component (glycone) from the saponin molecule, a step known as hydrolysis. This reaction helps in isolating sapogenins, which are the non-sugar part of saponins and serve as important precursors for synthesizing bioactive compounds, such as steroidal drugs (Vincken et al., 2007). The structure of sapogenin is presented in Fig. 2.

How Saponins Help in Synthesizing Sapogenins?

a. Structure of Saponins
   - Saponins are made of a sapogenin (steroidal or triterpenoid core) attached to sugar chains.
   - The sapogenin core is the biologically active part, which has various applications in pharmaceuticals.
b. Hydrolysis Process
   - To synthesize sapogenins, saponins undergo hydrolysis, where the sugar chains are cleaved off.
   - This can be done through acid hydrolysis or enzymatic treatment, depending on the method used (He et al., 2019).
c. Extraction of Sapogenins
   - After hydrolysis, the sapogenin is separated from the glycoside part.
   - This sapogenin can then be further purified and used as a precursor for synthesizing different compounds, such as steroid hormones or other bioactive molecules.
d. Applications of Sapogenins
   - Sapogenins are valuable in drug synthesis, particularly for the production of steroid-based drugs like cortisone and other anti-inflammatory agents (Richou et al., 1964).
   - They also have uses in cosmetics and health supplements due to their bioactive properties (de Costa et al., 2014).

Overall, the synthesis of immunomodulatory sapogenins in plants involves complex biochemical processes that contribute to their potential significance in human health and medicine (Fleck et al., 2006).

3.6 Inosine

Inosine, a biologically static purine nucleoside, is primarily made through the deamidation of adenosine, both intracellularly and extracellularly. Additionally, it can be generated by 5′-nucleotidase acting on 5′-inosinic acid. The structure of inosine is presented in Fig. 3. Recent research indicates that certain gut microbiota species also produce inosine, influencing host immune and proinflammatory functions. Inosine is further metabolized into hypoxanthine, xanthine, and uric acid. Transport across cell membranes is facilitated by equilibrative and concentrative nucleoside transporters, Inosine exerts its functions through receptor-dependent and receptor-independent mechanisms. Receptor-mediated

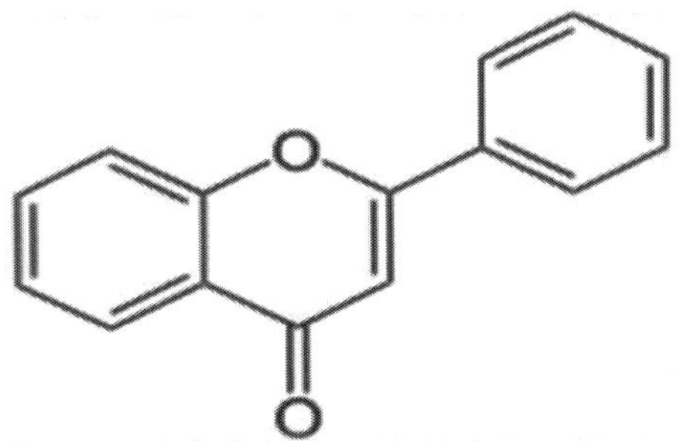

Fig. 2 Structure of sapogenin

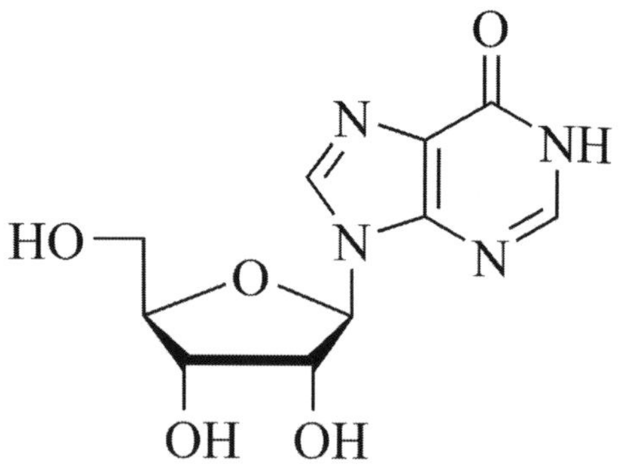

Fig. 3 Structure of inosine

actions involve adenosine receptor family members such as A1, A2A, A2B, and A3 G-seven transmembrane receptors While adenosine's role as a signaling molecule is well-established, the physiological and pathological implications of inosine in human health and disease contexts remain inadequately grasped.

Inosine is produced endogenously through three main processes:

A. 1 Adenosine deaminase (ADA) catalyzes the hydrolytic deamination of adenosine,
B. 25′-nucleotidase (5′NT) dephosphorylates inosine monophosphate (IMP), and
C. purine nucleoside phosphorylase (PNP) catalyzes the reaction of hypoxanthine and ribose-1p. Exogenous inosine can originate from two sources.

4 Synthetic Immunomodulators and Their Synthesis

4.1 Levamisole

Levamisole is widely employed for its immunomodulatory effects across different conditions, including rheumatoid arthritis, often co-administered with 5-fluorouracil. It has also demonstrated potential in treating colon cancer and melanoma. Furthermore, Levamisole has been used as an adjunct in radiotherapy for oral cancer, enhancing immunity by boosting lymphocyte subpopulations. The biosynthesis of levamisole is described in Fig. 4.

Levamisole, also known as phenylimidazoline (31.1.4), can be synthesized through several methods. One approach starts withbromoacetophenone, which reacts with thaizolidin-2-imine to form 3-(phenacyl)-2-imino-1,3-thiazolidine (31.1.1). This intermediate is then reacted with acetic anhydride to produce 3-(phenyacyl)-2-acetylimino-1,3-thiazolidine (31.1.2). The ketone group in this compound is reduced to an alcohol using sodium tetrahydroboride yielding (31.1.3). Subsequently, the hydroxyl group is substituted with chlorine using thionyl chloride. Finally, heating this product in acetic anhydride leads to the closure of the imidazole Another approach involves using styrene oxide as the starting material. Reacting phenylethylene oxide with thiazolidinimine yields 2-imino-3-(2-phenyl-2-hydroxyethyl)thiazolidine (31.1.5). This compound is then treated with thionyl chloride, followed by ethanoic acid, to produce levamisole (31.1.4) (Lorent et al., 2014).

Additionally, an alternative synthesis method using styrene oxide has been proposed. In this scheme, styrene oxide is reacted with aziridine to form 2-aziridion-1-phenylethanol-1 (31.1.6). This intermediate is then treated with potassium thiocyanate / thiourea to produce 3-(2-phenyl-2-hydroxyethyl)-2-amino-1,3-thiazolidine (31.1.5). Subsequent treatment with thionyl chloride and ethanoic anhydride yields the desired levamisole (31.1.4). ring, resulting in the formation of levamisole (31.1.4) (Juang & Liang, 2020).

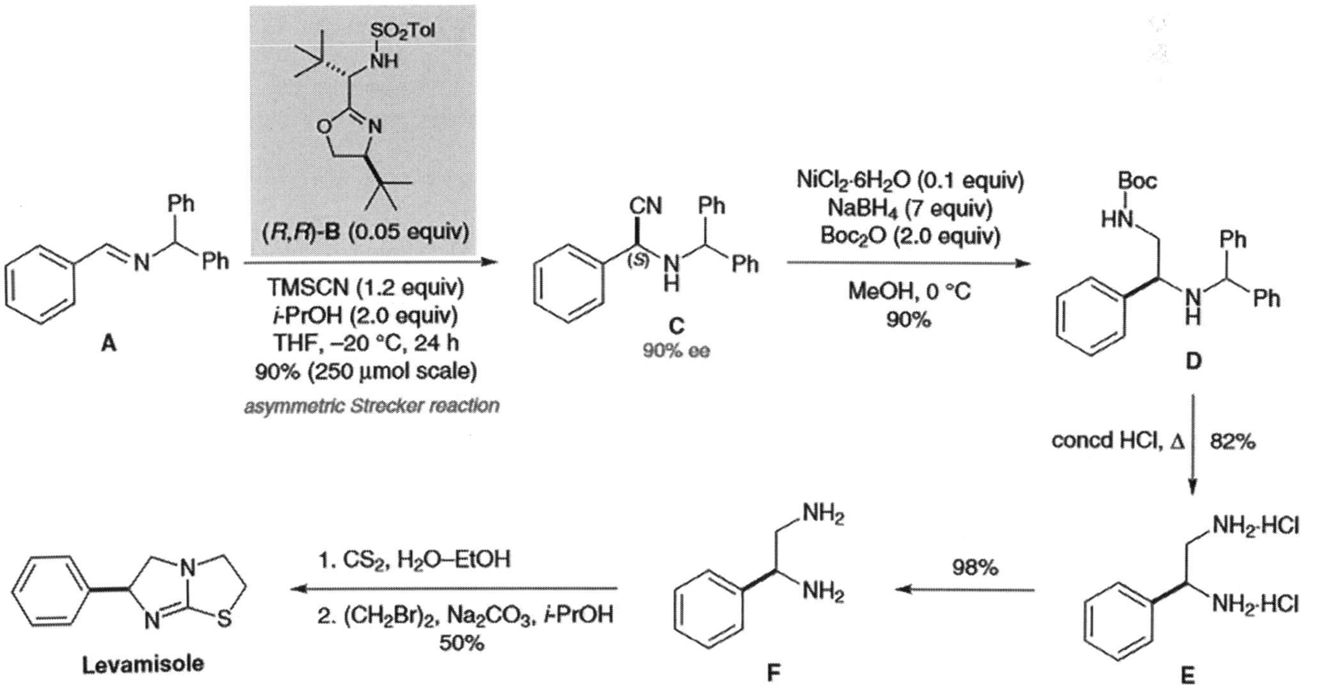

Fig. 4 Biosythesis of levamisole

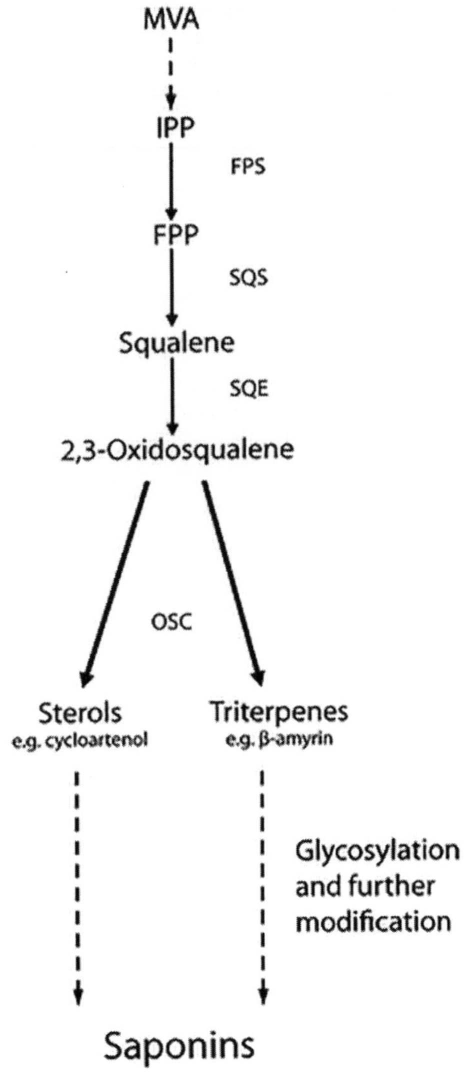

Fig. 5 Biosynthesis of saponin

4.2 Lenalidomide

An immunomodulatory agent known for its anti-tumor activity, primarily through anti-angiogenesis mechanisms. To enhance its pharmacological effectiveness, a sequence of Lenalidomide derivatives were plotted as inhibitors of tumor angiogenesis. These derivatives were screened for potential anti-angiogenesis targets using the reverse docking method with molecular docking software.

4.3 Astrgaloside IV

Astragaloside IV is obtained by extracting it from the roots of Astragalus membranaceus using a solvent. The extracted compound undergoes processes to increase its concentration, purify it from other substances, and ensure its quality and effectiveness before being used therapeutically. Astragaloside IV is derived from the roots of Astragalus membranaceus primarily through solvent extraction. The plant material, cleaned and dried, is often ground to enhance extraction efficiency. Ethanol or methanol is commonly employed as a solvent due to their ability to dissolve a broad spectrum of compounds, including astragalosides. The mixture undergoes agitation or heating to facilitate the dissolution of astragalosides, including Astragaloside IV. Following extraction, the crude extract containing astragalosides is concentrated to reduce volume and increase potency. Purification methods such as filtration, chromatography (like column or HPLC), or crystallization are then utilized to isolate Astragaloside IV from other components present in the extract. To confirm identity and purity, Astragaloside IV undergoes characterization using analytical techniques such as NMR spectroscopy and mass spectrometry. Rigorous quality control measures ensure the final product meets regulatory standards for purity and effectiveness. Once purified and verified, Astragaloside IV is formulated into pharmaceuticals or dietary supplements. Its renowned immunomodulatory properties make it a valuable component in both traditional medicine and modern healthcare applications.

5 Conclusion

This chapter delved into the detailed processes involved in the biosynthesis of immunomodulatory bioactive molecules. These molecules, including various interleukins, sapogenins, and synthetic agents like Lenalidomide derivatives, play vital roles in modulating the immune system and possess significant therapeutic potential. Biosynthesis Pathways: We examined the biosynthetic pathways of natural immunomodulators, focusing on the genetic and enzymatic processes involved. For example, the synthesis of interleukins encompasses gene activation, transcription, mRNA processing, translation, and post-translational modifications. The chapter covered the origins of these molecules from natural sources such as plants and gut microbiota, highlighting their complex biosynthetic mechanisms. Sapogenins, for instance, are derived from saponins through hydrolysis and genetic modifications. Synthetic Approaches: We explored synthetic methods for producing immunomodulatory molecules, such as the development of Lenalidomide derivatives. These methods use advanced techniques like reverse docking for target identification and chemical modifications to enhance bioactivity. Biological Activity and Mechanisms: The immunomodulatory effects of these molecules were discussed, including their interactions with immune cells and cytokine production modulation (Yoo et al., 2019). Both receptor-dependent and independent pathways of action were covered. Therapeutic Implications: The chapter underscored the therapeutic potential of these bioactive molecules in treating various diseases, such as cancer, rheumatoid arthritis, and immune disorders. The enhanced efficacy of synthetic derivatives compared to their natural counterparts was particularly noted. Future Directions the field of immunomodulatory bioactive molecules offers vast opportunities for further research and development. In summary, the biosynthesis of immunomodulatory bioactive molecules is a dynamic and promising area of research with significant implications for healthcare (Biocontrol Agents and Secondary Metabolites: Applications and Immunization for Plant Growth and Protection 2020).

References

Abbas, A. K., Lichtman, A. H., & Pillai, S. (2017). *Basic immunology: Functions and disorders of the immune system* (5th ed.). Elsevier.

Bafundo et al. (2021)

Bafundo, K. W., Duerr, I., Mcnaughton, J. L., & Johnson, A. B. (2021). The effects of a Quillaja and Yucca combination on performance and carcass traits of coccidia-vaccinated broilers exposed to an enteric disease challenge. *Poultry Science, 100*(11), 101391. https://doi.org/10.1016/j.psj.2021.101391.

Biocontrol Agents and Secondary Metabolites: Applications and Immunization for Plant Growth and Protection. (2020). Current trend and future prospects of secondary metabolite-based products from agriculturally important microorganisms.

Bioorganic Chemistry. (2023). Design, synthesis, antimicrobial screening and molecular modeling of novel 6,7 dimethylquinoxalin-2 (1H)-one and thiazole derivatives targeting DNA gyrase enzyme.

Brahmachari, G. (2008). Naturally occurring flavanones: An overview. *Natural Product Communications, 3*, 1934578X0800300820. https://doi.org/10.1177/1934578X0800300820.

Brás, J. P., Bravo, J., Freitas, J., Barbosa, M. A., Santos, S. G., Summavielle, T., et al. (2020). TNF-alpha-induced microglia activation requires miR-342: Impact on NF-kB signaling and neurotoxicity. *Cell Death & Disease, 11*(6), 1–15. https://doi.org/10.1038/s41419-020-2626-6.

Chávez-González, M. L., Sepúlveda, L., Verma, D. K., Luna-García, H. A., Rodríguez-Durán, L. V., Ilina, A., et al. (2020). Conventional and emerging extraction processes of flavonoids. *Processes, 8*, 434. https://doi.org/10.3390/pr804043.

de Costa, F., Yendo, A. C., Cibulski, S. P., Fleck, J. D., Roehe, P. M., Spilki, F. R., et al. (2014). Alternative inactivated poliovirus vaccines adjuvanted with Quillaja brasiliensis or Quil-A saponins are equally effective in inducing specific immune responses. *PLoS ONE, 9*(8), e105374. https://doi.org/10.1371/journal.pone.0105374.

Dias, M. C., Pinto, D. C. G. A., & Silva, A. M. S. (2021). Plant flavonoids: Chemical characteristics and biological activity. *Molecules, 26*, 5377. https://doi.org/10.3390/molecules26175377.

Fleck, J. D., Kauffmann, C., Spilki, F., Lencina, C. L., Roehe, P. M., & Gosmann, G. (2006). Adjuvant activity of Quillaja brasiliensis saponins onthe immune responses to bovine herpesvirus type 1 in mice. *Vaccine, 24*(37–39), 7129–7134. https://doi.org/10.1016/j.vaccine.2006.06.059.

Fiorucci, S., Santucci, L., & Distrutti, E. (2007). NSAIDs, coxibs, CINOD and H2S-releasing NSAIDs: What lies beyond the horizon. *Digestive and Liver Disease, 39*(12), 1043–1051. https://doi.org/10.1016/j.dld.2007.09.001.

Garofalo, S., Cocozza, G., Porzia, A., Inghilleri, M., Raspa, M., Scavizzi, F., et al. (2020). Natural killer cells modulate motor neuron-immune cell cross talk in models of amyotrophic lateral sclerosis. *Nature Communications, 11*(1), 1773.

He, Y., Hu, Z., Li, A., Zhu, Z., Yang, N., Ying, Z., et al. (2019). Recent advances in biotransformation of saponins. *Molecules, 24*(13), 2365. https://doi.org/10.3390/molecules24132365.

Heneka, M. T., Carson, M. J., Khoury, J. E. L., Landreth, G. E., Brosseron, F., Feinstein, D. L., et al. (2015). Neuroinflammation in Alzheimer's disease. *Lancet Neurology, 14*(4), 388–405. https://doi.org/10.1016/S1474-4422(15)70016-5.

Janeway, C. A., Travers, P., Walport, M., & Shlomchik, M. J. (2001). *Immunobiology: The immune system in health and disease* (5th ed.). Garland Science.

Jantan, I., Haque, M. A., Ilangkovan, M., & Arshad, L. (2019). An insight into the modulatory effects and mechanisms of action of Phyllanthus species and their bioactive metabolites on the immune system. *Frontiers in Pharmacology, 10*, 878. https://doi.org/10.3389/fphar.2019.00878.

Juang, Y. P., & Liang, P. H. (2020). Biological and pharmacological effects of synthetic saponins. *Molecules, 25*(21), 4974. https://doi.org/10.3390/molecules25214974.

Labh, S. N., & Shakya, S. R. (2014). Application of immunostimulants as an alternative to vaccines for health management in aquaculture. *International Journal of Fisheries and Aquatic Studies, 2*(1), 153–156.

Li, X., Liu, X., Cui, J., Song, W., Liang, Y., Hu, Y., et al. (2019). Epidemiological survey of antinuclear antibodies in healthy population and analysis of clinical characteristics of positive population. *Journal of Clinical Laboratory Analysis, 33*(6), e22965. https://doi.org/10.1002/jcla.22965.

Lorent, J. H., Quetin-Leclercq, J., & Mingeot-Leclercq, M. P. (2014). The amphiphilic nature of saponins and their effects on artificial and biological membranes and potential consequences for red blood and cancer cells. *Organic & Biomolecular Chemistry, 12*(43), 8803–8822. https://doi.org/10.1039/C4OB01652A.

Mukherjee, P. K. (2019). Bioactive phytocomponents and their analysis. In P. K. Mukherjee (Ed.), *Quality control and evaluation of herbal drugs* (pp. 237–328). Elsevier.

Nabavi, S. M., Šamec, D., Tomczyk, M., Milella, L., Russo, D., Habtemariam, S., et al. (2020). Flavonoid biosynthetic pathways in plants: Versatile targets for metabolic engineering. *Biotechnology Advances, 38*, 107316. https://doi.org/10.1016/j.biotechadv.2018.11.005.

Okoye, C. O., Jiang, H., Wu, Y., Li, X., Gao, L., Wang, Y., et al. (2023). Bacterial biosynthesis of flavonoids: Overview, current biotechnology applications, challenges, and prospects. *Journal of Cellular Physiology*, (Advance online publication). https://doi.org/10.1002/jcp.31006.

Pietta, P., Minoggio, M., & Bramati, L. (2003). Plant polyphenols: Structure, occurrence and bioactivity. In A. -U. Rahman (Ed.), *Studies in natural products chemistry* (Vol. 28, pp. 257–312). Elsevier.

Ramanadham, M., & Nageshwari, B. (2010). Anti-proliferative effect of levamisole on human myeloma cell lines in vitro. *Journal of Immunotoxicology, 7*(4), 327–332. https://doi.org/10.3109/1547691X.2010.514871.

Rehan, M. (2021). Biosynthesis of diverse class flavonoids via shikimate and phenylpropanoid pathway. In L. Q. Zepka, T. C. D. Nascimento, & E. Jacob-Lopes (Eds.), *Bioactive compounds (Ch. 6)*. IntechOpen. Retrieved from https://www.intechopen.com/.

Ren, W., Sun, H., Gao, G. F., Chen, J., Sun, S., Zhao, R., et al. (2020). Recombinant SARS-CoV-2 spike S1-fc fusion protein induced high levels of neutralizing responses in nonhuman primates. *Vaccine, 38*(37), 5653–5658. https://doi.org/10.1016/j.vaccine.2020.06.066.

Richou, R., Jensen, R., & Belin, C. (1964). Research on saponin, an adjuvant substance which stimulates immunity. *Revue D'immunologie Et De Therapie Antimicrobienne, 28*, 49–62.

Román-Albasini, L., Díaz-Véliz, G., Olave, F. A., Aguayo, F. I., García-Rojo, G., Corrales, W. A., et al. (2020). Antidepressant-relevantbehavioral and synaptic molecular effects of long-term fasudil treatment in chronically stressed male rats. *Neurobiology of Stress, 13*, 100234.https://doi.org/10.1016/j.ynstr.2020.100234.

Szebeni, J., Simberg, D., González-Fernández, Á., Barenholz, Y., & Dobrovolskaia, M. A. (2018). Roadmap and strategy for overcoming infusion reactions to nanomedicines. *Nature Nanotechnology, 13*(12), 1100–1108. https://doi.org/10.1038/s41565-018-0273-1.

Tariq, H., Asif, S., Andleeb, A., Hano, C., & Abbasi, B. H. (2023). Flavonoid production: Current trends in plant metabolic engineering and de novo microbial production. *Metabolites, 13*, 124. https://doi.org/10.3390/metabo13010124.

Tong, L. X., Sun, G. S., & Teng, J. M. (2015). Pediatric lichen sclerosis: A review of the epidemiology and treatment options. *Pediatric Dermatology, 32*(4), 593–599. https://doi.org/10.1111/pde.12615.

Vincken, J. P., Heng, L., de Groot, A., & Gruppen, H. (2007). Saponins, classification and occurrence in the plant kingdom. *Phytochemistry, 68*(3), 275–297. https://doi.org/10.1016/j.phytochem.2006.10.008.

Warshakoon, H. J., Hood, J. D., Kimbrell, M. R., Malladi, S., Wu, W. Y., Shukla, N. M., et al. (2009). Potential adjuvantic properties of innate immune stimuli. *Human Vaccines, 5*(6), 381–394. https://doi.org/10.4161/hv.5.6.8175.

Yendo, A. C., de Costa, F., Cibulski, S. P., Teixeira, T. F., Colling, L. C., Mastrogiovanni, M., et al. (2016). A rabies vaccine adjuvanted with saponins from leaves of the soap tree (Quillaja brasiliensis) induces specific immune responses and protects against lethal challenge. *Vaccine, 34*(19), 2305–2311. https://doi.org/10.1016/j.vaccine.2016.03.070.

Yoo, H. J., You, D. J., & Lee, K. W. (2019). Characterization and immunomodulatory effects of high molecular weight fucoidan fraction from the sporophyll of Undaria pinnatifida in cyclophosphamide-induced immunosuppressed mice. *Marine Drugs, 17*, 447. https://doi.org/10.3390/md17080447.

Zhou, X., Dong, Q., Kan, X., Peng, L., Xu, X., Fang, Y., et al. (2018). Immunomodulatory activity of a novel polysaccharide from Lonicera japonica in immunosuppressed mice induced by cyclophosphamide. *PLoS ONE, 13*(9), e0204152. https://doi.org/10.1371/journal.pone.0204152.

Biotechnological Application in the Production of Natural Sweeteners

C. Dhaarini, S. Tejashwini, and M. Gayathri

Abstract

Natural sweeteners are plant-based proteins with a sweet taste that resembles sugar. Natural sweeteners are non-nutritive but have more nutrients than refined sugar. Biotechnological methods such as fermentation and genetic engineering are used in their production. In addition to being healthier than artificial sweeteners, natural sweeteners offer other health advantages. The bacterium E. coli's mevalonate (MVA) and intrinsic methylerythritol 4-phosphate (MEP) pathways enable the eukaryotic S. cerevisiae to create isopentenyl diphosphate (IPP) and its isomer, dimethylallyl pyrophosphate (DMAPP), which are required for the manufacturing of plant-based isoprenoids. Both can consequently be used to create isoprenoid sweeteners. S. cerevisiae is regarded as a favorable host because it has an endoplasmic reticulum (ER) that permits it to perform the recombinant production of membrane-localized plant-derived P450s. As previously stated, P450s catalyze multiple steps in the synthesis of all three sweeteners. As a result, these microorganisms and enzymes facilitate the efficient synthesis of sweeteners, hence enhancing their availability. The purity, content, and thermal behavior of these sweeteners are measured via thermal analysis. The biosynthetic and biochemical procedures involved in the synthesis and characterization of naturally occurring sweetening chemicals will be the main topics of this review. We'll also go over the most current developments in the field of biotechnology.

Keywords

Fermentation • Genetic engineering • Isoprenoids • Thermal analysis • Chromatography

C. Dhaarini · S. Tejashwini · M. Gayathri (✉)
School of Biosciences and Technology, Vellore Institute of Technology, Tamil Nadu, Vellore Campus, Vellore, 632014, India
e-mail: gayathrigopinath@vit.ac.in

1 Introduction

Natural sweeteners: Natural sugars, sugar alcohols, terpenoid glycosides, certain amino acids, and polyphenols are sweet-tasting proteins derived from plants. A way to produce low-calorie, tooth-friendly products without the drawbacks of artificial additives is to use natural sweeteners that are strong but non-nutritive. Nature has around 100 different types of sweet chemicals classified under saccharides and non-saccharides, excluding simple sugars and polyols. Terpenoids, flavonoids, and proteins are their three primary types. These sweeteners can be further divided into three categories: semisynthetic (partially created by physico-chemical and enzymatic processes), artificial (synthetic, made solely from chemical components), and natural (extracted from plants or natural resources). Categorization of natural sugar substitutes is presented in Fig. 1 (Rayo-Mendez & Rodriguez-Llanos, 2023). Except for one, all of them have only been identified in green plants, not in any other type of life (Kinghorn and Compadre, 2001). Natural sweeteners in our diet can help lower the risk of acquiring chronic diseases linked to excessive sugar consumption and are also not absorbed by the human body. Natural sweeteners not only have a great taste, but they also increase vitamins and antioxidants, among other health benefits (Chéron et al., 2019). The use of natural sweeteners in the food business is crucial because it not only satisfies customer demand for healthier substitutes for regular sugar but also reflects the increased emphasis on environmental consciousness and sustainability. When compared to artificial sweeteners, natural sweeteners like honey, maple syrup, and stevia (***Stevia rebaudiana***) have many health benefits and preserve the natural flavors of the meal they sweeten. Sweet taste molecules span a wide chemical space with considerable structural diversity. Additionally, there is typically less processing involved in the growth and production of these natural sweeteners, which results in a lower carbon footprint and a more environmentally responsible method of

J. Devi (ed.), *Biotechnological Intervention in Production of Bioactive Compounds*,
Sustainable Landscape Planning and Natural Resources Management, https://doi.org/10.1007/978-3-031-76859-0_7

producing food (Philippe et al., 2014). Given these considerations, the use of natural sweeteners in the food business not only improves the general quality and healthfulness of products but also shows the industry's dedication to sustainability and innovation. Structure of natural sweet taste enhancers is provided in Fig. 2.

Because natural sweeteners are a more nutrient-dense option than refined sugars, they are essential for encouraging good eating habits. Using natural sweeteners lowers the consumption of empty calories and potentially hazardous substances, in contrast to artificial sweeteners which may pose health hazards. In certain situations, they can also assist in properly controlling blood sugar levels in those who have metabolic syndrome or diabetes. For example, because stevia contains anti-inflammatory and anti-immunomodulatory qualities, it would make a fantastic low-calorie substitute for regular sugar (Fry, 2012). Sugar overconsumption is frequently associated with weight gain and obesity. Thus, switching to natural sweeteners instead of refined sugar helps with weight management. These have more nutrients —like fiber, potassium, magnesium, and vitamins—than processed sweets, and they also improve oral health. Including these substitutes in our diets helps us lead a generally healthy lifestyle. Despite being safe to use, it's important to know where natural sweeteners fit into a balanced diet. Health benefits of natural sweeteners is presented in Table 1.

Due to significant health concerns regarding high sugar intake and the denial of chemical-based food additives, public interest in natural sweeteners has grown considerably in the past few years. Sweeteners made utilizing biotech techniques use less land and resources. In addition, it permits the incorporation of particular microorganisms into the soil, stimulates plant development, inhibits infections, lowers greenhouse gas emissions, and improves the fertility and structure of the soil. It safeguards natural resources and supports sustainable agriculture. All things considered, biotech techniques are essential to producing sustainable sweeteners since they conserve resources, improve soil health, and lessen environmental damage. Preferred host for the production of natural sweeteners is presented in Table 2.

2 Overview of the Biotechnological Methods

Natural sweeteners are produced using a variety of biotechnological techniques, such as genetic engineering, enzymatic synthesis, and fermentation. Fermentation, an

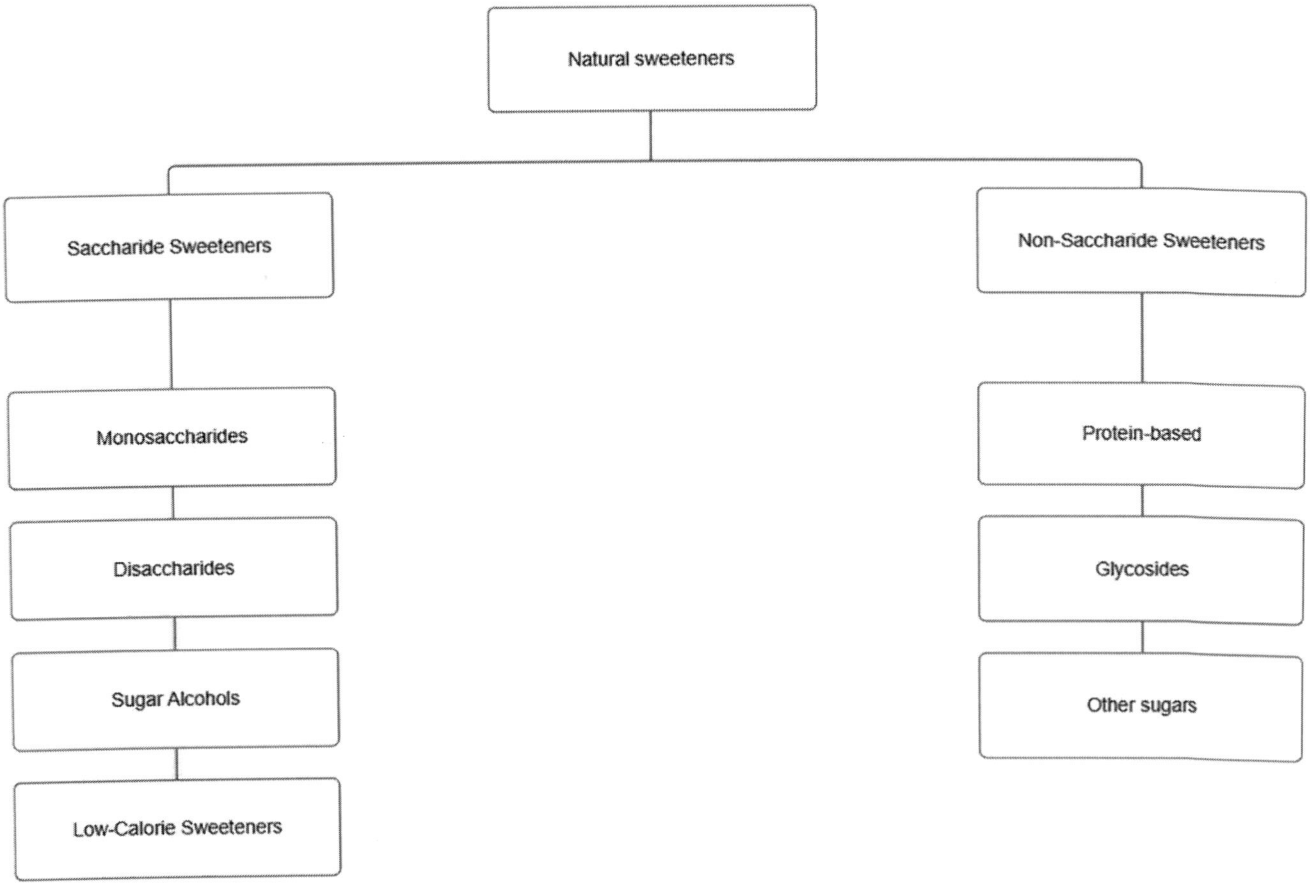

Fig. 1 Categorization of natural sugar substitutes

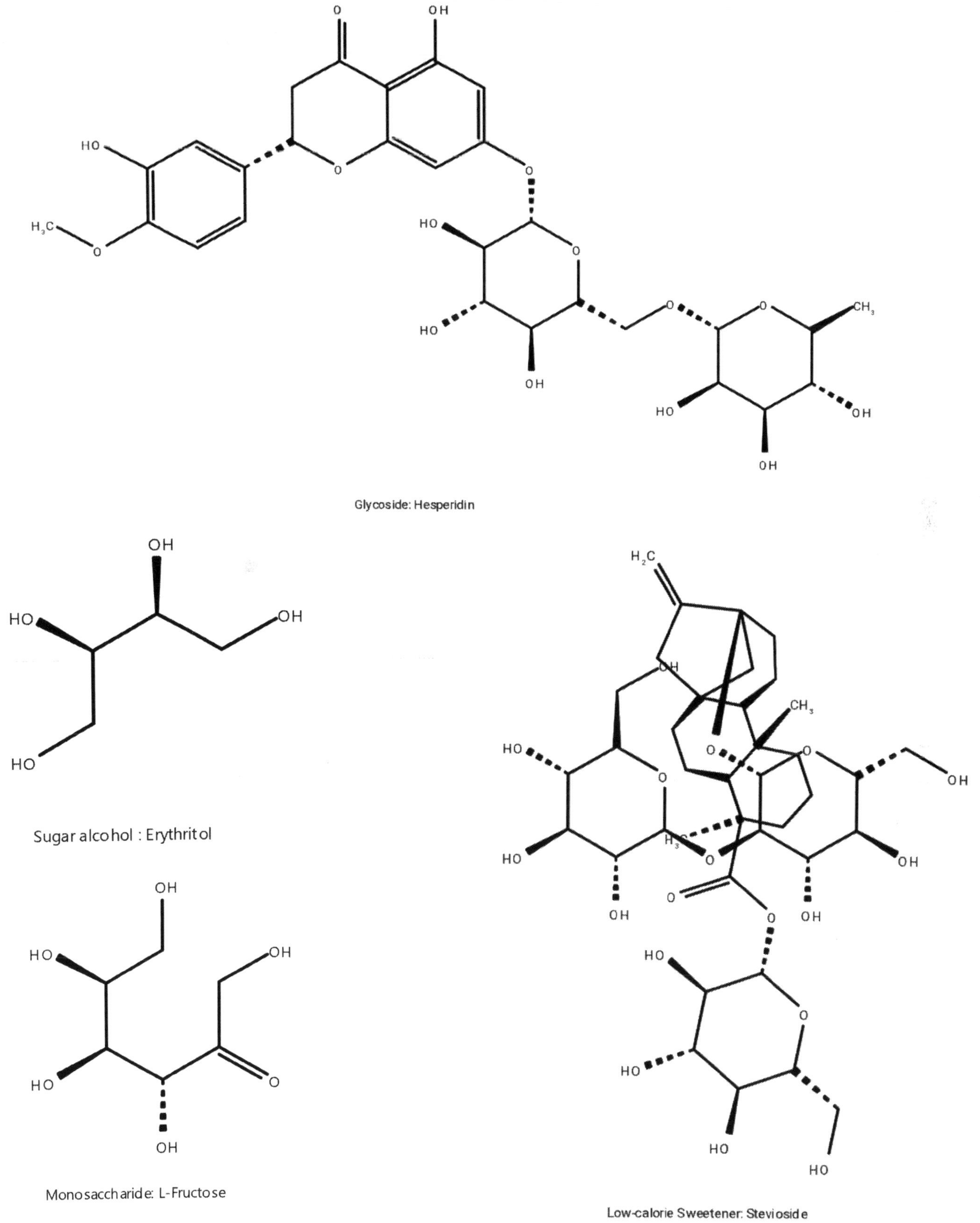

Fig. 2 Structure of natural sweet taste enhancers

Table 1 Health benefits of natural sweeteners

Natural sweeteners	Health benefits
Steviol glycosides	– Aids in blood sugar regulation – Can help with weight control
Maple syrup	– Contains antioxidants, which guard against cell damage – Provides essential minerals, including zinc and manganese
Glycyrrhizin	– Antiviral, antioxidant, anticancer, and anti-inflammatory properties – Promotes liver health
Erythritol	– Does not affect insulin or blood sugar levels – Anti-cariogenic qualities
Tagatose	– Helps with weight loss and doesn't cause glycemic spikes – Displays prebiotic, non-cariogenic, and anti-plaque potential
Mogrosides	– Anti-inflammatory and antioxidant properties – Anti-diabetic and anti-cancer effects

Table 2 Preferred host for the production of natural sweeteners

Natural sweeteners	Preferred hosts	Details
Steviol glycosides	*Stevia rebaudiana*	Steviol glycosides are extracted from the leaves of Stevia rebaudiana, a plant native to South America
Erythritol	*Yarrowia lipolytica*	Erythritol is produced using the yeast Yarrowia lipolytica, which is optimized for high yield and productivity
Monk fruit extract (Mogrosides)	*Siraitia grosvenorii*	Mogrosides are glycosides extracted from the fruit of Siraitia grosvenorii, known for their high-intensity sweetness
Xylitol	*Birch Trees*	Xylitol is commonly extracted from the bark of birch trees
Thaumatin	*Thaumatococcus daniellii*	Thaumatin is a protein sweetener extracted from the fruit of Thaumatococcus daniellii, a plant native to West Africa

age-old method, is now being used in modern kitchens to produce sugar substitutes. It enables the manufacturing of sweeteners with a lower environmental impact. Fermentation offers scalable output to satisfy increasing demand. Companies now create products that appeal to consumers who are health-conscious or who have dietary restrictions by using microbes to transform plant-based resources into chemicals that mimic the sweetness of sugar. The thriving industry of commercial stevia sweetener production by fermentation is demonstrated by these facilities. The combination of sweets and fermentation offers various advantages. Fermented sweeteners impart a unique characteristic to food products, meeting customer desire for varied taste profiles by improving flavor Texas Real Food (n.d.). Natural sweeteners produced through fermentation have a lower glycemic index and are therefore healthier than standard sweeteners. Fermentation requires less energy than other industrial processes and utilizes agricultural biowaste, so contributing to food security and waste reduction, making it a sustainable method. It is an economically viable solution for generating sweeteners due to its low energy requirements and potential for scaling. For example, Fermentation produces a variety of naturally occurring sweets, including: Date syrup is produced by fermenting dates, which keeps the fruit's natural nutrients in addition to producing a delicious product. Sorghum syrup is made from the fermented juices of the sorghum plant, this syrup is well-known for its strong flavor. We can also improve both the quantitative and qualitative properties of phytochemicals through genetic manipulation. Researchers have investigated genetic fidelity and diversity evaluation in *Stevia rebaudiana* using molecular markers (Shahnawaz et al., 2021). Rebaudioside M (Reb M), is a naturally occurring sweetener that has gained popularity due to its high sweetness and low-calorie content. Structure of Rebaudioside A is presented in Fig. 3. A study was conducted to increase Reb M production utilizing Saccharomyces cerevisiae, a typical microbe utilized in biotechnology. Stevia rebaudiana, a plant that contains steviol glycosides, is the source of Reb M (Buniowska et al., 2020). These substances, such as Reb M, have gained popularity as sugar substitutes since they are significantly sweeter than sugar and don't add calories to the diet. Nevertheless, Reb M content in Stevia plants is minimal, and low efficiency in prior attempts to manufacture Reb M in microbes has limited the yield. To address this issue, a study conducted at Jiangnan University genetically modified S. cerevisiae to more efficiently convert stevioside, another steviol glycoside, into Reb M. By first knocking down the

SCW2 gene in the yeast, they were able to stop the breakdown of the desired product (Esteve, 2020). This gene was responsible for breaking down glycosides. They then concentrated on the process of glycosylation, Sugar molecules are added to SGs' steviol chain. This is an important stage because it determines the final product's sweetness and sensory qualities. They discovered that UGT91D2 from stevia was the most effective glycosyltransferase after screening several glycosyltransferases, which are enzymes that link sugar molecules to other compounds (Li et al., 2024). The insertion of the silencing information regulator 2 (SIR2) into the yeast was one of the study's novel techniques. Because SIR2 has been shown to lengthen yeast cell life, it extended the altered S. cerevisiae's development cycle and allowed for a longer period during which Reb M could be produced (Howard, 2024). Starting with 10 g of stevioside, these genetic changes allowed for the generation of possible grams of Reb M per liter in a bioreactor. With a yield of 77.9%, steviol glycosides were converted to Reb M at the highest rate ever recorded, which is a major advancement for the sweetener's industrial production.

Enzymes are very specific, which makes them useful biocatalysts in the manufacturing of natural sweeteners (Carvalho and Fernandes, 2019). Fungi produce inulases, which are enzymes that have the potential to reduce manufacturing costs and increase the quality of syrup when in comparison to higher fructose corn syrup produced from starch. The food industry uses this enzyme to hydrolyze beta-2,1 glycosidic linkages found in polyfructans, such as inulin, found in a variety of plants, into glucose and fructose. When chicory or artichokes are hydrolyzed using inulinase, fructose syrup, and inulin oligosaccharides—which can be employed as a soluble fiber to support gut health or as a sweetener—are produced. As a result, the food business relies heavily on enzymes like inulases to produce natural sugars (Flores-Gallegos et al., 2018).

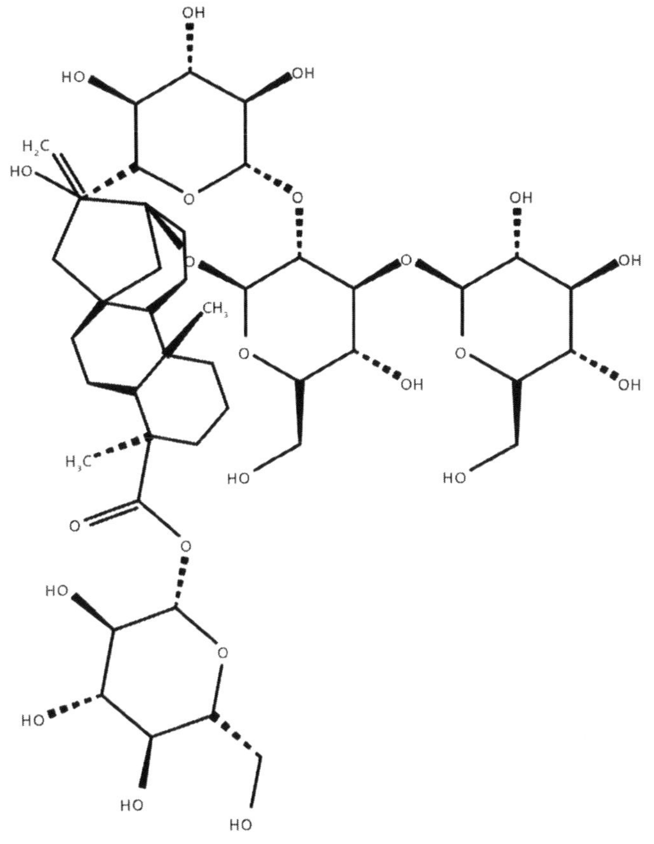

Fig. 3 Structure of Rebaudioside A (Reb-A)

3 Biosynthesis of Natural Sweeteners

Sweeteners are frequently produced by microorganisms such as bacteria or yeast, or by specific enzymes such as glucose isomerase or invertase. Swingle, or Grosvenorii, is an annual plant in the Cucurbitaceae family. Southern China is the natural home of S. grosvenorii, a plant whose fruit has traditionally been utilized as a sweetener and in traditional medicine to cure a few common illnesses like fever, colds, and other diseases. In China, S. grosvenorii is a protected species. Since living plant material is only cultivated in a limited number of places, it is illegal to export it. The component mogroside, a cucurbitane-type triterpenoid glucoside that is found in fresh fruit at a concentration of about 1%, is responsible for the fruit of S. grosvenorii's sweetness. Mogrosides vary in the kind and quantity of sugars they have attached, but they always contain mogrol as the triterpenoid aglycone. Structure of mogrosides is presented in Fig. 4. Approximately 20% of mogrosides, including mogroside V, which is sweeter than sucrose can be used by diabetic patients as a sugar substitute (Itkin et al., 2016). There are now numerous significant pharmacological effects that have been reported. The cyclization of 2,3-oxidosqualene by a class of biocatalysts known as oxidosqualene cyclases (OSCs) is the initial stage. It has been possible to characterize over 80 OSCs from plants. Cucurbitadienol synthase (CDS), an oxidosqualene cyclase, employs 2,3,22,23-diepoxysqualene as its substrate to create 24,25-epoxycucurbitadienol, which is one of the important features of this pathway. S. grosvenorii contains some genes that may express squalene epoxidases (SQEs), according to a genomic study. Of them, P450 (CYP87D18), epoxy hydrolases (EPHs), CDS, and All EPHs are significantly produced during the very first phases of fruit development and are thought to catalyze the next steps. Comparatively, CDS is formed by one gene. And functional investigations utilizing transgenic tobacco plants and S. cerevisiae revealed that it converts 2,3;22,23-diepoxysqualene into 24,25-epoxycucurbitadienol. This study's most significant finding was the technique underlying the hydroxylations at sites

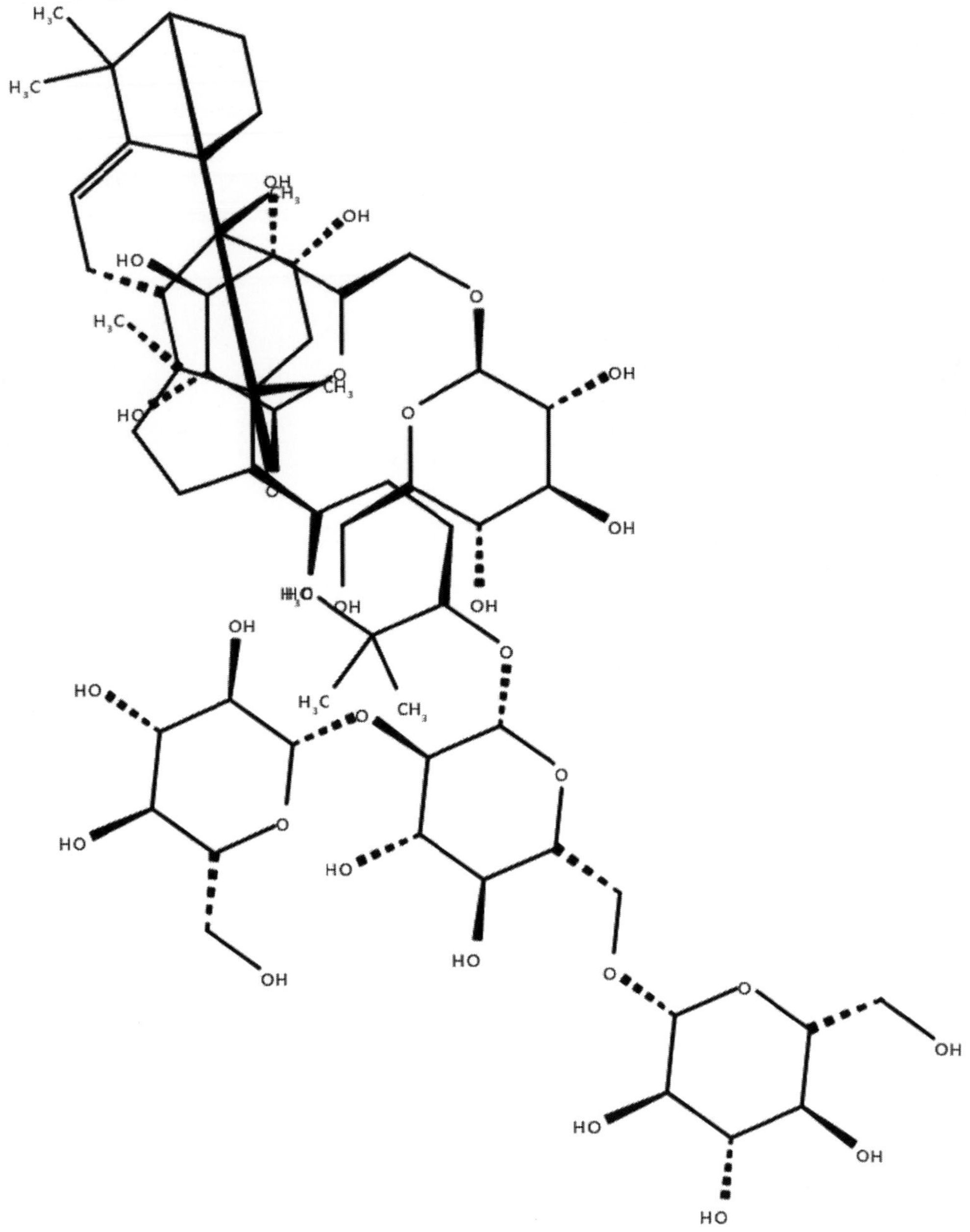

Fig. 4 Structure of mogrosides

C-24 and C-25. Initially, it was proposed that P450s catalyzed these hydroxylations by employing cucurbitadienol which was generated from 2,3-oxidosqualene as the substrate. Three of the eight genes encoding identified epoxide hydrolases found in the S. grosvenorii genome are highly expressed throughout the very first phases of fruit development. After examining these hydrolases, it was discovered that the 24,25-epoxycucurbitadienol substrate was converted to 24,25-dihydroxycucurbitadienol. In addition to catalyzing the C-11 hydroxylation of 24,25-dihydroxycucurbitadienol to produce 11,24,25-trihydroxycucurbitadienol—which is equivalent to the typical aglycone portion of mogrosides, mogrol—CYP87D18 is also implicated in the C-11 redox reaction of cucurbitadienol. Following this, five glucose molecules are added by glycosylation reactions—two at position C-3 and three at position C-24—to create mogroside V, an essential component of sweetness. It is known that the UGTs add one glucose molecule each at positions C-2′ and C-6′ of the C-24 glucose of mogroside IIE and transfer one glucose molecule at positions C-24 and C-3 of mogrol to create mogroside IIE using mogroside I-A1 to be a precursor. Among these, UGT720-269-1 is highly expressed during the early phases of fruit development, whereas UGT94-289-3 is the second UGT which is highly defined during the late phases of fruit development (Wang et al., 2017).

The creation of different plant-focused isoprenoids in S. cerevisiae has advanced remarkably. The semi-synthesis method of producing artemisinin at a cheap cost is made possible by the manufacture of artemisinic acid, which is an intermediate to the anti-malarial medication artemisinin. Numerous African nations are now receiving supplies of the artemisinin generated by this process. Future studies should concentrate on introducing the genes present in the metabolism of these natural sweeteners into bacteria, such as S. cerevisiae and E. coli, which are easier to cultivate as alternative hosts. The production of isopentenyl diphosphate (IPP) and its isomer, dimethylallyl pyrophosphate (DMAPP), which are required to make isoprenoids generated from plants, is achieved by the eukaryotic S. cerevisiae via the mevalonate (MVA) pathway and the intrinsic methylerythritol 4-phosphate (MEP) pathway in the bacterial E. coli. As a result, both can be used to create isoprenoid sweeteners (Seki et al., 2018). Because S. cerevisiae has an endoplasmic reticulum (ER) that enables the alternate production of membrane-localized plant-derived P450s, the organism is regarded as a favorable host. P450s catalyze several steps in the production of all three sweeteners, as was previously indicated. Consequently, these microbes and enzymes aid in the effective production of sweeteners, increasing their accessibility.

4 Characterization Techniques of Natural Sweeteners

Characterization techniques make understanding these sweeteners' composition, purity, and thermal behavior easier. Because natural sweeteners are low in calories and non-nutritive, they are becoming increasingly popular among health-conscious consumers. A variety of laboratory procedures are used in thermal analysis to measure the physical and chemical characteristics of materials concerning time and temperature. It sheds light on the composition and caliber of the items as well as the substrates. Thermogravimetric analysis (TGS) and differential scanning calorimetry (DSC) are the two most often utilized thermal analysis methods (Russo, 2023).

Using thermal analysis techniques like FTIR, and EDS, a study examined the concentration of cucurbitane glycosides from monk fruit and steviol glycosides from Stevia rebaudiana Bertoni in commercial samples of sweeteners. Understanding these materials' thermal resilience as well as the byproducts of thermal deterioration is crucial (Schmidt and Thomas, 2024). Before this, a Fourier-transform infrared spectrophotometer in an air atmosphere was coupled to a thermogravimeter (TG furnace) to study the thermal stability, behavior, and identification of the gaseous degradation compounds. A plate and frame heat exchanger (FT74X/HTST/UHT, Armfield, Inc.) with a nominal 66-s hold time tube was used for the studies. To reach the treatment conditions, the beverage was circulated through the heat exchanger after being deposited in a feed tank. Since fruit and vegetable blanching often occurs at a temperature of 60 °C, that number was selected. Similarly, pasteurizing liquid meals involves heating them to 88–99 °C for 15–30 s, which is standard procedure in the commercial sector. Every therapy was administered twice. After being chilled in an ice/water bath (FT61, Armfield, Inc.), the samples were packaged and refrigerated (4 ± 1 °C) until they were analyzed. The TG-DSC curves were obtained for two commercial powder formulations of stevia (S1, S2), an industrial example of pure steviol glycoside (PS), and a powder sample of dry leaves (LS) (Santana et al., 2022).

Several chromatographic methods, including Gas Chromatography (GC), Ion Chromatography (IC), Micellar Electrokinetic Chromatography (MEKC), Thin-Layer Chromatography (TLC), and High-Performance Liquid Chromatography (HPLC), are often used to determine low-calorie sweeteners. Faster techniques include Fourier-Transform Raman spectrometry, Fourier-Transform Infrared Spectroscopy (FTIR), and UV/Vis spectroscopy. The most frequently employed analytical technology for multi-analyte analysis is high-performance liquid chromatography (HPLC). Analyzing the affinities of the stationary phase column and the mobile phase forms the foundation of HPLC separation. Higher affinity compounds for the stationary phase will be segregated from lower affinities and held in the column. As a result, HPLC optimizes the gradient of the mobile phase to evaluate the content in only one pass. An appropriate detector is required to aid in recognizing and measuring non-nutritive sweeteners using HPLC (Gardana et al., 2010). Alternative detectors include light scattering, conductivity, mass spectrometry (MS), refractive index (RI), UV/Vis detectors, typically diode-array detectors (DAD), and others (Oktavirina et al., 2021). The primary steviol glycosides can be readily measured with good specificity, sensibility, and precision using the suggested HPLC approach.

The guidelines established by the regulating authorities concerning the manufacture, labeling, verification, and quality control of these natural sweeteners include the regulation of these sweeteners. To make sure that these sweeteners are used within acceptable bounds, each country has its own set of rules and guidelines. The Acceptable Daily Intake (ADI) value and maximum amount for high-intensity sweeteners in foodstuffs have been determined by several regulatory authorities around the world. On the other hand, these sweeteners are exempt from labeling regulations. While consuming non-nutritive sweeteners in moderation is better for energy intake than sugars, consuming these sweeteners in excessive quantities over an extended period may have negative consequences on human health.

While validation confirms whether the sweeteners comply with regulatory standards to ensure the desired product's constant quality and safety, quality control keeps an eye on the production process. In the food tech sector, which includes the sweetener production industry, it is imperative to underscore the significance of quality control and product validation. These measures guarantee the accessibility of safer goods for consumption and ensure that the anticipated flavor profile is delivered without any unfavorable side effects.

5 Comparison of Biotechnology and Traditional Methods

Natural sweeteners, as we all know, come from naturally occurring sources like plants. These consist of agave nectar, honey, and stevia (Stevia rebaudiana), which are all low in glycemic index, calories, and essential vitamins, minerals, and antioxidants. These sweeteners have a strong sweetness. A wide range of substances, including sugars, sugar alcohols, proteins, amino acids, terpenoid glycosides, and some polyphenols, are considered natural sweeteners. That being said, the market is only filled with commonly used sweeteners that have the necessary qualities, such as safety, pleasant flavor, high stability, good solubility, and affordable price. Except for fructose, which is the natural sugar with the highest sweetness and is abundant in fruit, agave nectar, honey, and some vegetables, other carbs have a steadily decreasing proportional sweetness potency in comparison to sugar. The two monosaccharides that the food industry uses most frequently as natural sweeteners are glucose and fructose. In many food products, fructose is used in place of sucrose due to its lower glycaemic index, strength as a sweetener, affordability, and capacity to enhance the flavor, color, texture, and shelf-life stability of the final product. Even though natural alternatives are thought to be healthier, these chemicals nevertheless need to be used in moderation (Sharififar et al., 2022).

Artificial sweeteners were also produced using conventional methods of sweetener synthesis. These sweeteners are calorie-free, chemically manufactured sugar substitutes that raise blood sugar levels. Aspartame, saccharin, erythritol, and sucralose are typical examples. Intense sweeteners offer sweetness in substantially greater amounts than sucrose and at varying potencies. Very few amounts are needed to achieve the appropriate sweetening effect because of their high sweetness potency. They therefore have the least negative impact on a product's energy value, which is ideal. Even though they are frequently referred to as "artificial sweeteners," these substances can be natural (such as rebaudioside and stevioside) or semisynthetic (such as neo hesperidin dihydrochalcone) (saccharin, aspartame, sucralose). Processed foods, particularly aerated and non-aerated drinks, tinned goods, baked goods, candies, and puddings, frequently include intense sweeteners.

Tagatose is a naturally found fructose isomer that is present in a variety of fruits and milk-based goods. It is thought to be a prebiotic and a taste booster. In the promotional sense, it is derived from lactose using a multi-stage biochemical procedure that also includes detoxification and distillation (Campbell et al., 2022). Its sweetness intensity, 0.92, is similar to that of sucrose but is digested differently, resulting in fewer calories and a milder glycemic response, as well as not favoring dental caries. When consumed in large quantities, tagatose can cause gas, diarrhea, and discomfort in the abdomen since it is only partially digested. Energy bars, breakfast cereals, chocolate gums, caramel, ice cream, soft beverages, and yogurt are all made with tagatose.

Enzymatic isomerization is a biotechnological method of producing tagatose, a natural sweetener preferred over chemical procedures (Oh, 2007). Biological methods for the synthesis of D-Tagatose is presented in Fig. 5 (Roy et al., 2018). Many enzymes can be used for biological D-tagatose manufacturing, such as L-arabinose isomerase (l-AI) EC 5.3.1.4, which, because of the similar substrate configurations, can catalyze the conversion of D-galactose into D-tagatose as well as L-arabinose into L-ribulose (Miao et al., 2023). However, biological D-tagatose synthesis is restricted because of the lower bioconversion efficiency of l-AI, the need for metal ions, the enzyme's low affinity for D-galactose, poor thermostability, and other factors. Therefore, by altering the functional characteristics of l-AI, it has been proposed that Utilizing genomic and protein engineering methods could increase the D-tagatose formation's transformation rate. It is possible to screen for particular mutants with higher catalytic activity by evaluating individual protein variants through the use of a highly defined selection method. When producing D-tagatose, the safety issue resulting from non-GRAS hosts' enzymes or cells is able to be rectified by introducing the L-arabinose isomerase gene to GRAS hosts such as C. glutamicum, Corynebacterium ammonagenes, and Bacillus megaterium (Roh et al., 2000). Ultimately, aside from the production and release of enzymes in a microbial host suitable for feeding. However, investigating requires research of novel sources of enzymes from GRAS microorganisms.

Biotechnological techniques have an advantage over conventional techniques in that they can overcome the constraints of plant-based manufacturing, guaranteeing a consistent supply and scalability of natural sweeteners. It's also been shown to improve sustainability by lessening environmental effects.

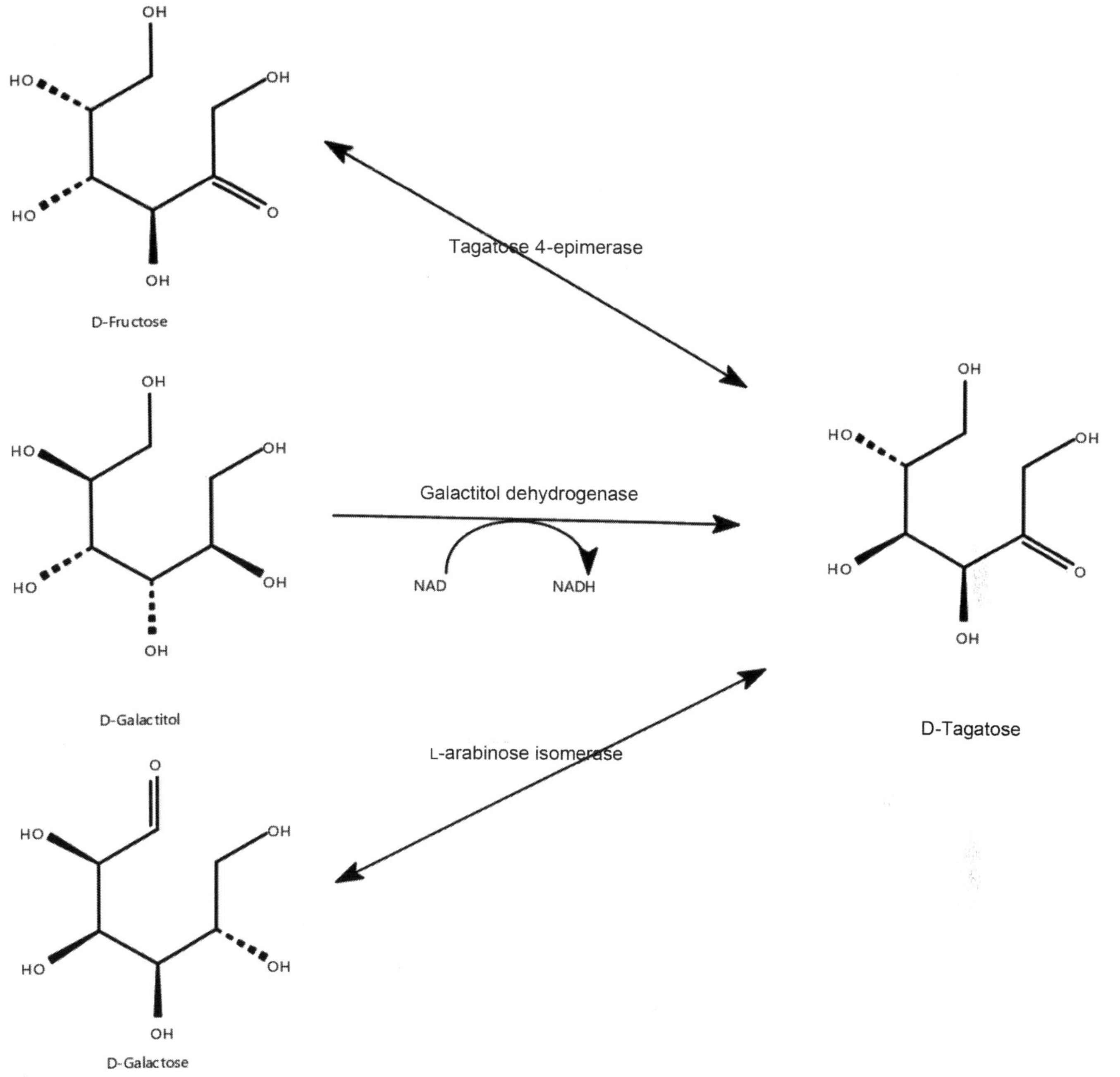

Fig. 5 Biological methods for the synthesis of D-tagatose

6 Applications and Potential Benefits

Natural sweeteners are used in a wide range of sectors. Soft drinks and fruit juices are sweetened with natural sweeteners such as agave nectar, honey, and stevia. Because natural sweeteners are thought to be healthier than artificial ones, consumers prefer them. As excipients, natural sweeteners help enhance taste perception and cover up the harshness of some medications. They are becoming more and more well-liked since they are less harmful than artificial sweeteners. For many liquid and oral medication formulations, they act as vehicles. Syrup bases are prepared using natural sweeteners. They are currently utilized in the pharmaceutical sector to make syrup bases, liquid oral preparations, pill and tablet coatings, and as a carrier for numerous medications. For those who are overweight, the low-calorie content makes it easy to ingest and helps avoid obesity (Majie, 2021).

A chronic illness that affects millions of individuals worldwide is diabetes. It is typified by elevated blood sugar levels, which can be attributed to either insufficient insulin production by the body or poor cell response to the produced insulin. Maintaining a healthy lifestyle and closely monitoring blood sugar levels are essential to managing diabetes. Swapping refined sugar for natural sweeteners is one strategy to help manage blood sugar levels. In addition to being tasty, natural sweeteners like honey, stevia, and maple syrup have less of an effect on blood sugar levels than processed sugar. The potential advantages of natural sweeteners for people with diabetes have drawn more attention in recent years. The usage of natural sweeteners including erythritol, xylitol, and stevia has been demonstrated in a study to have negligible effects on blood sugar levels in diabetics (Boesten et al., 2015). This is especially crucial as 422 million people worldwide are expected to have diabetes, an illness that is becoming more and more common. Another study discovered that using natural sweeteners in place of regular sugar can improve blood sugar control and lower the risk of problems for diabetics. The antioxidant and anti-inflammatory qualities of some natural sweeteners, which may be especially helpful for diabetics. In general, studies on natural sweeteners and diabetes indicate that including these substitutes in one's diet may provide a secure and practical means of sating sweet cravings without having an adverse influence on blood sugar levels.

Natural sweeteners effectively enhance the flavor character of produced goods without jeopardizing their chemical makeup or stability. Due to their antioxidant qualities, the phenolic chemicals that are contained in them are what keep them stable. These sweeteners aid in effectively preserving the proper pH levels in food goods, which is the only way to stop spoiling and increase the shelf life of the item. Longer shelf life is mostly dependent on two key factors: the sweetener's temperature stability and its compatibility with the biomolecules in food. Additionally, natural sweeteners have inherent antibacterial qualities that aid in the suppression of microorganisms that cause deterioration. The type of microorganisms present, their concentration, and the inherent qualities of the meal can all affect how effective they are at inhibiting microbial growth. Additionally, natural sweeteners can work in concert with other preservation techniques to broaden the range of microbial inhibition. This is particularly true in acidic pH situations, where the effects of these sweeteners can be more noticeable (Kroger et al., 2006).

Consumer behavior is greatly influenced by marketing initiatives and product labeling. Products labeled "no added sugar" or "made with natural sweeteners" may influence consumers to choose particular natural food items. Being open and honest about the sources of sweetness—whether it comes from monk fruit, stevioside, sugar beet, or reba—enables customers to make wise dietary decisions (Kemp and Lindley, 2007). Effect of Labeling: Customers who are empowered to choose items that meet their dietary needs and health goals can benefit from clear labels. Marketing Strategy can make use of the availability of natural sweeteners and their possible health advantages frequently influence customer buying decisions.

7 Current Research Trends and Prospects

The very sweet diterpene glycosides known as steviol glycosides are present in Stevia rebaudiana plants' foliage. The identification of the genes in charge of steviol glycoside production is a recent development. By adding genes for biosynthetic enzymes, scientists have modified microbial hosts (such as Saccharomyces cerevisiae) to generate steviol glycosides (Zhou et al., 2021). Mogrosides are triterpenoid glycosides of the cucurbitane family that are derived from monk fruit (Siraitia grosvenorii). Developments in our knowledge of the catalytic roles played by the biosynthetic enzymes that produce mogrosides. Microbial hosts can produce mogrosides efficiently thanks to biotechnological methods. Glycyrrhiza plants, sometimes known as licorice, are the source of glycyrrhizin, a triterpenoid glycoside of the oleanane class. The characterization of the enzymes involved in the manufacture of glycyrrhizin has advanced. Microbial manufacturing platforms improve sustainability and alleviate supply constraints.

Investigating natural sweeteners is more than simply a fad in the industry; it is a reflection of a continuous dedication to sustainability, innovation, and wellness. The environment, medical research, and the food business will all be significantly impacted by the difficulties and developments in this subject going forward. The market for natural sweeteners—which provide the sweetness of sugar without the negative health effects—is expanding. With their low-calorie contents and high sweetness profiles, monk fruit and thaumatin are emerging as viable substitutes. To stay ahead of consumer demands, the food sector is always finding new sources, such as sweet sorghum, and researching effective extraction techniques. Innovations in this field are fueled by the sweeteners' ability to preserve food in addition to their flavor and health benefits. The main factor influencing the use of natural sweeteners is consumer health (Saraiva et al., 2020). Research is closely examining how sweeteners affect long-term health, especially about metabolic conditions like type 2 diabetes. Researchers are becoming more interested in substances like mannitol, a sugar alcohol that lowers blood glucose levels and is helpful for diabetic patients, as they investigate the health impacts of sugars and sweeteners. Furthermore, research into the potential health effects of consuming specific amino acids and carbohydrates present in natural sweeteners is also ongoing.

Biotechnology presents viable pathways for the cost-effective production of natural sweeteners, guaranteeing supply, satisfying customer demand, and advancing sustainability.

8 Conclusion

Natural sweeteners produced biotechnologically provide sustainable and health-conscious substitutes for regular sugars. The efficiency and yield gains made possible by enzyme synthesis, genetic engineering, and the utilization of microorganisms have been emphasized in this chapter. Superior purity and consistency above conventional procedures are demonstrated by these sweeteners, whose quality and safety are guaranteed by advanced characterisation techniques such as high-performance liquid chromatography HPLC and mass spectrometry. The benefits of low glycemic index and antioxidant characteristics make them suitable for a wide range of uses, from food and beverages to pharmaceuticals. Current research trends indicate that this subject will continue to advance as they concentrate on finding new sources and streamlining production procedures.

Acknowledgment and Contributions The author (DC) gratefully acknowledges professor (GM), School of Biosciences and Technology, VIT Vellore Campus, Vellore, Tamilnadu for giving her valuable guidance during paper writing. The first author (DC) wrote the paper and the second author (TS) checked and edited the present manuscript.

References

Boesten, D. M. P. H. J., den Hartog, G. J. M., & de Cock, P. (2015). Health effects of erythritol. *Nutrafoods, 14*(3), 3–9. https://doi.org/10.1007/s13749-014-0067-5.

Buniowska, M., Carbonell-Capella, J. M., Znamirowska, A., Zulueta, A., Frígola, A., & Esteve, M. J. (2020). Steviol glycosides and bioactive compounds of a beverage with exotic fruits and *Stevia rebaudiana* Bert. As affected by thermal treatment. *International Journal of Food Properties, 23*(1), 255–268. https://doi.org/10.1080/10942912.2020.1719134.

Campbell, H. R., Alsharif, F. M., & Marsac, P. J. (2022). The development of a novel pharmaceutical formulation of D-tagatose for spray-drying. *Journal of Pharmaceutical Innovation, 17*(194–206). https://doi.org/10.1007/s12247-020-09507-4.

Carvalho, F., & Fernandes, P. (2019). Enzymes in sweeteners production. In B. Parameswaran, S. Varjani, & S. Raveendran (Eds.), *Green bio-processes. Energy, environment, and sustainability* (pp. 1–20). Springer. https://doi.org/10.1007/978-981-13-3263-0_9.

Chéron, J. -B., Marchal, A., & Fiorucci, S. (2019). Natural sweeteners. In L. Melton, F. Shahidi, & P. Varelis (Eds.), *Encyclopedia of food chemistry* (pp. 189–195). Academic Press. https://doi.org/10.1016/B978-0-08-100596-5.21620-X.

Esteve, M. J. (2020). Steviol glycosides and bioactive compounds of a beverage with exotic fruits and Stevia rebaudiana Bert. As affected by thermal treatment. *International Journal of Food Properties, 23* (1), 255–268. https://doi.org/10.1080/10942912.2020.1719134.

Flores-Gallegos, A. C., Veana, F., González-Herrera, S. M., Muñiz--Márquez, D. B., Sáenz-Galindo, A., & Rodríguez-Herrera, R. (2018). Fungal inulinases: An interesting option for food sweetener production. In M. Kuddus (Ed.), *Enzymes in food technology* (pp. 1–20). Springer. https://doi.org/10.1007/978-981-13-1933-4_5.

Fry, J. C. (2012). Natural low-calorie sweeteners. In D. Baines & R. Seal (Eds.), *Natural food additives, ingredients and flavourings* (pp. 41–75). Woodhead Publishing. https://doi.org/10.1533/9780857095725.1.41.

Gardana, C., Scaglianti, M., & Simonetti, P. (2010). Evaluation of steviol and its glycosides in *Stevia rebaudiana* leaves and commercial sweetener by ultra-high-performance liquid chromatography-mass spectrometry. *Journal of Chromatography A, 1217*(9), 1463–1470.

Howard, G. (2024, April 4). Making sweeter natural sweeteners with yeast and enzymes. *Natural Science News*. Retrieved June 29, 2024, from https://naturalsciencenews.com/article/1043

Itkin, M., Davidovich-Rikanati, R., Cohen, S., Portnoy, V., Doron-Faigenboim, A., Oren, E., et al. (2016). The biosynthetic pathway of the nonsugar, high-intensity sweetener mogroside V from *Siraitia grosvenorii. Proceedings of the National Academy of Sciences of the United States of America, 113*(47). https://doi.org/10.1073/pnas.1604828113.

Kemp, S., & Lindley, M. (2007). Developments in sweeteners. In A. Taylor & J. Hort (Eds.), *Modifying flavour in food* (pp. 185–201). Woodhead Publishing. https://doi.org/10.1533/9781845693367.185.

Kinghorn, A. D., & Compadre, C. M. (2001). Less common high-potency sweeteners. In O. L. O'Brien Nabors (Ed.), *Alternative sweeteners* (3rd ed., pp. 209–234). New York, NY: Marcel Dekker.

Kroger, M., Meister, K., & Kava, R. (2006). Low-calorie sweeteners and other sugar substitutes: A review of the safety issues. *Comprehensive Reviews in Food Science and Food Safety, 5*, 35–47.

Li, S., Luo, S., Zhao, X., Gao, S., Shan, X., Lu, J., et al. (2024). Efficient conversion of stevioside to rebaudioside M in *Saccharomyces cerevisiae* by engineering hydrolase system and prolonging the growth cycle. *Journal of Agricultural and Food Chemistry, 72* (14), 8140–8148. https://doi.org/10.1021/acs.jafc.4c01483.

Majie, A. (2021). Natural sweeteners and their use in the pharmaceutical industry. *ScienceOpen Posters*. https://doi.org/10.14293/S2199-1006.1.SOR-.PPCRINY.v1.

Miao, P., Wang, Q., Ren, K., Zhang, Z., Xu, T., Xu, M., et al. (2023). Advances and prospects of D-tagatose production based on a biocatalytic isomerization pathway. *Catalysts, 13*(11), 1437. https://doi.org/10.3390/catal13111437.

Oh, D. K. (2007). Tagatose: Properties, applications, and biotechnological processes. *Applied Microbiology and Biotechnology, 76*(1), 1–8. https://doi.org/10.1007/s00253-007-0981-1.

Oktavirina, V., Prabawati, N. B., Fathimah, R. N., Palma, M., Kurnia, K. A., Darmawan, N., et al. (2021). Analytical methods for determination of non-nutritive sweeteners in foodstuffs. *Molecules, 26*(11), 3135. https://doi.org/10.3390/molecules26113135.

Philippe, R. N., De Mey, M., Anderson, J., & Ajikumar, P. K. (2014). Biotechnological production of natural zero-calorie sweeteners. *Current Opinion in Biotechnology, 26*, 155–161. https://doi.org/10.1016/j.copbio.2014.01.004.

Rayo-Mendez, L. M., & Rodriguez-Llanos, J. H. (2023). Natural sweeteners. In G. A. Valencia (Ed.), *Natural Additives in Foods* (pp. 1–20). Cham: Springer. https://doi.org/10.1007/978-3-031-17346-2_5.

Roh, H. J., Kim, P., Park, Y. C., & Choi, J. H. (2000). Bioconversion of d-galactose into d-tagatose by expression of l-arabinose isomerase. *Biotechnology and Applied Biochemistry, 31*(1), 1–4.

Roy, S., Chikkerur, J., Roy, S. C., Dhali, A., Kolte, A. P., Sridhar, M., et al. (2018). Tagatose as a potential nutraceutical: Production, properties, biological roles, and applications. *Journal of Food Science, 83*(11), 2699–2709.

Russo, H. M. (2023, May). Thermal characterization by TG-DSC of commercial sweetener formulations and dried leaves of Stevia rebaudiana Bertoni. In *Paper Presented at the Proceedings of Brazilian Conference on Natural Products and Annual Meeting Macromolecular Evolution, Systematics and Ecology*.

Santana, N. D. S., Mothé, C. G., de Souza, M. N., & Setyaningsih, W. (2022). An investigation by thermal analysis of glycosidic natural sweeteners. *Journal of Thermal Analysis and Calorimetry, 147*(11), 13275–13287. https://doi.org/10.1007/s10973-022-11550-x.

Saraiva, A., Carrascosa, C., Raheem, D., Ramos, F., & Raposo, A. (2020). Natural sweeteners: The relevance of food naturalness for consumers, food security aspects, sustainability and health impacts. *International Journal of Environmental Research and Public Health, 17*(17), 6285. https://doi.org/10.3390/ijerph17176285.

Schmidt, S. J., & Thomas, L. C. (2024). Thermal analysis. In B. P. Ismail & S. S. Nielsen (Eds.), *Nielsen's food analysis*. Food science text series. Cham: Springer. https://doi.org/10.1007/978-3-031-50643-7_30.

Seki, H., Tamura, K., & Muranaka, T. (2018). Plant-derived isoprenoid sweeteners: Recent progress in biosynthetic gene discovery and perspectives on microbial production. *Bioscience, Biotechnology, and Biochemistry, 82*(6), 927–934. https://doi.org/10.1080/09168451.2017.1387514.

Shahnawaz, Pandey, D. K., & Konjengbam, M., et al. (2021). Biotechnological interventions of in vitro propagation and production of valuable secondary metabolites in *Stevia rebaudiana*. *Applied Microbiology and Biotechnology, 105*, 8593–8614. https://doi.org/10.1007/s00253-021-11580-9

Sharififar, F., Ashrafzadeh, A., & Kavirimanesh Khanaman, M. (2022). A review of natural peptide sweeteners. *International Journal of Peptide Research and Therapeutics, 28*, 158. https://doi.org/10.1007/s10989-022-10464-4.

Texas Real Food. (n.d.). Natural sweeteners and fermentation: How they enhance home cooking. Retrieved June 30, 2024, from https://discover.texasrealfood.com/natures-sweet-secrets/natural-sweeteners-and-fermentation.

Texas Real Food. (n.d.). Natural sweeteners: Exploring their efficacy in sustaining food freshness. Retrieved June 30, 2024, from https://discover.texasrealfood.com/natures-sweet-secrets/natural-sweetenersHoward.

Wang, R., Chiu, C. H., Lu, T. J., & Lo, Y. C. (2017). Biotransformation of mogrosides. In J. M. Merillon, & K. Ramawat (Eds.), *Sweeteners*. Reference series in phytochemistry. Cham: Springer. https://doi.org/10.1007/978-3-319-26478-3_13-1.

Zhou, X., Gong, M., Lv, X., et al. (2021). Metabolic engineering for the synthesis of steviol glycosides: Current status and future prospects. *Applied Microbiology and Biotechnology, 105*, 5367–5381. https://doi.org/10.1007/s00253-021-11419-3.

Role of Biotechnology in the Production of Health Promising Bioactive Molecules

Munish Sharma, Munit Sharma, Sumant Kumar Pandey, Deepak Sharma, and Munish Sharma

Abstract

Rapid advances in biotechnology have revolutionized the production of medically beneficial bioactive molecules. We aim in this chapter to provide an overview of the role of biotechnology in the synthesis and enhancement of bioactive compounds. Polyphenols, flavonoids, alkaloids, and terpenoids have powerful antioxidant, anti-inflammatory, antimicrobial, and anticancer properties. Low yield, environmental impact, and sustainability challenges often accompany traditional extraction methods. Through genetic engineering, metabolic engineering, and synthetic biology, biotechnology offers innovative solutions. Using genetic engineering techniques, such as gene cloning and expression, desired bioactive compounds can be produced in large quantities in microbial hosts. Through metabolic engineering, these production pathways are further optimized by altering the metabolic networks within the host organisms. The field of synthetic biology uses engineering principles to design and construct novel biological parts, devices, and systems, thus improving bioactive molecule production. Plant biotechnology is extensively investigated for the production of bioactive molecules. In addition, a number of different techniques, such as tissue culture, genetic alteration, or elicitation, are also used in this procedure. Biotechnological approaches have the potential to discover and produce novel bioactive compounds, ultimately improving human well-being and healthcare.

Keywords

Medicinal plants • Bioactive molecules • Biotechnology • Health benefits • Biosynthetic pathways

1 Introduction

Plants have been essential to human health and well-being since from centuries, making the basis of traditional medicine systems and modern pharmacology. Besides their medicinal properties, these plants are also used for cosmetics and aromatherapies because of their aromatic qualities. Their remarkable health benefits are primarily attributed to their rich reservoir of bioactive compounds, including alkaloids, flavonoids, terpenoids, phenolic compounds, and essential oils (Kaushal et al., 2020). In pharmaceuticals, nutraceuticals, and functional foods, MAPs exhibit a wide range of biological activities. So, it is important to understand these high-value bioactive compounds derived from medicinal and aromatic plants (Sharma et al., 2022a). This introduction aims to highlight the remarkable bioactive constituents of MAPs, their diverse therapeutic potentials and relevance in contemporary science.

Plants contain a variety of bioactive compounds. Despite meeting nutritional requirements, they promote health. Scientists have become increasingly interested in bioactive compounds due to their therapeutic properties, including

"Munish Sharma and Munish Sharma" Both two authors are different having same name and affiliation.

M. Sharma · M. Sharma (✉)
Department of Plant Sciences, Central University of Himachal Pradesh, Shahpur, Kangra, 176206, Himachal Pradesh, India
e-mail: munishptc@gmail.com

M. Sharma
e-mail: rs.munish.cuhp@gmail.com

M. Sharma
Department of Botany, Minerva P.G. College of Arts, Science and Commerce, Indora, Kangra, 176401, Himachal Pradesh, India

S. K. Pandey
School of Inter-Disciplinary and Trans-Disciplinary Studies (SOITS), Indira Gandhi National Open University, New Delhi, 110068, India

D. Sharma
Department of Plant Science, University of Manitoba, Winnipeg, MB R3T2N2, Canada

J. Devi (ed.), *Biotechnological Intervention in Production of Bioactive Compounds*,
Sustainable Landscape Planning and Natural Resources Management, https://doi.org/10.1007/978-3-031-76859-0_8

antioxidant, anti-inflammatory, anticancer, antimicrobial, and neuroprotective properties (Kumar et al., 2023). Bioactive compounds are becoming increasingly important in pharmacognosy, biotechnology, and functional foods. Bioactive compounds are used in pharmaceuticals, supplements, and nutraceuticals to improve human health. A bioactive compound study extends beyond its properties to its mechanism of action, biosynthesis, and potential synergies, providing insights into its safety and efficacy (Patil et al., 2009; Sharma et al., 2023, 2024a). The field of biotechnology has become one of the fastest growing fields in modern science, involving biological systems, organisms, and processes to produce innovative products and technologies. It has a profound impact on a wide range of fields, including agriculture, medicine, and environmental sciences (Kumar et al., 2024; Molina et al., 2015; Verma & Chandel, 2019).

Medicinal and aromatic plants have been used as raw materials for a variety of applications such as cosmetics, food, feed, environment protection, and pharmacology as well as an alternative to synthetic drugs. There are bioactive compounds found in plants and conventional extraction and synthesis methods face challenges related to sustainability, yield, and purity (Sharma et al., 2024d; Azmir et al., 2013).

In biotechnology, these challenges are addressed by cutting-edge techniques such as genetic engineering, metabolic engineering, and synthetic biology, enabling efficient and sustainable production. Researchers can enhance the biosynthesis of these compounds, enhance their stability and bioavailability, and even create new molecules with enhanced or unique properties using biotechnology. Pharmaceutical companies can develop new drugs based on this, agriculture can develop pest-resistant crops with nutritional enhancements, and the food industry can create functional foods promoting health (Ahuja et al., 2024; Kumar et al., 2020; Alamgir & Alamgir, 2018; Molina et al., 2015).

This chapter explores the dynamic interplay between biotechnology and the production of bioactive compounds, highlighting key techniques, advancements, and applications that are driving innovation and transforming industries worldwide.

2 Classifications of Bioactive Compounds

According to their chemical structure, biological origin, and functional properties, bioactive compounds are classified as naturally occurring chemicals with significant biological effects (Fig. 1) (Sharma et al., 2020a). There are many different types of alkaloids that are found in plants and are nitrogen-containing compounds. As a result of their strong pharmacological properties, they are able to relieve pain, prevent malaria, and prevent cancer. Among the most famous alkaloids are morphine, quinine, and caffeine, which are all examples of alkaloids. The flavonoids are a group of polyphenolic compounds that have been characterized for

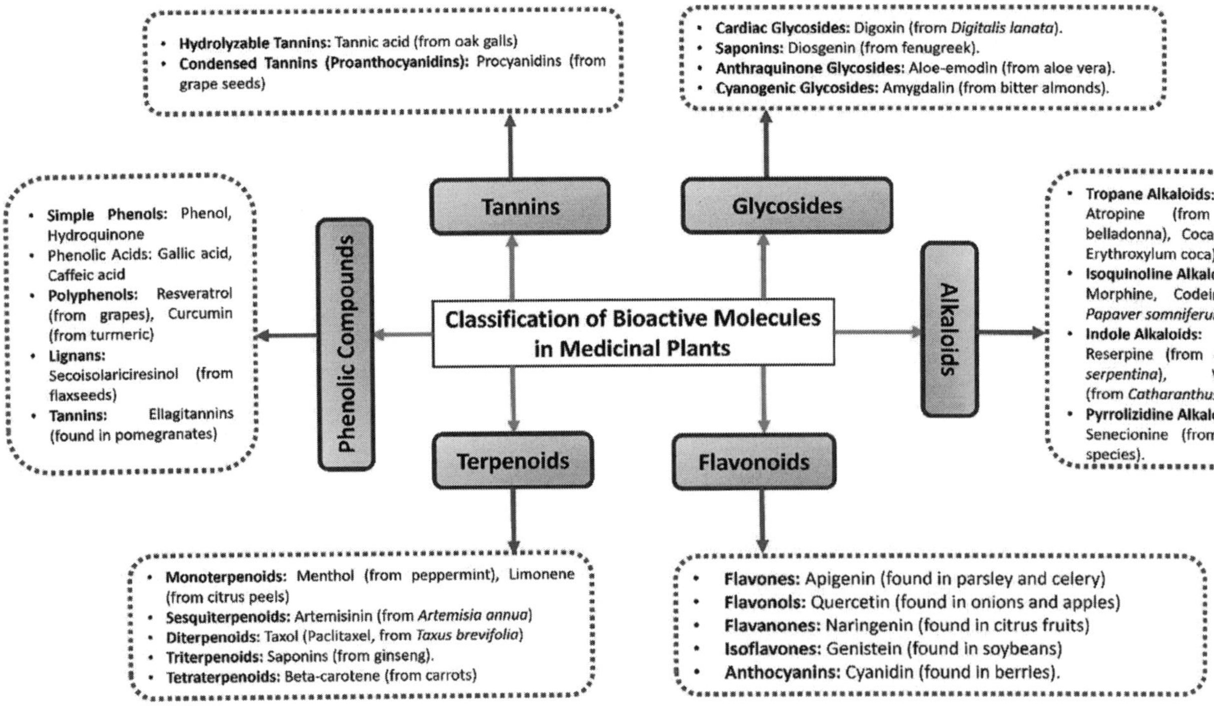

Fig. 1 Classification of bioactive compounds present in medicinal and aromatic plants

their antioxidant properties, as well as their role in protecting plants against UV radiation and pathogens; among them are quercetin, kaempferol, and catechin (Batool et al., 2023; Sharma et al., 2024c). Terpenoids are groups of organic compounds derived from five-carbon isoprene units which have a unique structure and include monoterpenoids such as menthol, sesquiterpenoids such as artemisinin, diterpenoids such as taxol, triterpenoids such as saponins, and tetraterpenoids such as vitamin A (Sharma et al., 2024d; Dhiman et al., 2023a, b).

Antioxidants are the organic compounds in this group that contain either one or more hydrocarbon groups attached to aromatic rings. An example of a phenolic compound is resveratrol, curcumin, and gallic acid, which are all phenolic compounds. There are a range of compounds known as glycosides, which contain a sugar molecule bonded to a moiety other than carbohydrate, such as digoxin and saponins, such as diosgenin. There are different kinds of saponins, such as steroidal saponins such as dioscin, and triterpene saponins such as ginsenoside. Their main characteristic is that they produce foam and that they have a variety of biological functions. It is important to differentiate between hydrolysable tannins, such as tannic acid, and condensed tannins, such as procyanidins, which are polyphenolic compounds with astringent properties (Dhiman et al., 2023a, b; Sharma et al., 2024e).

A polysaccharide is a long-chain carbohydrate molecule found in a wide variety of foods. For example, beta-glucans, found in oats and mushrooms, pectin, found in citrus fruits, and gums such as gum Arabic are examples. Antimicrobial peptides such as defensins, and antihypertensive peptides derived from the casein protein are examples of bioactive peptides and proteins that can have various physiological effects, including antimicrobial and antihypertensive properties. There are a variety of substances called steroids that have a characteristic four-ring structure, including steroidal hormones like oestrogen, phytosterols like beta-sitosterol, and steroidal saponins like dioscin. Coumarins are aromatic compounds that have the structure of benzopyrones and are known for their aromatic properties. There are many different types of coumarins, such as simple coumarins such as coumarin, furanocoumarins such as psoralen, and pyranocoumarins such as xanthyletin. Bioactive compounds are categorized highly in terms of their diversity and therapeutic potential, and this is of great significance when developing pharmaceuticals, nutraceuticals, and functional foods (Table 1) (Ahuja et al., 2024; Sharma et al., 2020b, 2022b).

Table 1 Some important medicinal plants having health promising bioactive compounds

Common name	Botanical name	Family	Bioactive compounds	Structure description	Structure
Aloe	*Aloe vera*	Asphodelaceae	Aloin, Emodin	Anthraquinone derivatives	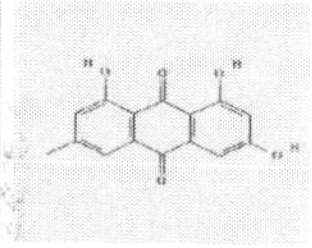
Ginger	*Zingiber officinale*	Zingiberaceae	Gingerol, Shogaol	Phenolic ketones	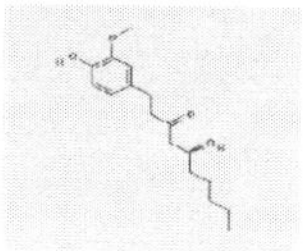
Turmeric	*Curcuma longa*	Zingiberaceae	Curcumin	Polyphenolic compound	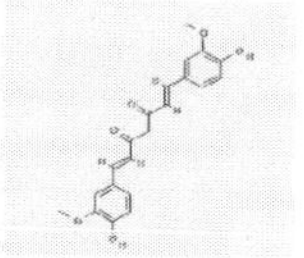
Garlic	*Allium sativum*	Amaryllidaceae	Allicin	Organosulfur compound	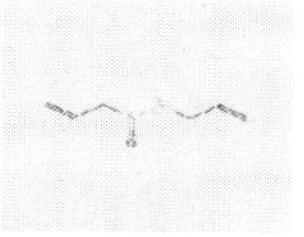
Ginseng	*Panax ginseng*	Araliaceae	Ginsenosides	Triterpene saponins	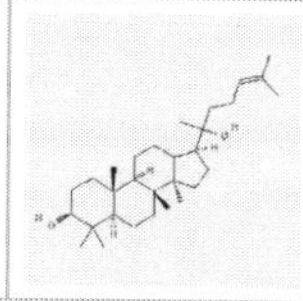

(continued)

Table 1 (continued)

Common name	Botanical name	Family	Bioactive compounds	Structure description	Structure
Echinacea	*Echinacea purpurea*	Asteraceae	Echinacoside	Phenolic compound	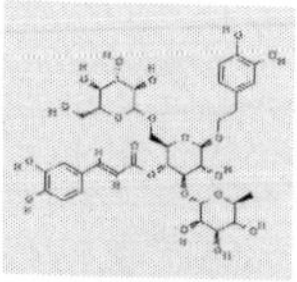
Neem	*Azadirachta indica*	Meliaceae	Azadirachtin	Tetranortriterpenoid	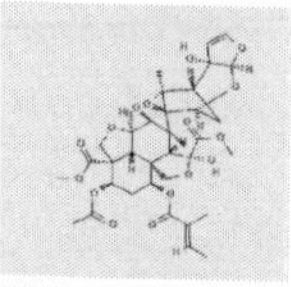
Mint	*Mentha piperita*	Lamiaceae	Menthol	Monoterpenoid	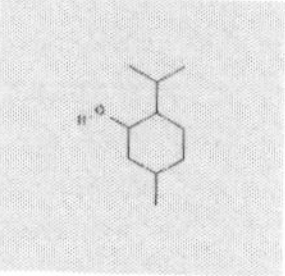
Chamomile	*Matricaria chamomilla*	Asteraceae	Apigenin	Flavonoid	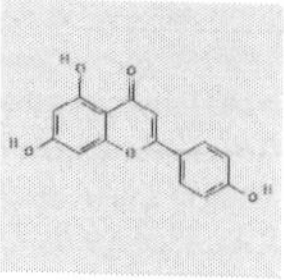
Lavender	*Lavandula angustifolia*	Lamiaceae	Linalool	Monoterpenoid alcohol	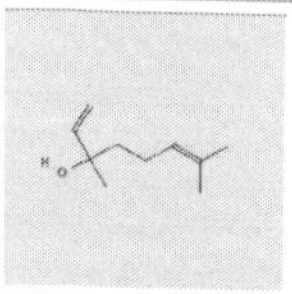
Rosemary	*Rosmarinus officinalis*	Lamiaceae	Rosmarinic acid	Polyphenolic compound	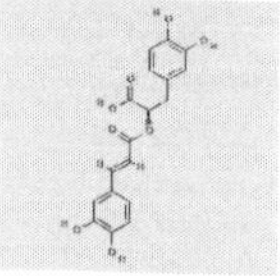
Sage	*Salvia officinalis*	Lamiaceae	Thujone	Monoterpenoid ketone	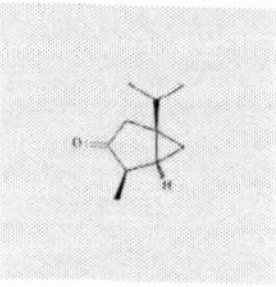
Basil	*Ocimum basilicum*	Lamiaceae	Eugenol	Phenylpropanoid	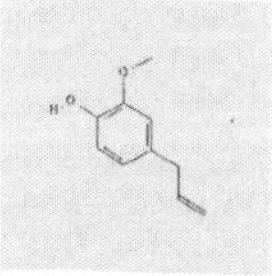
Thyme	*Thymus vulgaris*	Lamiaceae	Thymol	Monoterpenoid phenol	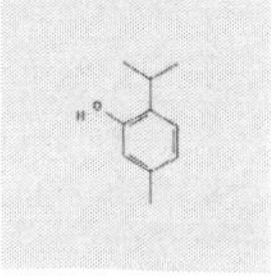
Ashwagandha	*Withania somnifera*	Solanaceae	Withanolides	Steroidal lactones	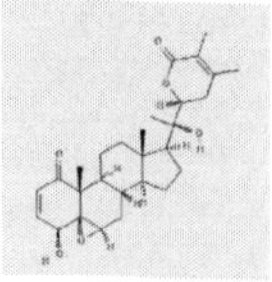

(continued)

Table 1 (continued)

Common name	Botanical name	Family	Bioactive compounds	Structure description	Structure
Cinnamon	*Cinnamomum verum*	Lauraceae	Cinnamaldehyde	Phenylpropanoid	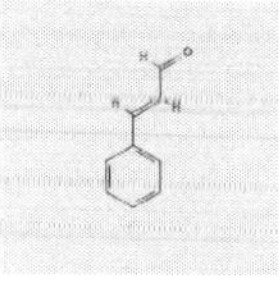
Clove	*Syzygium aromaticum*	Myrtaceae	Eugenol	Phenylpropanoid	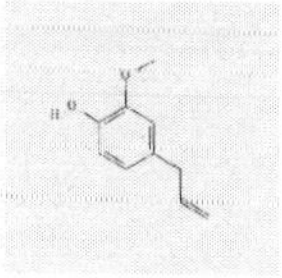
Dandelion	*Taraxacum officinale*	Asteraceae	Taraxasterol	Triterpenoid	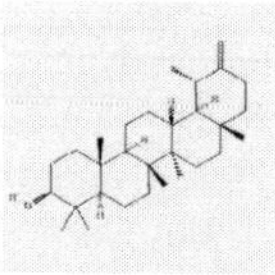
Holy Basil	*Ocimum sanctum*	Lamiaceae	Ursolic acid	Pentacyclic triterpenoid	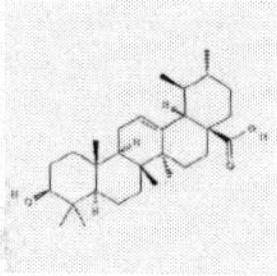
Licorice	*Glycyrrhiza glabra*	Fabaceae	Glycyrrhizin	Saponin glycoside	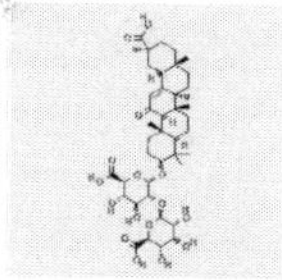
Ginkgo	*Ginkgo biloba*	Ginkgoaceae	Ginkgolides	Terpenoid lactones	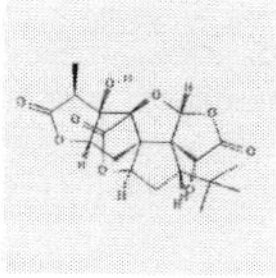
Green tea	*Camellia sinensis*	Theaceae	Epigallocatechin gallate	Polyphenolic compound	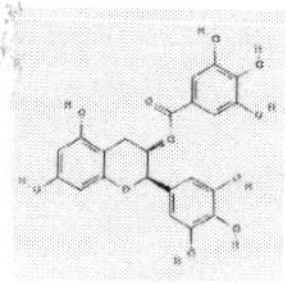
Fenugreek	*Trigonella foenum-graecum*	Fabaceae	Diosgenin	Steroidal saponin	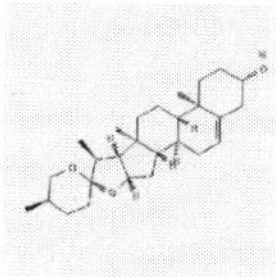
Milk Thistle	*Silybum marianum*	Asteraceae	Silymarin	Flavonolignan	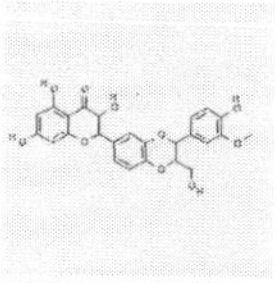
St. John's Wort	*Hypericum perforatum*	Hypericaceae	Hypericin	Anthraquinone derivative	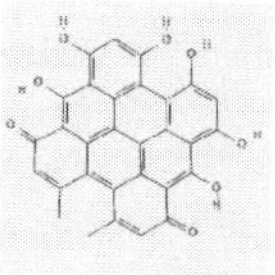

(continued)

Table 1 (continued)

Common name	Botanical name	Family	Bioactive compounds	Structure description	Structure
Valerian	*Valeriana officinalis*	Caprifoliaceae	Valerenic acid	Sesquiterpenoid	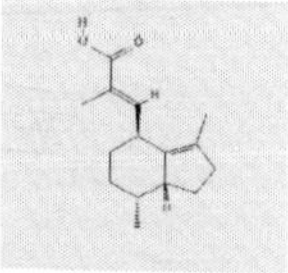
Cat's Claw	*Uncaria tomentosa*	Rubiaceae	Rhynchophylline	Indole alkaloid	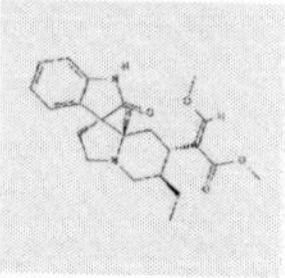
Evening Primrose	*Oenothera biennis*	Onagraceae	Gamma-linolenic acid	Fatty acid	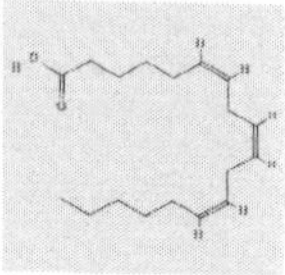
Feverfew	*Tanacetum parthenium*	Asteraceae	Parthenolide	Sesquiterpene lactone	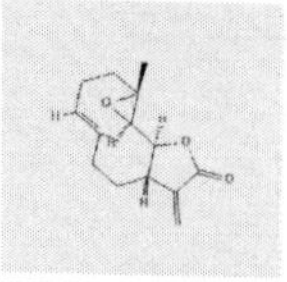
Devil's Claw	*Harpagophytum procumbens*	Pedaliaceae	Harpagoside	Iridoid glycoside	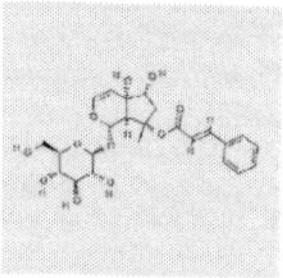
Goldenseal	*Hydrastis canadensis*	Ranunculaceae	Berberine	Isoquinoline alkaloid	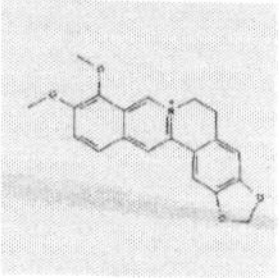
Red Clover	*Trifolium pratense*	Fabaceae	Isoflavones	Phytoestrogenic compounds	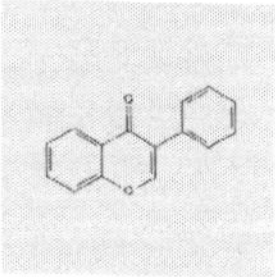
Black Cohosh	*Actaea racemosa*	Ranunculaceae	Actein, Cimicifugoside	Triterpene glycosides	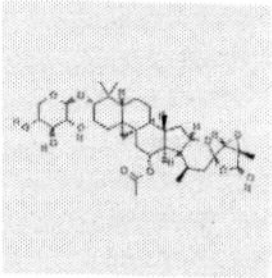
Saw Palmetto	*Serenoa repens*	Arecaceae	Beta-sitosterol	Phytosterol	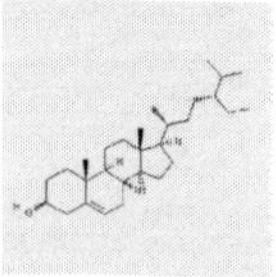
Elderberry	*Sambucus nigra*	Adoxaceae	Anthocyanins	Flavonoid pigments	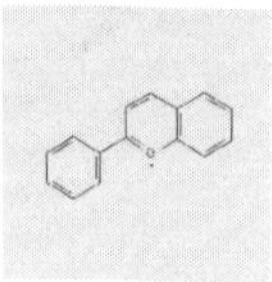

(continued)

Table 1 (continued)

Common name	Botanical name	Family	Bioactive compounds	Structure description	Structure
Cranberry	*Vaccinium macrocarpon*	Ericaceae	Proanthocyanidins	Condensed tannins	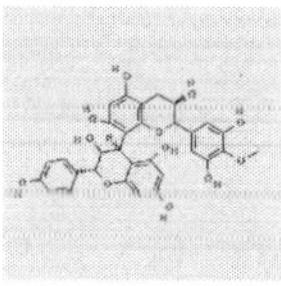
Hibiscus	*Hibiscus sabdariffa*	Malvaceae	Hibiscin	Flavonoid glycosides	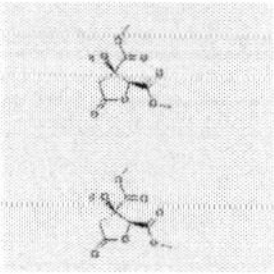
Burdock	*Arctium lappa*	Asteraceae	Arctiin, Arctigenin	Lignans	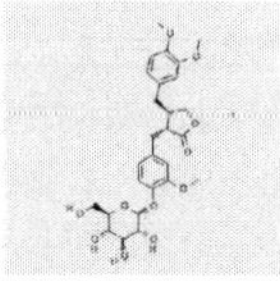
Yarrow	*Achillea millefolium*	Asteraceae	Achilline, Flavonoids	Sesquiterpene lactones, Polyphenolics	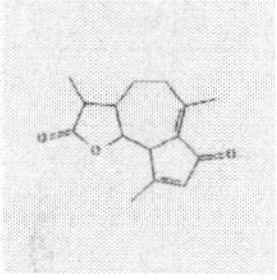
Gotu Kola	*Centella asiatica*	Apiaceae	Asiaticoside	Triterpene saponin	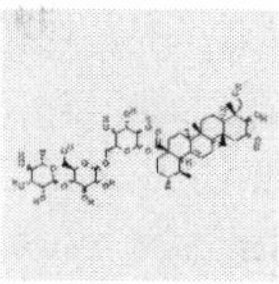
Nettle	*Urtica dioica*	Urticaceae	Scopoletin	Coumarin derivative	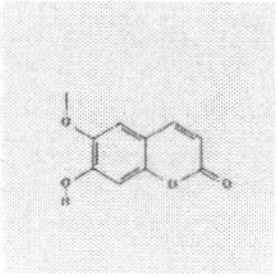
Passionflower	*Passiflora incarnata*	Passifloraceae	Passiflorine	Indole alkaloid	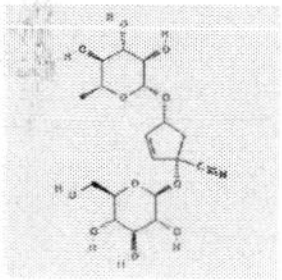
Peppermint	*Mentha × piperita*	Lamiaceae	Menthol, Menthone	Monoterpenoids	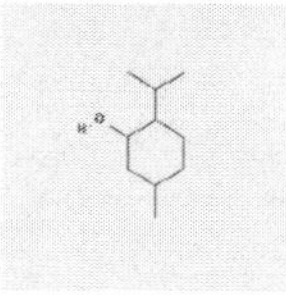
Skullcap	*Scutellaria baicalensis*	Lamiaceae	Baicalin	Flavone glycoside	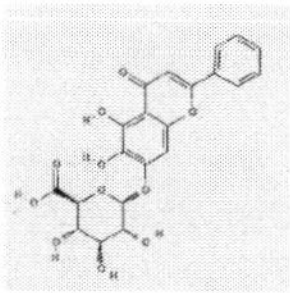
Soursop	*Annona muricata*	Annonaceae	Acetogenins	Polyketide derivatives	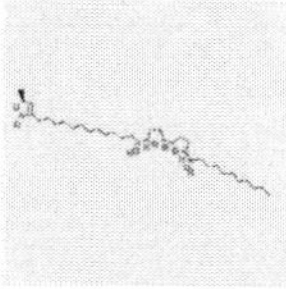

(continued)

Table 1 (continued)

Common name	Botanical name	Family	Bioactive compounds	Structure description	Structure
Willow	*Salix alba*	Salicaceae	Salicin	Phenolic glycoside	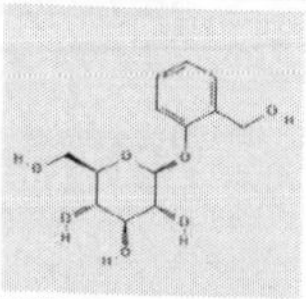

Source Structure original source and adapted from PubChem and Sharma et al. (2024a–g)

3 Biosynthesis of Bioactive Compounds

The biochemical pathways involved in the synthesis of bioactive compounds in plants are complex networks of enzymatic reactions that convert primary metabolites into a wide range of secondary metabolites exhibiting significant biological activities (Fig. 2). It is through these pathways that the plant uses to defend itself, grow, develop, and interact with the environment, all of which are vital to the plant's survival. In terms of biosynthetic pathways, alkaloids, flavonoids, terpenoids, phenolic compounds, glycosides, saponins, tannins, polysaccharides, peptides, proteins, steroids, and coumarin are just some of the classes of bioactive compounds that have distinct biosynthetic pathways (Sharma et al., 2024f).

During the biosynthesis of bioactive compounds in plants, a complex network of metabolic pathways is involved, each tightly regulated by the activity of specific enzymes. As a result of these metabolic pathways, primary metabolites become a diverse array of secondary metabolites, each of which exhibits a unique chemical structure and biological activity. Developing biotechnological applications for bioactive compounds to enhance their production for pharmaceutical, nutraceutical, and agricultural purposes is critical in order to research and understand their therapeutic potential as well as the development of biotechnological applications (Ahuja et al., 2024; Chung et al., 2016; Kumar & Singh, 2023; Uwineza & Waśkiewicz, 2020).

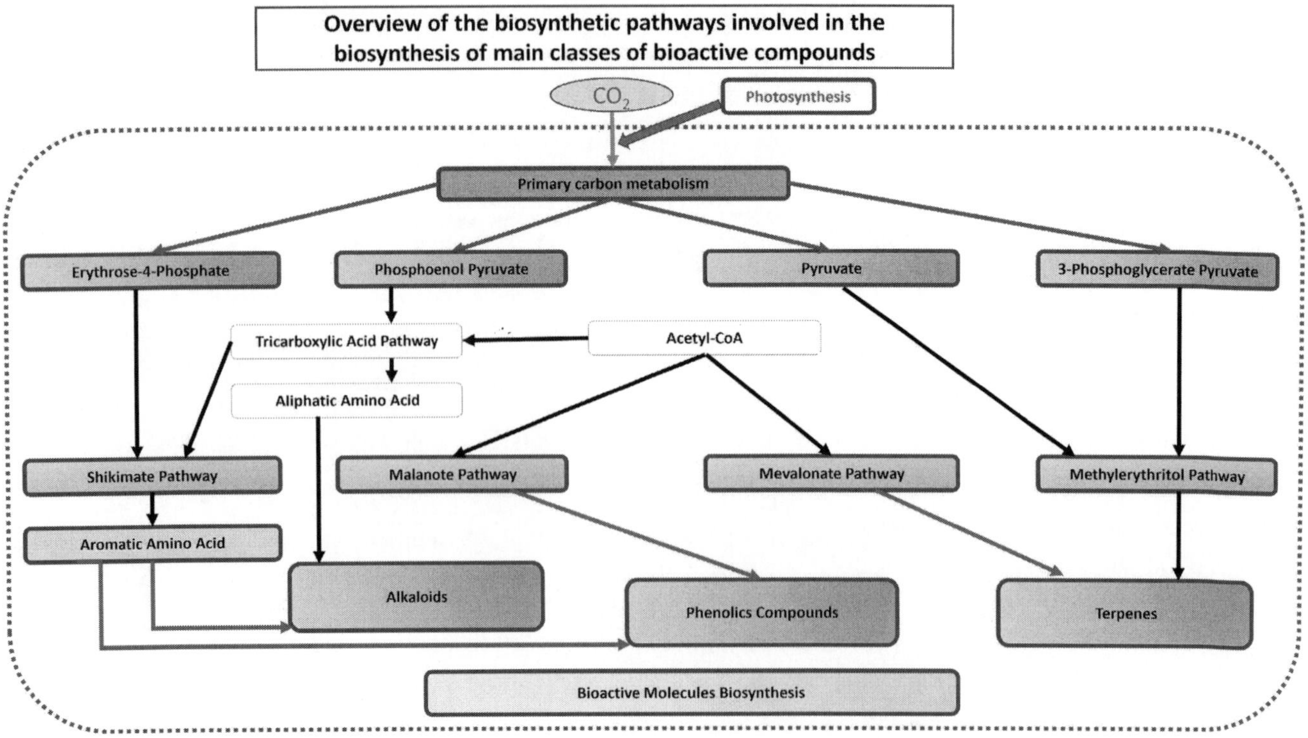

Fig. 2 Generalised biosynthetic pathways of bioactive compounds from medicinal and aromatic plants

3.1 Alkaloid Biosynthesis

As nitrogen-containing compounds, alkaloids are predominantly found in plants and show a number of pharmacological effects based on their chemical structure. The biosynthesis of alkaloids is usually initiated by amino acids, such as tyrosine, tryptophan, ornithine, and lysine, which are precursors of the alkaloid. A tyrosine molecule is converted into dopamine during the biosynthesis of morphine, which is derived from tyrosine. Throughout the enzymatic process of morphine synthesis, a number of enzymatic reactions are performed, including decarboxylation, methylation, and oxidation. These pathways are comprised of a number of enzymes, including decarboxylases, oxidases, and methyltransferases, each of which facilitates the conversion of amino acid precursors into complex alkaloid structures (Sharma et al., 2022c; Song et al., 2022).

3.2 Flavonoid Biosynthesis

There are a variety of polyphenolic compounds known as flavonoids and they play a role in the protection of plants through their antioxidant properties. Phenylpropanoid biosynthesis starts with the amino acid phenylalanine, which is the precursor to phenylpropanoid biosynthesis. An enzyme called phenylalanine ammonia lyase (PAL) is responsible for converting phenylalanine into cinnamic acid from phenylalanine. In order for naringenin chalcone, a key intermediate in the synthesis of naringenin, to be formed, a series of enzyme reactions is carried out, including chalcone synthase (CHS) and chalcone isomerase (CHI). As a result of further enzymatic modifications, a variety of flavonoid structures are produced, such as flavones, flavonols, flavanones, isoflavones, and anthocyanins, each of which has unique biological activities (Chen et al., 2023; Sharma et al., 2022d).

3.3 Terpenoid Biosynthesis

Terpenoids, which are also known as isoprenoids, are steroid compounds that are synthesized from units of isoprene derived from two primary pathways: that of the mevalonate (MVA) pathway and that of the methylerythritol phosphate (MEP) pathway. As a result of both pathways, the building blocks isopentenyl pyrophosphate (IPP) and dimethylallyl pyrophosphate (DMAPP) are produced. As a result of a series of reactions involving the HMG-CoA reductase enzyme, acetyl-CoA is converted into IPP and DMAPP as part of the MVA pathway. MEP is a pathway in which pyruvate and glyceraldehyde-3-phosphate are the starting materials, and the enzyme DXS is the key enzyme in the process. A wide variety of terpenoids can be formed by combining and modifying these isoprene units via terpene synthases, including monoterpenoids such as menthol, sesquiterpenoids such as artemisinin, diterpenoids such as taxol, triterpenoids like saponins and beta-carotene, among others (Ahuja et al., 2024; Li et al., 2023).

3.4 Phenolic Compound Biosynthesis

There are two pathways by which phenolic compounds are synthesized: the shikimate pathway and the phenylpropanoid pathway. The phenylalanine ammonia lyase (PAL) synthesizes cinnamic acid from phenylalanine, which is then converted into cinnamic acid by the shikimate pathway starting with simple sugars and producing aromatic amino acids such as phenylalanine. The formation of phenolic acids and polyphenols, including flavonoids, lignans, and tannins, is the outcome of this process. It has been well established that enzymes such as cinnamate 4-hydroxylase (C4H) and 4-coumarate-CoA ligase (4CL) play vital roles in the transformation of cinnamic acid derivatives into more complex phenolic structures (Hu et al., 2022; Sharma & Sharma 2021a).

3.5 Glycoside Biosynthesis

The formation of glycosides is based on the attachment of sugar moieties to non-carbohydrate aglycones, which may originate from flavonoids, terpenoids, or other secondary metabolites in plants. In glycosylation, a sugar unit is transferred from nucleotide-activated sugars into the aglycone by a process known as glycosyltransferases, which catalyze the sugar transfer. A steroid aglycone can be glycosylated to produce cardiac glycosides such as digoxin by glycosylating the steroid aglycone. By glycosylating the backbones of triterpenoid compounds, saponins are produced, resulting in highly active glycoside compounds that have a wide range of therapeutic properties (Kytidou et al., 2020; Sharma & Sharma, 2021b).

3.6 Saponin Biosynthesis

There are several enzymatic steps that are involved in the synthesis of saponins from their terpenoid backbones. This triterpenoid backbone is formed by the cyclization of oxidosqualene-by-oxidosqualene cyclase, starting out as oxidosqualene, which is cyclized by oxidosqualene cyclase. Upon oxidation, hydroxylation, and glycosylation of the backbone of the molecule, saponins can be synthesized. An enzyme called glycosyltransferase adds sugar units to the triterpenoid structure, which results in saponins being formed. Saponins are found to have diverse biological activities, ranging from antifungal properties to antiviral effects and cholesterol lowering properties (Sharma et al., 2021a; Upadhyay et al., 2018).

3.7 Tannin Biosynthesis

In the phenylpropanoid pathway, tannins are derived from polyphenolic compounds which have a polyphenolic structure. A tannin that can be hydrolyzed from phenolic acids, such as gallic acid, produces condensed tannins, also known as proanthocyanidins, which are polymers containing flavonoid units. In order to produce tannins, enzymes such as PAL, C4H, and 4CL are involved in the biosynthesis process, which facilitates the formation of phenolic precursor molecules. As a result of polymerization and further modification, the precursors can be converted into the diverse array of tannins found in plants. These tannins have many beneficial properties for the body, including antioxidant, antimicrobial, and astringent properties (Sharma et al., 2021b; He et al., 2015).

3.8 Polysaccharide Biosynthesis

There are several types of polysaccharides, which can be classified as long-chain carbohydrates composed of repeating monosaccharides. Their biosynthesis involves the polymerization of simple sugars such as glucose, fructose, and galactose, which are used as starting materials. During the formation of glycosidic bonds between sugar units, enzymes such as glycosyltransferases and polysaccharide synthases are involved. In order to synthesize beta-glucans, which have immunomodulatory and antitumor properties, glucose units are polymerized with 1-1-3 and 1-1-6 linkages through the use of reducing sugars (Sharma & Thakur, 2020; Yuan et al., 2019).

3.9 Peptide and Protein Biosynthesis

It is believed that bioactive peptides and proteins are produced through the process of ribosomal synthesis, where amino acids are assembled into polypeptides following genetic instructions. A number of post-translational modifications, including cleavage, phosphorylation, and glycosylation, contribute to the refinement and functionalization of these peptides and proteins. A number of antimicrobial peptides have been synthesized as pre-proteins, which are then cleaved into their bioactive forms, which play crucial roles in the plant's defense against pathogens (Akbarian et al., 2022; Sharma et al., 2020d).

3.10 Steroid Biosynthesis

The steroid hormones are synthesized from terpenoids in a chain reaction that begins with squalene, which is the starting material for lanosterol, which is produced by squalene epoxidase and oxido-squalene cyclase. There are many types of steroids that can be produced from lanosterol because of an enzymatic process called cyclization, hydroxylation, and oxidation which is catalyzed by various cytochrome P450 enzymes. It is important to note that plant sterols contribute to the structure of the cell membrane, and that steroidal hormones play a role in regulating a wide range of physiological processes (Sharma et al., 2020e; Bishop, 2007).

3.11 Coumarin Biosynthesis

The steps involved in the synthesis of coumarins are derived from cinnamic acid derivatives through the phenylpropanoid pathway. A number of enzymes, including the cinnamate 4-hydroxylase (C4H), hydrolyze cinnamic acid in order to form coumarin and this in turn is lactonized by the coumarin synthase enzyme to form coumarin. Various coumarin derivatives have been synthesized, including hydroxylated compounds and glycosylated compounds, which possess anticoagulant, anti-inflammatory, and anticancer properties (Zou et al., 2023; Bourgaud et al., 2006).

4 Medicinal Flora and Their Therapeutic Properties

Plants with medicinal and aromatic properties (MAPs) contain a wide range of bioactive compounds and are therefore highly sought after in pharmaceuticals, nutraceuticals, and cosmetics industries (Fig. 3). These MAPs are notable for having high-value bioactive compounds that are denoted by their high bioactivity (Batool et al., 2024; Jaspal et al., 2023; Singh et al., 2024; Thakur et al., 2020).

4.1 Rhododendron

Rhododendron trees contain a variety of phytochemicals, which are found in various parts of the tree. Flavonoids, which can prevent the formation of oxidants in the bloodstream, can be found in Rhododendron leaves. There is a substance called quercetin that can be found in leaf and flower extracts. In addition to rutin and sterol, the leaves are also rich in antioxidants. There are many compounds that can be found in the bark of this tree, including betulin, lupeol, and ursolic acid, among others. In comparison with synthetic antioxidant drugs, Rhododendron extracts have

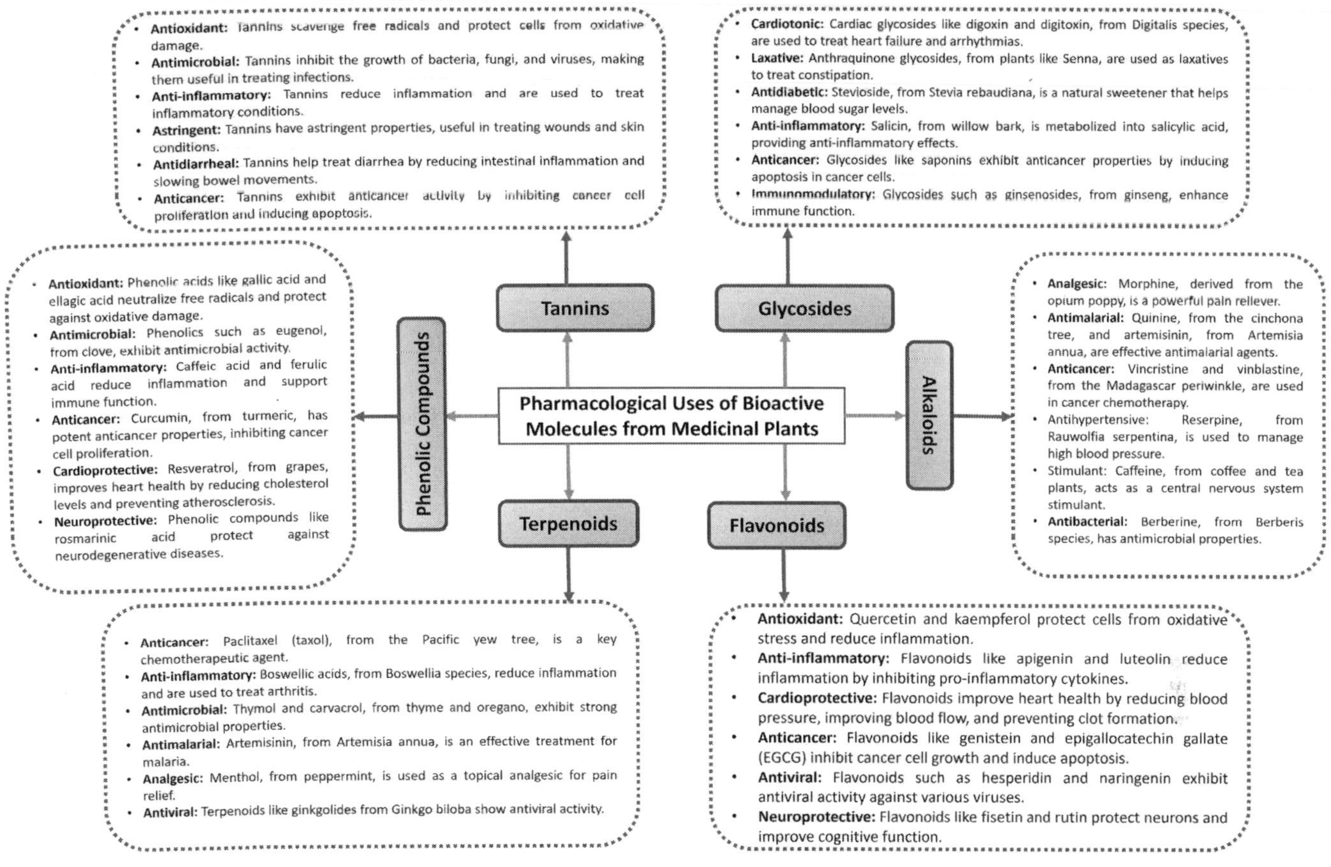

Fig. 3 Health promising benefits of bioactive compounds plants

been shown to exhibit high antioxidant properties (Prakash et al., 2007).

4.2 Aconitum

Aconitum species produce diterpene alkaloids and flavonoids in abundance due to the presence of this compound. As a result of the presence of the alkaloid aconitine within the tubers of the plant, the plant has medicinal properties. In addition to the tannic acid, starch, fatty acids and their glycerides, carbohydrates, and others that the plant contains, it also contains a wide range of other phytochemicals (Bahuguna et al., 2000). The use of aconite can be beneficial in the case of communicable diseases, such as measles, which can be transmitted by the air. Further, it has been found to be effective in treating diseases such as asthma, diabetes, leukoderma, leprosy, and convulsions. Antioxidants and antibacterial properties are attributed to plant components such as flavonoids and phenols. As a result, it is effective in treating coughs and congestion. There are also diuretic properties associated with this plant, so it is also both aphrodisiac and diuretic in nature (Beigh et al., 2006).

4.3 Bacopa Monnieri

This plant also contains a number of compounds that have therapeutic properties, including alkaloids, saponins, glycosides, flavonoids, and stigmasterols, in addition to a wide range of other compounds. Brahmi contains saponins known as bacosides, which are saponins found in plants. A number of flavonoids are found in it, including luteolin and apigenin. There are two alkaloids that have been isolated from Brahmi, namely herpes time and brahmine, which if taken at a high dosage may cause extreme headaches (Mitra et al., 2010). With Bacopa, it is possible to treat conditions like eczema, ulcerations, rashes, and rashes. A brain tonic, it is a popular medicine used to treat amnesia and Alzheimer's diseases. By detoxifying the blood, it removes toxins. In addition to preventing hair loss and premature graying, it is also an effective anti-hair loss and anti-hair loss herb. Hair follicles are rejuvenated and baldness is prevented (Bhandari et al., 2020). Anti-diabetic, anti-cancer, and anti-arthritic, it exhibits anti-diabetic and anti-cancer properties. A powerful antioxidant in Brahmi prevents free radical damage to the body (Ahuja et al., 2024; Rawat & Vashistha, 2011).

4.4 Glycyrrhiza

The root of liquorice contains two isoflavones, glabrene, and glabridin, both of which are phytoestrogens. Aside from saponin triterpenes (glycyrrhetinic acid, liquiric acid, and glycyrrhizin), the roots also contain flavonoids, coumarins, sugars, tannins, starch, choline, amino acids, and phytosterols, which are glycosides. There are many advantages to using Mulethi. It can reduce acid reflux and indigestion and can help to prevent ulcer formation by lowering stomach acid levels. It uses action on action to increase the flow of bile and lower the cholesterol level in the body. In terms of respiratory organ irritation and inflammation, this medicinal plant is useful for soothing and healing those symptoms (Thakur & Raj, 2017).

4.5 Berberis

Berberine, a quaternary iso-quinoline alkaloid found mostly in the stems and roots of this herbal plant, is its most important constituent. Roots and stems usually contain it. Quaternary ammonium salts of iso-quinoline alkaloids make up the bark. The leaves and stems of plants can also be used to extract lignans. *Berberis* species in the western Himalayas are rich in fibre, protein, fats and some minerals (Srivastava et al., 2001). Melasma is cured by Berberis extremely effectively. Counteracting stomach disorders is one of its benefits. Liver disorders can be prevented with its use. Due to its ability to prevent water loss, it is used as an occasional mild laxative (Joshi et al., 2011).

4.6 Aegle Marmelos

Kala et al. (2006) reported that this plant contains a variety of alkaloids, including aegline, marmesin, marmin, and marmelosin. In addition to diarrhea, dysentery, dyspepsia, malaria, fever, jaundice, and skin diseases like ulcers, urticaria, and eczema, nearly all parts of the tree are used to prepare herbal medicines (Sharma & Thokchom, 2014).

4.7 Hedychium

It has been reported that *H. spicatum* has antidiabetic properties, tranquillizing actions, pediculicidal effects, antimicrobial properties, antioxidant properties, antimalarial hepatoprotective actions, cytotoxic effects, anti-helminthic effects, stomachic effects, and tonic activities. Its anti-diabetic properties are attributed to 1, 8 cineole, a compound obtained from the rhizome. A study conducted by Chaudhary et al. (2012) demonstrated the anti-microbial properties of the essential oil. *H. acuminatum* is used most commonly for liver problems. Cataracts can be cured with *H. coronarium*. Indigestion is the best application of *H. marginatum* (Devi et al., 2014).

4.8 Picrorhiza Kurroa

This plant has been found to have beneficial effects as a laxative, liver stimulant, appetite and stimulant, and febrifuge in Ayurveda. Besides alleviating stomach aches, it is believed to stimulate appetite as well. Asthma and epidemic jaundice are also treated with the plant. In addition to treating disorders related to the chakra system, the herb is also useful for treating billow fever, urinary discharge, hiccups, blood troubles, burning sensations, and leukoderma (Sharma & Thokchom, 2014).

4.9 Taxus Wallichiana

Himalayan Yew is well known that *Taxus wallichiana* is well known for its therapeutic properties due to the presence of taxol, an anticancer substance. Taxol is found in the plant *Taxus wallichiana.* Taxol, which is found in the bark, needles, and seeds of the tree, is used for treating a variety of cancers, such as cancers of the lung, breast, and ovary. Due to its medicinal properties, it is an important ingredient in the pharmaceutical industry due to its economic importance (Bhuju & Gauchan, 2018; Sharma & Garg, 2015).

4.10 Juglans Regia

Due to the omega-3 fatty acids found in walnuts, walnut consumption has been linked to a number of health advantages, such as improved heart health due to its omega-3 fatty acid content, decreased inflammation, lowered cholesterol levels, and support for cognitive and mental health (Sharma et al., 2022a–d, 2024a–d). Juglone is a natural compound found in walnut trees (genus Juglans) and other plant species. Juglone can switch on pro-apoptotic proteins and hindrance anti-apoptotic proteins, thus inducing apoptosis, or programmed cell death, in cancer cells. Juglone can cause DNA damage in cancer cells. This can trigger a DNA repair response or lead to cell cycle arrest and apoptosis. Juglone has anti-angiogenic qualities, which means it can prevent the occurrence of fresh new blood vessels that provide tumours with nourishment, thereby restricting their growth. Tumour development and progression may be aided by persistent inflammation. Juglone anti-inflammatory qualities may be a factor in its anticancer characteristics. The NF-B and PI3K/Akt pathways, which are linked to cell survival, proliferation, and inflammation. It can be affected by juglone, as can other signalling pathways implicated in cancer. In a study of glioblastoma cancer cells, researchers discovered that juglone suppressed cell proliferation through activation of the p53 pathway, which controls apoptosis and cell cycle arrest (Sharma et al., 2022a–d).

4.11 Saussurea Costus

The perennial herbaceous plant Costus, also known as Kuth or Costus, belongs to the family Asteraceae. Most pharmacologically active components of the plant are its rhizomes, which offer wide-ranging therapeutic benefits. Known for its anti-inflammatory, analgesic, digestive, and expectorant properties, costus has long been used as a medicine. Aromatic properties of costus make it a popular ingredient in the manufacture of fragrances and incenses due to its scent properties. As a result of their characteristic musky odour, rhizomes of the plant can be used to make exotic perfumes owing to their ability to lend depth and richness to fragrances (Nadda et al., 2020; Ali & Venkatesalu, 2022).

4.12 Adhatoda Vasica

There are a variety of compounds that can be found in the various organs of adhatoda vasica, including alkaloids, phytosterols, polyphenolics, and glycosides. Its main constituents are quinazoline alkaloids of which vasicine is the chief compound. It consists mostly of quinazoline alkaloids. There are anticancer, antifungal, and antibacterial properties associated with the entire plant as a whole. The leaf product of *A. vasica* includes moderate antibacterial activity. The leaves of the plant have high antiulcer activity. The powder prepared from leave reported an antiulcer activity in rats observed in the ethanol-induced ulceration model compared to aspirin. Vasicine, the principal alkaloid of *Adhatoda*, showed anti-inflammatory activity. The vasicine alkaloid present in the *Adhatoda* plant has ureteric activity. Vasicine induces rhythmic contractions in non-pregnant and pregnant uterine muscle strips, producing an effect similar to oxytocin and mathergin. Vasicinone obtained from the roots produced transient hypotension in cats, the guinea pig while contraction of the intestine and depression of isolated heart thus shows good anticholinesterase activity (Shamsuddin et al., 2021).

4.13 Carum Carvi

The main components present in plants include carvacrol, carvone, limonen, α-pinene, linalool, carvenone, and p-cymene. The major component present in cumin includes cuminaldehyde, α and β-pinene, 1, 8 cineole, o- and p-cymene, α and γ terpinene, and safranal. Many flavonoids and phenolic compounds, lignin, and alkaloids are present in solvent-derived seed extracts. The seed and root of caraway also contain polyacetylene compounds. In a recent study, lipid transfer protein has been isolated from cumin seed. The flavonoid constituents of caraway include quercetin-3-glucuronides, isoquercetrin, quercetrin and kaempferol 3-glucoside (Sachan et al., 2016). Cumin and caraway are highly used against diabetes. In a glucose tolerance conducted test in rabbits, cumin increases the area under the glucose tolerance curve. A methanolic extract of Carum seed decreases blood sugar levels, inhibits glycosylated hemoglobin and creatinine, and improves serum insulin and glycogen content in diabetic rats (Jagtap et al., 2010). Cumin and caraway products have proved to have significant antioxidant activity in several test methods. A caraway root extract has also shown an important role against DPPH (1,1-diphenyl-2 picrylhydrazyl) radical activity (Johri, 2011). Both the seed of cumin and caraway oil were found to prevent the formation of colon-specific cancer. The plant is said to be estrogenic. An aqueous and ethanolic extract of Carum seed produces an antifertility effect by modulation of FSH and LH (luteinizing hormone) (Nalini et al., 2006).

4.14 Capparis Spinosa

Different flavonoids are present in caperbush i.e., quercetin 3-rutinoside, quercetin 7-rutenoside, quercetin 3-glucoside 7-rhamnoside, kaempferol 3-rhamnorutinoside. Caper contains more quercetin than any other plant per weight (Darwish & Aburjai, 2010). The petroleum ether, butanol, and hexane crude extract produced from the aerial part of the plant are most active against the growth of *Staphylococcus* epidermis and *Streptococcus faecalis* (Al-Snafi, 2015). Essential oil obtained from *C. spinosa* is active against *Micrococcus luteus*. Ethanolic extract of *C. spinosa* is active against fungal growth like *Alternaria sclerotium*. Root, bark and leaves may have anticarcinogenic activity. The hydrolysis product of indole 3-methyl glucosinolate has anti anti-carcinogenic effect (Arena & Bisgnano et al., 2008). Rutin is a powerful antioxidant bioflavonoid in the body and is a dietary supplement for capillary fragility (Lam et al., 2009).

4.15 Dactylorhiza Hatagirea

Its tuber contains glucoside, starch, albumen, volatile oil, and ashes. Chemically, it has dactylorhins A to E, dactyloses A and B, and lipids are found as a major constituent (Pant & Tsewang, 2012). The rhizomatous part of *D. hatagirea* has shown resistance against all gram +ve and gram –ve bacteria, and the aerial part has limited resistance properties. Part of the rhizome is more effective. An *E. coli* bacterium resistant to many drugs is highly susceptible to the extract of *D. Hatagirea*. It is also used to improve the functionality of sexual organisms and may be helpful in sexual behavior (Thakur & Dixit, 2007).

4.16 Dracocepallum Heterophyllum

Dracocephalum is rich in tannins, flavonoids, Saponins, Steroids, polyphenols, and alkaloids. The essential oil of *Dracocephalum* contains β-pinex and citronellol as the main compounds (Stapphen et al., 2015). The methanol extract was subjected to antioxidant essay using 2, 2,-diphenylpicry-1-hydrazy (DPPH), and different biochemical tests of aqueous and chloroform extract were performed. It shows the presence of flavonoid and tannins, which are a well-known natural antioxidant, so the antioxidant property is due to the fact of these constituents (Raj & Bala, 2010). The antibacterial testing by agar dilution shows more activity of *Dracocephalum* against the *staphylococcus aureus* as compared to the activity of *H. officinalis* (Stapphen et al., 2015). Essential oil obtained from the flowers of this plant show highly active property against the growth of fungi *Colletotrichum* species and was reported to be more toxic to *Aedes aegypti* larvae in larvicidal bio-assay (Stapphen et al., 2015). The presence of chemicals such as saponins, terpenoids, flavonoids, tannins, and steroids show anti-inflammatory action (Bever et al., 1980). Saponins and steroids are also responsible for central nervous system activity (Raj & Bala, 2010). Tannins present also inhibit the growth of fungi, yeast, bacteria, and viruses (Raj & Bala, 2010).

4.17 Ephedra Gerardiana

The major constituents include ephedrine, tannins, Catechin, Saponin, and essential oil. It is mostly taken as a decongestant and energy-enhancement supplement (Samenak et al., 2002). It is also used as a Bronchodilator and promotes relief in asthma. It serves as a vasoconstrictor and affects mucous membranes, rendering it an effective nasal decongestant. (Dewick, 2002).

4.18 Catharanthus Roseus

Vinblastine, a vinca alkaloid derived from *Catharanthus roseus*, was discovered as a natural compound. Primarily known for its effectiveness in treating Hodgkin's and Non-Hodgkin's lymphomas, breast cancer, sarcoma, and testicular cancer, its potential as a chemotherapeutic agent became apparent when it was tested in an experiment exploring the plant's potential anti-diabetic properties. Subsequently, these alkaloids have gained medical significance for their anti-tumor characteristics (Dhyani et al., 2022). Vinblastine causes disruption of microtubules by binding to them which causes cell death or apoptosis by arresting the cell cycle at the M phase (Trybus et al., 2022). Vinblastine disrupts microtubules, which are structural components of the cell's cytoskeleton. It attaches itself to the tubulin protein, which is responsible for microtubule formation and function. This disturbance obstructs the regular division of cells, which ultimately results in cell cycle arrest and a reduction in the growth of cancer cells (Jordan & Wilson, 1998). Vinblastine interferes with mitosis, the process of cell division, by interfering with microtubules. As a result, mitotic arrest occurs, which stops cancer cells from proliferating and dividing (Amos & Löwe, 1999). Cancer cells can engage in apoptosis when exposed to vinblastine. This process is aided by mitotic arrest and microtubule disruption (Dumontet & Jordan, 2010). To increase its efficacy, vinblastine is frequently used in conjunction with other chemotherapy medications. Combinations like these might work in concert to improve the patient's response to therapy as a whole (Bonate et al., 2012). Many diseases, such as lymphomas, leukaemia, testicular cancer, and certain solid tumours, can be treated with vinblastine. The kind and stage of cancer can affect vinblastine's efficacy. This study investigated the mechanism of action of vinblastine, a natural source-derived chemotherapeutic medication, in the treatment of aggressive juvenile brain cancers. It shown that vinblastine causes microtubule disruption, which results in apoptosis and cell cycle arrest (Lee et al., 2012). Vincristine is a natural vinca alkaloid which is also called as leurocristine has $C_{46}H_{56}N_4O_{10}$ formula is obtained from *Catharanthus roseus* is commonly used for the treatment of myeloma, leukemia, breast cancer, neck cancer and head cancer (Dhyani et al., 2022). This alkaloid blocks cell cycle in metaphase with its anti-microtubule activity (Schiller et al., 2018). Vincristine functions by inhibiting tubulin polymerization, resulting in the cessation of microtubule synthesis and promoting the breakdown of already formed tubules. It disrupts nucleic acid and protein synthesis by impeding the utilization of glutamic acid (Martino et al., 2018).

4.19 Taxus Baccata

Taxol is a first generation taxanes i.e., paclitaxel which is obtained from the leaf and bark of *Taxus baccata.* Taxanes are potent anti-cancer drugs that operate by attaching to microtubules, which play a crucial role in cell division. They are utilized in treating various cancers such as breast, lung and ovarian cancer. The binding of paclitaxel to ß-tubulin within microtubules reduces microtubule dynamics, leading to cell cycle arrest at the M-phase. Docetaxel, a semi-synthetic compound derived from *Taxus baccata*, is predominantly employed in therapies for breast, pancreatic, prostate, and lung cancers. Analogues of paclitaxel currently in clinical trials encompass larotaxel, ortataxel, anilataxel and teretaxel. Larotaxel is administered alone or in combination with other treatments for urethral bladder, lung, pancreatic, and breast cancers (Ojima et al., 2016; Xie & Zhou, 2017). The chemotherapy medication Taxol, also known as paclitaxel, is frequently used to stop the development and spread of cancer cells. It is a member of the taxanes pharmacological class and works against cancer in a number of different ways. Taxol binds to microtubules, which are essential structural components of the cell's cytoskeleton. Unlike vinca alkaloids, which destabilize microtubules, Taxol stabilizes them. This stabilization inhibits microtubule dynamics, preventing them from depolymerizing and disrupting the cell division process. As a result, cancer cells are apprehended in the G2/M phase of the cell division cycle, and this disruption of cell division leads to cell death (apoptosis) and inhibits cancer cell growth (Schiff et al., 1979). Additionally, Taxol possesses anti-angiogenic qualities. It obstructs angiogenesis, the process of forming fresh neo blood vessels that tumours need to get blood. Taxol helps starve the tumour by inhibiting the formation of new blood vessels, which limits the tumour's gain way in to nutrition and oxygen (Thigpen et al., 1994). Taxol has the ability to cause cancer cells to die. It stimulates biological pathways that result in programmed cell death by severing microtubules and interfering with the mitotic spindle (Yang et al., 2010).

4.20 Curcuma Longa

Curcumin is lead phytochemical obtained from *Curcuma longa* (Kocaadam & Şanlier, 2017; Zhang & Kitts, 2021). It modulates various nuclear and cellular factors, also upregulates the expression of various genes and their product which act against the growth of human glioblastoma cells (Vallianou et al., 2015). It also induces apoptosis as it efficiently releases cytochrome c which activates caspase 3 and caspase 9 resulting into cell death (Kuttikrishnan et al., 2019). The natural substance curcumin, which is present in the spice turmeric, has drawn a lot of interest due to its possible anticancer effects. Strong antioxidant curcumin can reduce oxidative stress and shield cells from DNA damage by neutralizing harmful reactive oxygen species (ROS) (Menon & Sudheer, 2007). Cancer is linked to chronic inflammation both during its initiation and progression. The anti-inflammatory qualities of curcumin may prevent the growth of cancer by lowering inflammation (Aggarwal et al., 2007). By starting pro-apoptotic proteins and hindrance anti-apoptotic proteins, curcumin leads cancer cells to go through programmed cell death (Jiao et al., 2009). Curcumin may limit the growth and spread of tumours by preventing the angiogenesis the process of creating fresh new blood vessel that provides the tumours with nutrition (Teiten et al., 2009). Curcumin has the ability to target multiple cancer-related signaling pathways, including those that control inflammation, cell survival, and proliferation, such as the PI3K/Akt, STAT3, and NF-κB passageway (Aggarwal et al., 2005; Shakibaei et al., 2007). Curcumin may affect the activation or suppression of genes implicated in the development of cancer by influencing the epigenetic regulation of gene expression (Yang et al., 2017). Curcumin is reported to increase the sensitivity of cancer cells to radiation, thereby increasing the effectiveness of radiation treatments (Kunnumakkara et al., 2007). Activation of telomerase is essential for immortalization of cancer cells, and curcumin has been demonstrated to inhibit the activity of telomerase (Shay & Wright, 2006). The properties of curcumin include its ability to control autophagy, a mode of metabolism that, depending on the situation, can either increase or decrease the survival capacity of cancer cells (Yamauchi et al., 2012). There has been a study that looked at the ability of curcumin to suppress the production of matrix metalloproteinase-2 (MMP-2) in laryngeal carcinoma cells, an enzyme implicated in tumour invasion and metastasis. As a result of that study, it was demonstrated that curcumin may be able to inhibit cancer cell invasion by suppressing the expression and activity of MMP-2 (Mitra et al., 2006).

4.21 Rauvolfia Serpentina

Serpine, which is a substance unique to the roots of this plant, is used in contemporary medicine to treat hypertension, which is a condition characterised by increased blood pressure, by decreasing sympathetic nervous system activity and vasodilation. Several studies have shown that *Rauvolfia serpentina* extract has antibacterial properties that make it effective against fungi and bacteria, as well as other microorganisms. It has also been found that some of the plant's alkaloids have antimalarial properties, suggesting that it may also be useful for treating malaria. *Rauvolfia serpentina* has traditionally been used as a sedative and anxiolytic. Alkaloids in the plant can reduce anxiety, tension, and sleeplessness by relaxing the central nervous system. There is enormous pharmaceutical potential in this plant because of its medicinal qualities and prospective applications in contemporary healthcare (Negi et al., 2014; Roy et al., 2023).

5 Biotechnological Approaches for Production of Bioactive Compounds

In order to improve yield, consistency, and sustainability of these valuable phytochemicals, biotechnological approaches are used to increase their bioavailability (Fig. 4). Plant tissue culture, genetic engineering, metabolic engineering, elicitation, and the use of microbial systems for production are some of the various methods available. Biotechnological approaches enable the production of valuable phytochemicals derived from medicinal plants to meet the growing demand for bioactive compounds. With cutting-edge methods (elicitation, genetic engineering, metabolic engineering, plant tissue culture, nanobiotechnology, and industrial nanobiotechnology), bioactive compounds can be produced in larger quantities, more consistently, and of higher quality (Table 2). This approach has been shown to support pharmaceutical, nutraceutical, and functional food development, while reducing reliance on wild harvesting of medicinal plants (Ahuja et al., 2024; Kumar et al., 2024; Wawrosch & Zotchev, 2021; Alamgir & Alamgir, 2018; Molina et al., 2015; De Ancos et al., 2015; Joe et al., 2015).

5.1 Plant Tissue Culture

In the field of plant tissue culture, it refers to the process of cultivating plant cells, tissues, or organs under sterile conditions on a media containing nutrients in an in vitro environment. It has been demonstrated through the use of this method that plants can be produced in large quantities and bioactive compounds extracted without having to be grown on a large scale in the field. There are *invitro* approaches exploited by various researchers are callus culture, cell suspension culture, organ culture, somatic embryogenesis (Ahuja et al., 2024; Wawrosch & Zotchev, 2021; Sharma & Thakur, 2020; Karuppusamy, 2009; Preil, 2003).

5.2 Genetic Engineering

Plant genomes can be altered through genetic engineering so that they produce more bioactive compounds by enhancing the production of proteins. It is possible to accomplish this by introducing new genes, modifying existing genes, or suppressing undesirable genes as a means to accomplish. Gene overexpression, RNA interference (RNAi),

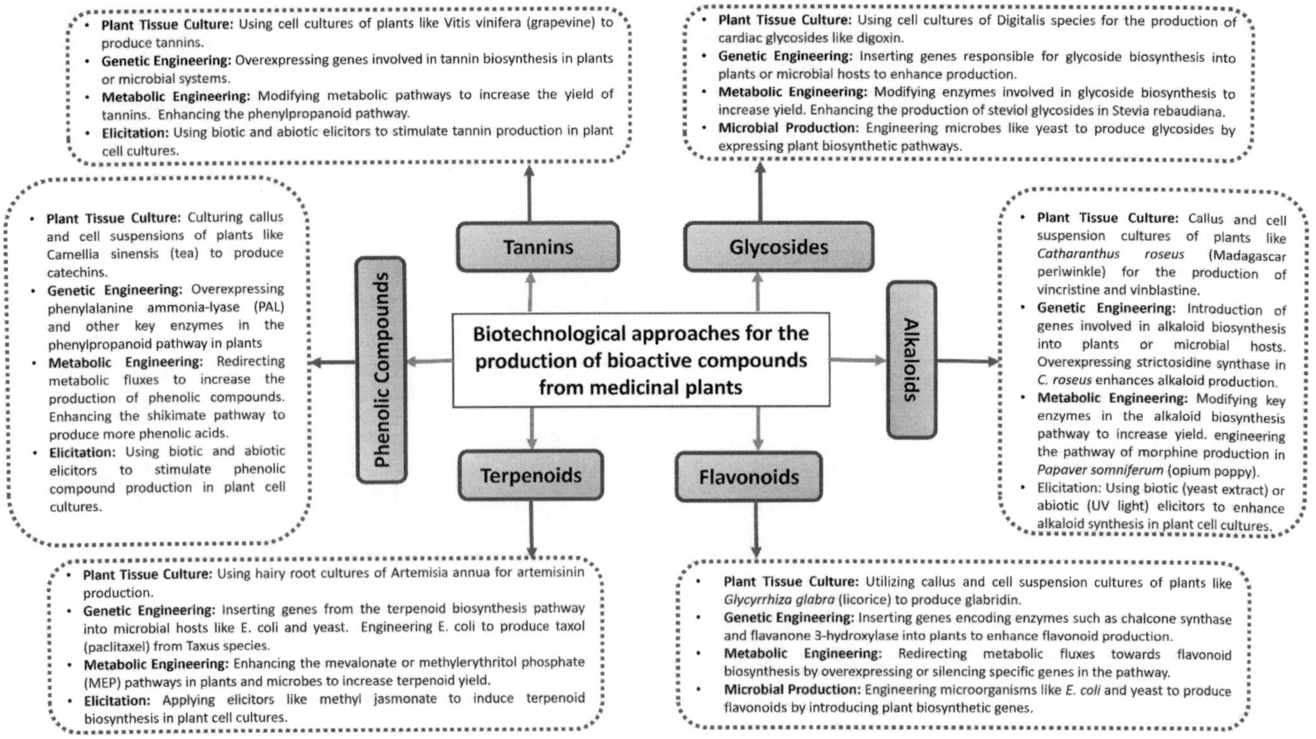

Fig. 4 Biotechnological approaches used for enhancing the production of bioactive compounds

Table 2 Biotechnological approaches used for the production of health promising bioactive compounds

Medicinal plant	Botanical name	Plant part used	Biotechnological approaches	Application	Mode of action
Turmeric	*Curcuma longa*	Rhizome	Tissue culture, Micropropagation	Anti-inflammatory, Antioxidant, Anticancer	Inhibition of NF-κB and COX-2, Antioxidant activity
Ginseng	*Panax ginseng*	Root	Callus culture, Hairy root culture	Immunomodulatory, Anti-fatigue	Modulation of immune response, Adaptogenic effects
Ginger	*Zingiber officinale*	Rhizome	Micropropagation, Callus culture	Antiemetic, Anti-inflammatory	Inhibition of prostaglandins and leukotrienes
Garlic	*Allium sativum*	Bulb	Cell culture, Genetic transformation	Antimicrobial, Cardioprotective	Inhibition of microbial enzymes, Antioxidant effects
Lavender	*Lavandula angustifolia*	Flowers	Micropropagation, Suspension culture	Anxiolytic, Sedative	Modulation of GABA receptors, Sedative effects
Echinacea	*Echinacea purpurea*	Roots, flowers	Callus culture, Hairy root culture	Immunostimulant, Antiviral	Stimulation of immune response, Inhibition of virus replication
Rosemary	*Rosmarinus officinalis*	Leaves	Micropropagation, Callus culture	Antioxidant, Cognitive enhancer	Free radical scavenging, Enhancement of cognitive function
Aloe vera	*Aloe barbadensis miller*	Leaves	Tissue culture, Genetic engineering	Wound healing, Moisturizing	Anti-inflammatory, Cell proliferation and collagen synthesis
Neem	*Azadirachta indica*	Seeds, leaves	Cell culture, Genetic transformation	Antimicrobial, Antimalarial	Inhibition of microbial enzymes, Modulation of immune response
Chamomile	*Matricaria chamomilla*	Flowers	Micropropagation, Suspension culture	Anti-inflammatory, Antispasmodic	Inhibition of COX-2 and TNF-α, Muscle relaxation
Ashwagandha	*Withania somnifera*	Roots	Callus culture, Hairy root culture	Adaptogenic, Antioxidant	Modulation of stress response, Free radical scavenging
Peppermint	*Mentha piperita*	Leaves	Micropropagation, Suspension culture	Antispasmodic, Analgesic	Inhibition of calcium channels, Modulation of pain receptors
Feverfew	*Tanacetum parthenium*	Leaves	Tissue culture, Cell suspension culture	Anti-migraine, Anti-inflammatory	Inhibition of serotonin release, Anti-inflammatory effects
Milk Thistle	*Silybum marianum*	Seeds	Callus culture, Genetic transformation	Hepatoprotective, Antioxidant	Inhibition of lipid peroxidation, Modulation of liver enzymes
St. John's Wort	*Hypericum perforatum*	Flowers, leaves	Micropropagation, Suspension culture	Antidepressant, Antiviral	Inhibition of reuptake of neurotransmitters, Antiviral effects
Ginkgo	*Ginkgo biloba*	Leaves	Tissue culture, Genetic engineering	Cognitive enhancer, Antioxidant	Improvement of blood flow, Free radical scavenging
Goldenseal	*Hydrastis canadensis*	Rhizome, roots	Callus culture, Hairy root culture	Antimicrobial, Anti-inflammatory	Inhibition of microbial enzymes, Modulation of immune response
Valerian	*Valeriana officinalis*	Roots	Micropropagation, Cell culture	Sedative, Anxiolytic	Modulation of GABA receptors, Sedative effects
Licorice	*Glycyrrhiza glabra*	Roots	Callus culture, Hairy root culture	Anti-inflammatory, Antiviral	Inhibition of prostaglandins and leukotrienes, Antiviral effects
Saffron	*Crocus sativus*	Stigma	Micropropagation, Suspension culture	Antioxidant, Antidepressant	Free radical scavenging, Modulation of neurotransmitters
Lemon Balm	*Melissa officinalis*	Leaves	Micropropagation, Callus culture	Antiviral, Anxiolytic	Inhibition of virus replication, Modulation of GABA receptors
Holy Basil	*Ocimum sanctum*	Leaves	Tissue culture, Genetic engineering	Adaptogenic, Antioxidant	Modulation of stress response, Free radical scavenging

(continued)

Table 2 (continued)

Medicinal plant	Botanical name	Plant part used	Biotechnological approaches	Application	Mode of action
Calendula	*Calendula officinalis*	Flowers	Callus culture, Hairy root culture	Wound healing, Anti-inflammatory	Promotion of cell proliferation, Inhibition of COX-2
Thyme	*Thymus vulgaris*	Leaves	Micropropagation, Cell culture	Antimicrobial, Antioxidant	Inhibition of microbial enzymes, Free radical scavenging
Sage	*Salvia officinalis*	Leaves	Micropropagation, Suspension culture	Antimicrobial, Cognitive enhancer	Inhibition of microbial enzymes, Enhancement of cognitive function
Cinnamon	*Cinnamomum verum*	Bark	Cell culture, Genetic engineering	Antioxidant, Antimicrobial	Free radical scavenging, Inhibition of microbial enzymes
Dandelion	*Taraxacum officinale*	Roots, Leaves	Callus culture, Suspension culture	Diuretic, Digestive aid	Promotion of urine production, Stimulation of digestive enzymes
Cat's Claw	*Uncaria tomentosa*	Bark, root	Callus culture, Hairy root culture	Anti-inflammatory, Immunomodulatory	Inhibition of NF-κB, Modulation of immune response
Elderberry	*Sambucus nigra*	Berries, flowers	Micropropagation, Suspension culture	Antiviral, Immunostimulant	Inhibition of virus replication, Stimulation of immune response
Eucalyptus	*Eucalyptus globulus*	Leaves	Tissue culture, Genetic engineering	Antimicrobial, Expectorant	Inhibition of microbial enzymes, Promotion of mucus clearance
Horsetail	*Equisetum arvense*	Stems	Callus culture, Hairy root culture	Diuretic, Anti-inflammatory	Promotion of urine production, Inhibition of COX-2
Gotu Kola	*Centella asiatica*	Leaves	Micropropagation, Suspension culture	Cognitive enhancer, Wound healing	Enhancement of cognitive function, Promotion of cell proliferation
Passionflower	*Passiflora incarnata*	Aerial parts	Callus culture, Hairy root culture	Anxiolytic, Sedative	Modulation of GABA receptors, Sedative effects
Black Cohosh	*Actaea racemosa*	Roots, rhizome	Tissue culture, Suspension culture	Menopausal symptom relief, Anti-inflammatory	Modulation of estrogen receptors, Inhibition of COX-2
Saw Palmetto	*Serenoa repens*	Berries	Micropropagation, Genetic engineering	Benign prostatic hyperplasia treatment	Inhibition of 5-alpha-reductase, Anti-inflammatory effects
Hawthorn	*Crataegus monogyna*	Berries, flowers	Callus culture, Hairy root culture	Cardioprotective, Antioxidant	Improvement of blood flow, Free radical scavenging
Bilberry	*Vaccinium myrtillus*	Berries	Micropropagation, Suspension culture	Antioxidant, Vision enhancement	Free radical scavenging, Improvement of ocular blood flow
Fenugreek	*Trigonella foenum-graecum*	Seeds	Tissue culture, Genetic engineering	Antidiabetic, Anti-inflammatory	Modulation of blood sugar levels, Inhibition of COX-2
Devil's Claw	*Harpagophytum procumbens*	Roots, tubers	Callus culture, Hairy root culture	Anti-inflammatory, Analgesic	Inhibition of COX-2 and TNF-α, Modulation of pain receptors
Licorice	*Glycyrrhiza glabra*	Roots	Callus culture, Hairy root culture	Anti-inflammatory, Antiviral	Inhibition of prostaglandins and leukotrienes, Antiviral effects
Red Clover	*Trifolium pratense*	Flowers	Micropropagation, Suspension culture	Menopausal symptom relief, Cardioprotective	Modulation of estrogen receptors, Improvement of blood flow
Evening Primrose	*Oenothera biennis*	Seeds	Tissue culture, Genetic engineering	Anti-inflammatory, Hormonal balance	Inhibition of prostaglandins, Modulation of hormonal balance
Green tea	*Camellia sinensis*	Leaves	Micropropagation, Suspension culture	Antioxidant, Weight management	Free radical scavenging, Modulation of metabolic rate
Bitter Melon	*Momordica charantia*	Fruits, seeds	Callus culture, Hairy root culture	Antidiabetic, Antiviral	Modulation of blood sugar levels, Inhibition of virus replication

(continued)

Table 2 (continued)

Medicinal plant	Botanical name	Plant part used	Biotechnological approaches	Application	Mode of action
Rhodiola	*Rhodiola rosea*	Root, rhizome	Callus culture, Hairy root culture	Adaptogenic, Cognitive enhancer	Modulation of stress response, Enhancement of cognitive function
Yohimbe	*Pausinystalia yohimbe*	Bark	Tissue culture, Genetic engineering	Aphrodisiac, Weight management	Alpha-2 adrenergic receptor antagonist, Modulation of metabolic rate
Moringa	*Moringa oleifera*	Leaves, seeds	Micropropagation, Suspension culture	Antioxidant, Anti-inflammatory	Free radical scavenging, Inhibition of COX-2 and TNF-α
Blueberry	*Vaccinium corymbosum*	Berries	Micropropagation, Suspension culture	Antioxidant, Vision enhancement	Free radical scavenging, Improvement of ocular blood flow
Astragalus	*Astragalus membranaceus*	Roots	Callus culture, Hairy root culture	Immunostimulant, Cardioprotective	Stimulation of immune response, Improvement of blood flow
Sea Buckthorn	*Hippophae rhamnoides*	Berries	Tissue culture, Genetic engineering	Antioxidant, Skin health	Free radical scavenging, Promotion of skin regeneration
Olive	*Olea europaea*	Leaves, fruit	Micropropagation, Suspension culture	Cardioprotective, Antioxidant	Improvement of blood flow, Free radical scavenging

Source And adapted from Sharma et al. (2024a–g)

CRISPR/Cas9, Agrobacterium-Mediated Transformation are some of the genetic engineering approaches exploited and reported time to time (Ahuja et al., 2024; Dincheva et al., 2015; Wawrosch & Zotchev, 2021).

5.3 Metabolic Engineering

A metabolic engineering approach involves taking steps to increase the production of bioactive compounds within plant cells by systematically changing metabolic pathways within the plant cell. The effectiveness of this approach will be determined by the ability to gain a deeper understanding of the plant's metabolic network (Ahuja et al., 2024; Wawrosch & Zotchev, 2021; Sharma & Sharma, 2021a; Moses et al., 2017; Dincheva et al., 2015).

5.4 Elicitation

Elicitation involves the use of biotic or abiotic agents to stimulate the production of secondary metabolites in plants. Elicitors can be physical, chemical, or biological in nature. Two types of elicitors, biotic elicitors and abiotic elicitors for are used production of secondary metabolites. (Ahuja et al., 2024; Wawrosch & Zotchev, 2021).

5.5 Microbial Systems for Production

Microbial systems, including bacteria, yeast, and fungi, can be engineered to produce plant-derived bioactive compounds. These systems offer several advantages, including rapid growth, ease of genetic manipulation, and scalability, Bioactive compounds can be produced by heterologous expression, fermentation technology as well as Co-Cultivation approaches (Wawrosch & Zotchev, 2021; Baadhe et al., 2015; Verma et al., 2020).

5.6 Nanobiotechnology

Using nanoparticles and nanoscale materials, nanobiotechnology can improve the delivery and efficacy of bioactive compounds by using low molecular weight compounds at a nanoscale level. As a result of utilizing this approach, phytochemicals can be made more stable, more solubilized, and more bioavailable. Nanoencapsulation, Nanocarriers and Nanobiosensors was developed to measure the release and production of bioactive compounds in real time (Ingle et al., 2020; Majumdar et al., 2020; Sharma et al., 2020a–e).

6 Conclusion

The use of natural or engineered molecules is crucial for treating illness. Several bioactive compounds are produced by biotechnology. The primary objective is to discover new medicines and health products from the many secondary metabolites found in these plants. It is essential to conserve and assess these natural resources in light of their conventional uses. Biotechnology contributes significantly to the development of bioactive molecules. Biotechnology has revolutionized bioactive chemical synthesis, giving scientists new ways to overcome traditional problems with natural sources. In synthetic biology, genetic engineering, and metabolic engineering, bioactive compounds are produced and enhanced. Gene engineering can be used to produce desired bioactives in microbes or plants. Additionally, it enhances their productivity. Metabolic engineering enhances the efficiency of host organisms by optimizing metabolic pathways. Synthetic biology integrates engineering concepts with biological systems. Using tissue culture, genetically modified plants, and elicitation techniques, plant bioactive chemicals can be produced more efficiently. Future biotechnological synthesis for bioactive compounds looks bright with technological advancement and multidisciplinary studies. CRISPR-Cas9 offers increased possibilities for developing bioactive compounds from microbes and plants. Bioactive molecules are produced through genomic, proteomic, and metabolomic technologies. Throughout this chapter, we emphasize the revolutionary potential of biotechnology in the production of bioactive molecules.

Acknowledgement and Contributions The authors thank Head, Department of Plant Sciences, Central University of Himachal Pradesh for providing necessary facilities i.e., access to the scientific journals. MS acknowledge the financial assistance received from Central University of Himachal Pradesh in the form of UGC-Non-Net fellowship. First, Second, Third and Fourth author (MS, MS, SKP, DS) wrote the paper and Corresponding author (MS) checked and edited the present chapter.

References

Aggarwal, B. B., Shishodia, S., Takada, Y., Banerjee, S., Newman, R. A., Bueso-Ramos, C. E., et al. (2005). Curcumin suppresses the Paclitaxel-Induced nuclear Factor-KB pathway in breast cancer cells and inhibits lung metastasis of human breast cancer in nude mice. *Clinical Cancer Research, 11*(20), 7490–7498. https://doi.org/10.1158/1078-0432.ccr-05-1192.

Aggarwal, B. B., Sundaram, C., Malani, N., & Ichikawa, H. (2007). Curcumin: The Indian solid gold. *Springer eBooks*, 1–75. https://doi.org/10.1007/978-0-387-46401-5_1.

Ahuja, A., Sharma, M., Mallubhotla, S., & Sharma, M. (2024). Bacosides neurotropic molecules production by tissue cultures of Bacopa monnieri (L.) Wettst. In *Elsevier eBooks* (pp. 353–373). https://doi.org/10.1016/b978-0-443-21818-7.00010-1.

Akbarian, M., Khani, A., Eghbalpour, S., & Uversky, V. N. (2022). Bioactive peptides: Synthesis, sources, applications, and proposed mechanisms of action. *International Journal of Molecular Sciences, 23*(3), 1445. https://doi.org/10.3390/ijms23031445.

Alamgir, A. N. M. (2018). Biotechnology, in vitro production of natural bioactive compounds, herbal preparation, and disease management (Treatment and Prevention). In *Progress in drug research* (pp. 585–664). https://doi.org/10.1007/978-3-319-92387-1_7.

Ali, S. I., & Venkatesalu, V. (2022). Botany, traditional uses, phytochemistry and pharmacological properties of Saussurea costus–an endangered plant from Himalaya-a review. *Phytochemistry Letters, 47*, 140–155. https://doi.org/10.1016/j.phytol.2021.12.008.

Al-Snafi, A. E. (2015). The chemical constituents and pharmacological effects of Capparis spinosa-an overview. *Indian Journal of Pharmaceutical Science and Research, 5*(2), 93–100.

Amos, L. A., & Lowe, J. (1999). How Taxol® stabilises microtubule structure. *Chemistry & Biology, 6*(3), R65–R69. https://doi.org/10.1016/s1074-5521(99)89002-4.

Arena, A., & Bisignano, G. (2008). Antiviral and immunomodulatory effect of some Saudi Arabian medicinal plants. *Agents Action, 17*(3–4), 383–384.

Azmir, J., Zaidul, I., Rahman, M., Sharif, K., Mohamed, A., Sahena, F., et al. (2013). Techniques for extraction of bioactive compounds from plant materials: A review. *Journal of Food Engineering, 117* (4), 426–436. https://doi.org/10.1016/j.jfoodeng.2013.01.014.

Baadhe, R. R., Potumarthi, R., Mekala, N. K., & Gupta, V. K. (2015). New trends in microbial production of natural complex bioactive isoprenoids. *Biotechnology of Bioactive Compounds: Sources and Applications*, 269–282.

Bahuguna, R., Purohit, M. C., Rawat, M. S. M., & Purohit, A. N. (2000). Qualitative and quantitative variations in alkaloids of Aconitum species from Garhwal Himalaya. *J. Plant Bio., 27*(2), 179–183.

Batool, Z., Singh, K., & Gairola, S. (2023). Medicinal plants traditionally used in the health care practices by the indigenous communities of the Trans-Himalayan region of Ladakh India. *Journal of Ethnopharmacology, 317*, 116837. https://doi.org/10.1016/j.jep.2023.116837.

Batool, Z., Singh, K., & Gairola, S. (2024). Nutraceutical and antioxidant potential of selected wild edible plants from the cold-arid desert of Ladakh. *India. Notulae Botanicae Horti Agrobotanici Cluj-Napoca, 52*(1), 13286–13286.

Beigh, S. Y., Nawchoo, I. A., & Iqbal, M. (2006). Cultivation and conservation of Aconitum heterophyllum: A critically endangered medicinal herb of the northwest Himalayas. *Journal of Herbs, Spices & Medicinal Plants, 11*(4), 47–56.

Bever, B. O. (1980). Oral hypoglycaemic plants in West Africa. *Journal of Ethnopharmacology, 2*(2), 119–127.

Bhandari, P., Sendri, N., & Devidas, S. B. (2020). Dammarane triterpenoid glycosides in Bacopa monnieri: a review on chemical diversity and bioactivity. *Phytochemistry, 172*, 112276. https://doi.org/10.1016/j.phytochem.2020.112276

Bhuju, S., & Gauchan, D. P. (2018). Taxus wallichiana (Zucc.), an endangered anti-cancerous plant: A review. *International Journal of Research, 5*(21), 10–21.

Bishop, G. J. (2007). Refining the plant steroid hormone biosynthesis pathway. *Trends in Plant Science, 12*(9), 377–380. https://doi.org/10.1016/j.tplants.2007.07.001.

Bonate, P. L., Strougo, A., Desai, A., Roy, M., Yassen, A., Van Der Walt, J. S., et al. (2012). Guidelines for the quality control of population pharmacokinetic-pharmacodynamic analyses: An industry perspective. *The AAPS Journal, 14*(4), 749–758. https://doi.org/10.1208/s12248-012-9387-9.

Bourgaud, F., Hehn, A., Larbat, R., Doerper, S., Gontier, E., Kellner, S., et al. (2006). Biosynthesis of coumarins in plants: A major pathway still to be unravelled for cytochrome P450 enzymes. *Phytochemistry Reviews, 5*(2–3), 293–308. https://doi.org/10.1007/s11101-006-9040-2.

Chaudhary, A. A., Dhar, U., Ahmad, A., Bhatt, I. D., Jugran, A., Kaur, G. (2012). Analysis of genetic diversity in Hedychium spicatum Buch. Ham. ex. D. Don in West Himalaya, India. *Journal of Medicinal Plants Research, 6*(23), 3984–3991.

Chen, S., Wang, X., Cheng, Y., Gao, H., & Chen, X. (2023). A review of classification, biosynthesis, biological activities and potential applications of flavonoids. *Molecules, 28*(13), 4982. https://doi.org/10.3390/molecules28134982.

Chung, H. H., Sung, Y. C., & Shyur, L. F. (2016). Deciphering the biosynthetic pathways of bioactive compounds in planta using omics approaches. *Medicinal Plants-Recent Advances in Research and Development*, 129–165.

Darwish, R. M., & Aburjai, T. A. (2010). Effect of ethnomedicinal plants used in folklore medicine in Jordan as antibiotic resistant inhibitors on Escherichia coli. *BMC Complementary And Alternative Medicine, 10*, 1–8.

De Ancos, B., Colina-Coca, C., González-Peña, D., & Sánchez-Moreno, C. 2015. Bioactive compounds from vegetable and fruit by-products. *Biotechnology of Bioactive Compounds: Sources and Applications*, 1–36.

Devi, N. B., Singh, P. K., & Das, A. K. (2014). Ethnomedicinal utilization of Zingiberaceae in the valley districts of Manipur. *Journal of Environmental Science Toxicology and Food Technology, 8*(2), 21–23.

Dewick, P. M. (2002). *Medicinal natural products: A biosynthentic approach* (2nd ed., p. 520). West Sussex, UK: Wiley and Son.

Dhiman, A., Sharma, M., Sharma, M. (2023). Ethnopharmacological Profile of Curcuma aromatica Salisb. *Journal of Mountain Research, 18*(1). https://doi.org/10.51220/jmr.v18i1.29.

Dhiman, R., Sharma, M., Sharma, M., & Sharma, M. (2023). Reviewing the medicinal potential of Valeriana jatamansi Jones: its traditional uses, phytochemistry, and pharmacological activities. *TMR Pharmacology Research, 3*(4), 23. https://doi.org/10.53388/pr202303023.

Dhyani, P., Quispe, C., Sharma, E., Bahukhandi, A., Sati, P., Attri, D. C., et al. (2022). Anticancer potential of alkaloids: A key emphasis to colchicine, vinblastine, vincristine, vindesine, vinorelbine and vincamine. *Cancer Cell International, 22*(1), 21–23. https://doi.org/10.1186/s12935-022-02624-9

Dincheva, I., Badjakov, I., & Kondakova, V. (2015). Metabolic engineering of bioactive compounds in berries. *Biotechnology of Bioactive Compounds: Sources and Applications*, 463–482.

Dumontet, C., & Jordan, M. A. (2010). Microtubule-binding agents: A dynamic field of cancer therapeutics. *Nature Reviews Drug Discovery, 9*(10), 790–803. https://doi.org/10.1038/nrd3253.

He, M., Tian, H., Luo, X., Qi, X., & Chen, X. (2015). Molecular progress in research on fruit astringency. *Molecules, 20*(1), 1434–1451. https://doi.org/10.3390/molecules20011434.

Hu, W., Sarengaowa, N., Guan, Y., & Feng, K. (2022). Biosynthesis of phenolic compounds and antioxidant activity in Fresh-Cut fruits and vegetables. *Frontiers in Microbiology*, 13. https://doi.org/10.3389/fmicb.2022.906069.

Ingle, A. P., Shende, S., Gupta, I., & Rai, M. (2020). Recent trends in the development of nano-bioactive compounds and delivery systems. In *Biotechnological production of bioactive compounds* (pp. 409–431). Elsevier.

Jagtap, A. G., & Patil, P. B. (2010). Antihyperglycemic activity and inhibition of advanced glycation end product formation by Cuminum cyminum in streptozotocin induced diabetic rats. *Food and Chemical Toxicology, 48*, 2030–2036.

Jaspal, N., Sharma, M., Prashar, D. B., Sharma, R., & Sharma, M. (2023). Jatropha curcas L.: A sustainable resource for biofuel feedstock with medicinal and commercial attributes. *Journal of Innovative Agriculture, 10*(3), 1–13.

Jiao, Y., Wilkinson IV, J., Di, X., Wang, W., Hatcher, H., Kock, N. D., et al. (2009). Curcumin, a cancer chemopreventive and chemotherapeutic agent, is a biologically active iron chelator. *Blood, The Journal of the American Society of Hematology, 113*(2), 462–469.

Joe, M. M., Benson, A., Ayyanar, M., & Sa, T. (2015). Bioactive compounds and medical significance of some endangered medicinal plants from the Western Ghats region of India. *Biotechnology of Bioactive Compounds: Sources and Applications*, 163–193.

Johri, R. K. (2011). Cuminum cyminum and Carum carvi: An update. *Pharmacognosy Reviews, 5*(9), 63.

Jordan, M. A., & Wilson, L. (1998). Microtubules and actin filaments: Dynamic targets for cancer chemotherapy. *Curr. Opi. Cell Biol., 10* (1), 123–130. https://doi.org/10.1016/s0955-0674(98)80095-1.

Joshi, P. V., Shirkhedkar, A. A., Prakash, K., & Maheshwari, V. L. (2011). Antidiarrheal activity, chemical and toxicity profile of Berberis aristate. *Pharmaceutical Biology, 49*(1), 94–100.

Kala, C. P., Dhyani, P. P., & Sajwan, B. S. (2006). Developing the medicinal plants sector in northern India: Challenges and opportunities. *Journal of Ethnobiology and Ethnomedicine, 2*, 1–15.

Karuppusamy, S. (2009). A review on trends in production of secondary metabolites from higher plants by in vitro tissue, organ and cell cultures. *Journal of Medicinal Plants Research, 3*(13), 1222–1239.

Kaushal, A., Sharma, M., Navneet, & Sharma, M. (2020). Ethnomedicinal, phytochemical, therapeutic and pharmacological review of the genus Erythrina. *International Journal of Botany Studies, 5*(6), 642–648.

Kocaadam, B., & Şanlier, N. (2017). Curcumin, an active component of turmeric (Curcuma longa), and its effects on health. *Critical Reviews in Food Science and Nutrition, 57*(13), 2889–2895. https://doi.org/10.1080/10408398.2015.1077195.

Kumar, P., Singh, K., Bhushan, A., Gupta, A. P., Babu, V., Gupta, P., et al. (2023). Botanical, chemical, and DNA barcode-based authentication of herbal ayurvedic medicines from the menispermaceae. *Revista Brasileira De Farmacognosia, 33*(2), 300–309. https://doi.org/10.1007/s43450-023-00369-8

Kumar, N., & Singh, R. S. (2023). *Biosynthesis of bioactive compounds in medicinal and aromatic plants: Manipulation by conventional and biotechnological approaches*. Springer Nature.

Kumar, P., Shaunak, I., & Verma, M. L. (2020). Biotechnological application of health promising bioactive molecules. In *Elsevier eBooks* (pp. 165–189). https://doi.org/10.1016/b978-0-444-64323-0.00006-0.

Kumar, S., Singh, A., Bist, C. M. S., & Sharma, M. (2024). Advancements in genetic techniques and functional genomics for enhancing crop traits and agricultural sustainability. Briefings in Functional Genomics. https://doi.org/10.1093/bfgp/elae017.

Kunnumakkara, A. B., Guha, S., Krishnan, S., et al. (2007). Curcumin potentiates antitumor activity of gemcitabine in an orthotopic model of pancreatic cancer through suppression of proliferation, angiogenesis, and inhibition of nuclear factor-κB–regulated gene products. *Cancer Research, 67*(8), 3853–3861. https://doi.org/10.1158/0008-5472.can-06-4257.

Kuttikrishnan, S., Siveen, K. S., Prabhu, K. S., et al. (2019). Curcumin induces apoptotic cell death via inhibition of PI3-Kinase/AKT pathway in B-precursor acute lymphoblastic leukemia. *Frontiers in Oncology, 9*, 484. https://doi.org/10.3389/fonc.2019.00484.

Kytidou, K., Artola, M., Overkleeft, H. S., & Aerts, J. M. F. G. (2020). Plant glycosides and glycosidases: A treasure-trove for therapeutics. *Frontiers in Plant Science*, 11. https://doi.org/10.3389/fpls.2020.00357.

Lam, S. K., Han, Q. F., et al. (2009). Isolation and characterization of a lectin with potentially exploitable activities from Caper seeds. *Bioscience Reports, 29*(5), 293–299.

Lee, K., Lee, Y., Ban, J. O., et al. (2012). The flavonoid resveratrol suppresses growth of human malignant pleural mesothelioma cells through direct inhibition of specificity protein. *International Journal of Molecular Medicine*. https://doi.org/10.3892/ijmm.2012.978.

Li, C., Zha, W., Li, W., Wang, J., & You, A. (2023). Advances in the biosynthesis of terpenoids and their ecological functions in plant resistance. *International Journal of Molecular Sciences, 24*(14), 11561.

Majumdar, M., Shivalkar, S., Pal, A., Verma, M. L., Sahoo, A. K., & Roy, D. N. (2020). Nanotechnology for enhanced bioactivity of bioactive compounds. In *Biotechnological production of bioactive compounds* (pp. 433–466). Elsevier.

Martino, E., Casamassima, G., Castiglione, S., et al. (2018). Vinca alkaloids and analogues as anti-cancer agents: Looking back, peering ahead. *Bioorganic & Medicinal Chemistry Letters, 28*(17), 2816–2826. https://doi.org/10.1016/j.bmcl.2018.06.044.

Menon, V. P., & Sudheer, A. R. (2007). Antioxidant and anti-inflammatory properties of curcumin. In *Springer eBooks* (pp. 105–125). https://doi.org/10.1007/978-0-387-46401-5_3.

Mitra, A., Chakrabarti, J., Banerji, A., et al. (2006). Curcumin, a potential inhibitor of MMP-2 in human laryngeal squamous carcinoma cells HEP2. *Journal of Environmental Pathology, Toxicology and Oncology, 25*(4), 679–690. https://doi.org/10.1615/jenvironpatholtoxicoloncol.v25.i4.70.

Mitra, P. K., Mitra, P., Das, A. P., Ghosh, C., Sarkar, A., & Chowdhury, D. (2010). Screening the efficacy of some East Himalayan medicinal plants on ethanol induced gastric ulcer in Albino rats. *Pleione, 4*(1), 69–75.

Molina, G., Pelissari, F. M., Pessoa, M. G., & Pastore, G. M. (2015). Bioactive compounds obtained through biotechnology. *Biotechnology of Bioactive Compounds: Sources and Applications*, 433–462.

Moses, T., Mehrshahi, P., Smith, A. G., & Goossens, A. (2017). Synthetic biology approaches for the production of plant metabolites in unicellular organisms. *Journal of Experimental Botany, 68*, 4057–4074. https://doi.org/10.1093/jxb/erx119.

Nadda, R. K., Ali, A., Goyal, R. C., Khosla, P. K., & Goyal, R. (2020). Aucklandia costus (syn. Saussurea costus): Ethnopharmacology of an endangered medicinal plant of the Himalayan region. *Journal of Ethnopharmacology, 263*, 113199.

Nalini, N., Manju, V., & Menon, V. P. (2006). Effect of spices on lipid metabolism in 1,2-dimethylhydrazine induced rat colon carcinogenesis. *Journal of Medicinal Food, 9*(2), 237–245.

Negi, J. S., Bisht, V. K., Bhandari, A. K., Bisht, D. S., Singh, P., & Singh, N. (2014). Quantification of reserpine content and antibacterial activity of Rauvolfia serpentina (L.) Benth. ex. Kurz. *African Journal of Microbiology Research, 8*(2), 162–166.

Ojima, I., Lichtenthal, B., Lee, S., et al. (2016). Taxane anticancer agents: A patent perspective. *Expert Opinion on Therapeutic Patents, 26*(1), 1–20. https://doi.org/10.1517/13543776.2016.1111872.

Pant, S., & Rinchen, T. (2012). Dactylorhiza hatagirea: A high value medicinal orchid. *Journal of Medicinal Plants Research, 6*(19), 3522–3524.

Patil, B. S., Jayaprakasha, G. K., Murthy, K. N. C., & Vikram, A. (2009). Bioactive Compounds: Historical perspectives, opportunities, and challenges. *Journal of Agriculture and Food Chemistry, 57* (18), 8142–8160. https://doi.org/10.1021/jf9000132.

Prakash, D., Upadhyay, G., Singh, B. N., Dhakarey, R., Kumar, S., & Singh, K. K. (2007). Free-radical scavenging activities of Himalayan rhododendrons. *Current Science*, 526–532.

Preil, W. (2003). Micropropagation of ornamental plants. In M. Laimer & W. Rücker (Eds.), *Plant tissue culture: 100 years since Gottlieb Haberlandt* (pp. 115–133). Springer Vienna.

Raj, J., & Bala, S. S. (2010). Antioxidative activity and phytochemical investigation on a high altitude medicinal plant Dracocephalum heterophyllum Benth. *Pharmacognosy Journal, 2*(6), 112.

Rawat, R., & Vashistha, D. P. (2011). Shortlisting of cultivable herbal plants in the Bhabhar region of the Garhwal Himalaya, Uttarakhand India. *International Journal of Medicinal and Aromatic Plants, 1* (1), 23–27.

Roy, A. C., Prasad, A., Priya, K., Das, P., Singh, S., Ghosh, C., et al. (2023). Anticancer effect of antioxidant-rich methanolic extract of Rauvolfia serpentina (L.) Benth. ex. Kurz leaves in HepG2 and HeLa cells: A mechanistic insight. *Biocatalysis and Agricultural Biotechnology, 50*, 102674. https://doi.org/10.1016/j.bcab.2023.102674.

Sachan, A. K., Das, D. R., & Kumar, M. (2016). Carum carvi-an important medicinal plant. *Journal of Chemical and Pharmaceutical Research, 8*(3), 529–533.

Samenuk, D., Link, M. S., Homoud, M. K., Contreras, R., Theohardes, T. C., Wang, P. J., et al. (2002). Adverse cardiovascular events temporally associated with ma huang, an herbal source of ephedrine. *Mayo Clinic Proceedings, 7*(1), 12–16.

Schiff, P. B., Fant, J., & Horwitz, S. B. (1979). Promotion of microtubule assembly in vitro by taxol. *Nature, 277*(5698), 665–667. https://doi.org/10.1038/277665a0.

Schiller, G. J., Damon, L. E., Coutre, S. E., et al. (2018). High-dose vincristine sulfate liposome injection, for advanced, relapsed, or refractory philadelphia chromosome-negative acute lymphoblastic leukemia in an adolescent and young adult subgroup of a phase 2 clinical trial. *Journal of Adolescent and Young Adult Oncology, 7* (5), 546–552. https://doi.org/10.1089/jayao.2018.0041.

Shakibaei, M., John, T., Schulze-Tanzil, G., et al. (2007). Suppression of NF-κB activation by curcumin leads to inhibition of expression of cyclo-oxygenase-2 and matrix metalloproteinase-9 in human articular chondrocytes: Implications for the treatment of osteoarthritis. *Biochemical Pharmacology, 73*(9), 1434–1445. https://doi.org/10.1016/j.bcp.2007.01.005.

Shamsuddin, T., Alam, M. S., Junaid, M., Akter, R., Hosen, S. M., Ferdousy, S., et al. (2021). Adhatoda vasica (Nees.): A review on its botany, traditional uses, phytochemistry, pharmacological activities and toxicity. *Mini Reviews in Medicinal Chemistry, 21*(14), 1925–1964.

Sharma, H., & Garg, M. (2015). A review of traditional use, phytoconstituents and biological activities of Himalayan yew, Taxus wallichiana. *Journal of Integrative Medicine, 13*(2), 80–90.

Sharma, M. N., & Sharma, M. (2021a). Ethnoveterinary studies of traditional medicinal plants in shivalik ranges of uttarakhand. *The Journal of Indian Botanical Society, 101*(3), 166–182.

Sharma, M., & Sharma, M. (2021b). *Plant tissue culture technique used in mining of high valued secondary metabolites from medicinal and aromatic plants* (pp. 670–673). JPS Scientific Publications, India.

Sharma, S., & Thokchom, R. (2014). A review on endangered medicinal plants of India and their conservation. *Journal of Crop and Weed, 10*(2), 205–218.

Sharma, M., Sharma, A. K., Thakur, R., & Sharma, M. (2020a). Dynamics of traditional information of medicinal plants from hilly terrains of ramban (J&K) India. *Indian Journal of Ecology, 47*(4), 1009–1013.

Sharma, M., Thakur, R., & Sharma, M. (2020b). Ethnomedicinal, phytochemical and pharmacological properties of Crocus sativus (saffron). *The Journal of Indian Botanical Society, 99*(3–4), 115–125. https://doi.org/10.5958/2455-7218.2020.00017.0.

Sharma, M., Thakur, R., & Sharma, M. (2020c). Ethno-botanical survey of medicinal plants of unexplored hilly areas of district Ramban (J&K). *International Journal of Botany Studies, 5*(3), 55–63.

Sharma, M., Thakur, R., Sharma, M., Sharma, A. K., & Sharma, A. K. (2020d). Changing scenario of medicinal plants diversity in relation to climate change: A review. *Plant Archives, 20*(2), 4389–4400.

Sharma, A. K., Sharma, A. K., Sharma, M., & Sharma, M. (2021a). Medicinal and aromatic plants: A potential source of novel bioactive compounds showing antiviral efficacy against coronavirus (SARS-CoV-2). *Indian Journal of Ecology, 48*(14), 7–16.

Sharma, M., Sharma, A. K., & Sharma, M. (2021b). Ethno-botanical study of medicinal plants from unexplored area of District Ramban (J&K) India. *Indian Journal of Agricultural Research, 55*(6), 702.

Sharma, M., Bithel, N., & Sharma, M. (2022b). Ethnobotanical study of medicinal plants among local tribes of Rajaji Tiger Reserve Haridwar. *Indian Journal of Ecology, 49*(3), 1197–1202.

Sharma, J., Singh, K., & Gairola, S. (2023). Plants used in female reproductive healthcare by the Gujjar tribe of the sub-Himalayan region of Uttarakhand, India. *Journal of Herbal Medicine, 42*, 100786. https://doi.org/10.1016/j.hermed.2023.100786.

Sharma, M., Yangzom, S., Sharma, M., & Sharma, M. (2024g). Ethnobotanical studies of high valued medicinal plants reported from Spiti valley in Himachal Pradesh. *Biology Bulletin, 51*(1), 104–114. https://doi.org/10.1134/s1062359023601568.

Sharma, M., & Thakur, R. (2020). Advanced biotechnology techniques used in conservation of threatened medicinal and aromatic plants. In *Compendium of research insights of life science students* (pp. 737–740).

Sharma, P., Guleria, P., & Kumar, V. (2020e). Green nanotechnology for bioactive compounds delivery. In *Biotechnological production of bioactive compounds* (pp. 391–407). Elsevier.

Sharma, A. K., Sharma, A. K., Sharma, M., & Sharma, M. (2022a). Assessment of land use change and climate change impact on biodiversity and environment. In *Springer proceedings in earth and environmental sciences* (pp. 73–89). https://doi.org/10.1007/978-3-031-05335-1_5.

Sharma, M., Sharma, M., Bithel, N., & Sharma, M. (2022c). Ethnobotany, phytochemistry, pharmacology and nutritional potential of medicinal plants from asteraceae family. *Journal of Mountain Research, 17*(2). https://doi.org/10.51220/jmr.v17i2.7.

Sharma, M., Sharma, M., & Sharma, M., 2022d. A comprehensive review on ethnobotanical, medicinal and nutritional potential of walnut (Juglans regia L.). *Proceedings of the Indian National Science Academy, 88*(4), 601–616. https://doi.org/10.1007/s43538-022-00119-9.

Sharma, M., Bithel, N., Sharma, K. K., & Sharma, M. (2024a). Potential of vegetables and plant metabolites in healthcare. In Apple Academic Press eBooks (pp. 3–31). https://doi.org/10.1201/9781032680125-2.

Sharma, M., Kashyap, N., Verma, M., Devi, B., Prashar, A., Sinha, S., & Sharma, M. (2024b). Role of natural products in cancer management: a comprehensive review. *Journal of Ethnopharmacology and Toxicology, 2*(1), 1–18. https://doi.org/10.37446/jet/ra/2.1.2024.1-18.

Sharma, M., Sharma, D., Sahu, S. C., Sharma, A., & Sharma, M. (2024c). Mapping the trends in global research productivity and conservation of saffron (Crocus sativus L.): Insight from bibliometric analysis during 1950–2022. *Biology Bulletin Reviews, 14*(2), 238–250. https://doi.org/10.1134/s2079086424020117.

Sharma, M., Sharma, A. K., Sharma, S., Sharma, A. K., Sharma, K. K., & Sharma, M. (2024d). Reviewing the insights of SARS-CoV-2: Its epidemiology, pathophysiology, and potential preventive measures in traditional medicinal system. *Clinical Traditional Medicine and Pharmacology*, 200147. https://doi.org/10.1016/j.ctmp.2024.200147.

Sharma, M., Sharma, M., Sahu, S. C., & Sharma, D. (2024e). Walnuts as functional food and nutraceutical: A bibliometric study of research trends on nutritional potential, phytochemistry and its health benefits. *Food and Humanity*, 100387. https://doi.org/10.1016/j.foohum.2024.100387.

Sharma, M., Sharma, M., Sharma, M., 2024f. Assessment of Variability and Genetic Factors among High Heritable Traits of Juglans regia (Walnut) from North Western Himalayan Regions, Agric. Res., https://doi.org/10.1007/s40003-024-00697-8.

Shay, J. W., & Wright, W. E. (2006). Telomerase therapeutics for cancer: Challenges and new directions. *Nature Reviews Drug Discovery, 5*(7), 577–584. https://doi.org/10.1038/nrd2081.

Singh, K., Kumar, P., Kumar, B., Sharma, J., Andrade-Cetto, A., Gupta, P., et al. (2024). Medicinal plants traditionally used in health care practices by inhabitants of Paddar region of Jammu and Kashmir, India. *Jounal of Ethnopharmacology, 334*, 118514.

Song, C., Ma, J., Li, G., Pan, H., Zhu, Y., Jin, Q., et al. (2022). Natural composition and biosynthetic pathways of alkaloids in medicinal dendrobium species. *Frontiers in Plant Science*, 13. https://doi.org/10.3389/fpls.2022.850949.

Srivastava, S. K., Khatoon, S., Rawat, A. K. S., Mehrotra, S., & Pushpangadan, P. (2001). Pharmacognostic evaluation of the root of Berberis aristata DC. *Natural Product Sciences, 7*(4), 102–106.

Stappen, I., Wanner, J., Tabanca, N., Wedge, D. E., Ali, A., Kaul, V. K., et al. (2015). Chemical composition and biological activity of essential oils of Dracocephalum heterophyllum and Hyssopus officinalis from Western Himalaya. *Natural Product Communications, 10*(1), https://doi.org/10.1177/1934578X1501000131.

Teiten, M. H., Gaascht, F., Eifes, S., et al. (2009). Chemopreventive potential of curcumin in prostate cancer. *Genes Nutrition, 5*(1), 61–74. https://doi.org/10.1007/s12263-009-0152-3.

Thakur, M., & Dixit, V. K. (2007). Aphrodisiac activity of Dactylorhiza hatagirea (D. Don) Soo in male albino rats. *Evidence-Based Complementary and Alternative Medicine, 4*, 29–31.

Thakur, A. K., & Raj, P. (2017). Pharmacological perspective of Glycyrrhiza glabra Linn: A mini-review. *Journal of Analytical & Pharmaceutical Research, 5*(5), 00156.

Thakur, R., Sharma, M., Sharma, M., & Sharma, A. K. (2020). Biodiversity status of major medicinal and aromatic plants (MAPs) In J&K, India. *Traditional Medicinal Plants, 2*, 81–112.

Thigpen, J., Blessing, J. A., Ball, H., et al. (1994). Phase II trial of paclitaxel in patients with progressive ovarian carcinoma after platinum-based chemotherapy: A gynaecologic oncology group study. *Clinical Oncology (Royal College of Radiologist), 12*(9), 1748–1753. https://doi.org/10.1200/jco.1994.12.9.1748.

Trybus, W., Trybus, E., & Krol, T. (2022). Emodin sensitizes cervical cancer cells to vinblastine by inducing apoptosis and mitotic death. *International Journal of Molecular Sciences, 23*(15), 8510. https://doi.org/10.3390/ijms23158510.

Upadhyay, S., Jeena, G. S., Shikha, N., & Shukla, R. K. (2018). Recent advances in steroidal saponins biosynthesis and in vitro production. *Planta, 248*(3), 519–544. https://doi.org/10.1007/s00425-018-2911-0.

Uwineza, P. A., & Waśkiewicz, A. (2020). Recent advances in supercritical fluid extraction of natural bioactive compounds from natural plant materials. *Molecules, 25*(17), 3847. https://doi.org/10.3390/molecules25173847.

Vallianou, N. G., Evangelopoulos, A., Schizas, N., & Kazazis, C. (2015). Potential anticancer properties and mechanisms of action of curcumin. *Anticancer Research, 35*(2), 645–651.

Verma, M. L., & Chandel, A. K. (2019). *Biotechnological production of bioactive compounds*. Elsevier.

Verma, M. L., Kishor, K., Sharma, D., Kumar, S., & Sharma, K. D. (2020). Microbial production of omega-3 polyunsaturated fatty acids. *Biotechnological Production of bioactive Compounds*, 293–326.

Wawrosch, C., & Zotchev, S. B. (2021). Production of bioactive plant secondary metabolites through in vitro technologies—status and outlook. *Applied Microbiology and Biotechnology, 105*(18), 6649–6668. https://doi.org/10.1007/s00253-021-11539-w.

Xie, S., & Zhou, J. (2017). Harnessing plant biodiversity for the discovery of novel anticancer drugs targeting microtubules. *Frontiers in Plant Science, 8*, 720. https://doi.org/10.3389/fpls.2017.00720.

Yamauchi, Y., Izumi, Y., Asakura, K., et al. (2012). Curcumin induces autophagy in ACC-MESO-1 cells. *Phytotherapy Research, 26*(12), 1779–1783. https://doi.org/10.1002/ptr.4645.

Yang, D., Yu, L., & Van, S. (2010). Clinically relevant anticancer polymer paclitaxel therapeutics. *Cancers, 3*(1), 17–42. https://doi.org/10.3390/cancers3010017.

Yang, J., Wang, C., Zhang, Z., et al. (2017). Curcumin inhibits the survival and metastasis of prostate cancer cells via the Notch-1 signaling pathway. *Acta Pathologica, Microbiologica Et Immunologica Scandinavica., 125*(2), 134–140. https://doi.org/10.1111/apm.12650.

Yuan, Y., Zhang, J., Kallman, J., Liu, X., Meng, M., & Lin, J. (2019). Polysaccharide biosynthetic pathway profiling and putative gene mining of Dendrobium moniliforme using RNA-Seq in different tissues. *BMC Plant Biology, 19*(1). https://doi.org/10.1186/s12870-019-2138-7.

Zhang, H. A., & Kitts, D. D. (2021). Turmeric and its bioactive constituents trigger cell signaling mechanisms that protect against diabetes and cardiovascular diseases. *Molecular and Cellular Biochemistry, 476*(10), 3785–3814. https://doi.org/10.1007/s11010-021-04201-6.

Zou, Y., Teng, Y., Li, J., & Yan, Y. (2023). Recent advances in the biosynthesis of coumarin and its derivatives. *Green Chemical Engineering, 5*(2), 150–154. https://doi.org/10.1016/j.gce.2023.04.003.

Role of Bioactive Compounds in Plant Disease Management

Jorge Miguel Heredia-Bátiz, Juan Pablo Manjarrez-Quintero, Octavio Valdez-Baro, María Magdalena Rivera-Salas, Heriberto Bayardo-Rosales, Luis Alfonso Jiménez-Ortega, and José Basilio Heredia

Abstract

Bioactive compounds can be obtained from different sources and methods since various plants and microorganisms can synthesize them. There is also a wide variety of methodologies for their production, where the isolation of these compounds is the most important stage. Medicinal plants are one of the most used sources for obtaining bioactive compounds of natural origin due to their high content of essential oils made up of volatile and aromatic compounds that largely belong to the group of terpenes. Bioactive compounds of natural origin play a very important role in agriculture since they can combat many diseases caused by biotic and abiotic factors. Among the most important factors are phytopathogenic microorganisms such as fungi and bacteria, which cause yearly losses worldwide. In addition to being effective against plant diseases, bioactive compounds help search for alternatives to chemical pesticides, which can harm health and the environment and represent great costs for producers. In this way, bioactive compounds of natural origin could be used as biopesticides, representing a great alternative within the integrated management of plant diseases.

Keywords

Bioactive compound • Extract • Essential oil • Biopesticide • Sustainable • Phytopathogen • Phytochemical

J. M. Heredia-Bátiz · J. P. Manjarrez-Quintero · O. Valdez-Baro · M. M. Rivera-Salas · H. Bayardo-Rosales · L. A. Jiménez-Ortega · J. B. Heredia (✉)
Department of Functional Foods and Nutraceuticals, Research Center for Food and Development, CONAHCYT, Culiacán, Sinaloa, México
e-mail: jbheredia@ciad.mx

1 Introduction

Agriculture is vital worldwide since food security depends on it, as well as a great diversity of industries dedicated to ornamentation, supplementation, and essential oils, among many others. Unfortunately, agricultural production can be affected by different factors, both biotic and abiotic, that can make crops sick. Among the most important biotic factors, we can find various phytopathogenic microorganisms such as fungi, viruses, bacteria, and nematodes, or larger ones such as insects and parasitic plants.

On the other hand, the abiotic factors comprise stresses like temperature, humidity, nutrition, or contamination (Agrios, 2009). These factors must be controlled since they can considerably reduce crop production yield. Due to this, integrated crop management comprises various strategies to give plants the optimal conditions for their growth and production. Among the multiple methods to achieve the highest production yields, the use of chemical pesticides is found to be one of the most implemented, which consists of the application of chemical substances in the soil, air, or on the plant that act in a systemic and contact manner to help the plant against diseases. Unfortunately, these chemical substances can be dangerous to health and the environment. Due to this, interest has emerged worldwide in seeking new alternatives for treating plant diseases. Some of these novel, safe, and ecological alternatives are biological control and, more recently, plant extracts.

Botanical extracts can be obtained by various methods depending on the plant matrix or the target compounds. Normally, aromatic plants are used due to the high amount of bioactive compounds that they possess, but there is also a high interest in the production of extracts from plant waste due to the global ecological trend (Freire & Santos, 2018; Ikram et al., 2023; Ismail et al., 2019; Mesa Vanegas et al., 2023; Sossou et al., 2022). Another way to obtain beneficial bioactive compounds for plants is by taking advantage of some microorganisms, generally fungi or bacteria, which

J. Devi (ed.), *Biotechnological Intervention in Production of Bioactive Compounds*,
Sustainable Landscape Planning and Natural Resources Management, https://doi.org/10.1007/978-3-031-76859-0_9

produce some secondary metabolites that can be recovered to be applied to crops, or these microorganisms can even be directly inoculated into crops as happens in biological control (Asad, 2022; Duchateau et al., 2024; Zhang et al., 2024). In this sense, bioactive compounds of plant and microbial origin are currently of great importance and interest, so in this chapter, we will talk about how bioactive compounds of plant and microbial origin can be obtained and examples of how these have been tested to be used as an alternative in the management of plant diseases.

2 Plant-Derived Bioactive Compounds

2.1 Extraction of Essential Oils

Essential oil extraction technologies are vital for obtaining high-quality oils used in various industries. These methods preserve the oil's integrity and potency, ensuring efficient extraction from different plant materials. Key techniques include hydrodistillation, steam distillation, supercritical fluid extraction, cold pressing, and solvent extraction.

Hydrodistillation: Plant materials are immersed in water and boiled. The water prevents overheating and preserves heat-sensitive compounds. The essential oil vapor is condensed and separated, allowing distillation below 100 °C (Tongnuanchan & Benjakul, 2014).

Steam Distillation: Plant material is placed on a perforated plate, and steam from an external boiler flows through a pipe. The essential oil vaporizes and is condensed for separation, avoiding thermal degradation by maintaining temperatures below 100 °C. Hydro-diffusion, a variation, uses steam introduced from the top, offering shorter processing time and higher productivity. Nutmeg oil is produced using hybrid steam distillation (Chakravarty et al., 2023).

Supercritical Fluid Extraction (SFE): SFE uses CO_2 in its supercritical state (31 °C, 7.38 MPa) for efficient extraction while preserving properties. Critical parameters include temperature, pressure, particle size, modifiers, and fluid flow rate. Higher pressures increase solvent power but may require fractionation to remove unwanted compounds. Co-solvents like ethanol enhance the solubility of polar compounds. SFE produces high-quality extracts without thermal degradation or toxic residues (Yousefi et al., 2019).

Cold Pressing: Mainly used for citrus fruits, this method involves breaking oil glands in the peel to release essential oils. The emulsion is centrifuged to separate the oil, avoiding thermal degradation of sensitive compounds (Stratakos & Koidis, 2016).

Solvent Extraction: Used for thermally labile materials like blossoms; this method immerses plant material in a solvent bath to dissolve the essential oil. The mixture is filtered and distilled to separate the oil from the solvent and impurities. Common solvents include alcohol, hexane, ethanol, petroleum ether, and methanol. This method operates at lower temperatures, minimizing heat-related chemical alterations, making it efficient for the perfume industry. Residual solvents limit their use in food products unless food-grade solvents are used (Stratakos & Koidis, 2016).

2.2 Obtaining Plant Extracts

Botanical extraction technologies have evolved significantly, incorporating both traditional and advanced methods to optimize the extraction of bioactive compounds.

Maceration: This traditional method involves soaking plant materials in a solvent at room temperature for an extended period. It is simple and cost-effective but time-consuming and solvent-intensive. Efficiency can be optimized by adjusting factors like solvent choice, solid-to-liquid ratio, and extraction time. For example, maceration of *Scirpus holoschoenus* with acetone for 24 h yielded 23.7%, with a total phenolic content of 182.29 mg GAE/g extract and antioxidant activity (DPPH IC50) of 32.4 mg/ml (Chuo et al., 2022).

Ultrasound-Assisted Extraction (UAE): UAE uses ultrasonic waves to create cavitation in the solvent, enhancing extraction by breaking plant cell walls and releasing active compounds. It offers higher yields, shorter extraction times, reduced solvent and energy consumption, and improved selectivity. However, it requires specialized equipment and precise control of conditions such as ultrasound frequency, power, and time. For example, UAE of camptothecin from *Nothapodytes nimmoniana* using methanol at 40 °C and a 75% duty cycle yielded 2.989 mg/g of plant (Chuo et al., 2022).

Microwave-Assisted Extraction (MAE): MAE uses microwaves to internally heat plant material, causing friction and heat that break cell walls and release bioactive compounds. Key factors influencing MAE performance include solvent type, solvent-to-solid ratio, microwave power, temperature, and extraction time. Polar solvents like water and aqueous ethanol are the most effective. MAE offers higher yields, reduced solvent use, shorter extraction times, and lower energy consumption than conventional methods. It is widely used in various industries to extract phenolic acids, flavonoids, alkaloids, polysaccharides, and essential oils. Careful optimization is required to maximize yields and minimize degradation (Yadav, 2020).

Deep Eutectic Solvents (DESs): DESs are green solvents formed by mixing hydrogen bond acceptors and donors to create eutectic mixtures with low melting points. They are low in toxicity, biodegradable, low in volatility, thermally and chemically stable, and cost-effective. DESs are used in green extraction processes, extracting phenolic compounds from natural sources. Their efficiency depends on the HBA-HBD molar ratio, extraction temperature, and water content. DESs provide a sustainable, efficient alternative to conventional solvents, supporting environmentally friendly advancements in industrial and research applications (Ali Redha, 2021).

2.3 Characterization and Purification

Characterizing and purifying bioactive compounds are essential in managing plant diseases. These techniques enable identifying and isolating specific components, contributing to developing effective plant disease management strategies. Fourier Transform Infrared (FTIR) Spectroscopy and various advanced purification methodologies are crucial in these processes.

Fourier Transform Infrared (FTIR) Spectroscopy is a fundamental analytical technique for identifying functional groups in bioactive compounds from medicinal plants. It is non-destructive, has minimal mechanical parts, and can quickly collect multiple scans per second, enhancing sensitivity and reducing noise. FTIR works by directing an infrared beam through a sample, which absorbs specific frequencies, producing a spectrum that serves as the sample's molecular fingerprint. Each absorption peak indicates the vibrational frequencies of the bonds within the molecules, enabling the identification of various functional groups. This method is widely used in phytochemical analysis to authenticate medicinal plants, evaluate phytochemical composition, and detect bioactive compounds in extracts, significantly contributing to understanding plant disease management by providing detailed molecular composition information (Wali et al., 2022).

High-Performance Liquid Chromatography (HPLC) separates compounds based on their interactions with stationary and mobile phases. Using a YMC-Gel column and ethanol at a flow rate of 2.5 ml/min, HPLC achieves high-purity capsaicin. Aqueous Two-Phase Extraction (ATPE) and Simulated Moving Bed Chromatography (SMBC) enhance purity through partitioning and adsorption processes, respectively. High-Speed Counter-Current Chromatography (HSCCC) and Sequential Centrifugal Partition Chromatography (sCPC) utilize liquid–liquid partitioning for high-purity capsaicin extraction. Supercritical Fluid Extraction (SCFE), employing CO_2 under controlled conditions, achieves high yields and is suitable for industrial-scale production (Wang et al., 2022).

There have been successfully isolated and purified bioactive compounds from the stem bark of *Jatropha podagrica*. The process involved methanol extraction and fractionation using hexane and ethyl acetate. The ethyl acetate extract, noted for its strong antioxidant activity, was purified through column chromatography and analyzed using thin-layer chromatography. Gas chromatography-mass spectrometry (GC-MS) identified methyl gallate, gallic acid, fraxetin, and tomentin as key bioactive constituents. Further confirmation was obtained through electrospray ionization-mass spectrometry (ESI-MS), liquid chromatography-electrospray ionization mass spectrometry (LC-ESI-MS), and NMR spectroscopies. This effective purification process highlights the potential of *Jatropha podagrica* as a source of significant bioactive compounds for managing plant diseases (Minh et al., 2019). A representation of the extraction process can be seen in Fig. 1.

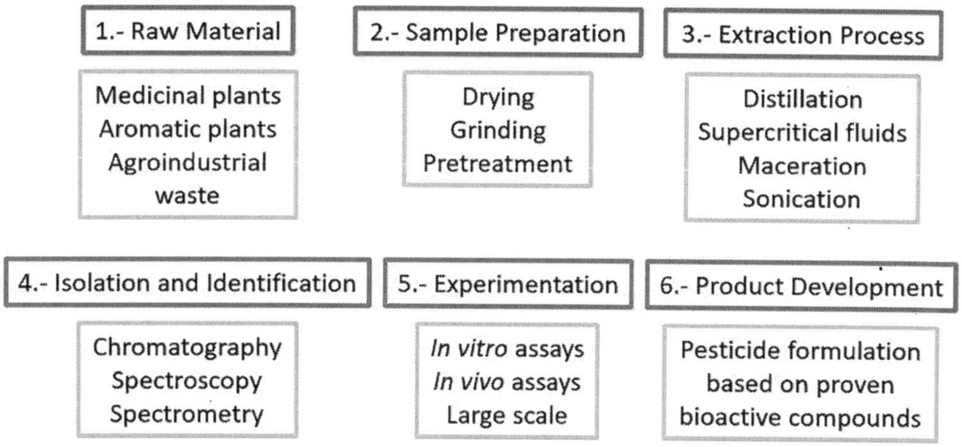

Fig. 1 An overview on the process of obtaining bioactive compounds

3 Microbial-Derived Bioactive Compounds

3.1 Biosynthesis of Bacterial Compounds

Bacteria are the largest group of microorganisms that generate bioactive compounds used in treating diseases since about two-thirds of the antibiotics used in treating diseases in humans, animals, and plants are manufactured from bacteria of the genus *Streptomyces*. Specifically, bacterial species belonging to the genera *Streptomyces, Bacillus, Pseudomonas, Serratia, Burkholderia, Pantoea,* and *Staphylococcus* have been studied for their ability to produce secondary metabolites with potential for the control of diseases in plants. Secondary metabolites are low molecular weight compounds produced during the stationary phase of bacterial growth (Fig. 2). These compounds are made in a group of biosynthetic genes that mainly encode polyketides and peptides of non-ribosomal synthesis.

Numerous bacterial compounds such as phenazines, antibiotics, biosurfactants, siderophores, and volatile organic compounds play an important role in direct antagonism against plant pathogenic agents as well as in stimulating plant immunity (Duchateau et al., 2024; Ruparelia et al., 2022). Phenazines are nitrogenous heterocyclic compounds produced mainly by *Pseudomonas, Streptomyces, Burkholderia,* and *Pantoea species*. The mode of action of phenazines is due to their ability to interfere with electron transport. At the same time, biosurfactants are compounds of an amphiphilic nature characterized by reducing surface tension. Among the bioactive compounds with biosurfactant properties are glycolipids and lipopeptides. Species of Bacillus, Pseudomonas, Streptomyces, Burkholderia, and Serratia produce lipopeptides.

The antimicrobial activity of lipopeptides is due to their ability to interact with cell membranes. Antimicrobial lipopeptides produced by *Bacillus* species are mainly classified into three families: iturin, surfactin, and fengycin. Lipopeptides are a heterogeneous group of amino acids (7–10 amino acid residues) and variable-length fatty acid chains. Lipopeptides are synthesized in enzyme complexes called non-ribosomal peptide synthetases (NRPS), sometimes with the help of polyketide synthases (PKS). Non-ribosomal peptide synthetases are multidomain enzymatic complexes since, during the synthesis of lipopeptides, each of the domains is responsible for the incorporation of a particular amino acid (Duchateau et al., 2024; Fira et al., 2018).

In addition to suppressing the growth of plant pathogenic agents, volatile organic compounds produced by bacteria can also improve plant growth. Volatile organic compounds are metabolites with low molecular weight, low polarity, and high vapor pressure, which allows them to diffuse into the air. These compounds are produced through catabolic pathways such as glycolysis, fermentation, sulfur metabolism and mainly in the biosynthesis of fatty acids. Among the volatile organic compounds produced by bacteria are dimethyl disulfide, 1-butanol, decanone, nonane, benzothiazole, which inhibit the growth of different plant pathogenic fungi such as *Fusarium oxysporum, Aspergillus niger, Phytophora infestans, Phytium ultimum, Verticillium*

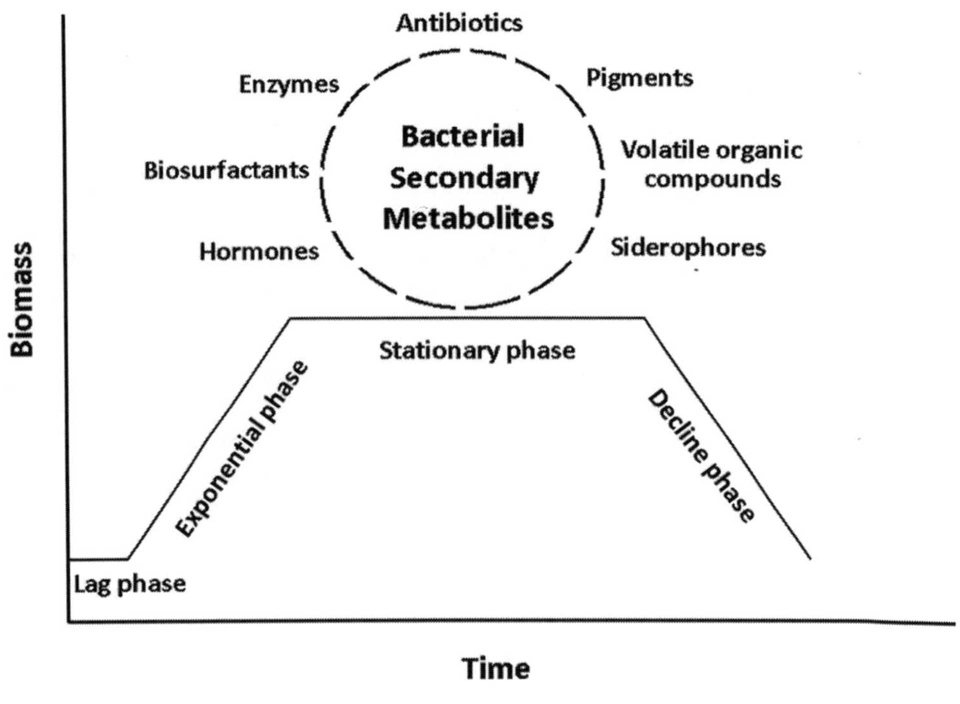

Fig. 2 Bacterial growth curve with production of secondary metabolites

longisporum, Rhizoctonia solani, Sclerotinia sclerotiorum, among others (Nguyen et al., 2023; Rana et al., 2024).

3.2 Biosynthesis of Fungal Compounds

Fungi can produce various biologically active compounds to control plant diseases; among these compounds of fungal origin are antibiotics, enzymes, and secondary metabolites such as peptaibols, gliotoxins, glycolipids, terpenes, and volatile organic compounds. Regarding fungal enzymes, various species of *Trichoderma* secrete extracellular chitinases and proteases that degrade the walls of plant pathogenic fungi and oomycetes such as *Rhizoctonia solani, Sclerotinia sclerotiorum, Botrytis cinerea, Fusarium oxisporum* and *Pythium ultimum*. These fungal enzymes can also inhibit the germination of spores of *Alternaria alternata, Magnaporthe grisea,* and *Fusarium verticillioides*.

Another bioactive compound that has potent antibacterial and antifungal activity by inhibiting the growth of microorganisms such as *Escherichia coli, Pseudomonas aeruginosa, Streptococcus aureus, and Alternaria solani*, is terrain, which is a secondary metabolite produced by the marine fungus *Aspergillus terres*. Mycotoxins also have antimicrobial, antiviral, and immunomodulatory properties. Sulfur mycotoxins such as gliotoxins are produced by species of *Gliocladium*, *Aspergillus*, *Eurotium*, *Trichoderma*, and *Penicillium*, where the transcription of gliotoxin biosynthesis genes is affected by factors such as temperature, pH, aeration and composition of the growing medium (Asad, 2022; Daguerre et al., 2014; Zaehle et al., 2014; Zhang et al., 2024).

Glycolipids with antifungal activity can alter the membrane of plant pathogenic fungi, such as powdery mildew. Fungal species such as *Pseudozyma flocculosa, Pseudozyma fusiformata, Pseudozyma gramicola,* and *Ustilago maydis* are characterized by producing glycolipids with antimicrobial activity. Some of the terpenic compounds produced by *Trichoderma* species present antifungal activity. Similarly, fungal enzymes such as glucose oxidase secreted by *Aspergillus* and *Penicillium* species have activity against pathogenic fungi such as *Sclerotinia sclerotiorum, Rhizoctonia solani,* and *Verticillium dahliae*. The antifungal properties of glucose oxidase are due to the production of hydrogen peroxide during the oxidation of glucose by the action of the enzyme (Daguerre et al., 2014).

Regarding volatile organic compounds with antifungal properties that positively affect plants, these compounds are mainly produced by yeasts such as *Saccharomyces cerevisiae*. The main volatile organic compounds yeast produces are alcohols, aldehydes, ketones, phenols, and esters (Di Francesco et al., 2024). Other promising bioactive compounds are antimicrobial peptides; among these peptaibols are the largest group. Peptaibols are linear polypeptides that exhibit activity against fungi, bacteria, and viruses; fungi of the genus Trichoderma, Gliocladium, and Emericellopsis produce these compounds. The biological properties of peptaibols derive from their ability to form pores in lipid membranes and the systemic resistance they induce in plants. Peptaibols are also synthesized in multifunctional enzymatic domains called non-ribosomal peptide synthetases; these protein complexes are formed by multiple modules: initiation module, elongation modules, and termination module, in this last module, the peptide is released from the enzyme complex (Daguerre et al., 2014; Pereira-Dias et al., 2023).

4 Bioactive Compounds for Diseases Caused by Biotic Factors

4.1 Bioactive Compounds Against Fungi

Essential oils are useful for controlling phytopathogenic fungi that cause diseases in plants. These metabolites can be effective in most in vitro cases. However, the trend toward replacing or reducing chemical fungicides has increased interest in conducting in vivo tests to validate their fungicidal effects and develop botanical-based formulations. It's important to consider the challenges associated with using essential oils for controlling diseases caused by phytopathogenic fungi, as these oils tend to have low stability and high volatility. Therefore, finding the best way to formulate them is crucial to ensure their effectiveness is not compromised by environmental conditions (Raveau et al., 2020).

The most commonly represented fungicidal activity by essential oils is against major phytopathogenic fungi such as *Fusarium*, *Alternaria*, *Botrytis*, *Rhizoctonia*, *Penicillium*, and *Aspergillus*, being the plant genera *Thymus*, *Origanum*, *Syzygium*, *Mentha*, *Ocinum*, *Cinnamomum*, *Rosmarinus*, and *Piper* the main aromatic plants that produce essential oil with fungicide potential; this for the protection of crops of high economic value such as pineapple, banana, watermelon, papaya, guava, mango, and orange (Mohd Israfi et al., 2022; Raveau et al., 2020). It must be considered that essential oils are apolar, so they must be properly formulated to be adequately dispersed. Upon contact with fungal cells, they can exhibit fungicidal effects, allowing assays to measure inhibition of mycelial growth, spore germination, and inhibition zones. This enables the determination of EC_{50}, MIC, and MFC values. Lower necessary concentrations indicate better efficacy of the essential oil. The next step involves identifying and testing individual components individually and in mixtures to maximize the effect of these molecules (Chang et al., 2022; Raveau et al., 2020).

Essential oils can have a direct or indirect fungicidal effect, as the same molecules in essential oils may be responsible for their fungicidal effects through various mechanisms of action (cell wall alteration, membrane rupture, gene expression modification, ergosterol biosynthesis inhibition, mitochondrial damage, cell division inhibition, nucleic acid, and protein synthesis inhibition, and efflux pump inhibition). Additionally, these molecules can induce systemic resistance in plants (antioxidant enzyme activity, histone modification of resistance genes, increased production of phytohormones and phenolic compounds), promoting protection against fungal phytopathogens (Kesraoui et al., 2022; Raveau et al., 2020).

Plant extracts against fungi are highly diverse, as they depend significantly on the extraction method, solvent, process conditions, plant species, and the type of fungus. The key lies in the extracted metabolites that possess fungicidal potential. These include phenolic acids, tannins, flavonoids, alkaloids, terpenes, and saponins, along with all their subgroups derived from these phytochemical groups. The variation in fungicidal phytochemicals stems from the different mechanisms of action they can exert on fungal cells. These molecules exhibit distinct physicochemical properties. Among the most studied mechanisms of plant extracts are membrane and cell wall disruption, inhibition of ergosterol biosynthesis, increased lipid peroxidation, oxidative stress, nucleic acid and enzyme inhibition, and metabolic alterations (such as inhibition of cellular respiration, sugars, pyruvate, Krebs cycle, etc.).

Additionally, they promote systemic resistance in host plants by enhancing enzymes like superoxide dismutase, peroxidase, catalase, polyphenol oxidase, phenylalanine ammonia-lyase, β-1,3-glucanase, and chitinase, as well as phytohormones like auxins, abscisic acid, salicylic acid, gibberellins, cytokinins, ethylene, and jasmonic acid (Ehiobu et al., 2022; Elshafie et al., 2023; Jiang et al., 2023a, b). There are multiple extracts with fungicidal effects; however, the most frequently reported ones are derived from hydrophilic extracts such as ethanol, methanol, and aqueous extracts. Additionally, there are extracts using acetone, ethyl acetate, petroleum ether, chloroform, and hexane. These extracts come from various plants, including oregano, thyme, clove, cinnamon, pepper, citrus, rosemary, and mint. However, their potential depends on the vegetative part from which the fungicidal metabolites are extracted and the target fungi. The most evaluated fungi are *Alternaria* spp., *Aspergillus* spp., *Botritys* spp., *Colletotrichum* spp., *Curvularia* spp., *Fusarium* spp., *Mucor* spp., *Penicillium* spp., *Phytophthora* spp., *Pythium* spp., *Puccinia* spp. *Rhizoctonia* spp., and *Rhizopus* spp. (Achar et al., 2021; Bhatta, 2022; Deresa & Diriba, 2023).

We can elucidate that using non-volatile plant extracts, similar to essential oils, holds significant potential for reducing or replacing chemical fungicides and addressing the growing demand for environmentally friendly alternatives. However, several factors must be considered. While research is thriving, there remains a lack of depth in areas such as optimizing extraction methods to maximize fungicidal effects, isolating and purifying fungicidal components from mixed extracts containing multiple inert molecules, identifying and quantifying novel fungicidal molecules of botanical origin, evaluating synergistic effects among phytochemicals and even with existing fungicides, exploring mechanisms of action through gene expression, stability testing, encapsulation, and formulation for maximum efficacy in open-field conditions, and assessing scalability to high production levels. These efforts aim to establish these phytochemicals as viable and competitive options against current solutions (Deresa & Diriba, 2023; Elshafie et al., 2023; Jiang et al., 2023a, b).

Microbial compounds have emerged as a promising alternative for controlling phytopathogenic fungi, offering a more sustainable and less environmentally detrimental option than conventional chemical fungicides. One of the most promising strategies involves the use of bacteria such as *Bacillus* spp. or *Pseudomonas* spp., these bacteria not only promote plant growth but also act as biological control agents against phytopathogens. For instance, *Bacillus* spp. has demonstrated inhibition of fungi such as *Alternaria* and *Fusarium* by producing antifungal lipopeptides like fengycin and surfactin (Santos et al., 2024). Volatile compounds have also emerged as a viable alternative. These compounds can inhibit the growth of phytopathogens through various molecular mechanisms. For example, certain VOCs from *Pseudomonas* have been shown to cause DNA damage to the pathogen and structural damage to fungal hyphae, and some compounds can affect the expression of specific genes in phytopathogens (Almeida et al., 2023).

Endophytic fungi possess the ability to inhibit phytopathogens through antibiosis, producing a variety of antimicrobial chemical compounds, especially secondary metabolites such as alkaloids, flavonoids, isocoumarins, lignans, peptides, phenols, phenylpropanoids, quinones, steroids, terpenoids, and volatile compounds (Priyashantha et al., 2023), which have potent antimicrobial properties. *Trichoderma* spp., for example, produces a variety of hydrolytic enzymes and antibiotics that inhibit the growth of phytopathogenic fungi and enhance the systemic resistance of host plants (Enespa & Chandra, 2017). Although microbial compounds offer a promising and sustainable alternative for the control of phytopathogenic fungi, they still face significant challenges that limit their application on a large scale, such as variability in efficacy under different environmental conditions and the complexity of mass production and formulation. However, these strategies will likely be optimized with ongoing advances in biotechnological research and a better understanding of microbial interactions. This could overcome current obstacles and fully leverage the potential of biological control agents for more sustainable agriculture.

4.2 Bioactive Compounds Against Viruses

The potential of essential oils against other phytopathogens, such as fungi, bacteria, oomycetes, and nematodes, has been extensively studied, and formulations are already replacing or reducing chemical products. However, research is scarce for phytopathogenic viruses due to the greater difficulty in controlling viral diseases than other pathogens. The most effective ways to prevent disease using essential oils include eradicating the vector, inducing systemic plant resistance (increasing enzymes PAL, SOD, CAT, and POD, boosting hormones like salicylic acid, and expressing genes ICS1, NPR1, and PR-1a), and directly destroying viral particles. Strictly speaking, applying these essential oils preventively yields better results when using these phytochemicals.

Currently, research on essential oil use against phytopathogenic viruses has primarily focused on mosaic viruses, which are among the most common. However, investigations are still limited, highlighting the need for greater interest in botanical solutions. The effects vary depending on the virus and the essential oil composition, with aromatic plant essential oils containing phenolic monoterpenes like carvacrol and thymol (from oregano and thyme) being some of the most studied and effective (Biniaz et al., 2024; Carezzano et al., 2023; Jiang et al., 2023a, b; Paduch-Cichal et al., 2023; Taglienti et al., 2022). One must consider the concentration of molecules with antiviral effects to inhibit viral particles using essential oils. Even among isomers, significant differences in antiviral effects can exist. Therefore, not all plants with essential oils can combat these viruses.

Additionally, the timing and dosage of essential oil application play a crucial role. There are differences when inoculating before or after applying the extract, with superior effects observed when applying the essential oil after inoculation (Paduch-Cichal et al., 2023). One way to control virus-transmitted diseases is by using essential oils to manage the vectors of these viruses. For instance, essential oils can reduce the oviposition of *Bemisia tabaci* (whitefly) or have repellent and mortality effects against aphids. By keeping insect vectors, such as whiteflies (vectors of geminivirus) or aphids (vectors of mosaic viruses), in check, we can prevent virus spread in plants. However, caution must be exercised regarding the concentration used to avoid phytotoxicity. Moreover, these vectors often remain on plant surfaces for extended periods. Therefore, the use of encapsulated essential oils is crucial for virus control. Encapsulation enhances the residual effect of essential oils on plant parts compared to non-encapsulated applications. Without encapsulation, residual effects are minimal, only affecting what comes into contact shortly after application. Over time, this leaves the plant vulnerable, allowing other vectors to introduce the virus (Peres et al., 2020; Taglienti et al., 2023).

Plant extracts, like essential oils, also hold a particular interest in testing their antiphytoviral effects against mosaic viruses in various plants, primarily tobacco. However, thousands of compounds are present when extracting metabolites from any part of the plant. Only a select group of metabolites exhibit antiphytoviral effects. Therefore, one of the most suitable strategies involves fractionating and purifying these compounds and testing them separately to identify molecules with high potential (Islam et al., 2018). These extracts can exhibit antiphytoviral effects through multiple pathways by reducing infection or preventing viral particle replication. However, the timing of extract administration and the subsequent days after application are crucial. Over time, the effect may diminish due to the loss of active molecules influenced by various environmental and physicochemical factors. Therefore, considering these factors is essential to enhance the effectiveness of these extracts (Shen et al., 2008).

Plant extracts against phytopathogenic viruses are categorized as any other than essential oils. This includes extracts obtained using solvents and various extraction techniques and residues from essential oil hydrodistillation, such as hydrosols or residual water. Hydrosols, in particular, have potential against phytopathogenic viruses. However, their effectiveness may vary compared to essential oils due to the chemical composition unique to each aromatic plant species. While they share many compounds, the concentration differs. Additionally, hydrosols undergo chemical changes during heating. Despite this, their advantage lies in obtaining larger quantities than essential oils, making them a valuable byproduct often overlooked (Vuko et al., 2021a, b). The evaluation method for antiphytoviral activity must also consider whether the hydrosol is applied before or simultaneously with the inoculum. The effect is more pronounced when using the hydrosol first, as it may activate systemic resistance mechanisms, thereby reducing disease severity. Additionally, the effect varies over time, depending significantly on residuality and tissue absorption (Nazlić et al., 2021; Vuko et al., 2022).

Plant diseases caused by viruses cause significant losses in agricultural production because there are currently few effective agents for controlling viral diseases in plants. Among the bioactive compounds of microbial origin used as antiviral agents in plants is ningnanmycin, an antimicrobial agent obtained from fermentation with Streptomyces. Its activity against tobacco mosaic virus (TMV) is related to the inhibition of the coat proteins assembly, used by viral particles to protect the viral RNA, which results in a loss of TMV pathogenicity. In addition, ningnanmycin can also induce systemic plant resistance against papaya ringspot virus (PRSV) and cucumber mosaic virus (MCV) (Sun et al., 2022; Wang et al., 2023a, b; Zhao et al., 2017).

Microorganisms produce antibiotics, alkaloids, cytotoxins, and various secondary metabolites, so developing biogenic pesticides from living organisms to control TMV is feasible since these metabolites have low toxicity, high efficiency, and high biocompatibility. The tobacco mosaic virus (TMV) has become the representative agent of virology due to its wide host range and the fact that it can cause enormous losses in the agricultural sector worldwide (Guo et al., 2020; Sun et al., 2022). Among the agents that inhibit TMV is cytosinepeptidemycin, which is a secondary metabolite derived from *Streptomyces ahygroscopicus*, which can also inhibit the infection of potato virus Y (PVY), rice black dwarf virus (SRBSDV) and cucumber mosaic virus (CMV) (Guo et al., 2020; Zhao et al., 2017).

Another compound that provides defense against viral diseases is chitosan, a natural compound obtained from chitin, which is present in the cell walls of fungi and the exoskeleton of arthropods. Chitosan has activity against various plant viruses such as PRSV, TMV, CMV and PVY. Chitosan interacts with DNA and proteins, interfering with the key processes of virus multiplication (transcription and translation) (Fan et al., 2022). Other bioactive compounds with antiviral activity are anthraquinone derivatives, which different species of Aspergillus can produce. The anti-TMV activity of anthraquinones produced from *Aspergillus oryzae* is because these compounds act as inducers of systemic resistance, in addition to increasing the tolerance of plants to TMV (Yang et al., 2023).

In addition to ningnanmycin, cytosinepeptidemycin, chitosan, and anthraquinones, lectins derived from cyanobacteria can potentially be antiviral agents. Lectins are proteins or glycoproteins that can reversibly bind to free sugars or carbohydrates; in the case of lectins obtained from cyanobacteria, these lectin-carbohydrate interactions can be used to control diseases (Singh et al., 2017).

4.3 Bioactive Compounds Against Bacteria

The control of phytopathogenic bacteria is also a very important topic where alternatives to conventional control methods are also sought. Below are some studies in which plant and microbial-origin bioactive compounds showed promising results. To begin with bioactive compounds of plant origin, we will talk about essential oils, which are volatile and nonpolar fractions extracted from plants that contain compounds with very high bioactivities. It has been shown that the essential oil of different plants can help against diseases caused by *Ralstonia solanacearum*, especially oils that contain antibacterial compounds such as p-Cymene, farnsul, saponins, terpene hydrocarbons, phenolic compounds, tannins, beta caryophyllin, geranyl acetate, cytronolal, eugenol, flavonoids, citral, geraniol, alkaloids, terpenoids, and steroids. For example, different essential oils have been tested against the wilt disease caused by *R. solanacearum* in hermes and sponta potato cultivars.

Pimpinella anisum L. essential oil was able to reduce the severity of wilt disease by up to 82%, possibly due to its phenolic compounds such as anethole, linoleic acid, limonene, and fenchone; *Thymus vulgaris* L. oil up to 77.6% possibly due to its high content of monoterpenes such as y-terpinene and β-cymene. *Cymbopogon flexuosus* L. oil contains alkaloids and phenols such as α-terpinene, γ-terpinene, linalool, and α-pinene, which could help achieve the 69.4% reduction. In the case of *Syzigium aromaticum* L., this reduced up to 60.1%, possibly due to the content of carvacrol, eugenol, eugenol acetate, and caryophyllene present in its essential oil. Similarly, the essential oils of *Foeniculum vulgare* L. were used, reducing the severity up to 55.7%, a reduction attributed to its content of trans-anethole, fenchone, limonene, and estragole (Abd-Elrahim et al., 2022). Potato soft rot is another disease that can appear in this very important crop and is caused by *Pectobacterium carotovorum* subsp. *carotovorum* (Pcc).

It has been shown that the essential oils of *Myrtus communis* and *Ferula gummosa* can reduce the incidence of the disease from 31.9–56.6%, where the major compounds were oxygenated monoterpenes, α -pinene, 1,8-cineole, limonene, linalool and linalyl acetate in *M. communis* and β-pinene, γ -elemene, germacrene D, terpinen-4-ol and p-cymene in *F. gummosa* (Hajian-Malekiet al., 2021). On the other hand, the essential oils of *Cymbopogon winterianum* and *Citrus aurantium* can be effective against the cause of citrus canker, *Xanthomonas citri* subsp. *citri*, with MIC values ≥ 0.487 µl/ml and ≥ 0.47 µl/ml, respectively. The majority content of these essential oils was citronellal, geraniol, citronellol, geranial, limonene, and isopulegol for *C. winteranium*, and linalyl acetate and linalool for *C. aurantium* (Sauer et al., 2015). Another important disease is fire blight caused by *Erwinia amylovora* in the Rosaceae family. The oils of *Apium graveolens* and *Curcuma longa* have been tested against this microorganism, demonstrating that they can reduce its pathogenicity by up to 41.7 and 30.17% in unripe pear fruits, an activity attributed to the phytochemical content of these oils (Akhlaghi et al., 2020).

Essential oils may be the most studied extracts. However, other extracts, such as aqueous ones, can be obtained through various methods; therefore, they differ in composition from essential oils. These other plant extracts also play a very important role in the search for alternatives for managing plant diseases, so some cases are mentioned below where these have shown promising results against phytopathogenic bacteria. *Streptomyces scabies* is a microorganism of economic importance since it attacks tubers such as potatoes; hence, ethanolic extracts of myrrh, garlic, lavender, wild indigo root, frankincense, barberry, and cayenne have been tested, showing that they can inhibit its growth up to 75% or more. Furthermore, the flavonoids sophoraflavanone G, aceosidin, baicalein, and quercetin showed MICs of 6.8 ± 0.4 μM, 100.0 ± 2.1 μM, 202.9 ± 5.3 μM, and 285.2 ± 6.8 μM, respectively (Gutierrez et al., 2022).

In another study, the hydroalcoholic extract of *Larrea tridentata* was tested against *Clavibacter michiganensis* sbsp. *michiganensis*, *Pseudomonas syrngae* and *Xanthomonas campestris*. In addition to testing the hydroalcoholic extract, an aqueous fraction and another obtained with ethyl acetate were tested. The latter was the most bioactive, demonstrating MICs of 3.12 mg/mL, 3.12 mg/mL, and 0.39 mg/mL against *C. michiganensis* sbsp. *michiganensis*, *P. syringae,* and *X. campestris*, respectively. Demonstrating that the nature of the solvent varies the phytochemical composition and, therefore, the bioactivity of the extract (Morales-Ubaldo et al., 2021).

In that sense, the effectiveness of an essential oil and an extract has also been compared. The herb *Teucrium polium* and the phytopathogenic bacteria *Pectobacterium cartovorum*, *Xanthomonas campestris*, *Ralstonia solanacearum*, *Streptomyces scabies*, *Rhizobium vitis*, *Rhizobium radiobacter*, *Brenneria nigrifluens*, *Pantoea agglomerans* and *Pseudomonas aeruginosa* were used. In this study, the essential oil presented MIC of 12.5 uL against *X. campestris*, *R. vitis*, *R. solanacearum,* and *S. scabies*, and MIC of 25 uL against *R. radiobacter*, *P. agglomeranse,* and *B. nigrifluens*, where the majority compounds were α-pinene and myrcene. In the case of the methanolic extract, it presented MIC of 6.25 uL against *S. scabies*, 12.5 uL against *B. nigrifluens* and *R. radiobacter*, and 25 uL against *P, carotovorum*, *X, campestris*, *R. vitis*, *P. agglomeranse* and *R solanacearum*, and sinapic acid and eugenol as its major compounds (Purnavab et al., 2015).

Secondary metabolites are low molecular weight compounds produced by bacteria, although they are unnecessary for their growth and replication. These bacterial compounds provide them with a competitive advantage by having antimicrobial properties. Secondary metabolites that exhibit a wide range of antibacterial activities against plant pathogenic bacteria include streptomycin, tetracycline, chloramphenicol obtained from *Streptomyces* spp., and other antibiotics isolated from *Bacillus* spp., such as bacillomycins and bacilenes.

The mode of action of antibiotics depends on their type and focuses mainly on the inhibition of protein synthesis and the cell wall (Ruparelia et al., 2022). Bacterial endophytes synthesize various bioactive compounds with antagonistic activity against plant pathogens and, therefore, can inhibit the growth of plant pathogenic bacteria. An example is *Pseudomonas fluorescens*, which produces pyoluteorin, an antibiotic that inhibits root rot caused by *Thielaviopsis basicola* (Nguyen et al., 2023). Volatile compounds produced by *Mucor racemosus* inhibit the growth of *Erwinia carotovora*, while volatile compounds such as nonanal produced by *Pseudomonas aurantica* inhibit the bacterial growth of *Ralstonia solanacearum* and *Clavibacter michiganensis* (Tahir et al., 2017; Wang et al., 2023a, b).

4.4 Bioactive Compounds Against Nematodes

Essential oils are gaining attention as natural nematicides due to the environmental and health risks of traditional chemical nematicides. Several studies highlighted their potential against various nematode species. Essential oils from 29 medicinal plants have been evaluated against pinewood nematode (*Bursaphelenchus xylophilus*). Oils from *Paeonia moutan*, *Perilla frutescens*, *Boswellia carterii*, and *Schizonepeta tenuifolia* achieved 100% mortality at 2 mg/ml, with LC50 values from 0.21–0.57 mg/ml. These oils, rich in monoterpenoids, phenols, and sesquiterpenes, showed strong nematicidal properties, suggesting potential as biocontrol agents. This study highlights the effectiveness of these oils and the need for further research on their mode of action, formulation, and safety (Choi et al., 2007). Other essential oils have also been evaluated against stem and bulb nematode (*Ditylenchus dipsaci*).

Cinnamomum cassia oil achieved 100% mortality within 4 h at 2000 ppm, while Frankincense and Dill's oils showed significant effects after 24 h. Oils from European Silver Fir and Mountain Pine also demonstrated substantial nematicidal effects, with *Thymus vulgaris* and thymol showing considerable activity. This study emphasizes the potential of these oils in nematode management and highlights the need for standardization and field trials (Douda et al., 2022). In the same way, the impact of ten essential oils on root gall formation by *Meloidogyne incognita* in tomato plants has been studied. Oils from *Artemisia absinthium*, *Lavandula officinalis*, *Mentha arvensis*, *Thymus serpyllum*, *Piper nigrum*, *Citrus bergamia*, and *Eucalyptus citriodora* effectively reduced root galls and egg masses, with higher concentrations showing better results. These oils also promoted plant growth, highlighting their potential as sustainable alternatives to chemical nematicides. The study found a linear relationship between the effectiveness of essential oils in reducing egg masses and root gall formation, indicating their dual benefit in nematode control and plant health promotion (Ozdemir & Gozel, 2018).

Artemisia annua essential oil (AEO) has been evaluated against citrus nematode (*Tylenchulus semipenetrans*). AEO showed 100% mortality at 250 ppm after 48 h and significantly reduced egg hatching. AEO also exhibited repellent properties, preventing nematode infestation. Greenhouse experiments revealed that AEO treatments reduced nematodes in soil and roots while promoting plant growth. These findings suggest AEO as a promising alternative for managing citrus nematodes, though further field trials are necessary (Asadi et al., 2024). These studies demonstrate the potential of essential oils as eco-friendly nematicides, emphasizing the need for additional research to optimize their use in integrated pest management strategies. The significant nematicidal properties highlight the promise of essential oils as sustainable solutions for agricultural nematode management.

In addition to essential oils, plant extracts are useful against diseases caused by nematodes. It has been shown that alcoholic or phenolic compounds such as kojic acid, eugenol, and punicalagin 1; aldehyde compounds such as furfural, salicylaldehyde, and cinnamic aldehyde; ketone compounds such as pulegone, fervenulin, and spectinabilin; acid compounds such as malic, linoleic, caffeic, chlorogenic, and rosmarinic acids; ester compounds such as ethyl cinnamate, xanthoxol, and bergapten; alkaloid compounds such as betaine, drupacine, walterone, and protopinium; and terpenoid compounds such as limonene, β-pinene, and cordionic acid; among many other compounds, they can be used to control nematodes and the diseases they can cause (Chen & Song, 2021).

Methanolic extracts of *Cucurbita maxima*, *Prunus africana*, *Pelargoniumídoides*, *Croton sylvaticus*, *Solanum aculeastrum*, *Vernonia colorata*, *Searsia lancea,* and *Merwilla plumbea* are effective against the root-knot nematode *Meloidogyne incognita*, demonstrating reductions in the number of nematodes ranging from 50–97%, proving to contain compounds such as alkaloids, terpenoids, phenols, flavonoids, tannins, steroids, saponins, carbohydrates and anthraquinones (Sithole et al., 2021). Similarly, extracts from 13 different medicinal plants were tested against *Meloidogyne incognita*, where *Solidago* sp. and *Catharanthus roseus* showed the best results, with J2 mortality of 72–100% and the presence of phenols, flavonoids, triterpenoids, coumarins, alkaloids, and glycosides in their composition (Kesba et al., 2021). In another study where ethanolic extracts of 20 wild plants were tested against *Nacobbus aberrans*, *Cosmos sulphureus* obtained the highest mortality rate at 72 h. In this study, the content of total phenols and total flavonoids was evaluated, where this plant presented values of 22.46 ± 0.08 mg GAE/g DT and 100.03 ± 1.06 mg CE/g DT (Vázquez-Sánchez et al., 2018).

Various studies have identified a wide range of microorganisms with nematicidal activity, including notable bacteria such as *Bacillus* spp. and *Pseudomonas* spp., among others (Deng et al., 2024), as well as different fungi like *Trichoderma* spp., mycorrhizae, and other endophytic fungi (Poveda et al., 2020). These microorganisms produce bioactive compounds capable of immobilizing, inhibiting, or killing nematodes. For instance, *Bacillus thuringiensis* can produce crystal proteins that induce cytotoxicity in various nematodes (Vázquez-Pineda et al., 2012), while *Pseudomonas* spp. produce antibiotics that suppress these phytoparasitic organisms (Timper et al., 2009). Furthermore, fungi of the genus *Trichoderma* are known to secrete hydrolytic enzymes such as chitinases and proteases, which degrade the nematode cuticle and lead to their demise. Some fungi produce toxic compounds, such as mycotoxins, which affect nematodes. An example is *Pleurotus ostreatus*, which produces paralyzing mycotoxins for these organisms (Satou et al., 2008). Another example is trans-2-decenoic acid derived from linoleic acid, which exhibits activity against nematodes, insects, and other fungi (Bhat et al., 2023).

Current research in this field is extensive, driven by the adverse effects of chemical nematicides. However, developing and optimizing these biological strategies is crucial, as natural compounds tend to be less consistent and effective than traditional synthetic chemicals. Although many microorganisms show promising nematicidal results in vitro, not all have proven equally effective under field conditions. It is, therefore, essential to continually improve performance and understand the complex biotic and abiotic interactions to maximize the benefits of bionematicides in sustainable agricultural systems (Abd-Elgawad, 2021). The transition to more sustainable methods in agriculture involves exploring and harnessing the potential of these biological agents. Besides their role in nematode control, some of these microorganisms can offer additional benefits, such as enhancing soil health and microbial biodiversity, which are crucial for the long-term sustainability of agricultural systems. This approach aims not only to reduce dependence on synthetic chemical products but also to maintain and enhance crop productivity in an environmentally responsible manner.

4.5 Bioactive Compounds Against Parasitic Plants

Essential oils have shown potential as natural herbicides against parasitic plants like *Cuscuta* and *Striga*. Oils from *Anethum graveolens*, *Cymbopogon citratus*, and *Ocimum gratissimum* inhibit weed germination and growth while inducing suicidal germination in Striga, reducing its seed bank. However, careful management is needed to minimize phytotoxic effects on host crops.

There is an investigation of the essential oil extracted from *Anethum graveolens* (dill), which contains carvone (51.69%) and limonene (39.88%). The study assessed the oil's effects on the germination and early seedling growth of *Cuscuta campestris*, *Cuscuta epithymum*, and their host plants, alfalfa, and red clover. For *Cuscuta campestris*, germination inhibition ranged from 67% at 0.01% to 94% at 1%. Seedling growth inhibition varied from 5% at 0.01% to 96% at 1%. For *Cuscuta epithymum*, germination inhibition was 100% at 1% and 0.5%, 80% at 0.1%, and 67% at 0.01%. Seedling growth inhibition ranged from 4% at 0.01% to 100% at 1%.

Host plants also exhibited significant phytotoxic responses. Red clover germination was inhibited by 35% at 0.01% and 70–100% at higher concentrations. Seedling growth inhibition mirrored these results. Alfalfa showed 5% germination inhibition at lower concentrations and 36–100% at higher concentrations, with a corresponding trend in seedling growth inhibition. Dill essential oil shows considerable herbicidal potential against *Cuscuta* species but requires careful management to prevent crop damage. Further research should validate these results in field trials and refine application techniques to mitigate adverse effects on crops (Saric-Krsmanovic et al., 2023).

The allelopathic properties of essential oils from ten local aromatic plants on *Striga hermonthica* germination, a major African parasitic weed, have been studied. Oils were extracted by steam distillation and tested on *Striga* seeds under various dilutions (10, 0.2, and 0.4%) in controlled laboratory conditions. The inhibition bioassay results showed that essential oils significantly affected *Striga* seed germination. *Cymbopogon citratus* oil at 10% dilution inhibited germination (100%). Oils from *Ocimum gratissimum*, *Lippia multiflora*, and *Cymbopogon nardus* also demonstrated substantial inhibitory effects, with inhibition rates of 92.60%, 92.12%, and 74.54%, respectively. These oils contain metabolites that act as bio-herbicides, potentially useful for controlling *Striga* and improving cereal yields. Field use of these oils could limit environmental pollution and combat climate change. The stimulation bioassay revealed that lower concentrations of some essential oils could stimulate *Striga* seed germination. The 0.2% dilutions of oils from *Cymbopogon citratus*, *Ocimum gratissimum*, *Hyptis suaveolens*, *Ocimum americanum*, and *Cymbopogon schoenanthus* induced germination rates ranging from 60.64% to 0.97%. This suggests these oils might be used for ecological weed management through suicidal germination, where stimulated *Striga* seeds germinate but fail to survive without a host, reducing the seed bank (Sawadogo-Ilboudo et al., 2021).

On the other hand, speaking of extracts, it has been shown that the compounds sarmentosine and sarmentine obtained from ethanolic extracts of *Piper sarmentosum* can be useful against *Echinochloa crusgalli* and *Amaranthus retroflexus*, showing inhibition rates greater than 80% at concentrations of 5 mg/mL (Feng et al., 2019). In the search for bioactive compounds for managing plant diseases, attempts have also been made to use residues, such as the residues of *Brucea javanica* (L.) Merr., which demonstrated herbicidal potential against *Eleusine indica* L. and *Bidens pilosa* L. in a study where different extracts were used. Among these, the ethyl acetate extract had the best results at a concentration of 5 mg/mL, with an inhibition rate of 100% for both plants in greenhouse conditions. From this extract, the compounds charisphthalate, vanillic acid, protocatechuic acid, methyl ester of 4-(9H-b-carbolin-1-yl)-4-oxobut-2-enoic acid, pythyriactrin, and bruecins D-F were identified (Huang et al., 2021).

In the sense of comparing essential oils and extracts, the essential oils of *Origanum syriacum*, *Origanum onites,* and *Origanum majorana* have been used against *Thlaspi arvense*, *Amaranthus retroflexus*, *Rumex cripus,* and *Lactuca serriola*. The extracts tested were acetone and n-hexane, which did not obtain results as good as those of essential oils. In general, *O. onites* oil obtained the best mortality rates (73.33–90%) against the different herbs, in which its major compounds were carvacrol, thymol, linalool, p-cymene, and B-caryophyllene, respectively (Kordali et al., 2022).

Allelopathy is one of the principal interactions that has captured researchers' attention in developing suitable replacements for current synthetic products (Saxena, 2014). It can be defined as the chemical interaction of microbial and plant secondary metabolites with organisms to affect the growth and development of biological systems. The secondary metabolites involved in allelopathy are termed allelochemicals (Rice, 1984).

Considering that seed germination is a critical phase for the development and infestation of parasitic plants, an additional approach has been proposed for managing these weeds using secondary metabolites from certain microorganisms (Graniti, 1991). Environmentally friendly strategies are based on biological control using phytopathogenic agents and/or the phytotoxic metabolites they produce (Evidente et al., 2013). Various metabolites with the potential for weed and crop disease control have been discovered in developing herbicides from microbial and natural compounds. Among bacteria, Tabtoxin produced by *Pseudomonas syringae* var. *tabaci*, Phaseolotoxin by *Pseudomonas syringae* pv. *phaseolicola*, and Coronatine by *Pseudomonas coronafaciens* have demonstrated significant phytotoxic activities by inhibiting key plant enzymes (Saxena, 2014). Actinomycetes, particularly *Streptomyces*, produce Bialaphos, Herbimycin, and Pyrizadocidin, which act against a broad range of both broadleaf and grassy weeds. In the realm of cyanobacteria, compounds like Cyanobacterin and Dihydroactiniodiolide have shown herbicidal activity by interfering with key photosynthetic processes. Finally, among fungi, AF-toxins from *Alternaria fragariae* and Maculosin from *Alternaria alternata* are specific to certain plant species and offer a targeted control approach (Saxena, 2014).

The use of microbial herbicides has been explored for many years, yet numerous challenges remain to be overcome before they can be widely applied in agricultural lands, forests, grasslands, or waterways. These challenges include improving the efficacy of microbial activity, the durability of the suppressive compound, delivery systems, and preventing damage to non-target organisms. Other important considerations are interactions with chemical herbicides, regulations, commercialization and mass production, as well as economic feasibility (Alvarez-Fernandez, 2012). Some bioactive compounds with possible use for plant disease management are listed in Table 1.

5 Bioactive Compounds for Diseases Caused by Abiotic Factors

5.1 Bioactive Compounds for Temperature Stress

Changes in the climate in the primary agricultural regions of the world directly affect crop productivity and yields, potentially reducing food security. Presently, intensive production systems rely on agronomic management and protected agriculture to mitigate the impacts of abiotic stress. When used alongside biofertilizers, biostimulants, and bioprotectors, they play a crucial role in maintaining and boosting productivity because they possess bioactive effects, as those mentioned in Table 2. Agronomic management, protected agriculture, and biotechnology have mitigated temperature stress. However, worldwide climate change calls for exploring additional tools to ensure food security. One strategy involves using phytochemicals and other natural products that boost the antioxidant system. Below are described some successful examples of this approach.

One defense mechanism against temperature stress is the increase in the production of secondary metabolites (phytochemicals). These compounds work alongside antioxidant enzymes to help minimize the effects of excessive production of reactive oxygen species (ROS), which is triggered by high or low temperatures. For instance, in a review article, Mishra et al. (2023), mention that levels of flavonoids and total phenols increase under heat stress conditions. Interestingly, even under heat stress, the roots produce excessive amounts of flavonoids, making micronutrients from inaccessible sources soluble, thereby facilitating their absorption. This indicates that flavonoids have a dual effect, serving as antioxidants and as agents that enable the absorption of micronutrients, thus contributing to the survival of crops.

It is worth noting that applying biofertilizers externally during periods of oxidative heat stress can enhance thermal resistance and aid in the retention of nutrients. An illustrative example is X. Zhang et al. (2023), who evaluated the foliar application of sodium nitroprusside (100 μM) and gibberellic acid (5 μg/mL), in wheat crops subjected to heat stress (40 °C for 6 h per day for 15 days). Concomitant application was found to enhance plant growth. There was a 4.48% increase in plant height, nearly a 30% increase in fresh weight, and an 87% increase in dry weight. Additionally, the rate of photosynthesis increased by almost 40%. Furthermore, also had a stimulating effect on the antioxidant system, leading to an increase in the enzymes superoxide dismutase (SOD), catalase (CAT), ascorbate peroxidase (APX), and glutathione reductase. The content of proline and glycine betaine was also observed to increase. The antioxidant effects are attributed to sodium nitroprusside, which is classified as a nitric oxide donor.

Nitric oxide is a signaling molecule that participates in the stress signaling cascade. The application of sodium nitroprusside acts as a modulator of the endogenous antioxidant system. Additionally, stimulating the production of osmoprotectors such as proline and glycine betaine improves protection against oxidative stress and dehydration, maintaining the stability of the cell membrane. Gibberellic acid has been shown to stimulate the endogenous antioxidant system by increasing antioxidant enzymes and synthesizing osmoprotectants (Zhang et al., 2023).

Guo et al. (2022), evaluated the effects of foliar administration of gibberellic acid on heat-stressed tomato crops (var. Ahmar and Roma). It was found that 75 mg/L of gibberellic acid improved the root and shoot biomass yield. Additionally, improvements were observed in physiological parameters such as greenness index, photosynthetic range, and transpiration. The synthesis of osmoprotectants like proline and nutrients such as phosphorus, nitrogen, and potassium increased. Furthermore, the highest application of 100 mg/L mitigated oxidative stress, stimulating the activity of SOD, peroxidase, and CAT, and reducing markers such as malondialdehyde and H_2O_2. Gibberellic acid, a diterpenoid-type phytohormone, plays a role in physiological and biochemical processes, regulating the cell cycle, photosynthesis, and growth.

Endophytic bacteria can colonize plants without damaging plant tissue and induce plant tolerance to biotic and abiotic stress. Various endophytic bacteria such as *Bacillus,*

Table 1 Bioactive compounds with possible use against plant diseases caused by biotic factors

Source of bioactive compounds	Disease-causing agent	Attributed bioactive compounds	References
Bacillus spp.	*Alternaria*, *Fusarium*	Fengycin, surfactin	Santos et al. (2024)
Streptomyces spp.	Tobacco mosaic virus, Papaya ringspot virus, Cucumber mosaic virus	Ningnanmycin	Sun et al. (2022), Wang et al. (2023a, b), Zhao et al. (2017)
Streptomyces ahygroscopicus	Tobacco mosaic virus, Potato virus Y, Rice black dwarf virus, Cucumber mosaic virus	Cytosinepeptidemycin	Guo et al. (2020), Zhao et al. (2017)
Fungi	PRSV, TMV, CMV, PVY	Chitosan	Fan et al. (2022)
Aspergillus oryzae	TMV	Anthraquinone derivatives	Yang et al. (2023)
Pimpinella anisum L.	*Ralstonia solanacearum*	Anethole, linoleic acid, limonene, fenchone	Abd-Elrahim et al. (2022)
Thymus vulgaris L.		γ-Terpinene, β-cymene	
Cymbopogon flexuosus L.		α-terpinene, γ-terpinene, linalool, α-pinene	
Syzigium aromaticum L.		Carvacrol, eugenol, eugenol acetate and caryophyllene	
Foeniculum vulgare L.		Trans-anethole, fenchone, limonene and estragole	
Myrtus communis	*Pectobacterium carotovorum* subsp. *carotovorum*	α-pinene, 1,8-cineole, limonene, linalool, linalyl acetate	Hajian-Maleki et al. (2021)
Ferula gummosa		β-pinene, γ-elemene, germacrene D, terpinen-4-ol, p-cymene	
Cymbopogon winterianum	*Xanthomonas citri* subsp. *citri*	Citronellal, geraniol, citronellol, geranial, limonene, isopulegol	Sauer et al. (2015)
Citrus aurantium		Linalyl acetate, linalool	
Teucrium polium	*Pectobacterium cartovorum*, *Xanthomonas campestris*, *Ralstonia solanacearum*, *Streptomyces scabies*, *Rhizobium vitis*, *Rhizobium radiobacter*, *Brenneria nigrifluens*, *Pantoea agglomerans*	α-pinene, myrcene, sinapic acid, eugenol	Purnavab et al. (2015)
Pseudomonas fluorescens	*Thielaviopsis basicola*	Pyoluteorin	Nguyen et al. (2023)
Pseudomonas aurantica	*Ralstonia solanacearum*, *Clavibacter michiganensis*	Nonanal	Wang et al. (2023a, b)
Thymus vulgaris	*Ditylenchus dipsaci*	Thymol	Douda et al. (2022)
Anethum graveolens	*Cuscuta campestris*, *Cuscuta epithymum*	Carvone, limonene	Saric-Krsmanovic et al. (2023)
Piper sarmentosum	*Echinochloa crusgalli*, *Amaranthus retroflexus*	Sarmentosine, sarmentine	Feng et al. (2019)
Brucea javanica (L.) Merr. residues	*Eleusine indica* (L.) Gaertn, *Bidens pilosa* L.	Charisphthalate, vanillic acid, protocatechuic acid, methyl ester of 4-(9H-b-carbolin-1-yl)-4-oxobut-2-enoic acid, pythyriactrin, bruecins D-F	Huang et al. (2021)
Origanum onites	*Thlaspi arvense*, *Amaranthus retroflexus*, *Rumex cripus*, *Lactuca serriola*	Carvacrol, thymol, linalool, p-cymene, β-carioplhyllene	Kordali et al. (2022)

Table 2 Biostimulants and plant-derived bioactive compounds to mitigate abiotic stress

Stress	Specific condition	Crop	Biostimulant/bioactive compound	Mechanism of action	References
Temperature	Cold stress	*H. vulgare*	*S. angustifolium* and *H. pluvialis* extract	Improve the osmoprotectant content and enhance the photosynthetic parameters	(Babazadeh et al., 2023)
		C. pepo	Allantoin, ascorbate, salicylic acid, amino acids, and sources of Se, Mo and Co	Stimulation of photosynthetic efficiency and antioxidant	(Navarro-Morillo et al., 2024)
	Heat stress	*G. max*	*A. nodosum* extract	Improve the antioxidant status	(Repke et al., 2022)
		S. lycopersicum	*A. nodosum* extract	Up-regulated the gene transcription of protective heat shock proteins	(Carmody et al., 2020)
Nutrient	NPK deficiency	*S. lycopersicum*	Humic acid and amino acids	Improve the osmoprotectant content and enhance the nutraceutical content	(Koleška et al., 2017)
Hydric	Drought stress	*B. juncea*	Orthosilic acid and seaweed extract	Improve the osmoprotectant content and enhance the antioxidant status	(Sujata Goyal et al., 2023)
		T. aestivum	Aqueous extract of *S. platensis*	Improve the gas exchange and metabolism	(Elnajar et al., 2024)
		Saccharum spp. hybrids	Extract of seaweed *A. nodosum*	Mitigate oxidative stress and improve the enzymatic antioxidant activity	(Jacomassi et al., 2022)
Pollution	Heavy metals-contaminated saline soil	*C. annuum*	Moringa and licorice extracts	Enhancing antioxidative endogenous systems	(Desoky et al., 2019)
	Copper toxicity	*Z. mays*	Gallic acid	Enhancing antioxidative endogenous systems	(Yetişsin & Kurt, 2020)
	Cadmium and manganese toxicity	*E. densa*	Anthocyanins	Enhancing antioxidative endogenous systems	(Maleva et al., 2018)

Burkholderia, Enterobacter, Paenibacillus, Microbacterium, Azospirillum, Acinetobacter, Streptomyces, Nocardia, Pseudomonas, and *Klebsiellla* can promote plant growth through the synthesis of indole acetic acid and improve resistance to abiotic stress. Plant-endophyte interactions have biostimulant actions in plants, such as the induction of root growth, induction of the antioxidant system of plants, and production and release of volatile hormones such as methyl salicylate, methyl jasmonate, and ethylene mainly. Sunflower plants under water stress showed that endophytic bacteria can promote plant growth by producing salicylic acid (Nguyen et al., 2023). In stress due to drought and high temperatures, plants promote the emission of isoprene, which is important in protecting against reactive oxygen and nitrogen species formed under stress conditions. Various bacteria can produce isoprene; however, bacteria of the *Bacillus* genus are considered the group of bacteria with the highest production of isoprene (Rani et al., 2023). Once applied to plants, bioactive compounds produced by fungi and bacteria, such as terpenes, phenolics, and volatile hormones, can improve nutrient absorption, crop quality, and increase protection against thermal, mechanical, chemical, and oxidative stresses. The decrease in abiotic stress of plants is a desired effect since stressors can reduce crop yields (Ciriello et al., 2024).

5.2 Bioactive Compounds for Moisture Stress

Hydric stress is the main abiotic stress that affects the productivity and quality of crops. It occurs when there is limited or scarce water available to the roots, leading to increased transpiration and systemic dehydration. This results in the loss of metabolic functionality. Factors such as changes in rainfall patterns, soil moisture retention capacity, and the specific water needs of the crop can impact productivity. Additionally, the growing global population and demand for drinking water further exacerbate this issue (Caicedo-Lopez et al., 2020). However, exogenous applications of both microbial and non-microbial biostimulants (chitosan, fulvic and humic acids, protein hydrolysates, phosphates, algae extracts) can contribute to the growth, uptake, and translocation of micro and macronutrients (Bhupenchandra et al., 2022).

An example is the one presented by Ali et al. (2019), who supplied α-tocopherol (0.001, 0.01, and 0.1 mM) via foliar to wheat plants exposed to water deficit. The stress treatment was subjected to only two irrigations during the experimental period, while the control group was irrigated four times. Administration with α-tocopherol increased the spike length (7.67 ± 0.05 cm), the plant fertile tillers (5.33 ± 0.33), the number of spikelets/spike (356 ± 19), the weight in 100 g (3.50 ± 0.29 g), the yield per plant (14.27 ± 1.18 g), likewise, the level of carotenoids in the leaves increased by 0.25 mg/g. At the same time, markers of oxidative stress, such as malondialdehyde and H_2O_2, decreased, and there was an increase in the levels of antioxidant enzymes, including SOD, APX, CAT, and peroxidase. Additionally, the treatment led to higher levels of total phenolic compounds and flavonoids when administered at a dose of 0.01 mmol, indicating that tocopherol stimulates the antioxidant response, resulting in increased levels of enzymatic markers and the production of phytochemicals like flavonoids and carotenoids. These biochemical changes ultimately contribute to an improved redox status.

Hanif et al. (2021), examined the impact of spraying rice crops with different concentrations of proline (10, 20, and 30 mM) under hydric stress. The application of 30 mM proline across all treatments resulted in the highest activation of antioxidant enzymes (SOD, POD, CAT), leading to a reduction in oxidative stress markers like malondialdehyde (MDA). Additionally, this concentration significantly boosted the chlorophyll content in the leaves compared to the other levels. The application of 30 mM proline also elevated the levels of osmoprotectants such as glycine betaine and proline, which are essential for combating water stress. On the other hand, the application of chitosan and its semisynthetic derivatives (N-succinyl and N, O-dicarboxymethylated) at a rate of 0.5 mg per plant has been proven to enhance the antioxidant response and productivity of corn hybrids in conditions of hydric stress. Chitosan treatments were found to reduce MDA levels and boost the activity of antioxidant enzymes such as SOD, CAT, APX, and GPX. Additionally, chitosan treatments increased total phenolic compounds and osmoprotective compounds like proline. Furthermore, the quality of the grain, specifically the content of soluble sugars, amino acids, and starch, was higher with chitosan treatments. These positive effects can be attributed to the structural properties of chitosan, which interacts with genes related to antioxidant response, leading to improved plant performance under stress (Rabêlo et al., 2019).

Certain microorganisms produce a variety of secondary metabolites that function as growth modulators and protect against water stress. Plant Growth-Promoting Bacteria (PGPB) affect hormonal balance by producing and modulating hormones such as auxins and abscisic acid (ABA). Auxins enhance growth and root formation, improving water absorption and root structure. ABA aids plants in tolerating dehydration by promoting stomatal closure (Bittencourt et al., 2023). Bacteria such as *Azospirillum brasilense* provide exogenous ABA, enhancing the plant's ability to cope with abiotic stress (Hungria et al., 2010). Some bacteria possess the enzyme ACC deaminase, which reduces ethylene levels, thereby promoting growth under adverse conditions (Gamalero & Glick, 2015). Additionally, PGPBs synthesize osmolytes such as glycine betaine, soluble sugars, trehalose, and proline, which, when secreted, act synergistically with plant osmolytes to function as osmoprotectants, enhancing drought tolerance (Ozturk & Turkyilmaz Unal, 2021).

Inoculation with PGPBs like *Arthrobacter sp.* and *Bacillus spp.* has increased proline levels in plants, enhancing their capacity to withstand water stress by maintaining cellular turgor and membrane stability (Sziderics et al., 2007). Moreover, exopolysaccharides (EPS) are high molecular weight, heterogeneous polymers that are water-soluble and secreted by microorganisms into the surrounding environment (Ghosh et al., 2019). EPS synthesis is a common self-protection mechanism in bacteria, benefiting bacteria and plants under stress conditions. Bacteria from the genera *Bacillus*, *Pseudomonas*, and *Azospirillum* secrete EPS under water stress conditions, which increases water retention in the soil and enhances abiotic stress tolerance in plants. Furthermore, volatile organic compounds (VOCs) are lipophilic, low molecular weight molecules emitted by plants and microorganisms in response to stress. Bacteria such as *Burkholderia*, *Pantoea*, *Serratia*, and *Chromobacterium* synthesize these compounds, promoting plant growth and modulating essential physiological processes (Bittencourt et al., 2023).

5.3 Bioactive Compounds for Nutrient Stress

Crops need at least 17 nutrients to grow properly and be productive, with 14 of them as essential minerals that must be supplied correctly. This includes considering doses, application times, absorption, and supply routes specific to each crop. If these nutrients are not available in the right form or quantity, it can lead to nutrient stress, affecting the plant's growth, development, and productivity (Islam et al.,

2022). The symptoms caused include chlorosis, necrosis, and stunted growth. While applying fertilizers is essential for crop maintenance, it must be carefully controlled and regulated.

Plants have defense mechanisms such as morphological modifications at the shoot, root, and leaf levels and biochemical changes. These include upregulating nutrient transporters, increasing phosphatases, secreting organic acids, accumulating flavonoids, and releasing ROS (Francis et al., 2023). An interesting article was published by Fawzy et al. (2023), whom investigated the effects of different biostimulants on green beans grown in nutrient-deficient soil. The biostimulants—chitosan, a mixture of amino acids, humic acid, or yeast extracts—were applied as foliar sprays at concentrations of 2 mL/L or 2 g/L. The results indicated that chitosan applications mainly increased vegetative parameters such as the fresh and dry weight of leaves, the number of leaves per plant, and plant size (height and weight). The nutritional content also improved significantly after the application of chitosan, with an 18% increase in crude protein and nitrogen, phosphorus, and potassium (NPK). The authors suggested that the biostimulants worked by stimulating the antioxidant system, regulating nutrient uptake genes, and promoting the production of primary and secondary metabolites.

Another example is published by Koleška et al. (2017), whom evaluated the effects of administering a commercial biostimulant, which is a mixture of organic matter, polysaccharides, proteins, polypeptides, amino acids, vitamins, and humic acid, on tomato crops experiencing NPK deficiency. The authors reported that applying the biostimulant led to a 14% increase in total yield and improved the quality of the fruits. The biostimulant significantly reduced the total soluble solids while increasing the total acidity. Additionally, the biostimulant prevented a decrease in lycopene and total carotenoids and also prevented an increase in ascorbic acid.

Microbial-derived bioactive compounds have proven to be effective tools in mitigating nutrient stress in plants, promoting their growth and resilience under adverse conditions. These compounds, produced by bacteria and fungi, include phytohormones, siderophores, antibiotics, and secondary metabolites, which collectively improve nutrient availability and soil health, thereby contributing to agricultural sustainability. One of the key mechanisms by which microorganisms assist plants in coping with nutrient stress is through the production of siderophores. These compounds have a high affinity for iron, allowing plants to access this essential micronutrient even in soils with low iron availability. Certain bacteria are known for their ability to produce siderophores, enhancing iron absorption and thus promoting plant growth under nutrient-deficient conditions (Vaishnav et al., 2017).

Furthermore, microorganisms promote plant growth by synthesizing phytohormones such as auxins and cytokinins. These plant hormones play a crucial role in root development and nutrient absorption. Auxins, in particular, are a class of small molecules capable of inducing plant growth responses (Teale et al., 2006). They are involved in most of the quantitative growth changes that occur throughout a plant's life cycle. Auxins are the most important plant hormones produced by *Azospirillum*, *Pseudomonas* spp., and *Bacillus megaterium* (Keswani et al., 2020). Another significant group of bioactive compounds are antibiotics and secondary metabolites, which suppress soil pathogens and improve plant nutrient availability. *Streptomyces* spp., for instance, produce antibiotics that inhibit soil pathogens, reducing competition for nutrients and favoring plant growth (Vurukonda et al., 2018). Volatile organic compounds (VOCs) produced by bacteria also play a crucial role in water stress. For example, *Bacillus subtilis* is known for releasing VOCs that can induce systemic resistance in plants, enhancing their tolerance and promoting more efficient nutrient use (Hashem et al., 2019).

5.4 Bioactive Compounds for Pollution Stress

Crops and plants are exposed to various environmental pollutants, including UV radiation, ozone, nitrogen oxides, heavy metals, pesticides, microplastics, persistent organic pollutants, excessive salinity, and other toxic substances. Heavy metals such as Pb and Cd significantly affect crops' growth, development, and productivity. Symptoms crops may exhibit include chlorosis, impaired photosynthesis, blocked respiration, and reduced growth, which can modify the antioxidant machinery and lead to oxidative stress due to ROS (Mareri et al., 2022). Some plant defense mechanisms include activating the antioxidant system and facilitating cellular detoxification through antioxidant enzymes.

The plant also undergoes physiological changes at the leaf level to regulate the stomatal aperture (Oksanen & Kontunen-Soppela, 2021). Using biostimulants and natural substances can help alleviate the impacts of stress caused by contamination from foreign substances such as heavy metals. They work through various mechanisms, including enhancing the antioxidant system, boosting the accumulation of osmoprotectants, regulating the buildup and movement of foreign substances, influencing gene expression, and improving the photochemical system, among other mechanisms (Feng et al., 2023). Below are some examples of bioactive compounds that have been shown to mitigate pollution stress, particularly heavy metals. Flavonoids such as quercetin and naringenin have been shown to reduce the toxic effects of heavy metals in various crops, including corn, *A. thaliana*, and tomato. They achieve this by stimulating the plants' natural antioxidant system and increasing the levels of chlorophyll and secondary metabolites such as carotenoids and flavonoids (Singh et al., 2021).

For instance, Aslam et al. (2022) evaluated the effects of different concentrations of quercetin (15, 25, 40 μM) on Fenugreek (*T. corniculata* L.) plants that were exposed to chromium stress. It was evident that exposure to chromium had toxic effects on the plants. However, when treated with quercetin, there was an improvement in plant growth. Specifically, the treatment with 25 μM of quercetin improved chlorophyll a and b levels, total carotenoids, photosynthetic rate, and transpiration rate. This treatment also led to an increase in gas exchange. Furthermore, under the same treatment, there was an increase in proline content and a decrease in H_2O_2 compared to the control group without quercetin. These effects suggest that quercetin may modulate the endogenous antioxidant system as it also led to an increase in the enzymatic activity of SOD, POD, CAT, and APX by 28%, 22%, 29%, and 33%, respectively. The detoxifying effects of quercetin are attributed to its antioxidant capacity and the stimulation of endogenous antioxidant enzymes.

Arikan et al. (2023), evaluated the impact of quercetin and kaempferol (25 μM) on wheat plants exposed to arsenic stress. They found that wheat plants treated with arsenic, without flavonoids, showed a decrease in relative growth rate. However, when both flavonoids were applied together, the negative effects of arsenic stress were completely mitigated, and the growth rate returned to normal levels similar to those of the control group without stress. Additionally, applying both flavonoids restored the osmotic potential in the plants. The researchers also observed that both flavonoids were present inside the plant cells, specifically in guard cells, and this presence influenced the gene expression of proteins involved in the photosystem, such as PsaA, PsaB, PsbA, and PsbD. Moreover, the application of both flavonoids led to a 20% decrease in intracellular ROS (H_2O_2), and an increase in the activity of antioxidant enzymes such as SOD, POX, APX, and GPX, indicating a regulatory role of the plant's endogenous antioxidant system. The same research group recently published advancements in evaluating the exogenous application of hesperidin and chlorogenic acid in corn plants under arsenic stress. Their research aims to understand the mechanisms of action involved in regulating the antioxidant and photosynthetic systems and the synthesis of fatty acids (Arikan et al., 2022; Elbasan et al., 2024).

Microbial-derived compounds have garnered significant attention as a sustainable approach to mitigating environmental pollution stress. These compounds, produced by bacteria and fungi, play a crucial role in bioremediation processes, helping to detoxify and restore contaminated ecosystems. The ability of microorganisms to produce a wide range of bioactive compounds, such as enzymes, biosurfactants, and secondary metabolites, makes them invaluable tools in combating pollution. Enzymes are the most efficient bioremediation tools, facilitating various chemical changes in contaminants. The specificity of enzymes is typically broad enough to act on different molecules with similar structures. Additionally, enzymes can be modified to enhance their stability and efficiency under specific conditions or with particular substrates (Mousavi et al., 2021).

The microbial enzymes most commonly involved in bioremediation are cytochrome P450s, dehydrogenases, laccases (Lac), proteases, hydrolases, lipases, and dehalogenases. These enzymes have shown promise and efficiency in the degradation and transformation of various pollutants, including agrochemicals, aromatic hydrocarbons, dyes, polymers, detergents, and halogenated substances (Narayaran et al., 2023). Biosurfactants are another class of compounds derived from microorganisms that have shown great potential in pollution mitigation. These amphiphilic compounds contain hydrophobic and hydrophilic domains (Singh et al., 2020).

Natural surfactants or biosurfactants can be produced extracellularly or as part of the cell membrane by various microorganisms, such as bacteria, fungi, and yeasts (Sharma et al., 2021). Examples include *Pseudomonas aeruginosa* (produces rhamnolipids), *Bacillus subtilis* (produces a lipopeptide called surfactin), *Nocardia amarae*, and *Saccharomyces lipolytica* (Bustamante et al., 2012). Secondary metabolites produced by microorganisms also contribute significantly to mitigating pollution stress. These metabolites, including organic acids, alcohols, and phenolic compounds, can chelate heavy metals and other toxic elements, reducing their bioavailability and toxicity. For instance, the production of citric acid by *Aspergillus niger* has been shown to enhance the leaching of heavy metals from contaminated soils, aiding in their removal and recovery (Ren et al., 2009).

6 Conclusions

A wide variety of bioactive compounds can be obtained from different plant matrices and microorganisms, the most common being aromatic plants and bacteria, respectively. In addition to the diversity of sources of these compounds, there are different methodologies for obtaining them, which will vary depending on the source, the compounds of interest, and personal limitations. It has been shown that these compounds can be useful in crop management since they present bioactivity against a large number of agents that cause plant diseases, such as fungi, viruses, bacteria, nematodes, and parasitic plants. Besides, bioactive compounds can favor the plant when it suffers some abiotic stresses related to temperature, moisture, nutrients, and pollution. In addition to these benefits, using bioactive compounds to manage plant diseases is very helpful to health and the environment since they reduce or replace synthetic substances.

References

Abd-Elgawad, M. M. M. (2021). Optimizing safe approaches to manage plant-parasitic nematodes. *Plants, 10*(9), 1911.

Abd-Elrahim, R., Tohamy, M. R. A., Atia, M. M., Elashtokhy, M. M. A., & Ali, M. A. S. (2022). Bactericidal activity of some plant essential oils against Ralstonia solanacearum infection. *Saudi Journal of Biological Sciences, 29*(4), 2163–2172. https://doi.org/10.1016/j.sjbs.2021.11.045.

Achar, P. N., & Sreenivasa, M. Y. (2021). Current perspectives of biocontrol agents for management of fusarium verticillioides and its fumonisin in cereals—a review. *Journal of Fungi, 7*(9), 776.

Agrios, G. N. (2009). Plant pathogens and disease: General introduction. In M. Schaechter (Ed.) *Encyclopedia of microbiology* (3rd Edn.). Oxford: Academic Press.

Akhlaghi, M., Tarighi, S., & Taheri, P. (2020). Effects of plant essential oils on growth and virulence factors of Erwinia amylovora. *Journal of Plant Pathology, 102*(2), 409–419. https://doi.org/10.1007/s42161-019-00446-9.

Ali, Q., Ali, S., Iqbal, N., Javed, M. T., Rizwan, M., Khaliq, R., et al. (2019). Alpha-tocopherol fertigation confers growth physio-biochemical and qualitative yield enhancement in field grown water deficit wheat (Triticum aestivum L.). *Scientific Reports, 9*(1), 12924. https://doi.org/10.1038/s41598-019-49481-7.

Ali Redha, A. (2021). Review on extraction of phenolic compounds from natural sources using green deep eutectic solvents. *Journal of Agricultural and Food Chemistry, 69*(3), 878–912. https://doi.org/10.1021/acs.jafc.0c06641.

Almeida, O. A. C., de Araujo, N. O., Dias, B. H. S., de Sant'Anna Freitas, C., Coerini, L. F., Ryu, C. -M., et al. (2023). The power of the smallest: The inhibitory activity of microbial volatile organic compounds against phytopathogens. *Frontiers in Microbiology*, 13. https://doi.org/10.3389/fmicb.2022.951130.

Alvarez-Fernandez, R. (2012). *Herbicides—environmental impact studies and management approaches*. InTech. https://doi.org/10.5772/1206

Arikan, B., Ozfidan-Konakci, C., Yildiztugay, E., Zengin, G., Alp, F. N., & Elbasan, F. (2022). Exogenous hesperidin and chlorogenic acid alleviate oxidative damage induced by arsenic toxicity in Zea mays through regulating the water status, antioxidant capacity, redox balance and fatty acid composition. *Environmental Pollution, 292*, 118389. https://doi.org/10.1016/j.envpol.2021.118389.

Arikan, B., Yildiztugay, E., & Ozfidan-Konakci, C. (2023). Protective role of quercetin and kaempferol against oxidative damage and photosynthesis inhibition in wheat chloroplasts under arsenic stress. *Physiologia Plantarum, 175*(4), e13964. https://doi.org/10.1111/ppl.13964.

Asad, S. A. (2022). Mechanisms of action and biocontrol potential of Trichoderma against fungal plant diseases-a review. *Ecological Complexity, 49*, 100978. https://doi.org/10.1016/j.ecocom.2021.100978.

Asadi, S. Z., Jamali, S., Ghadamyari, M., & Mottaghitalab, V. (2024). Control of citrus nematode Tylenchulus semipenetrans (Tylenchida: Tylenchulidae) using plant-based products under in vitro and in vivo conditions. *Journal of Plant Protection Research, 64*(2), 106–114. https://doi.org/10.24425/jppr.2024.150246.

Aslam, M. A., Ahmed, S., Saleem, M., Shah, A. A., Shah, A. N., Tanveer, M., et al. (2022). Quercetin ameliorates chromium toxicity through improvement in photosynthetic activity, antioxidative defense system; and suppressed oxidative stress in Trigonella corniculata L. *Frontiers in Plant Science*, 13.

Babazadeh, B. A., Sadeghzadeh, N., & Hajiboland, R. (2023). The impact of algal extract as a biostimulant on cold stress tolerance in barley (Hordeum vulgare L.). *Journal of Applied Phycology, 35*(6), 2919–2933. https://doi.org/10.1007/s10811-023-03107-8.

Bhat, A. A., Shakeel, A., Waqar, S., Handoo, Z. A., & Khan, A. A. (2023). Microbes vs. Nematodes: Insights into Biocontrol through Antagonistic organisms to control root-knot nematodes. *Plants, 12* (3). https://doi.org/10.3390/plants12030451.

Bhatta, U. K. (2022). Alternative management approaches of citrus diseases caused by penicillium digitatum (Green Mold) and penicillium italicum (Blue Mold). *Frontiers in Plant Science*, 12. https://doi.org/10.3389/fpls.2021.833328.

Bhupenchandra, I., Chongtham, S. K., Devi, E. L., Sahoo, M. R., Choudhary, A. K., Salam, M. D., et al. (2022). Role of biostimulants in mitigating the effects of climate change on crop performance. *Frontiers in Plant Science*, 13.

Biniaz, Y., Kavoosi, G., & Afsharifar, A. (2024). Antiviral activity and tobacco-induced resistance, mediated by essential oil nano-emulsions from Zataria multiflora and Satureja bakhtiarica. *International Journal of Pest Management, 70*(3), 432–444. https://doi.org/10.1080/09670874.2021.1985653.

Bittencourt, P. P., Alves, A. F., Ferreira, M. B., da Silva Irineu, L. E., Pinto, V. B., & Olivares, F. L. (2023). Mechanisms and applications of bacterial inoculants in plant drought stress tolerance. *Microorganisms, 11*(2). https://doi.org/10.3390/microorganisms11020502.

Bustamante, M., Durán, N., & Diez, M. C. (2012). Biosurfactants are useful tools for the bioremediation of contaminated soil: A review. *Journal of Soil Science and Plant Nutrition, 12*, 667–687.

Caicedo-Lopez, L. H., Contreras-Medina, L. M., Guevara-Gonzalez, R. G., Perez-Matzumoto, A. E., & Ruiz-Rueda, A. (2020). Effects of hydric stress on vibrational frequency patterns of capsicum annuum plants. *Plant Signaling & Behavior, 15*(7), 1770489. https://doi.org/10.1080/15592324.2020.1770489.

Carezzano, M. E., Reyna, P. G., Accotto, E., Giordano, W., Oliva, M. D. L. M., Rodriguez Pardina, P., et al. (2023). Plant-derived essential oils and aqueous extract as potential ingredients for a biopesticide: Phytotoxicity in soybean and activity against soybean mosaic virus. *Processes, 11*(8), 2265.

Carmody, N., Goñi, O., Łangowski, Ł., & O'Connell, S. (2020). Ascophyllum nodosum extract biostimulant processing and its impact on enhancing heat stress tolerance during tomato fruit set. *Frontiers in Plant Science*, 11.

Chakravarty, I., Parmar, V. M., & Mandavgane, S. A. (2023). Current trends in essential oil (EO) production. *Biomass Conversion and Biorefinery, 13*(17), 15311–15334. https://doi.org/10.1007/s13399-021-01963-3.

Chang, Y., Harmon, P. F., Treadwell, D. D., Carrillo, D., Sarkhosh, A., & Brecht, J. K. (2022). Biocontrol potential of essential oils in organic horticulture systems: From farm to fork. *Frontiers in Nutrition*, 8. https://doi.org/10.3389/fnut.2021.805138.

Chen, J.-X., & Song, B.-A. (2021). Natural nematicidal active compounds: Recent research progress and outlook. *Journal of Integrative Agriculture, 20*(8), 2015–2031. https://doi.org/10.1016/S2095-3119(21)63617-1.

Choi, I.-H., Park, J.-Y., Shin, S.-C., Kim, J., & Park, I.-K. (2007). Nematicidal activity of medicinal plant essential oils against the pinewood nematode (Bursaphelenchus xylophilus). *Applied Entomology and Zoology, 42*(3), 397–401. https://doi.org/10.1303/aez.2007.397.

Chuo, S. C., Nasir, H. M., Mohd-Setapar, S. H., Mohamed, S. F., Ahmad, A., Wani, W. A., et al. (2022). A glimpse into the extraction methods of active compounds from plants. *Critical Reviews in Analytical Chemistry, 52*(4), 667–696. https://doi.org/10.1080/10408347.2020.1820851.

Ciriello, M., Fusco, G. M., Woodrow, P., Carillo, P., & Rouphael, Y. (2024). Unravelling the nexus of plant response to non-microbial biostimulants under stress conditions. *Plant Stress, 11*, 100421. https://doi.org/10.1016/j.stress.2024.100421.

Daguerre, Y., Siegel, K., Edel-Hermann, V., & Steinberg, C. (2014). Fungal proteins and genes associated with biocontrol mechanisms of soil-borne pathogens: A review. *Fungal Biology Reviews, 28*(4), 97–125. https://doi.org/10.1016/j.fbr.2014.11.001.

Deng, Y. J., Chen, Z., Chen, Y. P., Wang, J. P., Xiao, R. F., Wang, X., et al. (2024). Lipopeptide C_{17} Fengycin B exhibits a novel antifungal mechanism by triggering metacaspase-dependent apoptosis in fusarium oxysporum. *Journal of Agricultural and Food Chemistry, 72*(14), 7943–7953. https://doi.org/10.1021/acs.jafc.4c00126.

Deresa, E. M., & Diriba, T. F. (2023). Phytochemicals as alternative fungicides for controlling plant diseases: A comprehensive review of their efficacy, commercial representatives, advantages, challenges for adoption, and possible solutions. *Heliyon, 9*(3). https://doi.org/10.1016/j.heliyon.2023.e13810.

Desoky, E.-S.M., Elrys, A. S., & Rady, M. M. (2019). Integrative moringa and licorice extracts application improves capsicum annuum fruit yield and declines its contaminant contents on a heavy metals-contaminated saline soil. *Ecotoxicology and Environmental Safety, 169*, 50–60. https://doi.org/10.1016/j.ecoenv.2018.10.117.

Di Francesco, A., Moret, E., Cignola, R., Garagozzo, L., Torelli, E., & Di Foggia, M. (2024). Yeasts volatile organic compounds (VOCs) as potential growth enhancers and molds biocontrol agents of mushrooms mycelia. *Fungal Biology, 128*(4), 1859–1867. https://doi.org/10.1016/j.funbio.2024.05.007.

Douda, O., Zouhar, M., & Manasova, M. (2022). Effect of plant essential oils on the mortality of Ditylenchus dipsaci (Kuhn, 1857) nematode under in vitro conditions. *Plant, Soil and Environment, 68*(9), 410–414.

Duchateau, S., Crouzet, J., Dorey, S., & Aziz, A. (2024). The plant-associated Pantoea spp. as biocontrol agents: Mechanisms and diversity of bacteria-produced metabolites as a prospective tool for plant protection. *Biological Control, 188*, 105441. https://doi.org/10.1016/j.biocontrol.2024.105441.

Ehiobu, J., Idamokoro, E., & Afolayan, A. (2022). Biofungicides for improvement of potato (Solanum tuberosum L.) production. *Scientifica, 2022*(1), 1405900. https://doi.org/10.1155/2022/1405900.

Elbasan, F., Arikan, B., Ozfidan-Konakci, C., Tofan, A., & Yildiztugay, E. (2024). Hesperidin and chlorogenic acid mitigate arsenic-induced oxidative stress via redox regulation, photosystems-related gene expression, and antioxidant efficiency in the chloroplasts of Zea mays. *Plant Physiology and Biochemistry, 208*, 108445. https://doi.org/10.1016/j.plaphy.2024.108445.

Elnajar, M., Aldesuquy, H., Abdelmoteleb, M., & Eltanahy, E. (2024). Mitigating drought stress in wheat plants (Triticum Aestivum L.) through grain priming in aqueous extract of spirulina platensis. *BMC Plant Biology, 24*(1), 233. https://doi.org/10.1186/s12870-024-04905-z.

Elshafie, H. S., Camele, I., & Mohamed, A. A. (2023). A comprehensive review on the biological. *Agricultural and Pharmaceutical Properties of Secondary Metabolites Based-Plant Origin., 24*(4), 3266.

Enespa, & Chandra, P. (2017). Microbial volatiles as chemical weapons against pathogenic fungi. In D. K. Choudhary, A. K. Sharma, P. Agarwal, A. Varma, & N. Tuteja (Eds.), *Volatiles and food security: Role of volatiles in agro-ecosystems* (pp. 227–254). Singapore: Springer Singapore.

Evidente, A., Andolfi, A., & Cimmino, A. (2013). Microbial and plant metabolites as potential herbicides for the control of parasitic plants. *ACS Symposium Series, 1141*, 153–166. https://doi.org/10.1021/bk-2013-1141.ch011.

Fan, H., Yan, X., Fu, M., Liu, D., Awan, A. W., Chen, P., et al. (2022). Interactive effect of biological agents chitosan, lentinan and ningnanmycin on papaya ringspot virus resistance in papaya (Carica papaya L.). *Molecules, 27*(21), 7474. https://doi.org/10.3390/molecules27217474.

Fawzy, Z. F., El-Ramady, H., Azab, M. A., & Mahdy, H. A. A. (2023). Can foliar application of natural biostimulants reduce nitrate and fiber content in fresh green bean under soil nutrient deficiency? *Bulletin of the National Research Centre, 47*(1), 165. https://doi.org/10.1186/s42269-023-01135-5.

Feng, G., Chen, M., Ye, H. C., Zhang, Z. K., Li, H., Chen, L. L., et al. (2019). Herbicidal activities of compounds isolated from the medicinal plant Piper sarmentosum. *Industrial Crops and Products, 132*, 41–47. https://doi.org/10.1016/j.indcrop.2019.02.020.

Feng, D., Wang, R., Sun, X., Liu, L. N., Liu, P., Tang, J., et al. (2023). Heavy metal stress in plants: Ways to alleviate with exogenous substances. *Science of the Total Environment, 897*, 165397. https://doi.org/10.1016/j.scitotenv.2023.165397.

Fira, D., Dimkić, I., Berić, T., Lozo, J., & Stanković, S. (2018). Biological control of plant pathogens by Bacillus species. *Journal of Biotechnology, 285*, 44–55. https://doi.org/10.1016/j.jbiotec.2018.07.044.

Francis, B., Aravindakumar, C. T., Brewer, P. B., & Simon, S. (2023). Plant nutrient stress adaptation: A prospect for fertilizer limited agriculture. *Environmental and Experimental Botany, 213*, 105431. https://doi.org/10.1016/j.envexpbot.2023.105431.

Freire, M., & Santos, C. (2018). Reação de espécies vegetais a Meloidogyne enterolobii e eficiência de seus extratos aquosos no controle do patógeno. *Semina: Ciências Agrárias, 39*, 2385. https://doi.org/10.5433/1679-0359.2018v39n6p2385.

Gamalero, E., & Glick, B. R. (2015). Bacterial Modulation of Plant Ethylene Levels. *Plant Physiology, 169*(1), 13–22. https://doi.org/10.1104/pp.15.00284.

Ghosh, D., Gupta, A., & Mohapatra, S. (2019). A comparative analysis of exopolysaccharide and phytohormone secretions by four drought-tolerant rhizobacterial strains and their impact on osmotic-stress mitigation in Arabidopsis thaliana. *World Journal of Microbiology and Biotechnology, 35*(6), 90. https://doi.org/10.1007/s11274-019-2659-0.

Graniti, A. (1991). Phytotoxins and their involvement in plant diseases. Introduction. *Experientia, 47*(8), 751–755. https://doi.org/10.1007/BF01922453.

Guo, Y., Dong, Y., Xu, C., Xie, Q., Xie, Y., Xia, Z., et al. (2020). Novel combined biological antiviral agents Cytosinpeptidemycin and Chitosan oligosaccharide induced host resistance and changed movement protein subcellular localization of tobacco mosaic virus. *Pesticide Biochemistry and Physiology, 164*, 40–46. https://doi.org/10.1016/j.pestbp.2019.12.006.

Guo, T., Gull, S., Ali, M. M., Yousef, A. F., Ercisli, S., Kalaji, H. M., et al. (2022). Heat stress mitigation in tomato (Solanum lycopersicum L.) through foliar application of gibberellic acid. *Scientific Reports, 12*(1), 11324. https://doi.org/10.1038/s41598-022-15590-z.

Gutierrez, J., Bakke, A., Vatta, M., & Merrill, A. R. (2022). Plant natural products as antimicrobials for control of streptomyces scabies: A causative agent of the common scab disease. *Frontiers in Microbiology*, 12. https://doi.org/10.3389/fmicb.2021.833233.

Hajian-Maleki, H., Baghaee-Ravari, S., & Moghaddam, M. (2021). Herbal essential oils exert a preservative effect against the potato soft rot disease. *Scientia Horticulturae*, 285. https://doi.org/10.1016/j.scienta.2021.110192.

Hanif, S., Saleem, M. F., Sarwar, M., Irshad, M., Shakoor, A., Wahid, M. A., et al. (2021). Biochemically triggered heat and drought stress tolerance in rice by proline application. *Journal of Plant Growth Regulation, 40*(1), 305–312. https://doi.org/10.1007/s00344-020-10095-3.

Hashem, A., Tabassum, B., & Fathi Abd Allah, E. (2019). Bacillus subtilis: A plant-growth promoting rhizobacterium that also impacts biotic stress. *Saudi Journal of Biological Sciences, 26*(6), 1291–1297. https://doi.org/10.1016/j.sjbs.2019.05.004.

Huang, Y., Chen, L., He, B., Liu, S. Y., Zeng, D. Q., Wang, Y. H., et al. (2021). Herbicidal activity and bioactive components of Brucea javanica (L.) Merr. residue. *Arabian Journal of Chemistry, 14*(7). https://doi.org/10.1016/j.arabjc.2021.103228.

Hungria, M., Campo, R. J., Souza, E. M., & Pedrosa, F. O. (2010). Inoculation with selected strains of Azospirillum brasilense and A. lipoferum improves yields of maize and wheat in Brazil. *Plant and Soil, 331*(1), 413–425. https://doi.org/10.1007/s11104-009-0262-0.

Ikram, M., Singh, S., Ansari, M. J., Islam, J., Shariq, M., Sulaiman Alharbi, R., et al. (2023). Evaluation of botanicals for the management of Meloidogyne incognita infecting carrot and volatile nematicidal metabolite profiling. *Journal of King Saud University-Science, 35*(9), 102911. https://doi.org/10.1016/j.jksus.2023.102911.

Islam, W., Qasim, M., Noman, A., Tayyab, M., Chen, S., & Wang, L. (2018). Management of tobacco mosaic virus through natural metabolites. *Records of Natural Products*, 12. https://doi.org/10.25135/rnp.49.17.10.178.

Islam, W., Tauqeer, A., Waheed, A., & Zeng, F. (2022). MicroRNA mediated plant responses to nutrient stress. *International Journal of Molecular Sciences, 23*(5). https://doi.org/10.3390/ijms23052562.

Ismail, M., Fayyaz, S., Azad, M., Javed, S., Ali, I., Ali, S., et al. (2019). Nematicidal properties of some medicinal plants from selected flora of gilgit-baltistan. *Journal of Animal and Plant Sciences, 29*(4), 1182–1187.

Jacomassi, L. M., Viveiros, J. D. O., Oliveira, M. P., Momesso, L., de Siqueira, G. F., & Crusciol, C. A. C. (2022). A seaweed extract-based biostimulant mitigates drought stress in sugarcane. *Frontiers in Plant Science*, 13.

Jiang, Y., Pan, X., Li, Y., Yang, Y., Jia, Y., Lei, B., et al. (2023a). Linalool induces resistance against tobacco mosaic virus in tobacco plants. *Plant Disease, 107*(7), 2144–2152. https://doi.org/10.1094/pdis-09-22-2246-re.

Jiang, M., Wang, T., Simal-Gandara, J., Siddaiah, C. N., Dai, X. F., Chen, J. Y., et al. (2023b). Overview of the control of plant fungal pathogens by natural products derived from medicinal plants. *Plant Protection Science, 59*(4), 303–316. https://doi.org/10.17221/17/2023-pps.

Kesba, H., Abdel-Rahman, A., Sayed, S., & Al-Sayed, A. S. (2021). Screening the nematicidal potential of indigenous medicinal plant extracts against Meloidogyne incognita under lab. and greenhouse conditions. *Egyptian Journal of Biological Pest Control, 31*(1). https://doi.org/10.1186/s41938-021-00429-y.

Kesraoui, S., Andrés, M. F., Berrocal-Lobo, M., Soudani, S., & Gonzalez-Coloma, A. (2022). Direct and indirect effects of essential oils for sustainable crop protection. *Plants, 11*(16), 2144.

Keswani, C., Singh, S. P., Cueto, L., García-Estrada, C., Mezaache-Aichour, S., Glare, T. R., Borriss, R., Singh, S. P., Blázquez, M. A., & Sansinenea, E. (2020). Auxins of microbial origin and their use in agriculture. *Applied Microbiology and Biotechnology, 104*(20), 8549–8565. https://doi.org/10.1007/s00253-020-10890-8

Koleška, I., Hasanagić, D., Todorović, V., Murtić, S., Klokić, I., Parađiković, N., et al. (2017). Biostimulant prevents yield loss and reduces oxidative damage in tomato plants grown on reduced NPK nutrition*. *Journal of Plant Interactions, 12*(1), 209–218. https://doi.org/10.1080/17429145.2017.1319503.

Kordali, S., Kabaagac, G., Şen, İ., Yilmaz, F., & Najda, A. (2022). Phytotoxic effects of three origanum species extracts and essential oil on seed germinations and seedling growths of four weed species. *Agronomy, 12*, 2581. https://doi.org/10.3390/agronomy12102581.

Maleva, M., Garmash, E., Chukina, N., Malec, P., Waloszek, A., & Strzałka, K. (2018). Effect of the exogenous anthocyanin extract on key metabolic pathways and antioxidant status of Brazilian elodea (Egeria densa (Planch.) Casp.) exposed to cadmium and manganese. *Ecotoxicology and Environmental Safety, 160*, 197–206. https://doi.org/10.1016/j.ecoenv.2018.05.031.

Mareri, L., Parrotta, L., & Cai, G. (2022). Environmental stress and plants. *International Journal of Molecular Sciences, 23*(10). https://doi.org/10.3390/ijms23105416.

Mesa Vanegas, A. M., Wagner Arenas, J., Ocampo Jiménez, O. O., & Monsalve Fonnegra, I. Z. (2023). Nematicidal activity and in vitro radical scavenging from Piper cumbricola and Piper eriopodon. *Biocatalysis and Agricultural Biotechnology, 47*, 102595. https://doi.org/10.1016/j.bcab.2022.102595.

Minh, T. N., Xuan, T. D., Tran, H.-D., Van, T. M., Andriana, Y., Khanh, T. D., et al. (2019). Isolation and purification of bioactive compounds from the stem bark of jatropha podagrica. *Molecules, 24*(5). https://doi.org/10.3390/molecules24050889.

Mishra, S., Spaccarotella, K., Gido, J., Samanta, I., & Chowdhary, G. (2023). Effects of heat stress on plant-nutrient relations: An update on nutrient uptake, transport, and assimilation. *International Journal of Molecular Sciences, 24*(21). https://doi.org/10.3390/ijms242115670.

Mohd Israfi, N. A., Mohd Ali, M. I. A., Manickam, S., Sun, X., Goh, B. H., Tang, S. Y., et al. (2022). Essential oils and plant extracts for tropical fruits protection: From farm to table. *Frontiers in Plant Science*, 13. https://doi.org/10.3389/fpls.2022.999270.

Morales-Ubaldo, A. L., Rivero-Perez, N., Avila-Ramos, F., Aquino-Torres, E., Prieto-Méndez, J., Hetta, H. F., et al. (2021). Bactericidal activity of larrea tridentata hydroalcoholic extract against phytopathogenic bacteria. *Agronomy-Basel, 11*(5). https://doi.org/10.3390/agronomy11050957.

Mousavi, S. S., Sundararaju, S., & Goh, P. S. (2021). Recent advancements in enzyme immobilization techniques for environmental applications. *Journal of Environmental Chemical Engineering, 9*(5), 106197. https://doi.org/10.1016/j.jece.2021.106197

Narayaran, M., Ali, S. S., & El-Sheekh, M. (2023). A comprehensive review on the potential of microbial enzymes in multipollutant bioremediation: Mechanisms, challenges, and future prospects. *Journal of Environmental Management, 334*, 117532. https://doi.org/10.1016/j.jenvman.2023.117532

Navarro-Morillo, I., Navarro-León, E., Atero-Calvo, S., Rios, J. J., Ruiz, J. M., & Blasco, B. (2024). Biostimulant-induced mitigation of cold and drought stresses in zucchini plants. *Scientia Horticulturae, 331*, 113114. https://doi.org/10.1016/j.scienta.2024.113114.

Nazlić, M., Fredotović, Ž., Vuko, E., Fabijanić, L., Kremer, D., Stabentheiner, E., et al. (2021). Wild species veronica officinalis L. and veronica saturejoides Vis. ssp. saturejoides—biological potential of free volatiles. *Horticulturae, 7*(9), 295.

Nguyen, T.-D., Nguyen, T.-T., Pham, M.-N., Duong, H.-N., Pham, T.-T., Nguyen, T.-P., et al. (2023). Relationships between endophytic bacteria and medicinal plants on bioactive compounds production. *Rhizosphere, 27*, 100720. https://doi.org/10.1016/j.rhisph.2023.100720.

Oksanen, E., & Kontunen-Soppela, S. (2021). Plants have different strategies to defend against air pollutants. *Current Opinion in*

Environmental Science & Health, 19, 100222. https://doi.org/10.1016/j.coesh.2020.10.010.

Ozdemir, E., & Gozel, U. (2018). Nematicidal activities of essential oils against Meloidogyne incognita on tomato plant. *Fresenius Environmental Bulletin, 27*(6), 4511–4517.

Ozturk, M., & Turkyilmaz Unal, B. (2021). Osmoregulation and its actions during the drought stress in plants. *Physiologia Plantarum, 172*(2), 1321–1335. https://doi.org/10.1111/ppl.13297.

Paduch-Cichal, E., Mirzwa-Mróz, E., Wojciechowska, P., Bączek, K., Kosakowska, O., Węglarz, Z., et al. (2023). Antiviral activity of selected essential oils against cucumber mosaic virus. *Plants, 12*(1), 18.

Pereira-Dias, L., Oliveira-Pinto, P. R., Fernandes, J. O., Regalado, L., Mendes, R., Teixeira, C., et al. (2023). Peptaibiotics: Harnessing the potential of microbial secondary metabolites for mitigation of plant pathogens. *Biotechnology Advances, 68*, 108223. https://doi.org/10.1016/j.biotechadv.2023.108223.

Peres, M. C., de Souza Costa, G. C., dos Reis, L. E. L., da Silva, L. D., Peixoto, M. F., Alves, C. C. F., et al. (2020). In natura and nanoencapsulated essential oils from Xylopia aromatica reduce oviposition of Bemisia tabaci in Phaseolus vulgaris. *Journal of Pest Science, 93*(2), 807–821. https://doi.org/10.1007/s10340-019-01186-6.

Poveda, J., Abril-Urias, P., & Escobar, C. (2020). Biological control of plant-parasitic nematodes by filamentous fungi inducers of resistance: Trichoderma, mycorrhizal and endophytic fungi. *Frontiers in Microbiology*, 11. https://doi.org/10.3389/fmicb.2020.00992.

Priyashantha, A. K. H., Karunarathna, S. C., Lu, L., & Tibpromma, S. (2023). Fungal endophytes: An alternative biocontrol agent against phytopathogenic fungi. *Encyclopedia, 3*(2), 759–780.

Purnavab, S., Ketabchi, S., & Rowshan, V. (2015). Chemical composition and antibacterial activity of methanolic extract and essential oil of Iranian Teucrium polium against some of phytobacteria. *Natural Product Research, 29*(14), 1376–1379. https://doi.org/10.1080/14786419.2014.1000320.

Rabêlo, V. M., Magalhães, P. C., Bressanin, L. A., Carvalho, D. T., Reis, C. O. D., Karam, D., et al. (2019). The foliar application of a mixture of semisynthetic chitosan derivatives induces tolerance to water deficit in maize, improving the antioxidant system and increasing photosynthesis and grain yield. *Scientific Reports, 9*(1), 8164. https://doi.org/10.1038/s41598-019-44649-7.

Rana, A., Sudakov, K., Carmeli, S., Miyara, S. B., Bucki, P., & Minz, D. (2024). Volatile organic compounds of the soil bacterium Bacillus halotolerans suppress pathogens and elicit defense-responsive genes in plants. *Microbiological Research, 281*, 127611. https://doi.org/10.1016/j.micres.2024.127611.

Rani, A., Rana, A., Dhaka, R. K., Singh, A. P., Chahar, M., Singh, S., et al. (2023). Bacterial volatile organic compounds as biopesticides, growth promoters and plant-defense elicitors: Current understanding and future scope. *Biotechnology Advances, 63*, 108078. https://doi.org/10.1016/j.biotechadv.2022.108078.

Raveau, R., Fontaine, J., & Lounès-Hadj Sahraoui, A. (2020). Essential oils as potential alternative biocontrol products against plant pathogens and weeds: A review. *Foods, 9*(3), 365.

Ren, W.-X., Li, P.-J., Geng, Y., & Li, X.-J. (2009). Biological leaching of heavy metals from a contaminated soil by Aspergillus niger. *Journal of Hazardous Materials, 167*(1), 164–169. https://doi.org/10.1016/j.jhazmat.2008.12.104.

Repke, R. A., Silva, D. M. R., dos Santos, J. C. C., & de Almeida Silva, M. (2022). Increased soybean tolerance to high-temperature through biostimulant based on Ascophyllum nodosum (L.) seaweed extract. *Journal of Applied Phycology, 34*(6), 3205–3218. https://doi.org/10.1007/s10811-022-02821-z.

Rice, E. L. (1984). In *Allelopathy* (2nd Edn). Academic Press, pp. 368. ISBN 0125870558.

Ruparelia, J., Rabari, A., Mitra, D., Panneerselvam, P., Das-mohapatra, P. K., & Jha, C. K. (2022). Efficient applications of bacterial secondary metabolites for management of biotic stress in plants. *Plant Stress, 6*, 100125. https://doi.org/10.1016/j.stress.2022.100125.

Santos, A. M., Soares, A., Luz, J., Cordeiro, C., Sousa Silva, M., Dias, T., et al. (2024). Microbial interactions as a sustainable tool for enhancing PGPR antagonism against phytopathogenic fungi. *Sustainability, 16*(5). https://doi.org/10.3390/su16052006.

Saric-Krsmanovic, M., Gajic-Umiljendic, J., Radivojevic, L., Santric, L., Đorđević, T., & Đurović-Pejčev, R. (2023). Sensitivity of cuscuta species and their hosts to anethum graveolens essential oil. *Pesticidi i Fitomedicina, 38*, 33–39. https://doi.org/10.2298/PIF2301033S.

Satou, T., Kaneko, K., Li, W., & Koike, K. (2008). The toxin produced by Pleurotus ostreatus reduces the head size of nematodes. *Biological & Pharmaceutical Bulletin, 31*(4), 574–576. https://doi.org/10.1248/bpb.31.574

Sauer, A. V., Santos, E. M., Gonçalves-Zuliani, A. M. O., Nocchi, P. T. R., Nunes, W. M. C., & Bonato, C. M. (2015). Bacteriostatic and bactericidal activity in vitro of different essential oils as alternative treatments to control xanthomonas citri subsp. citri. In *Paper Presented at the XII International Citrus Congress-International Society of Citriculture*.

Sawadogo-Ilboudo, T. C., Yonli, D., Traoré, H., & Boussim, J. I. (2021). Use of essential oils from local plants as potential bio-herbicides to deplete Striga hermonthica seedbank. *Journal of Research in Weed Science, 4*(1), 57–69. https://doi.org/10.26655/JRWEEDSCI.2021.1.5.

Saxena, S. (2014). Microbial metabolites for the development of ecofriendly agrochemicals. *Allelopathy Journal, 33*, 1–24.

Sharma, J., Sundar, D., & Srivastava, P. (2021). Biosurfactants: Potential agents for controlling cellular communication, motility, and antagonism. *Frontiers in Molecular Biosciences*, 8. https://doi.org/10.3389/fmolb.2021.727070.

Shen, J.-G., Zhang, Z.-K., Wu, Z.-J., Ouyang, M.-A., Xie, L.-H., & Lin, Q.-Y. (2008). Antiphytoviral activity of bruceine-D from Brucea javanica seeds. *Pest Management Science: Formerly Pesticide Science, 64*(2), 191–196. https://doi.org/10.1002/ps.1465.

Singh, R. S., Walia, A. K., Khattar, J. S., Singh, D. P., & Kennedy, J. F. (2017). Cyanobacterial lectins characteristics and their role as antiviral agents. *International Journal of Biological Macromolecules, 102*, 475–496. https://doi.org/10.1016/j.ijbiomac.2017.04.041.

Singh, S., Kumar, V., Singh, S., Dhanjal, D. S., Datta, S., Sharma, D., et al. (2020). 16-Biosurfactant-based bioremediation. In V. C. Pandey & V. Singh (Eds.), *Bioremediation of Pollutants* (pp. 333–358). Elsevier.

Singh, P., Arif, Y., Bajguz, A., & Hayat, S. (2021). The role of quercetin in plants. *Plant Physiology and Biochemistry, 166*, 10–19. https://doi.org/10.1016/j.plaphy.2021.05.023.

Sithole, N. T., Kulkarni, M. G., Finnie, J. F., & Van Staden, J. (2021). Potential nematicidal properties of plant extracts against Meloidogyne incognita. *South AfricaN Journal of Botany, 139*, 409–417. https://doi.org/10.1016/j.sajb.2021.02.014.

Sossou, B., Izuogu, N., Anifowose, A., & Ahamefule, H. (2022). Controlling root-knot nematode Meloidogyne incognita in tomatoes using modified effective microorganisms-fermented plant extract and compost manure. https://doi.org/10.30486/IJROWA.2021.1937252.1307.

Stratakos, A. C., & Koidis, A. (2016). Chapter 4-methods for extracting essential oils. In V. R. Preedy (Ed.), *Essential oils in food preservation, flavor and safety* (pp. 31–38). Academic Press.

Sujata Goyal, V., Baliyan, V., Avtar, R., & Mehrotra, S. (2023). Alleviating drought stress in Brassica juncea (L.) Czern & Coss. by

Foliar application of biostimulants—orthosilicic acid and seaweed extract. *Applied Biochemistry and Biotechnology, 195*(1), 693–721. https://doi.org/10.1007/s12010-022-04085-2.

Sun, Y., Wu, H., Zhou, W., Yuan, Z., Hao, J., Liu, X., et al. (2022). Effects of indole derivatives from Purpureocillium lilacinum in controlling tobacco mosaic virus. *Pesticide Biochemistry and Physiology, 183*, 105077. https://doi.org/10.1016/j.pestbp.2022.105077.

Sziderics, A. H., Rasche, F., Trognitz, F., Sessitsch, A., & Wilhelm, E. (2007). Bacterial endophytes contribute to abiotic stress adaptation in pepper plants (Capsicum annuum L.). *Canadian Journal of Microbiology, 53*(11), 1195–1202. https://doi.org/10.1139/w07-082.

Taglienti, A., Donati, L., Ferretti, L., Tomassoli, L., Sapienza, F., Sabatino, M., et al. (2022). In vivo antiphytoviral activity of essential oils and hydrosols from origanum vulgare, thymus vulgaris, and rosmarinus officinalis to control Zucchini yellow mosaic virus and tomato leaf curl New Delhi virus in cucurbita pepo L. *Frontiers in Microbiology*, 13. https://doi.org/10.3389/fmicb.2022.840893.

Taglienti, A., Donati, L., Dragone, I., Ferretti, L., Gentili, A., Araniti, F., et al. (2023). In vivo antiphytoviral and aphid repellency activity of essential oils and hydrosols from mentha suaveolens and foeniculum vulgare to control Zucchini yellow mosaic virus and its vector aphis gossypii. *Plants, 12*(5), 1078.

Tahir, H. A. S., Gu, Q., Wu, H., Niu, Y., Huo, R., & Gao, X. (2017). Bacillus volatiles adversely affect the physiology and ultra-structure of Ralstonia solanacearum and induce systemic resistance in tobacco against bacterial wilt. *Scientific Reports, 7*(1). https://doi.org/10.1038/srep40481.

Teale, W. D., Paponov, I. A., & Palme, K. (2006). Auxin in action: Signalling, transport and the control of plant growth and development. *Nature Reviews Molecular Cell Biology, 7*(11), 847–859. https://doi.org/10.1038/nrm2020.

Timper, P., Koné, D., Yin, J., Ji, P., & McSpadden Gardener, B. B. (2009). Evaluation of an antibiotic-producing strain of Pseudomonas fluorescens for suppression of plant-parasitic nematodes. *Journal of Nematology, 41*(3), 234–240.

Tongnuanchan, P., & Benjakul, S. (2014). Essential oils: Extraction, bioactivities, and their uses for food preservation. *Journal of Food Science, 79*(7), R1231–R1249. https://doi.org/10.1111/1750-3841.12492.

Vaishnav, A., Varma, A., Tuteja, N., & Choudhary, D. K. (2017). Characterization of bacterial volatiles and their impact on plant health under abiotic stress. In D. K. Choudhary, A. K. Sharma, P. Agarwal, A. Varma, & N. Tuteja (Eds.), *Volatiles and food security: Role of volatiles in agro-ecosystems* (pp. 15–24). Singapore: Springer Singapore.

Vázquez-Pineda, A., Bravo-de-la-Parra, A., Mendoza-de-Gives, P., Liébano-Hernández, E., Hernández-Linares, I., Yáñez-Pérez, N., Aguilar-Marcelino, L., Ramírez-Vargas, G., Hernández-Castro, E., Gutiérrez-Segura, I., & López-Arellano, M. E. (2012). Use of products derived from *Bacillus thuringiensis* as an alternative control in nematodes of veterinary importance: Review. *Mexican Journal of Livestock Sciences, 3*(1), 77–88.

Vázquez-Sánchez, M., Medina-Medrano, J. R., Cortez-Madrigal, H., Angoa-Perez, M. V., Munoz-Ruiz, C. V., & Villar-Luna, E. (2018). Nematicidal activity of wild plant extracts against second-stage juveniles of nacobbus aberrans. *Nematropica, 48*(2), 136–144.

Vuko, E., Dunkić, V., Maravić, A., Ruščić, M., Nazlić, M., Radan, M., et al. (2021). Not only a weed plant—biological activities of essential oil and hydrosol of dittrichia viscosa (L.) Greuter. *Plants, 10*(9), 1837.

Vuko, E., Dunkić, V., Ruščić, M., Nazlić, M., Mandić, N., Soldo, B., et al. (2021). Chemical composition and new biological activities of essential oil and hydrosol of hypericum perforatum L. ssp. veronense (Schrank) H. Lindb. *Plants, 10*(5), 1014.

Vuko, E., Radman, S., Jerković, I., Kamenjarin, J., Vrkić, I., & Fredotović, Ž. (2022). A plant worthy of further study—volatile and non-volatile compounds of portenschlagiella ramosissima (Port.) tutin and its biological activity. *Pharmaceuticals, 15*(12), 1454.

Vurukonda, S. S. K. P., Giovanardi, D., & Stefani, E. (2018). Plant growth promoting and biocontrol activity of streptomyces spp. as endophytes. *International Journal of Molecular Sciences, 19*(4). https://doi.org/10.3390/ijms19040952.

Wali, S., Gharge, S., Naik, A., Kadgaonkar, M., Bhandurge, P., & Suryawanshi, S. S. (2022). Isolation, characterization and identification of bioactive compounds from herbal medicines and medicinal plant extracts by fourier transform infrared spectroscopy: A brief review. *International Journal of Ayurvedic Medicine, 13*(3), 570–578. https://doi.org/10.47552/ijam.v13i3.2753.

Wang, F., Xue, Y., Fu, L., Wang, Y., He, M., Zhao, L., & Liao, X. (2022). Extraction, purification, bioactivity and pharmacological effects of capsaicin: A review. *Critical Reviews in Food Science and Nutrition, 62*(19), 5322–5348. https://doi.org/10.1080/10408398.2021.1884840

Wang, C., Ma, G., Zhang, S., Zhao, K., & Li, X. (2023a). Study on the binding of ningnanmycin to the helicase of Tobamovirus virus. *Pesticide Biochemistry and Physiology, 194*, 105494. https://doi.org/10.1016/j.pestbp.2023.105494.

Wang, K., Ke, S., Fang, W., Liu, F., & Zhang, Z. (2023b). Agroactive volatile organic compounds from microbes: Chemical diversities and potentials of application in crop protection. *Advanced Agrochem, 2*(1), 39–57. https://doi.org/10.1016/j.aac.2022.12.004.

Yadav, N., Sharma, S., Joys, J. S., & Kumar, S. (2020). Microwave assisted extraction of bioactive compounds: A brief review. *Journal of the Indian Chemical Society,* 97(10a), 1751–1756.

Yang, F.-X., Ma, Y.-Y., Wu, Y.-P., Zhao, G.-K., Li, Y.-P., Li, Z.-J., et al. (2023). Extraction and characterization of anti-virus anthraquinones from Nicotiana tabacum-derived Aspergillus oryzae YNCA1220. *Pesticide Biochemistry and Physiology, 196*, 105613. https://doi.org/10.1016/j.pestbp.2023.105613.

Yetişsin, F., & Kurt, F. (2020). Gallic acid (GA) alleviating copper (Cu) toxicity in maize (Zea mays L.) seedlings. *International Journal of Phytoremediation, 22*(4), 420–426. https://doi.org/10.1080/15226514.2019.1667953.

Yousefi, M., Rahimi-Nasrabadi, M., Pourmortazavi, S. M., Wysokowski, M., Jesionowski, T., Ehrlich, H., & Mirsadeghi, S. (2019). Supercritical fluid extraction of essential oils. *TrAC Trends in Analytical Chemistry, 118*, 182–193. https://doi.org/10.1016/j.trac.2019.05.038.

Zaehle, C., Gressler, M., Shelest, E., Geib, E., Hertweck, C., & Brock, M. (2014). Terrein biosynthesis in aspergillus terreus and its impact on phytotoxicity. *Chemistry & Biology, 21*(6), 719–731. https://doi.org/10.1016/j.chembiol.2014.03.010.

Zhang, L., Yang, Q., Jiang, Y., Yu, L., Song, N., Zhao, D., et al. (2024). Antifouling and antibacterial bioactive metabolites of marine fungus (terrein)/polyamide thin-film composite reverse osmosis membranes for desalination applications. *Desalination, 572*, 117140. https://doi.org/10.1016/j.desal.2023.117140.

Zhang, X., Ma, M., Wu, C., Huang, S., & Danish, S. (2023). Mitigation of heat stress in wheat (Triticum aestivum L.) via regulation of physiological attributes using sodium nitroprusside and gibberellic acid. *BMC Plant Biology, 23*(1), 302. https://doi.org/10.1186/s12870-023-04321-9.

Zhao, L., Feng, C., Wu, K., Chen, W., Chen, Y., Hao, X., et al. (2017). Advances and prospects in biogenic substances against plant virus: A review. *Pesticide Biochemistry and Physiology, 135*, 15–26. https://doi.org/10.1016/j.pestbp.2016.07.003.

Bioactive Compounds: Industrial and Agricultural Applications

Janvika Varma, Arun Kumar Kulshrestha, Pranay Punj Pankaj, Vijay Jagdish Upadhye, and Anupama Shrivastav

Abstract

Bioactive substances boost crop output and enable sustainable agricultural methods by acting as organic insecticides, growth regulators, and disease inhibitors. Additionally, they have a significant impact on the creation of nutritional supplements and functional foods that advance human health. They also support environmentally friendly agricultural practices by improving plant growth and soil health. This chapter explores the various applications of bioactive chemicals in agriculture and industry. Because of their innate antibacterial, antioxidant, and anti-inflammatory qualities, these compounds are essential in the formulation of medicines, cosmetics, and nutraceuticals, meeting the growing demand for natural and healthful goods. The chapter also covers current developments in bioengineering, biotechnology, and extraction methods, providing creative ways to maximize the potential of bioactive substances and highlights the importance of these substances in tackling issues related to industry and agriculture, as well as in fostering sustainability and better resource management, all of which lead to the well-being of ecosystems and economies.

J. Varma · A. K. Kulshrestha
Department of Medical Laboratory Technology, Faculty of Life Health and Allied Sciences, ITM Vocational University, Ravaal, Waghodia, Vadodara, 391760, Gujarat, India

P. P. Pankaj
Department of Zoology, Nagaland University, HQRS: Lumami, Zunheboto, 798627, Nagaland, India

V. J. Upadhye
Research and Development Cell (RDC), Parul Institute of Applied Sciences (PIAS), Parul University (DSIR- SIRO Recognized), Limda, Waghodia, Vadodara, 391760, Gujarat, India

A. Shrivastav (✉)
Department of Microbiology, Faculty of Life Health and Allied Sciences, ITM Vocational University, Ravaal, Waghodia, Vadodara, 391760, Gujarat, India
e-mail: anupamashrivastav@gmail.com; anupamashrivastav@yahoo.co.in; anupamam@itmvu.in

Keywords

Bioactive compounds • Industrial applications • Agricultural applications sustainable farming • Extraction techniques

1 Introduction

Bioactive compounds include secondary metabolites, enzymes, polysaccharides, proteins, peptides, and lipids that significantly impact biological systems. Examples of bioactive compounds include vitamins, minerals, and phytochemicals. Research is being conducted on their ability to enhance health, prevent illness, and promote well-being simultaneously (Garcia-Marti et al., 2023).

In the food industry, bioactive compounds are used to impart flavor, antioxidants, and carotenoids into food. However, these compounds often have limited efficiency; thus, encapsulation has become a method for enhancing their quality. Bioactive substances can exist in gaseous, solid, or liquid form and may be coated with either a single or double layer. Encapsulation techniques include extrusion, fluid bed coating, coacervation, spray drying, and nanoemulsions. Starch, chitosan, modified starch, proteins, and natural polymers are used to encapsulate bioactives such as fish oil, flavors, lycopene, and β-carotene in the spray drying process. Coacervation involves encapsulating polyunsaturated fatty acids, essential oils, fragrances, and enzymes using gums, proteins, starch, maltodextrin, xanthan gum, and pectin. Extrusion is used to encapsulate flavors, carotenoids, probiotic microbes, and oils with sodium alginate, maltodextrin, modified starch, and pectin. In fluid bed coating, probiotic microbes, vitamins, and lactic acid carotenoids are encapsulated using cellulose, maltodextrin, gums, and proteins.

Both the agricultural and industrial sectors make extensive use of bioactive compounds derived from plants. These

J. Devi (ed.), *Biotechnological Intervention in Production of Bioactive Compounds*, Sustainable Landscape Planning and Natural Resources Management, https://doi.org/10.1007/978-3-031-76859-0_10

compounds can stimulate or inhibit receptors, activate or inactivate enzymes, and increase or decrease gene expression.

Bioactive compounds exhibit a diverse array of biological activities in crops and can be either hydrophilic or hydrophobic. Additionally, they have the potential to enhance bioavailability. The consumption of meat is an important component of the human diet because it provides the body with crucial nutrients such as protein, minerals, and vitamins. However, the composition of fatty acids, especially in ruminant species, has raised concerns due to its potential impact on cardiovascular health. Researchers have explored altering feed elements such as grass, linseed, and algae, as well as enriching non-ruminant meats like pork with n-3 fatty acids from sources such as linseed, fish oils, and algae, to optimize the fatty acid profile. Furthermore, maintaining adequate levels of bioactive compounds like vitamin E, along with appropriate packaging and processing methods, is essential to ensure product quality (Wood, 2023).

Opuntia species have long been used as traditional therapeutic treatments for chronic ailments and as dietary resources due to their nutritional value and biological functionalities. The plant holds significant economic potential and exhibits variations in phytochemical composition between its wild and domesticated varieties. The phenolic content, antioxidants, pigments, and other components found in both the aerial and underground parts of Opuntia species contribute to their beneficial properties. The ongoing global trend of antimicrobial resistance is reducing the availability of antimicrobial substances derived from microorganisms. As a result, there is increasing interest in exploring plants such as Opuntia as potential sources of new antimicrobial compounds.

Macromolecular antioxidants enhance the overall biological functionality of dietary matrices. Opuntia seeds, cladodes, and fruits are underappreciated reservoirs of natural compounds. Methods for isolating chemicals from Opuntia residues have been developed to analyze and identify useful constituents. A study conducted a systematic review and bibliometric analysis to explore research gaps, trends, and advancements in the recovery and utilization of betanin and vitexin. The study focused on sustainable extraction techniques and the biological properties of these compounds. The primary research areas included antioxidant activity, betalain, betanin, and betacyanin. Demand for betanin and vitexin has increased due to their beneficial biological properties and their use in pharmaceutical applications and nutraceutical food products. Sustainable extraction methods for betanin and vitexin include ultrasound-assisted extraction, pulsed electric field extraction, pressurized liquid extraction, and subcritical water extraction. Bioactive compounds derived from plants, fruits, and vegetables have been used in pharmaceuticals for a significant period of time. Over the past 12 years, bibliometric analysis has been used to evaluate progress in green extraction technologies and the biological properties of betanin and vitexin (De Freitas Marinho et al., 2023).

Food waste production is a major issue in the food life cycle, with households in developed countries contributing 42% of it. The zero-waste economy concept seeks to use waste as raw material for new products. Agro-industrial residues can be repurposed for use in the food, cosmetics, and pharmaceutical industries. Pretreatment is crucial for breaking down agricultural lignocellulosic materials, which can then be bioconverted through enzymatic hydrolysis and fermentation. Bioactive compounds are extracted from natural sources via solid–liquid extraction using organic solvents in heat-reflux systems. Extraction techniques include Soxhlet extraction, maceration, hydrodistillation, solvent extraction, solar solvent extraction, pressurized liquid extraction, supercritical fluid extraction, microwave-assisted extraction, ultrasound-assisted extraction, and pulsed electric field extraction.

Horticulture faces the challenge of providing healthy food for the growing global population, and sustainable food production is essential to combat hunger and malnutrition. To increase agricultural productivity, researchers are exploring organic biostimulants (PDBs) as eco-friendly alternatives to synthetic chemicals. Regulation 2019/1009 governs the use of biostimulants, which stimulate plant nutrition processes and improve nutrient use efficiency. Plant extracts play a vital role in improving agricultural sustainability, particularly regarding crop quality and yield. While synthetic plant protection products are being studied, evidence on their use remains limited (Godlewska et al., 2021).

Lignin nanocapsules, derived from plant cell wall components, offer a sustainable and efficient method for controlled-release systems in agriculture, medicine, and pharmaceuticals. These non-toxic, biodegradable, and economical nanomaterials are ideal carriers for gene delivery systems, drug carriers, and heavy metal absorbers. Synthesis methods include acid precipitation, ultra-sonication, and interfacial crosslinking, demonstrating the versatility and potential of lignin-based nanomaterials in promoting food security and enhancing nutrient delivery.

Thus, bioactive compounds play a crucial role in both the agricultural and commercial sectors due to their natural origins. These compounds have numerous industrial applications and have been extensively studied for their positive effects on health. Several industries, such as pharmaceuticals, nutraceuticals, and cosmeceuticals, benefit from these applications. When utilized as natural pesticides and growth promoters, they significantly improve the quality and productivity of agricultural production. By enhancing the health and resilience of plants, bioactive compounds play a vital role in promoting sustainable agriculture. Furthermore, these compounds are essential in maintaining the quality of post-harvest products, leading to innovative product development, promoting sustainability, and protecting the environment.

2 Sources of Bioactive Compounds

Interest in bioactive substances obtained from natural sources has increased dramatically in recent years. These substances are diverse and abundant among plant species and are primarily secondary metabolites. They have numerous advantageous properties, including antibacterial, immunomodulatory, anti-inflammatory, and antioxidant activities. These bioactive substances fall into several important categories, such as terpenes, flavonoids, alkaloids, coumarins, and stilbenes. To enhance their potential uses, research into these compounds and understanding their mechanisms of action is crucial.

Many living organisms, such as plants, animals, algae, fungi, and bacteria, are sources of bioactive compounds, which possess important biological properties like antioxidant, antibacterial, anti-inflammatory, and anticancer effects. These compounds can be found in various foods, including fruits, vegetables, grains, nuts, spices, medicinal herbs, probiotics, as well as in saponins, alkaloids, vitamins, and minerals (Adefegha, 2021).

More than 60% of the anti-cancer medications available today are derived from plants, and over 3,000 plant species are known to have anticancer potential (Dwarka et al., 2022). Researchers are also investigating plants and marine organisms as potential sources of safer and more effective antidiabetic drugs, as bioactive compounds play a key role in addressing metabolic disorders like diabetes (Ahmed et al., 2023). Therefore, the current scientific focus is on exploring novel bioactive compounds and understanding their unknown effects, which are essential to meet human needs across various fields.

Typically, these compounds are present in trace amounts in food matrices, particularly in those derived from plants (phytocompounds). Although they are not essential for life, like proteins, lipids, carbohydrates, vitamins, and minerals, their anti-inflammatory and antioxidant properties make them beneficial for health.

These factors collectively drive the growth of multiple research directions involving bioactive compounds, from their extraction to their application in various industries, including the food, cosmetic, and pharmaceutical sectors. Consequently, many experimental methods involve extracting these compounds using diverse techniques and materials, including waste, particularly from the agri-food industry.

Alkaloids, flavonoids, terpenoids, and phenolic acids are among the bioactive molecules most frequently produced by plants. These compounds have been shown to have significant health benefits and are critical for plant defense mechanisms. Flavonoids and phenolic acids are abundant in fruits like berries (strawberries, blueberries) and vegetables (broccoli, spinach). These compounds possess antioxidant properties that help reduce inflammation and oxidative stress (Scalbert et al., 2005). Various culinary herbs and spices, including ginger, garlic, and turmeric, are excellent sources of bioactive compounds like allicin, curcumin, and gingerol, which have anti-inflammatory, antibacterial, and anticancer properties.

Certain bioactive substances with considerable medicinal potential, such as terpenoids and carotenoids, are produced by marine sources like algae and invertebrates. For example, fucoidans, which have antiviral and anticoagulant properties, are found in brown algae (Wijesinghe & Jeon, 2012).

Microorganisms such as fungi and bacteria are also prolific producers of bioactive compounds. Antibiotics like tetracycline and streptomycin, which are produced by soil bacteria, particularly from the genus Streptomyces, have revolutionized modern medicine. A vast array of bioactive compounds, including immunosuppressants, statins, and penicillin, are produced by fungi, including yeasts and molds. These compounds are widely used in pharmaceuticals to treat infections, reduce cholesterol, and prevent organ transplant rejection.

Bioactive peptides and proteins from animal sources, such as collagen and gelatin, contribute to joint health and wound healing (Song et al., 2019). Bee products, such as honey, propolis, and royal jelly, are well-known for their antioxidant, antibacterial, and anti-inflammatory properties (Pasupuleti et al., 2017).

Dairy and non-dairy fermented foods provide beneficial bioactive compounds. Probiotics that promote gut health can be found in sauerkraut, kimchi, and miso, while peptides with antihypertensive properties are present in yogurt and kefir.

3 Extraction of Bioactive Compounds

The term "extraction of bioactive compounds" refers to the process of separating and isolating materials from a solid or liquid mixture using a suitable solvent, such that each component is either more soluble or less soluble than the other ingredients in the mixture. Extraction techniques can be categorized into two types based on the state of the two phases in the mixture: solid–liquid and liquid–liquid extraction.

The principle underlying the extraction process is molecular diffusion, which is defined as the equalization of the concentration of dissolved materials between systems in contact with one another. The amount of material diffused (per unit area and time) is inversely proportional to the diffusion distance, relative to particle size, and directly proportional to the diffusion coefficient and the concentration difference at the boundary layer.

In general, the particle diameter is roughly inversely proportional to thc diffusivity in the liquid phase. Diffusion significantly increases with a rise in system temperature; however, increased temperature also dramatically affects the solution by lowering its viscosity.

3.1 Traditional Extraction Methods

Liquid–liquid extraction, commonly referred to as solvent extraction, involves two immiscible liquid phases—an aqueous phase and an organic phase. The organic phase dissolves the target analyte, and the phases are combined in a separatory funnel where, after shaking, they form distinct layers. The analyte partitions between the phases based on its relative solubility in the two immiscible liquids (Wells, 2003). Despite being suitable for temperature-sensitive materials and azeotropic mixtures, liquid–liquid extraction has disadvantages, such as the need for high volumes of organic solvents, emulsion formation, and being a labor-intensive process.

Solid-phase extraction (SPE) offers an alternative, where the chemical component is retained on a solid sorbent and later recovered through elution (Murakami et al., 2020). This method requires complete solute elution from the solid particles and maximum, consistent solute absorption by the sorbent. Despite its advantages, such as solvent efficiency and cost-effectiveness over liquid–liquid extraction (Poole, 2003), SPE has drawbacks, including the limited sorption capacity of the sorbent and the need for filtration to remove fine particles (Poole). Solid-phase microextraction (SPME) simplifies the extraction process by dispersing the solid phase in a small amount of extracting phase and exposing the sorbent to the sample for a set period. This technique is advantageous due to its automation and ability to analyze bioactive chemicals at low concentrations. Conventional methods such as maceration and Soxhlet extraction have been used to extract phenolic compounds. While maceration is inexpensive and simple, it takes longer and yields less because it operates at lower temperatures and requires less agitation.

3.2 Modern Extraction Techniques

Ultrasound-assisted extraction (UAE) employs sound frequencies ranging from 18 to 100 kHz to enhance extraction yields by improving mass transfer and rupturing cellular matrices to release desired compounds. This technique utilizes acoustic cavitation, where pressure variations cause microbubbles to form, oscillate, and rapidly collapse. According to Alarcon-Rojo et al. (2015), high-power ultrasound (10–1000 $W.cm^{-2}$) is highly effective in the food industry, enhancing cavitation effects and physicochemical properties to promote the release of extractable compounds. UAE allows rapid heat and mass transfer during the extraction of bioactive compounds, especially when combined with other technologies (Sumere et al., 2018; Mounir, 2021).

Instant controlled pressure drop (DIC) technology combines hydro-thermo-mechanical processes and rapid pressure changes to optimize the extraction of organic products. Key components of DIC technology include a controlled pressure-drop valve, a vacuum machine, an extraction vessel, and a condensate recovery trap. DIC is more effective than traditional drying techniques for strawberries and can improve the nutritional content and shelf life of heat-sensitive food products such as apples and onions. According to Toepfl et al. (2006), pulsed electric field (PEF) extraction enhances compound extraction by subjecting bio cells to a brief external electrical field that increases cell membrane permeability. PEF, being a non-thermal process, is an energy-efficient method used to extract compounds from plant cells, such as carotenoids and polyphenols. This technology can improve the yield and quality of juices and oils, making it a valuable alternative to traditional thermal processes (PEF Pretreatment-assisted extraction) (Puértolas et al., 2015).

Enzyme-assisted extraction (EAE) uses enzymes to break down cell walls, allowing solvents to access and release bound phytochemicals. Common enzymes used to aid the extraction of polyphenols from plant materials include lipase, pectinase, and protease. EAE is an environmentally friendly method that uses water as a solvent and operates at moderate temperatures to prevent degradation. However, the extraction process can take up to 48 h, requiring patience for completion.

Microwave extraction employs electromagnetic waves to quickly and efficiently heat and extract compounds. Microwave-assisted extraction (MAE) is a promising technique for obtaining bioactive compounds from plant and animal sources, offering faster extraction times, higher yields, and reduced solvent consumption compared to conventional methods (Pérez-Martínez et al., 2020). Recent advancements in green microwave extraction methods, such as microwave hydrodiffusion and gravity (MHG) and microwave-assisted distillation (MAD), have further improved the process, offering better repeatability and low energy consumption (Saini & Keum, 2018).

4 Applications of Bioactive Compounds

Bioactive compounds are defined as molecules that influence living organisms, tissues, or cells. Due to their potential applications in various industries, particularly in agriculture and pharmaceuticals, bioactive compounds have garnered significant attention in recent years. Their versatility stems from their ability to interact with biological systems, which allows for a wide range of uses, from crop protection to the development of pharmaceuticals. Figure 1 illustrates the key applications of bioactive compounds across different sectors.

4.1 Industrial Applications

Applications of bioactive compounds in the pharmaceutical and nutraceutical industries are among the most well-known. These substances are crucial in the development of pharmaceuticals and nutritional supplements. For instance, flavonoids—a class of bioactive compounds—have shown promise in treating cardiovascular diseases due to their antioxidant properties. Additionally, polyphenols, another class of bioactive compounds, are widely used in nutraceuticals for their anti-inflammatory and anti-carcinogenic properties. Bioactive compounds sourced from microalgae, plants, and other organisms exhibit a wide range of bioactivities, including antibacterial, antiviral, anticancer, and immunomodulatory properties, demonstrating immense potential for pharmaceutical applications. These compounds, found in fruits, vegetables, and medicinal plants, have been employed by the pharmaceutical industry to develop plant-based drugs that prevent and treat conditions such as cancer and cardiovascular diseases.

Seventy percent of Earth's ecosystems are aquatic, hosting a vast array of species that provide essential ecosystem services such as raw materials, seafood, tourism, and cultural heritage. Seaweeds, or macroalgae, have long been utilized as a nutrient-rich food source and in traditional medicine. However, advancements in technology over recent decades have enabled the characterization and utilization of the biological properties of macroalgal compounds for biotechnological applications. Seaweeds contain unique compounds with properties that hold great potential in medicine and industry. Due to their affordability, ease of cultivation, palatability, and non-toxic nature, macroalgae are excellent alternatives to synthetic chemicals.

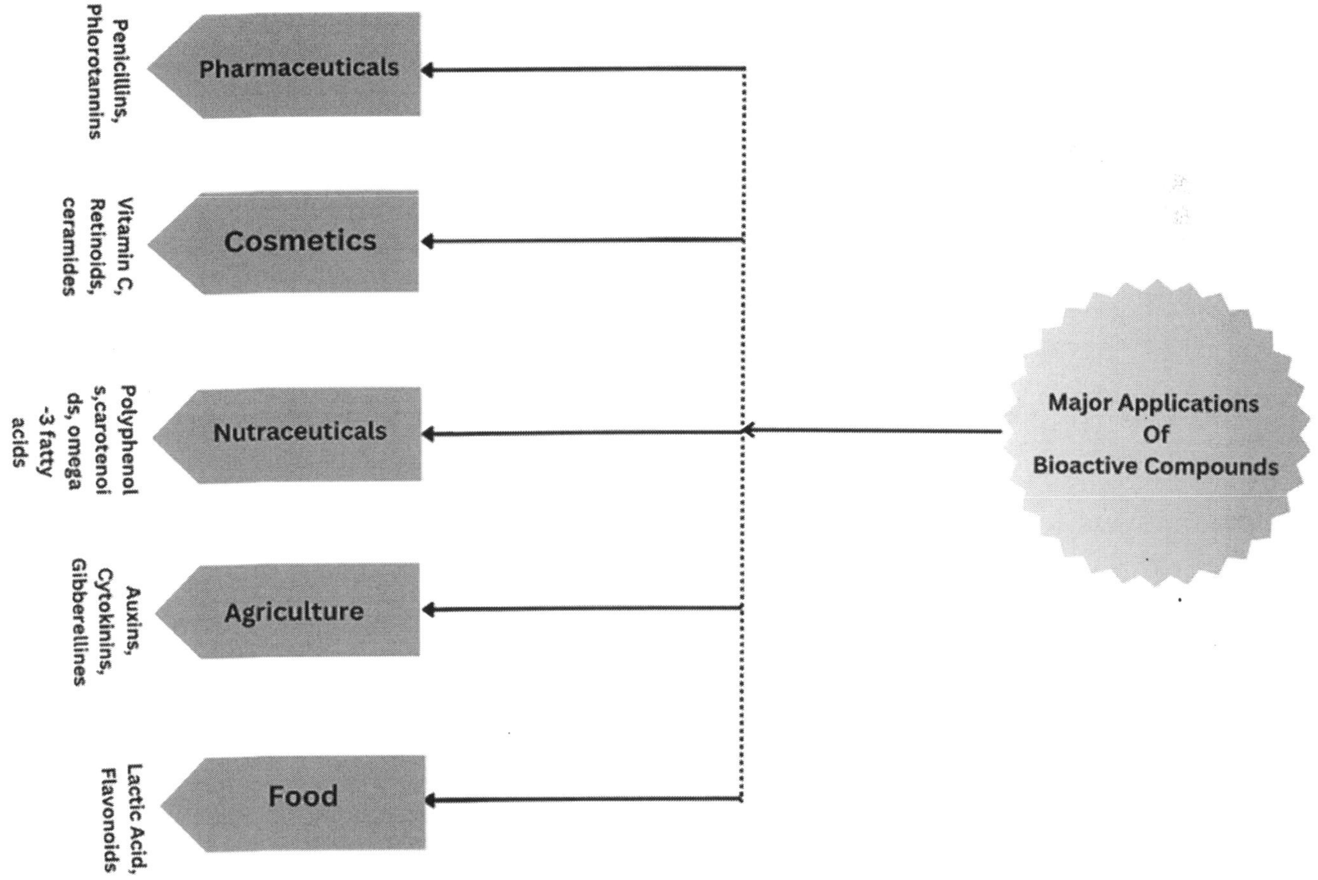

Fig. 1 Major application of bioactive compounds

Bioactive compounds derived from plants have also garnered significant attention in pharmaceutical applications due to their medicinal properties. These compounds, including phenolic acids, alkaloids, flavonoids, and terpenoids, exhibit a variety of biological activities such as antibacterial, anticancer, anti-inflammatory, and cardioprotective effects. The therapeutic potential of plant-derived bioactive compounds is increasingly recognized in the development of novel drugs and therapies for various diseases. For instance, flavonoids have shown potential in reducing the risk of chronic conditions such as cardiovascular diseases and certain types of cancer due to their antioxidant properties (Shahidi & Ambigaipalan, 2015). Similarly, alkaloids like quinine and morphine have been used for their analgesic and antimalarial properties. The growing body of evidence supporting the efficacy of these compounds highlights the need for further research to fully realize the potential of plant-based medicines in improving human health.

To accelerate the drug development process, pharmaceutical companies are investing heavily in advanced technologies like high-throughput screening (HTS). They are particularly focused on plant-derived molecules for the treatment of serious conditions like heart diseases. HTS enables the rapid testing of biological activity in hundreds to millions of samples, facilitating the identification of potential therapeutic candidates from complex plant extracts. Moreover, the integration of HTS with other technologies such as computational biology and genomics has provided a deeper understanding of these compounds' mechanisms of action, thereby speeding up the development of these substances into effective treatments. The commitment of pharmaceutical companies to leveraging these cutting-edge technologies underscores the growing awareness of the potential of plant-derived compounds in the development of new medications for heart diseases and other critical health conditions.

Similarly, the use of bio-based polymers in drug delivery systems has emerged as a viable strategy for pharmaceutical formulations due to their inherent advantages over conventional synthetic polymers in terms of biodegradability and biocompatibility. Bio-based polymers, derived from natural materials like proteins, polyesters, and polysaccharides, hold significant promise in improving the efficacy and safety of drug delivery. Their biodegradability and compatibility with biological systems make them ideal for the development of advanced drug delivery systems. For example, the natural polysaccharide chitosan has been extensively researched for its mucoadhesive properties and its ability to enhance drug bioavailability. Similarly, polylactic acid (PLA), a bio-based polyester, is well-known for its use in controlled-release formulations due to its non-toxic nature and predictable degradation profile. The growing interest in bio-based polymers is driving innovations in drug delivery technologies.

4.1.1 Pharmaceutical and Nutraceutical

The goal is to achieve a more focused and prolonged release of treatments, which will enhance patient outcomes and minimize side effects. This highlights the importance of further research in the area of bio-based polymers to fully unlock their potential for medicinal applications. Innovative biodegradable surfactants derived from renewable resources are rapidly emerging, offering safer alternatives to conventional surfactants made from petrochemicals for industrial and pharmaceutical use. These bio-based surfactants, sourced from natural materials like sugars, proteins, and plant oils, offer significant ecological advantages due to their biodegradability and low toxicity.

Because renewable resources are used in their production, these industrial processes are more sustainable and have a reduced environmental impact. For instance, several bacteria produce glycolipid surfactants called rhamnolipids, which have been shown to possess excellent surface-active properties and biodegradability, making them valuable in bioremediation, cosmetics, and the pharmaceutical industries. Similarly, another class of microbial surfactants, sophorolipids, has demonstrated promise in improving the solubility and bioavailability of hydrophobic drugs, offering a safer alternative to petroleum-based formulations. The shift towards renewable and biodegradable surfactants aligns with global efforts to promote sustainable development and reduce the environmental footprint of industrial activities. Continued research and innovation in this field are essential for advancing the application of bio-based surfactants, leading to long-term environmental and economic benefits.

In the nutraceutical sector, bioactive compounds play a critical role due to their wide range of functional properties and health benefits. These compounds are abundantly found in various foods such as fruits, vegetables, seeds, and microorganisms. Examples include polyphenols, carotenoids, bioactive peptides, and vitamins. Polyphenols, found in berries, tea, and wine, are known for their strong anti-inflammatory and antioxidant properties, which help reduce the risk of chronic diseases such as cancer and cardiovascular conditions (Scalbert et al., 2005). Carotenoids, present in leafy greens, tomatoes, and carrots, have been linked to oxidative stress reduction and improved eye health, while also providing essential provitamin A activity (Kazaz et al., 2022).

Bioactive peptides, derived from protein hydrolysates of fish, milk, and soy, exhibit antihypertensive, antibacterial, and immunomodulatory properties, enhancing their potential in functional foods and dietary supplements (Pihlanto, 2016). Vitamins, essential micronutrients found in a wide range of food sources, are crucial for maintaining metabolic processes and preventing deficiency-related diseases (Gupta, 2022). The integration of these bioactive compounds into nutraceutical products underscores their importance in disease prevention and health promotion, driving industry innovation and growth.

Bioactive compounds, with their antibacterial, anticancer, anti-inflammatory, and antihypertensive actions, are key in managing diseases and promoting overall well-being. Researchers have shown that antioxidants, such as the polyphenols in fruits and vegetables, neutralize free radicals, reducing oxidative stress and the risk of chronic diseases. Their anticancer potential has been linked to their ability to inhibit cancer cell growth and induce apoptosis. Preclinical studies have shown promising results for compounds like curcumin and resveratrol (Perales-Salinas et al., 2024).

Anti-inflammatory bioactive compounds, such as flavonoids and omega-3 fatty acids, regulate inflammatory pathways, reducing inflammation-related conditions. Essential oils and peptides also possess antibacterial properties, providing novel methods to combat bacterial infections and drug resistance. Bioactive compounds are also involved in managing metabolic disorders, as some polyphenols and alkaloids have demonstrated antidiabetic effects by lowering blood glucose levels and enhancing insulin sensitivity. Furthermore, bioactive peptides from dietary proteins have been shown to possess antihypertensive properties by inhibiting the angiotensin-converting enzyme (ACE), aiding in blood pressure regulation.

These wide-ranging health benefits underscore the potential of bioactive compounds in nutraceuticals for promoting health and preventing disease. The growing popularity of bioactive compounds in nutraceutical products is due to their natural origins and lower side effect profiles, making them preferable to synthetic ingredients. This trend reflects the increasing demand from consumers for natural ingredients in dietary supplements and functional foods.

Nutraceuticals enriched with bioactive compounds are experiencing a surge in the market, as consumers seek products that not only enhance overall health and prevent diseases but also provide nutritional benefits. Natural bioactive compounds with cardioprotective, anti-inflammatory, and antioxidant properties, such as polyphenols, carotenoids, and omega-3 fatty acids, are particularly favored. Moreover, natural compounds often have fewer side effects compared to synthetic chemicals, making them more appealing to health-conscious consumers (Shahidi & Ambigaipalan, 2015).

The move towards natural bioactive compounds is also driven by the goal of promoting environmentally friendly and sustainable practices in food production and supplement manufacturing (Liu, 2023). As more research uncovers the therapeutic potential of these compounds, they are likely to be increasingly incorporated into nutraceutical products to meet consumer demand for natural, health-promoting ingredients.

Technological advancements, such as microencapsulation and nanoencapsulation, have significantly improved the preservation, solubility, and bioavailability of bioactive compounds, broadening their applications in the food, pharmaceutical, and nutraceutical industries. These encapsulation techniques involve enclosing bioactive molecules in microscopic or nanoscopic carriers, which enhances the stability and controlled release of the compounds while protecting them from degradation during processing and storage (Rao & Naidu, 2016). For example, nanoencapsulation has been shown to improve the solubility of hydrophobic bioactive compounds, thereby increasing their efficacy and absorption in the human body (Neyestani & Shariatifar, 2021). Additionally, this technique allows for targeted delivery, ensuring the release of bioactive compounds at specific sites in the body, which enhances their therapeutic efficacy and reduces the likelihood of side effects (Lanzas et al., 2019).

According to the incorporation of bioactive compounds into various products has been facilitated by these advanced delivery systems, opening new avenues for the development of functional foods and nutraceuticals with enhanced health benefits. As encapsulation technologies continue to evolve, they have the potential to revolutionize the efficient and effective application of bioactive compounds across a wide range of industries.

4.1.2 Cosmetics

Bioactive compounds are increasingly finding their way into the cosmetics industry due to their beneficial properties and the growing demand from consumers for natural products. Examples of bioactive compounds from marine bioresources, such as hydrolyzed collagen, glycogen, and hyaluronic acid, are becoming more popular as sustainable and safe alternatives in cosmetic formulations. Hydrolyzed collagen, known for its excellent moisturizing and anti-aging properties, enhances skin elasticity and reduces the appearance of wrinkles. Glycogen, a natural polysaccharide, provides energy to skin cells, promoting cellular renewal and improving skin texture. Hyaluronic acid, a powerful humectant, can hold significant amounts of water, keeping the skin hydrated and supple.

These marine-derived bioactive compounds are not only effective but also align with the growing trend towards sustainability and eco-friendly practices in the cosmetics industry. Their incorporation into skincare products addresses consumer preference for natural and ethically sourced ingredients, driving innovation and growth in the market. Continued research into marine bioresources is likely to uncover even more potent bioactive ingredients, further expanding their application in the cosmetics sector. These bioactive substances are essential in addressing skin aging concerns because they influence collagen formation, act as cellular regulators and antioxidants, and reduce free radical levels in tissues. Free radicals are a primary cause of oxidative stress and skin aging. Antioxidants, such as those found in marine-derived components, help neutralize these radicals (Ganguly et al., 2022).

Additionally, these substances function as cellular regulators, influencing various signaling pathways that support skin healing and regeneration. These bioactive ingredients are crucial for developing anti-aging skincare products that meet the increasing demand for safe and effective cosmetics due to their multifunctional properties. Further research is being conducted on phytochemicals, which are derived from plants and used as flavorings and antioxidants in cosmetics and personal care products, for their medicinal and cosmeceutical benefits. Flavonoids, tannins, and terpenoids are natural substances with potent antioxidant properties that protect the skin from oxidative stress and environmental damage.

For instance, flavonoids from fruits and vegetables, such as quercetin and catechins, neutralize free radicals and reduce inflammation, promoting healthier and more resilient skin (Middleton et al., 2022). Tannins, abundant in tea and some berries, have astringent properties that help tighten pores and firm the skin, improving overall skin appearance. According to de Castro et al. (2019), terpenoids, including substances like menthol and limonene, are popular in lotions, creams, and perfumes as natural aroma agents and potent antioxidants. As consumer preferences for natural and sustainable ingredients grow, the integration of phytochemicals into cosmeceutical products supports environmental responsibility and skin health simultaneously.

The personal care and cosmetics industry is experiencing innovation due to ongoing research into plant-based bioactive chemicals, which continues to reveal new benefits and applications (de Castro et al., 2019). The use of biofunctional molecules in cosmetics emphasizes their wide range of pharmacological properties and highlights their importance for product development and innovation across various sectors. These bioactive substances, including polysaccharides, enzymes, and peptides, have anti-aging, anti-inflammatory, and skin-repairing properties that make them invaluable in cosmetic formulations (Maleki et al., 2022).

For example, biofunctional peptides, such as matrixyl and argireline, which are well-known for their effects on collagen formation and wrinkle reduction, help promote youthful skin (Lupo & Souyoul, 2023). Enzymes like superoxide dismutase, which are powerful antioxidants, neutralize free radicals and protect the skin from oxidative damage (Ganguly et al., 2022). Polysaccharides, such as hyaluronic acid, enhance skin hydration and flexibility due to their superior moisture retention properties. Incorporating these biofunctional compounds into cosmetic products improves their efficacy while satisfying consumer demand for multi-functional, scientifically backed skincare products.

The use of bioactive (co)oligoesters as controlled delivery systems for substances like p-anisic acid shows great promise in the cosmetic market due to their precise release profiles and biocompatibility. These advanced delivery systems are designed to enhance the effectiveness and stability of active ingredients, ensuring prolonged release and targeted action on the skin. P-anisic acid, known for its antibacterial and anti-inflammatory properties, retains its potency for a longer duration due to its encapsulation in (co) oligoesters, which protect it from degradation and allow for slow release (Steinberg, 2020).

The biocompatibility of these oligoesters ensures their safety in cosmetic formulations by reducing the risk of irritation and adverse reactions. A consistent therapeutic effect is made possible by the controlled release mechanism, which improves the overall effectiveness of skincare products (Di Marzio et al., 2020). The development of bioactive (co)oligoesters as delivery systems represents a major breakthrough in cosmetic research, providing a versatile platform for incorporating various active compounds to enhance product efficacy and customer satisfaction.

The use of bioactive compounds in cosmetics reflects the increasing popularity of natural, sustainable ingredients that are environmentally friendly and therapeutically beneficial. The demand for bioactive compounds derived from natural sources is being driven by consumers' growing preference for products that are not only effective but also ethically sourced and environmentally sustainable (Messire et al., 2023). These compounds, which include naturally occurring peptides, plant extracts, and marine-derived ingredients, are known for their powerful biological activities, such as anti-aging, anti-inflammatory, and antioxidant effects (Afandi, 2022).

For example, ingredients like aloe vera and green tea extract are widely known for their restorative and revitalizing properties, making them popular choices in skincare formulations (Godić et al., 2020). The use of bioactive compounds also supports sustainable practices by reducing the environmental impact of cosmetic products, as many of these substances are renewable and biodegradable (Abd Gani, 2020). By enhancing product performance and safety, the trend of incorporating bioactive substances is revolutionizing the cosmetics industry while promoting a more sustainable and responsible approach to personal care and beauty.

4.1.3 Food Sector

The food sector benefits greatly from bioactive substances. As previously mentioned, bioactive chemicals required by the human body are found in food, nutrient supplements (Talekar et al., 2018), food coloring (Chhikara et al., 2019), meat and animal products (Pogorzelska-Nowicka et al., 2018), and other food products. They are also added to food to enhance its quality. For instance, ascorbic acid is frequently added to prevent oxidation; vanillin and cinnamaldehyde are used as flavoring agents; and carotenoids, curcumin, and anthocyanins are used as coloring agents. One of the primary areas within the food industry that generates a significant amount of bioactive compounds is fermentation (Sadh et al., 2018). Examples include single-cell proteins, which serve as an alternative protein source; lactic acid, used for acidulation and preservation; xanthan, employed as an emulsifier, thickener, and stabilizer; laccase, utilized in baking and the beverage industry as a stabilizer (Mayolo-Deloisa et al., 2020); astaxanthin, a coloring agent (Gwaltney-Brant, 2021); citric acid, used as a preservative and flavoring agent; fumaric acid, which serves as an acidulant (Karaffa & Kubicek, 2021); and many others.

Fruits, wine industry wastes, and plant, fruit, and vegetable wastes—such as peels, seeds, and pomace—are the primary sources of these bioactive chemicals (Shirahigue & Antonini, 2020). Apples, mangoes, plums, bananas, and citrus fruits are examples of common fruits that contain phenolic acids, flavanols, carotenoids, anthocyanins, and lipids. Common vegetables like potatoes, carrots, beetroot, and broccoli contain carotenoids, flavonoids, phenolic acids, and carbohydrates. Other sources of carotenoids include Neochloris oleoabundans, C. anñuum baydgi, and Helianthus annuus L. (sunflower) wastes (petals, florets). Anthocyanins can be found in chokeberry fruit, strawberries, raspberries, blueberries, blackberries, and Tannat grape pomace, as well as in the papery peel of Allium cepa L. var. Ascalonicum, Phyllostachys pubescens (bamboo), Tannat grape pomace, Oryza sativa (brown rice), Thymus atlanticus (Moroccan thyme), and bitter gourd. The extraction tables provide a detailed list of these sources.

Due to their chemical and physical properties, industrial processes generate a lot of coproducts that are difficult to preserve. Traditionally, these byproducts have been used to produce compost or animal feed. However, their composition suggests they contain highly valuable compounds that, with the right conversion, could be developed into commercially viable products or used as ingredients in new food products, capitalizing on the abundance of potentially valuable compounds they contain.

Byproducts from the food industry include fruits, seeds, skins, and leftover membranes. These byproducts contain numerous bioactive compounds such as dietary fiber (pectin, cellulose, hemicellulose, and lignin), minerals (potassium, calcium, magnesium, and selenium), organic acids (citric, oxalic, and malic acids), vitamins (thiamine, riboflavin, and niacin), phenolic acids (chlorogenic, ferulic, and sinapic acids), flavonoids (hesperidin, narirutin, didymin, hesperetin, and diosmin), terpenes (limonene), and carotenoids (lutein, β-carotene, and zeaxanthin) (Zhang et al., 2020; Multari et al., 2020). These bioactive compounds are associated with numerous health benefits, including anti-inflammatory, antibacterial, antioxidant, anti-hypertensive, neuroprotective, and antiallergenic properties (Manchanda et al., 2023; Ruviaro et al., 2020). As a result, the food sector is increasingly interested in producing various products using agro-industrial waste byproducts.

Bioactive compounds, especially phenolics, are critical to plants' defense against various pathogens and stresses due to their antibacterial properties (Xie, 2024). Many fruits and vegetables are rich in flavonoids, a subclass of phenolics essential to both human health and plant defense. Proanthocyanidin-rich grape seed extracts have demonstrated strong antibacterial and antioxidant properties, which make them useful for extending the shelf life of food products such as cooked chicken. Similarly, the potent antibacterial compounds found in citrus peels and mango seed extracts have been shown to suppress the growth of various pathogens, thereby enhancing food safety and preservation (Lee & Kim, 2022).

Moreover, avocado peels and seeds exhibit effective antibacterial properties due to their high phenolic content, suggesting their potential use in food preservation. These findings highlight the potential for using bioactive components from food waste as natural preservatives, offering a waste-free and sustainable approach to food preservation (Ferreira-Lazarte & Villamiel, 2023). Fruit pomace, a byproduct of fruit processing, holds significant potential for enhancing the nutritional and functional properties of various food products. Incorporating pomace into baked goods, meat, and dairy products can improve dietary fiber content, emulsion stability, and water retention while also enhancing flavor and aroma. Studies have shown that apple pomace can be effectively used in baked snacks and meat products, improving textural qualities and sensory attributes when incorporated up to certain levels (Iqbal et al., 2021). Similarly, mango peel powder and red grape pomace have demonstrated comparable benefits in biscuits and pork burgers, respectively, by enhancing fiber content and maintaining sensory quality (Younis & Ahmad, 2015).

4.2 Agriculture Application

Bioactive substances generated from horticultural crops are important in agricultural applications because they provide a natural and sustainable alternative to manufactured chemicals. These substances, which include polyphenols, flavonoids, and carotenoids, possess strong bioactive properties that can enhance plant growth, pest and disease resistance, and overall crop health. Encapsulation technologies, such as modern micro- and nanoencapsulation methods, have been employed to increase the potency and stability of these bioactive compounds. These encapsulation strategies protect bioactive substances from degradation and regulate their release, ensuring long-term and targeted delivery to plants (Raddatz & Menezes, 2021). For example, nanoencapsulation can improve the solubility and bioavailability of hydrophobic chemicals, making them more effective in agricultural applications (Malerba & Cerana, 2022). Furthermore, the use of bioactive compounds from horticultural crops aligns with the growing demand for environmentally friendly and sustainable farming practices, reducing reliance on harmful synthetic chemicals and enhancing environmental health. Continued research and development in this area have the potential to further integrate natural bioactive substances into sustainable agriculture, boosting crop production and sustainability.

Moreover, bioactive substances derived from plants and microorganisms have shown promising insecticidal and repellent properties against various pests, offering a sustainable alternative to conventional pesticides. Compounds such as essential oils, alkaloids, and terpenoids exhibit strong bioactivity and can effectively regulate pest populations while minimizing environmental damage (Isman, 2020). For example, neem oil, obtained from the neem tree, contains azadirachtin, which inhibits insect growth and reproduction, making it an excellent natural pesticide. Similarly, microbial bioactive substances such as spinosad, produced by the soil bacterium Saccharopolyspora spinosa, have been successfully used as insecticides due to their unique mode of action and low toxicity to non-target organisms (Tiwari et al., 2021).

Nanotechnology has the potential to significantly enhance the potency and stability of these bioactive substances. Applications of nanotechnology can improve the distribution, controlled release, and targeted action of these substances, increasing their efficacy while reducing the required dosages. This combination of bioactive compounds and advanced nanotechnological methods holds great promise for developing next-generation, environmentally friendly pest management solutions in agriculture. The use of bioactive substances from plants and microorganisms in pest management not only enhances agricultural sustainability but also promotes ecosystem health and reduces the environmental impact of conventional chemical pesticides. These natural substances, including essential oils, flavonoids, and microbial metabolites, offer effective pest management while being biodegradable and less harmful to non-target organisms, thus preserving biodiversity (Isman, 2020).

The integration of bioactive substances supports integrated pest management strategies that focus on long-term prevention and ecological balance, reducing the need for synthetic pesticides that can pollute soil and water and lead to pesticide resistance. Farmers can achieve effective pest management while minimizing environmental concerns, resulting in healthier soils and ecosystems. Additionally, the implementation of such sustainable practices aligns with global efforts to reduce agricultural carbon footprints and promote environmentally friendly farming methods (Ghorbani et al., 2020). Research and development of bioactive substances and their delivery technologies, such as nanotechnology, are advancing their efficacy and applicability, paving the way for more sustainable and resilient agricultural practices.

For instance, the compound bio-fertilizer additive, including FistulainA and salinomycin, enhances plant growth, disease resistance, nutrient delivery, and crop yield without harming the environment, making it a practical and effective agricultural application strategy. Similarly, the EM active beneficial bacteria mother liquor bio-organic liquid fertilizer promotes crop growth, nutrient absorption, disease resistance, and odor reduction in poultry, thereby increasing agricultural productivity and sustainability. Bioactive substances derived from plants and microorganisms hold significant potential for pest management in agriculture, offering an environmentally friendly alternative to synthetic pesticides. These natural compounds, including alkaloids, terpenoids, and microbial metabolites, possess inherent pesticidal properties that can be used to manage a range of agricultural pests. The application of nanotechnology enhances the efficacy and stability of these bioactive substances, allowing for controlled release and targeted delivery to pest populations. Nanotechnology improves the solubility and bioavailability of these substances, ensuring their long-term effectiveness and reducing application rates, thereby minimizing environmental impact (López-Cabeza, 2020). This synergy between bioactive substances and nanotechnology not only improves crop protection but also boosts food production by reducing crop losses and increasing output quality. Continued research and deployment of these advanced technologies have the potential to revolutionize agricultural practices, making them more sustainable and resilient in the face of pest threats. Table 1 presents sources, mechanisms, benefits, and names of various bioactive molecules along with their applications in different sectors.

Table 1 Bioactive compounds: source, target organism, mode of action, practical benefits

Sr. No.	Bioacive molecule	Source	Target organesm/condition	Mechanism of action	Practical benefits
Nutraceutical					
1	Curcumin	Curry powder (Curcuma longa)	Cancer, inflammation	Antioxidant, anti-inflammatory, and cell signaling pathway modulator	Padmanaban and Rangarajan (2016)
2	Resveratrol	Peanuts, berries, and grapes	Cardiovascular diseases, aging, antioxidant	Activation of sirtuins, and anti-inflammatory, antioxidant	Pagaret al. (2019)
3	Fatty Acids Omega-3	Chia seeds, flaxseed, and fish oil	Heart conditions, inflammation	Lipid-level modulating, blood clotting-reducing, and anti-inflammatory	Calder (2019)
Antimicrobial					
4	Allicin	Garlic (Allium sativum)	Heart disease, infections	Antimicrobial, blood pressure-lowering, and antioxidant	Chidinma (2019)
5	Thymol	Thymus vulgaris	Thymol infections, plant disease management	Antifungal and antibacterial	Salehi et al. (2018)
6	Berberine	Barberry (Berberis vulgaris), and goldenseal (Hydrastis canadensis)	Infections and metabolic conditions	Antiviral, antifungal, and antibacterial	Imenshahidi and Hosseinzadeh (2019)
Anti-inflammatory					
7	Quercetin	Onions, apples, berries	Allergies, inflammation	Antioxidant, anti-inflammatory, inhibits histamine release	Qi et al. (2022)
8	Boswellia acid	As Boswellia serrata, Frankincense	Joints and inflammation	Lowers inflammation and promotes the health of joints	Chaudhary and Kumari (2021)
Agriculture					
9	Azadirachtin	The Azadirachta indica, or neem tree	Termite management	Growth regulator for insects that interferes with molting and reproduction	Agbo et al. (2019)
10	Chitosan	Exoskeletons of shellfish	Encouraging plant growth and managing pests	Boosts plant immunity and has antifungal properties	Younes and Rinaudo (2015)
11	Salicylic acid	Bark from Willow (Salix spp.)	Resistance to plant diseases and growth regulation	Triggers the plant's systemic acquired resistance (SAR)	Dempsey and Klessig (2017)
Interstitial applications					
12	Humic acid	Organic matter that has decomposed	Fertility of soil and water retention	Promotes nutrient uptake and improves soil structure	Bayat et al. (2021)
13	Saponins	Different plants (Quillaja saponaria, for example)	Biopesticides, enhancers for drug delivery	Disrupts cell membranes of pests, enhances solubility of drugs	Francis et al. (2021)
14	Fulvic acid	Organic matter that has decomposed	Absorption of nutrients and soil health	Increases soil microbial activity and chelates minerals	Lumactud et al. (2022)
15	Phytosterols	Different plants (such as corn and soybean)	Control of cholesterol and health of crops	Lowers human cholesterol absorption and maintains the integrity of plant cells	Farquhar (2020)

5 Conclusion

Bioactive chemicals derived from plants and microorganisms have significant potential to revolutionize several industries, particularly agriculture. Utilizing these natural substances can enhance pest management, increase crop yields, and promote sustainable farming practices. Combining bioactive substances with advanced technologies such as nanotechnology amplifies their efficacy and application. This integration enables regulated release and targeted delivery, improving the stability and effectiveness of bioactive compounds. This approach not only provides effective pest management but also reduces environmental impact by decreasing the reliance on synthetic pesticides, which can pollute soil and water and contribute to pesticide resistance. The application of bioactive compounds also supports international initiatives that advocate for environmentally friendly farming methods and aim to reduce the agricultural sector's carbon footprint. These sustainable practices lead to more robust and productive agricultural systems, enhancing ecological balance and soil health. In summary, ongoing research and development in this field are paving the way for the next generation of environmentally friendly pest control methods and sustainable farming techniques. By harnessing bioactive chemicals, we can make significant strides in increasing agricultural productivity, protecting the environment, and ensuring global food security.

References

Abd Gani, S. S. (2020). New trends in cosmetics: The potential use of red pitaya and its by-products as cosmetic active ingredients. *Journal of mechanics of continua and mathematical Sciences*, (1). https://doi.org/10.26782/jmcms.spl.9/2020.05.00014.

Adefegha, S. A. (2021). Requirements of bioactive compounds for health claims. *Food Bioactives and Health*, 337–367. https://doi.org/10.1007/978-3-030-57469-7_10.

Afandi, N. (2022). Natural active ingredients used in topical cosmetic formulations for anti- ageing: A systematic review. *International Journal of Pharmaceuticals, Nutraceuticals and Cosmetic Science, 5*(1), 67–78. https://doi.org/10.24191/ijpnacs.v5i1.05.

Agbo, E., Nta, A. I., & Ajaba, M. O. (2019). *Bio-pesticidal Properties of Neem (Azadirachta indica)*. Science Domain International.

Ahmed, M. N., Dable-Tupas, G., Rahmatullah, M., Ratta, K., Michael, M., & Ghosh, R. (2023). Bioactive compounds in diabetes care and prevention. *Role of Nutrigenomics in Modern-Day Healthcare and Drug Discovery*, 387–438. https://doi.org/10.1016/b978-0-12-824412-8.00014-x.

Alarcon-Rojo, A., Janacua, H., Rodriguez, J., Paniwnyk, L., & Mason, T. (2015). Power ultrasound in meat processing. *Meat Science, 107*, 86–93. https://doi.org/10.1016/j.meatsci.2015.04.015.

Bayat, H., Shafie, F., Aminifard, M. H., & Daghighi, S. (2021). Comparative effects of humic and fulvic acids as biostimulants on growth, antioxidant activity and nutrient content of yarrow (Achillea millefolium L.). *Scientia Horticulturae, 279*, 109912. https://doi.org/10.1016/j.scienta.2021.109912.

Calder, P. C. (2019). Is increasing microbiota diversity a novel anti-inflammatory action of marine n–3 fatty acids? *The Journal of Nutrition, 149*(7), 1102–1104. https://doi.org/10.1093/jn/nxz043.

Chaudhary, G., & Kumari, I. (2021). Boswellia serrata ROXB. EX COLEBR. (Salai): An ayurvedic herb with anti-inflammatory potential. *International Journal of Pharmaceutical Sciences Review and Research, 69*(1). https://doi.org/10.47583/ijpsrr.2021.v69i01.024.

Chhikara, N., Kushwaha, K., Sharma, P., Gat, Y., & Panghal, A. (2019). Bioactive compounds of beetroot and utilization in food processing industry: A critical review. *Food Chemistry, 272*, 192–200. https://doi.org/10.1016/j.foodchem.2018.08.022.

Chidinma, O. (2019). Therapeutic effects of garlic: A review. *Scientific Journal of Biology & Life Sciences, 1*(1). https://doi.org/10.33552/sjbls.2019.01.000502.

de Castro, R. D. (2019). *Chemical and antifungal analysis of essential oils and phytochemicals against Candida albicans*. https://doi.org/10.31487/j.cmr.2018.01.005.

De Freitas Marinho, L., Sganzerla, W. G., Ferreira, V. C., Jimenez Moreno, J. A., Rostagno, M. A., & Forster-Carneiro, T. (2023). Advances in green extraction methods, biological properties, and applications of betanin and vitexin: An updated review and bibliometric analysis. *Biocatalysis and Agricultural Biotechnology, 51*, 102744. https://doi.org/10.1016/j.bcab.2023.102744.

Dempsey, D. A., & Klessig, D. F. (2017). How does the multifaceted plant hormone salicylic acid combat disease in plants and are similar mechanisms utilized in humans? *BMC Biology, 15*(1). https://doi.org/10.1186/s12915-017-0364-8.

Di Marzio, N., Eglin, D., Serra, T., & Moroni, L. (2020). Bio-fabrication: Convergence of 3D bioprinting and nano-biomaterials in tissue engineering and regenerative medicine. *Frontiers in Bioengineering and Biotechnology*, 8. https://doi.org/10.3389/fbioe.2020.00326.

Dwarka, D., Baijnath, H., & Mellem, J. J. (2022). Bioactive compounds as therapeutic intervention in cancer therapy. *Therapeutic Use of Plant Secondary Metabolites*, 84–110. https://doi.org/10.2174/9789815050622122010007.

Farquhar, J. W. (2020). Plant sterols: Their biological effects in humans. *Handbook of Lipids in Human Nutrition*, 101–105. https://doi.org/10.1201/9781003068099-10.

Ferreira-Lazarte, A., & Villamiel, M. (2023). Green technologies for sustainable food production and preservation: Ultrasounds. *Sustainable Food Science-A Comprehensive Approach*, 139–157. https://doi.org/10.1016/b978-0-12-823960-5.00057-3.

Francis, G., Kerem, Z., Makkar, H. P. S., & Becker, K. (2021). Reflections on 'The biological action of saponins in animal systems: A review.' *British Journal of Nutrition, 127*(7), 1034–1036. https://doi.org/10.1017/s0007114521004852.

Ganguly, B., Hota, M., & Pradhan, J. (2022). *Skin aging: Implications of UV radiation, reactive oxygen species and natural antioxidants*. Reactive Oxygen.

Garcia-Marti, M., Jafari, S. M., Rashidinejad, A., Xiao, J., & Simal-Gandara, J. (2023). An overview of different food bioactive ingredients. *Handbook of Food Bioactive Ingredients*, 1–27. https://doi.org/10.1007/978-3-030-81404-5_1-1.

Ghorbani, R., et al. (2020). Sustainable agriculture: The role of the diverse bioactive naturally occurring chemicals. *Sustainable Agriculture Reviews, 39*, 267–284.

Godić, A., Pojšak, B., Adamič, M., & Dahmane, R. G. (2020). Corrigendum to "The role of antioxidants in skin cancer prevention and treatment." *Oxidative Medicine and Cellular Longevity, 2020*, 1–1. https://doi.org/10.1155/2020/1969760.

Godlewska, K., Ronga, D., & Michalak, I. (2021). Plant extracts-importance in sustainable agriculture. *Italian Journal of Agronomy, 16*(2). https://doi.org/10.4081/ija.2021.1851.

Gupta, K. (2022). The vitamins: Fundamental aspects in nutrition and health. *Journal of Nutrition Education and Behavior, 54*(12), 1132–1133. https://doi.org/10.1016/j.jneb.2022.07.003.

Gwaltney-Brant, S. M. (2021). Nutraceuticals in hepatic diseases. *Nutraceuticals*, 117–129. https://doi.org/10.1016/b978-0-12-821038-3.00008-2.

Imenshahidi, M., & Hosseinzadeh, H. (2019). Berberine and barberry (Berberis vulgaris): A clinical review. *Phytotherapy Research, 33* (3), 504–523. https://doi.org/10.1002/ptr.6252.

Iqbal, A., Schulz, P., & Rizvi, S. S. (2021). Valorization of bioactive compounds in fruit pomace from agro-fruit industries: Present Insights and future challenges. *Food Bioscience, 44*, 101384. https://doi.org/10.1016/j.fbio.2021.101384.

Isman, M. B. (2020). Botanical insecticides in the twenty-first century—fulfilling their promise? *Annual Review of Entomology, 65*(1), 233–249. https://doi.org/10.1146/annurev-ento-011019-025010.

Karaffa, L., & Kubicek, C. P. (2021). Production of organic acids by fungi. *Encyclopedia of Mycology*, 406–419. https://doi.org/10.1016/b978-0-12-809633-8.21066-2.

Kazaz, S., Miray, R., Lepiniec, L., & Baud, S. (2022). Plant monounsaturated fatty acids: Diversity, biosynthesis, functions and uses. *Progress in Lipid Research, 85*, 101138. https://doi.org/10.1016/j.plipres.2021.101138.

Lanzas, C., Farthing, T., & Dawson, D. (2019). High-resolution location data: Modeling and applications in veterinary medicine. *Frontiers in Veterinary Science*, 6. https://doi.org/10.3389/conf.fvets.2019.05.00010.

Lee, S., & Kim, H. J. (2022). Antioxidant activities of premature and mature mandarin (Citrus unshiu) peel and juice extracts. *Food Science and Biotechnology, 31*(5), 627–633. https://doi.org/10.1007/s10068-022-01064-5.

Liu, R. H. (2023). Health benefits of dietary phytochemicals in whole foods. *Nutritional Health*, 177–190. https://doi.org/10.1007/978-3-031-24663-0_14.

López-Cabeza, R. (2020). Methods for understanding the fate of nanopesticides in soil and water. *Nanopesticides*, 111–136. https://doi.org/10.1007/978-3-030-44873-8_5.

Lumactud, R. A., Gorim, L. Y., & Thilakarathna, M. S. (2022). Impacts of humic-based products on the microbial community structure and functions toward sustainable agriculture. *Frontiers in Sustainable Food Systems*, 6. https://doi.org/10.3389/fsufs.2022.977121.

Lupo, M. P., & Souyoul, S. (2023). Cosmeceuticals in dermatologic conditions. *Cosmeceutical Science in Clinical Practice*, 173–178. https://doi.org/10.1201/9781315165905-14.

Maleki, M. H., Nezhad, H. J., Bahadori, N. K., Daneshniya, M., & Latifi, Z. (2022). Antimicrobial activity of bioactive peptides and their applications in food safety. *Bioactive Peptides from Food*, 415–426. https://doi.org/10.1201/9781003106524-26.

Malerba, M., & Cerana, R. (2022). Chitosan nanoparticle-based delivery systems for sustainable agriculture. *Role of Chitosan and Chitosan-Based Nanomaterials in Plant Sciences*, 197–211. https://doi.org/10.1016/b978-0-323-85391-0.00022-8.

Manchanda, P., Kaur, H., Mankoo, R. K., Kaur, J., Kaur, M., & Sidhu, G. S. (2023). Effect of solvent ratio, temperature and time on extraction of bioactive compounds and their antioxidant potential from callus, leaf and peel extracts of Citrus pseudolimon Taraka. *Journal of Food Measurement and Characterization, 17*(6), 6180–6190. https://doi.org/10.1007/s11694-023-02111-3.

Mayolo-Deloisa, K., González-González, M., & Rito-Palomares, M. (2020). Laccases in food industry: Bioprocessing, potential industrial and biotechnological applications. *Frontiers in Bioengineering and Biotechnology*, 8. https://doi.org/10.3389/fbioe.2020.00222.

Messire, G., Serreau, R., & Berteina-Raboin, S. (2023). Antioxidant effects of catechins (EGCG), andrographolide, and curcuminoids compounds for skin protection, cosmetics, and dermatological uses: An update. *Antioxidants, 12*(7), 1317. https://doi.org/10.3390/antiox12071317.

Middleton, F. M., McGregor, R., Webb, R. H., Wilson, N. J., & Moreland, N. J. (2022). Cytokine imbalance in acute rheumatic fever and rheumatic heart disease: Mechanisms and therapeutic implications. *Autoimmunity Reviews, 21*(12), 103209. https://doi.org/10.1016/j.autrev.2022.103209.

Mounir, S. (2021). Review for "A review on instant controlled pressure drop technology-a strategic tool for extraction of bioactive compounds". https://doi.org/10.1111/ijfs.15408/v1/review1.

Multari, S., Licciardello, C., Caruso, M., & Martens, S. (2020). Monitoring the changes in phenolic compounds and carotenoids occurring during fruit development in the tissues of four citrus fruits. *Food Research International, 134*, 109228. https://doi.org/10.1016/j.foodres.2020.109228.

Murakami, H., Omiya, M., Miki, Y., Umemura, T., Esaka, Y., Inoue, Y., et al. (2020). Evaluation of the adsorption properties of nucleobase-modified sorbents for a solid-phase extraction of water-soluble compounds. *Talanta, 217*, 121052. https://doi.org/10.1016/j.talanta.2020.121052.

Neyestani, M., & Shariatifar, N. (2021). Food ingrdient encapsulation using nanoliposome technology: A review study. *Journal of Food Safety and Hygiene*. https://doi.org/10.18502/jfsh.v5i4.5697.

Padmanaban, G., & Rangarajan, P. N. (2016). Curcumin as an adjunct drug for infectious diseases. *Trends in Pharmacological Sciences, 37*(1), 1–3. https://doi.org/10.1016/j.tips.2015.09.007.

Pagar, K. R., Khandbahale, S. V., & Phadtare, D. G. (2019). The therapeutic potential of resveratrol: A review of clinical trials. *Asian Journal of Pharmaceutical Research, 9*(3), 193. https://doi.org/10.5958/2231-5691.2019.00031.5..

Pasupuleti, V. R., Sammugam, L., Ramesh, N., & Gan, S. H. (2017). Honey, propolis, and royal jelly: A comprehensive review of their biological actions and health benefits. *Oxidative Medicine and Cellular Longevity, 2017*.

Perales-Salinas, V., Purushotham, S. S., & Buskila, Y. (2024). Curcumin as a potential therapeutic agent for treating neurodegenerative diseases. *Neurochemistry International, 178*, 105790. https://doi.org/10.1016/j.neuint.2024.105790.

Pérez-Martínez, B. T., Aboudzadeh, M. A., Schubert, U. S., Leiza, J. R., & Tomovska, R. (2020). Microwave irradiation versus conventional heating assisted free-radical copolymerization in solution. *Chemical Engineering Journal, 399*, 125761. https://doi.org/10.1016/j.cej.2020.125761.

Pihlanto, A. (2016). Milk proteins: Bioactive peptides. *Encyclopedia of Dairy Sciences*, 935–941. https://doi.org/10.1016/b978-0-08-100596-5.00932-x.

Pogorzelska-Nowicka, E., Atanasov, A. G., Horbańczuk, J., & Wierzbicka, A. (2018). Bioactive compounds in functional meat products. *Molecules, 23*(2), 307. https://doi.org/10.3390/molecules23020307.

Poole, C. F. (2003). New trends in solid-phase extraction. *TrAC Trends in Analytical Chemistry, 22*(6), 362–373. https://doi.org/10.1016/s0165-9936(03)00605-8.

Puértolas, E., Luengo, E., Álvarez, I., & Raso, J. (2012). Improving mass transfer to soften tissues by pulsed electric fields: Fundamentals and applications. *Annual Review of Food Science and Technology, 3*(1), 263–282. https://doi.org/10.1146/annurev-food-022811-101208.

Qi, W., Qi, W., Xiong, D., & Long, M. (2022). Quercetin: Its antioxidant mechanism, antibacterial properties and potential application in prevention and control of toxipathy. *Molecules, 27* (19), 6545. https://doi.org/10.3390/molecules27196545.

Raddatz, G. C., & de Menezes, C. R. (2021). Microencapsulation and co-encapsulation of bioactive compounds for application in food: Challenges and perspectives. *Ciência Rural, 51*(3). https://doi.org/10.1590/0103-8478cr20200616.

Rao, P. J., & Naidu, M. M. (2016). Nanoencapsulation of bioactive compounds for nutraceutical food. *Sustainable Agriculture Reviews*, 129–156. https://doi.org/10.1007/978-3-319-39306-3_4.

Ruviaro, A. R., Barbosa, P. de P. M., Martins, I. M., De Ávila, A. R. A., Nakajima, V. M., Dos Prazeres, A. R., et al. (2020). Flavanones biotransformation of citrus by-products improves antioxidant and ACE inhibitory activities in vitro. *Food Bioscience, 38*, 100787. https://doi.org/10.1016/j.fbio.2020.100787.

Sadh, P. K., Kumar, S., Chawla, P., & Duhan, J. S. (2018). Fermentation: A boon for production of bioactive compounds by processing of food industries wastes (By-Products). *Molecules, 23* (10), 2560. https://doi.org/10.3390/molecules23102560.

Saini, R. K., & Keum, Y.-S. (2018). Carotenoid extraction methods: A review of recent developments. *Food Chemistry, 240*, 90–103. https://doi.org/10.1016/j.foodchem.2017.07.099.

Salehi, B., Mishra, A. P., Shukla, I., Sharifi-Rad, M., del Contreras, M. M., Segura-Carretero, A., et al. (2018). Thymol, thyme, and other plant sources: Health and potential uses. *Phytotherapy Research, 32* (9), 1688–1706. https://doi.org/10.1002/ptr.6109.

Scalbert, A., Johnson, I. T., & Saltmarsh, M. (2005). Polyphenols: Antioxidants and beyond. *The American Journal of Clinical Nutrition, 81*(1). https://doi.org/10.1093/ajcn/81.1.215s.

Shahidi, F., & Ambigaipalan, P. (2015). Phenolics and polyphenolics in foods, beverages and spices: Antioxidant activity and health effects–a review. *Journal of Functional Foods, 18*, 820–897. https://doi.org/10.1016/j.jff.2015.06.018.

Shirahigue, L. D., & Ceccato-Antonini, S. R. (2020). Agro-industrial wastes as sources of bioactive compounds for food and fermentation industries. *Ciência Rural, 50*(4). https://doi.org/10.1590/0103-8478cr20190857.

Song, E., Kim, T., Chung, H., & Kwak, S. (2019). Collagen peptides from animal by-products: A review on the processing, biological, and functional aspects. *Marine Drugs, 17*(9), 467.

Steinberg, D. C. (2020). Global regulation of preservatives and cosmetic preservatives. *Cosmetic Microbiology*, 195–205. https://doi.org/10.1201/9780429113697-13.

Sumere, B. R., De Souza, M. C., Dos Santos, M. P., Bezerra, R. M. N., Da Cunha, D. T., Martinez, J., et al. (2018). Combining pressurized liquids with ultrasound to improve the extraction of phenolic compounds from pomegranate peel (Punica granatum L.). *Ultrasonics Sonochemistry, 48*, 151–162. https://doi.org/10.1016/j.ultsonch.2018.05.028.

Talekar, S., Patti, A. F., Singh, R., Vijayraghavan, R., & Arora, A. (2018). From waste to wealth: High recovery of nutraceuticals from pomegranate seed waste using a green extraction process. *Industrial Crops and Products, 112*, 790–802. https://doi.org/10.1016/j.indcrop.2017.12.023.

Tiwari, S., Lata, C., & Prasad, V. (2021). Application of nanobiotechnology in overcoming salinity stress. *Nanobiotechnology*, 375–398. https://doi.org/10.1007/978-3-030-73606-4_16.

Toepfl, S., Mathys, A., Heinz, V., & Knorr, D. (2006). Review: Potential of high hydrostatic pressure and pulsed electric fields for energy efficient and environmentally friendly food processing. *Food Reviews International, 22*(4), 405–423. https://doi.org/10.1080/87559120600865164.

Wells, M. J. M. (2003). Principles of extraction and the extraction of semivolatile organics from liquids. *Sample Preparation Techniques in Analytical Chemistry*, 37–138. https://doi.org/10.1002/0471457817.ch2.

Wijesinghe, W., & Jeon, Y.-J. (2012). Biological activities and potential industrial applications of fucose rich sulfated polysaccharides and fucoidans isolated from brown seaweeds: A review. *Carbohydrate Polymers, 88*(1), 13–20. https://doi.org/10.1016/j.carbpol.2011.12.029.

Wood, J. D. (2023). Meat composition and nutritional value. *Lawrie*, 665–685. https://doi.org/10.1016/b978-0-323-85408-5.00012-1.

Xie, L. (2024). Biofoundries for plant-derived bioactive compounds. *Engineering Biology for Microbial Biosynthesis of Plant-Derived Bioactive Compounds*, 257–283. https://doi.org/10.1016/b978-0-443-15558-1.00005-9.

Younes, I., & Rinaudo, M. (2015). Chitin and chitosan preparation from marine sources. Structure, properties and applications. *Marine Drugs, 13*(3), 1133–1174. https://doi.org/10.3390/md13031133.

Younis, K., & Ahmad, S. (2015). Waste utilization of apple pomace as a source of functional ingredient in buffalo meat sausage. *Cogent Food & Agriculture, 1*(1), 1119397. https://doi.org/10.1080/23311932.2015.1119397.

Zhang, Y., Liao, J., & Qi, J. (2020). Functional and structural properties of dietary fiber from citrus peel affected by the alkali combined with high-speed homogenization treatment. *Lwt, 128*, 109397.

Cancer-Fighting Bioactive Agents Derived from Microbes

Kaushika Olymon, Eshan Abbas, Shivanshi Madeshiya, and Aditya Kumar

Abstract

Cancer is a leading global cause of death, with projections estimating 16.4 million deaths annually by 2040. The complexity of cancer, driven by both internal and external factors, necessitates innovative research focused on prevention, early detection, and personalized treatments. Natural products, especially those from plants, environments, and microorganisms, have been crucial in developing anticancer drugs, with over 60% of current treatments derived from these sources. Recent efforts have intensified in identifying bioactive compounds from microbes due to increasing cancer cell resistance to conventional treatments and severe side effects. Microbes, with their vast species diversity, produce a wide range of bioactive products that are often more cost-effective to extract than to synthesize chemically. These microbial agents can modulate the immune response, inhibit cell division, and induce apoptosis, offering potential solutions to multidrug resistance and toxicity issues. Microorganisms, particularly those from extreme environments, are rich sources of unique anticancer metabolites. Understanding the mechanisms by which microbial-derived agents exert anticancer effects—such as inhibiting signaling pathways, inducing necrosis and apoptosis, reducing angiogenesis, and interfering with translation and splicing—is essential for their therapeutic application. This chapter explores the diverse categories of microbial-derived anticancer compounds, their mechanisms of action, practical applications, and the challenges they face. It highlights recent advancements and promising outcomes in translating these compounds from laboratory research to clinical use, emphasizing their significant potential to improve cancer treatment outcomes.

K. Olymon · E. Abbas · S. Madeshiya · A. Kumar (✉)
Department of Molecular Biology and Biotechnology, Tezpur University, Tezpur, Assam 784028, India
e-mail: aditya@tezu.ernet.in

Keywords

Cancer therapy • Microbial-derived anti-cancer agents • Anticancer compounds • Natural products • Anti-cancer drugs

1 Introduction

Cancer is a leading cause of death globally, with an estimated 16.4 million deaths per year by 2040 (Law et al., 2020). Both internal and external factors can trigger the complex cellular and molecular changes that result in cancer (Wainwright & Scaffidi, 2017). The primary goals in cancer research include prevention, early detection, and the development of innovative treatments, such as personalized therapies that consider the tumor's molecular biology and the host's immune system (Lukasiewicz & Fol, 2018).

Natural products from plants, environments, and microorganisms have been instrumental in the fight against cancer, with over 60% of current anticancer drugs derived from these sources (Law et al., 2020). The search for bioactive compounds from microbes has intensified in recent decades due to increasing resistance of cancer cells to apoptosis-inducing treatments (Tan et al., 2019). Ideally, chemotherapeutic drugs should target only cancer cells, but most current treatments also harm normal cells (Ser et al., 2015). Microbial anticancer drugs may offer solutions to issues like multidrug resistance and severe side effects (Tan et al., 2015).

Microbes, with their species diversity, can produce a vast range of bioactive products. Extracting these compounds from microbes can be more cost-effective than chemical synthesis. Numerous microbes have been studied for their potential to produce cancer-fighting agents. These agents can modulate the immune response, prevent cell division, and induce apoptosis (Martinez-Montiel et al., 2016). Microorganisms, especially those in extreme environments, are

J. Devi (ed.), *Biotechnological Intervention in Production of Bioactive Compounds*, Sustainable Landscape Planning and Natural Resources Management, https://doi.org/10.1007/978-3-031-76859-0_11

valuable sources of natural anticancer metabolites due to their unique compounds (Safarpour et al., 2019).

Understanding the mechanisms behind the anticancer properties of microbial-derived bioactive agents is crucial for their application in cancer therapy. These substances work by inhibiting critical signaling pathways, causing necrosis and apoptosis, reducing angiogenesis, and interfering with translation and splicing (Baindara & Mandal, 2020). Their diverse mechanisms suggest the potential for combination therapies targeting multiple aspects of cancer progression.

This chapter explores the various categories of microbial-derived compounds with anticancer properties, their mechanisms of action, practical applications, and the challenges they face. It also highlights recent advancements and promising outcomes in translating these compounds from laboratory research to real-world applications, emphasizing their significance and potential impact on cancer treatment.

2 The Evolution of Microbes as Anticancer Agents: A Historical Perspective

The use of microorganisms in cancer treatment dates to the 1800s. In 1867, German physician Dr. Wilhelm Busch observed cancer improvement in a patient with erysipelas, now known as Streptococcus pyogenes. This observation marked an early understanding of the potential for bacterial infections to impact cancerous tissues (Broadway & Scharf, 2019). Dr. William Coley, later known as the pioneer of immunotherapy, took these findings further, by developing what came to be known as "Coley's toxins"—a concoction of killed bacterial microbes including Streptococcus pyogenes and Serratia marcescens, which he used to treat cancer patients, in the late 19th and early 20th centuries. His work led to significant tumor regression in many cases, and although his methods were controversial and not universally accepted at the time, they laid the groundwork for modern immunotherapy (Bickels et al., 2002).

In 1928, Alexander Fleming made a groundbreaking discovery that revolutionized medicine. He observed that a mold infection on a Petri dish containing Staphylococcus aureus inhibited bacterial growth. This mold, Penicillium notatum, produced penicillin, the first antibiotic. Penicillin's ability to target bacterial infections without harming human cells made it a critical drug during World War II, saving countless lives from bacterial infections (Pham et al., 2023). This discovery also highlighted the potential of microbial products in therapeutic applications beyond bacterial infections, setting the stage for the exploration of microbes in cancer treatment.

Viral therapy has a similarly intriguing history in cancer treatment (Fig. 1). Unlike bacteria, viruses were identified much later, with their characterization as particulate matter occurring in 1917 (Kelly & Russell, 2007). The potential of viral infections to cause tumor regression was noted as early as 1904, when Dock observed improvement in a leukemia patient following an influenza infection (Pelner et al., 1958), suggesting that viral infections could provoke an immune response strong enough to impact cancer cells. With advancements in genetic engineering, the potential of viral therapy expanded significantly. For instance, researchers began to harness viruses like the herpes simplex virus to specifically target and kill cancer cells, as demonstrated in glioma treatments (Forbes et al., 2018).

Research into microbial secondary metabolites started with primary focus on their antimicrobial properties, which later started to include focus also on therapeutic potential for treating diseases such as cancer. Actinobacteria, particularly the genus Streptomyces, emerged as prolific producers of bioactive compounds. Streptomyces species have been responsible for the discovery of several significant anticancer drugs, such as dactinomycin, bleomycin, doxorubicin, and mitomycin C (Demain & Vaishnav, 2011). These drugs work through various mechanisms, such as intercalating DNA, inhibiting topoisomerase II, or cross-linking DNA, leading to the death of cancer cells.

Marine environments, which make up 95% of Earth's biosphere, have proven to be a rich and largely untapped source of bioactive compounds (Jimeno et al., 2004). Marine bacteria, cyanobacteria, actinobacteria, and fungi have yielded numerous compounds with potent anticancer properties. For example, sarcodictyin, derived from marine bacteria, disrupts microtubule function, which is essential for cell division. Bryostatins, produced by the bacterial symbiont of the bryozoan Bugula neritina, have shown promise in modulating protein kinase C, a key player in cell growth and differentiation (Gerwick & Fenner, 2013).

Symbiotic bacteria associated with marine animals have also been identified as sources of novel anticancer agents. These bacteria often produce toxins that benefit their hosts by deterring predators or competitors, and these same toxins can have powerful anticancer effects. For instance, macrolactin-A, produced by bacteria associated with the dinoflagellate *Noctiluca scintillans*, has been found to inhibit melanoma cell growth. Similarly, bryostatins from bryozoan symbionts have demonstrated potential in cancer treatment by inducing apoptosis and inhibiting cell proliferation (Hale et al., 2002; Simidu et al., 1990).

Advancements in biotechnology and molecular biology have enabled the genetic modification of microorganisms, transforming them from agents of infection to sophisticated tools in cancer therapy. Engineered bacteria and viruses are now being designed to target cancer cells, deliver therapeutic agents directly to tumors, and stimulate the immune system to attack malignancies. This progress highlights the versatility of microbes and their potential in developing

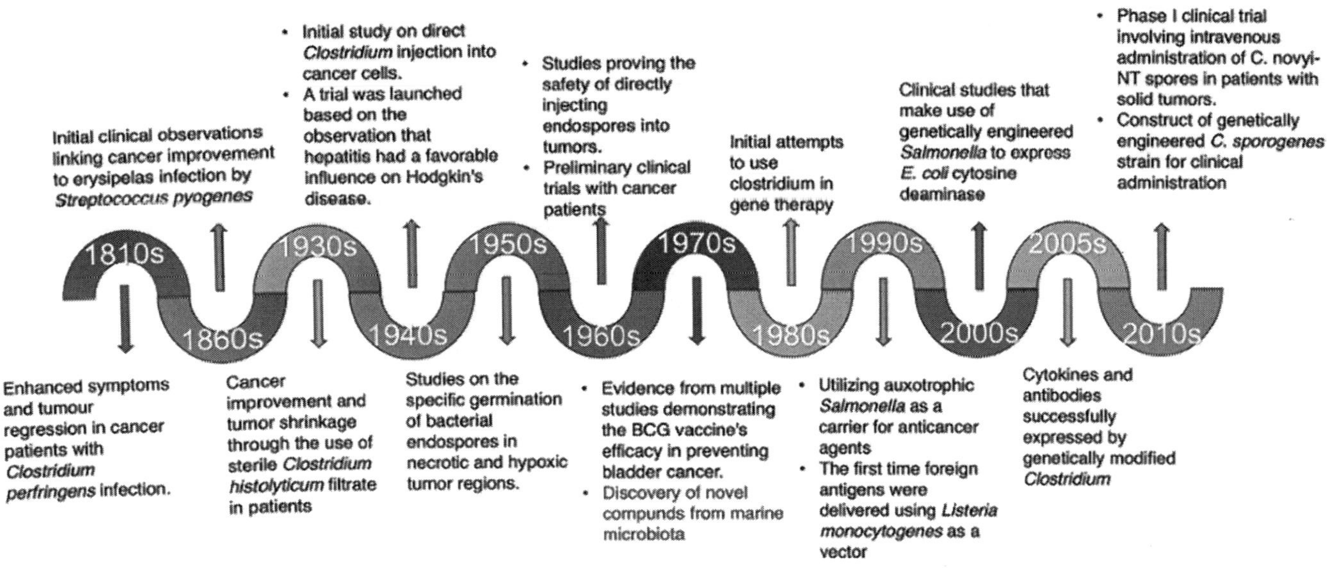

Fig. 1 Timeline of early observations on the microbe-based methods for cancer treatment

next-generation cancer treatments. As research deepens our understanding of the interactions between microbes and cancer cells, these innovations offer promising prospects for more effective and targeted therapies.

3 Mechanisms of Anticancer Activity: How Microbes Combat Cancer

Microbial infections activate the immune system, eliciting both innate and adaptive responses that fight bacterial pathogens and cancer cells. This coincidental benefit has led researchers to explore microbes for their anticancer activity. Anticancer drugs derived from microbes, their classifications, and mechanisms of action, treatment applications and side effects are tabulated in Table 1. Identifying microbial strains with minimal risk to the host is crucial in developing microbe-based cancer therapies (Nallar et al., 2017).

3.1 Bacterial Therapy

Bacterial therapy can combat cancer through several mechanisms.

3.1.1 Direct Oncolysis

Bacteria can directly destroy cancer cells by producing exotoxins or through nutrient competition. Exotoxins are proteins secreted by bacteria that can damage cell membranes, disrupt cellular functions, or inhibit protein synthesis in cancer cells. By competing for nutrients, bacteria can deprive cancer cells of the essential resources they need to grow and proliferate.

3.1.2 Inducing Apoptosis

Genetically modified bacteria can be engineered to inhabit the tumor microenvironment and induce apoptosis (programmed cell death) in cancer cells. These bacteria can be designed to express specific molecules that trigger apoptotic pathways within the cancer cells, leading to their death without harming normal cells (Kasinskas & Forbes, 2007).

3.1.3 Immune Sensitization

Some bacteria can enhance the immune system's ability to recognize and attack cancer cells. These immune-sensitizing bacteria can present cancer-specific antigens to the immune system, effectively "teaching" immune cells to identify and destroy cancer cells. This approach can be particularly effective in treating metastatic diseases, where cancer has spread to multiple locations in the body (Wood & Paterson, 2014).

3.2 Viral Therapy

Viral therapies utilize viruses to introduce therapeutic genes into cancer cells, activating the immune system or directly causing cell death.

3.2.1 Gene Introduction

Viruses can be engineered to deliver genes that encode proteins to stimulate an immune response or induce cell lysis in cancer cells. For example, certain viruses can introduce genes that produce cytokines, which are signaling molecules that enhance the immune system's ability to attack cancer cells.

Table 1 Anticancer drugs derived from microbes, their classifications, and mechanisms of action, treatment applications and side effects

Drugs	Source	Class	Mechanism of action	Treatment	Side effects	Citations
Bleomycin	*Streptomyces verticillus*	Glycopeptides	Induces DNA strand breaks by generating free radicals, causing apoptosis in cancer cells	Treats Hodgkin's lymphoma, non-Hodgkin's lymphoma, testicular cancer, and squamous cell carcinomas	Pulmonary toxicity, skin reactions, and fever	Kawai and Akaza (2003)
Doxorubicin	*Streptomyces peucetius*	Anthracycline antibiotic	Intercalates into DNA, inhibiting topoisomerase II, preventing DNA replication and transcription	Used against breast cancer, lymphomas, leukemias, and multiple myeloma	Cardiotoxicity, myelosuppression, nausea, vomiting, and hair loss	Rivankar (2014), Sohail et al. (2021)
Mitomycin C	*Streptomyces caespitosus*	Antitumor antibiotics	Acts as a bioreductive alkylating agent that cross-links DNA, inhibiting DNA synthesis	Treats upper gastrointestinal cancers, anal cancer, and bladder cancer	Bone marrow suppression, nausea, vomiting, and renal damage	Bradner (2001), Patil et al. (2016)
Salinosporamide A	*Salinispora tropica*	Proteasome inhibitors	Inhibits proteasomes, disrupting protein degradation and leading to apoptosis	Investigated for multiple myeloma and solid tumors	Fatigue, gastrointestinal issues, and neurotoxicity	Gulder and Moore (2010)
Didemnin B	*Trididemnum solidum*	Cyclic depsipeptide	Inhibits protein synthesis and induces apoptosis	Tested in clinical trials for various cancers, including multiple myeloma and solid tumors	Severe toxicity halted its development, but derivatives are being explored	Shin et al. (1994), Xu et al. (2012)
Plitidepsin	*Aplidium albicans*	Cyclic depsipeptides	Induces apoptosis by binding to eEF1A2, leading to oxidative stress and cell death	Under investigation for multiple myeloma and other solid tumors	Includes myalgia, fatigue, and gastrointestinal symptoms	Leisch et al. (2019), Miyake et al. (2021)
Bryostatin 1	*Bugula neritina*	Macrocyclic lactones	Modulates protein kinase C (PKC) activity, affecting cell growth and differentiation	Explored for use in lymphoma, leukemia, and Alzheimer's disease	Myalgia, gastrointestinal issues, and fatigue	Kollar et al. (2014), Kortmansky and Schwartz (2003)
Trabectedin	*Ecteinascidia turbinata*	Alkaloids	Binds to the minor groove of DNA, disrupting transcription and causing cell death	Approved for the treatment of soft tissue sarcoma and ovarian cancer	Neutropenia, liver enzyme elevation, and fatigue	Boccia et al. (2024), Imai et al. (2024)
Lurbinectedin	*Derived from Trabectedin*	Alkylating agents	Similar to Trabectedin, binds to DNA and inhibits transcription	Approved for small cell lung cancer and under investigation for other cancers	Neutropenia, nausea, and fatigue	Cheng et al. (2024)

(continued)

Table 1 (continued)

Drugs	Source	Class	Mechanism of action	Treatment	Side effects	Citations
Calicheamicin	*Micromonospora echinospora ssp. Calichensis*	Carbohydrate	Binding to minor groove of double-stranded DNA	Acute myeloid leukemia (AML) cells	Myelosuppression, hepatotoxicity, infusion-related reactions, increased infection risk, gastrointestinal toxicity, fatigue	Ricart (2011)
Epothilone B	*Sporangium cellulosum*	Polyketidemacro lactone	Bind to tubulin and induces apoptosis	Metastatic or locally advanced breast cancer	Peripheral Neuropathy, Fatigue, Musculoskeletal Pain and Gastrointestinal Toxicity	Griffin et al. (2003), Harrison and Swanton (2008)
Geldanamycin	*Streptomyces hygroscopicus*	Macrocyclic polyketide	Targets HSP90 and suppresses tumorigenic proteins	Breast cancer, melanoma, multiple myeloma, solid tumors and hematologic malignancies	Hepatotoxicity, Gastrointestinal Toxicity, Cardiotoxicity and Fatigue	Blagosklonny (2002), Neckers (2006)
Actinomycin D	*Streptomyces spp.*	Peptolides	Inhibits transcription by inhibiting progression of RNA polymerase	Wilms' Tumor, Rhabdomyosarcoma, Ewing's Sarcoma, Gestational Trophoblastic Neoplasia and Testicular Cancer	Alopecia, Hepatotoxicity, Myelosuppression and Fatigue	Liu et al. (2016)
Prodigiosin	*Serratia marcescens*	Prodiginines	Targets Wnt/β-catenin signaling and retriggers p53	Colon Cancer, Breast Cancer, Lung Cancer, Gastric Cancer and hematological Malignancies	Cytotoxicity, gastrointestinal toxicity, myelosuppression, hepatotoxicity, nephrotoxicity, and cardiotoxicity	Li et al. (2018), Wang et al. (2016)
Staurosporine	*Streptomyces staurosporeus*	Alkaloid	Inhibits tyrosine kinases	Acute myeloid leukemia (AML) with FLT3 mutations, and advanced systemic mastocytosis	Cytotoxicity, myelosuppression, hepatotoxicity, gastrointestinal toxicity, cardiotoxicity, and neurotoxicity	Gescher (1998), Yadav et al. (2015)

3.2.2 Cancer Vaccines

These vaccines encode and express unique cancer antigens, which are then presented to the immune system. This helps the immune system recognize and target cancer cells specifically. Virotherapies, on the other hand, release endogenous antigens through cell lysis, further stimulating the immune response against the cancer (Elsedawy & Russell, 2013).

3.2.3 Oncolytic Vaccines

Hybrid therapies known as oncolytic vaccines use oncolytic viruses to encode tumor antigens in oncolytic vectors. These vaccines stimulate the immune system by presenting tumor-specific antigens, effectively combining the benefits of virotherapy and cancer vaccination. Clinical trials for oncolytic viruses are ongoing, exploring their efficacy in treating various types of cancer (Forbes et al., 2018).

3.3 Fungal Therapy

Fungi, renowned for their nutritional content, have also shown significant potential in regulating cancer immunotherapies.

3.3.1 Modulating Immune Responses

Fungal β-glucan, a polysaccharide found in the fungal cell walls, can modulate the immune system. β-glucan enhances the immune response by activating macrophages, dendritic cells, and natural killer (NK) cells. This heightened immune activity can help in identifying and destroying cancer cells (Tian et al., 2013).

3.3.2 Polysaccharides with Immunotherapeutic Effects

Polysaccharides extracted from fungi like *Agaricus blazei* and *Coriolus versicolor* have demonstrated significant immunotherapeutic effects against tumors. These polysaccharides can stimulate the immune system, increase the production of cytokines, and promote the activity of immune cells that target cancer cells (Liu et al., 2015).

3.4 Detailed Mechanisms of Microbial Anti-cancer Activity

The diverse mechanisms by which microbes exert anticancer effects present a promising research avenue for developing more targeted and effective cancer therapies with fewer side effects compared to traditional methods like chemotherapy and radiation (Fig. 2). Microbial therapies are particularly valuable for treating metastatic diseases, as microbes can selectively target cancer cells, modulate the immune system, and disrupt critical cellular processes. This multifaceted approach offers significant potential in the fight against cancer.

3.4.1 Direct Microbial Cytotoxicity

- **Release of Lytic Enzymes**: Certain bacteria secrete lytic enzymes that degrade the cell walls of cancer cells, leading to their destruction.
- **Production of Anticancer Metabolites**: Microbes produce various metabolites with anticancer properties. These metabolites can inhibit cell proliferation, induce apoptosis, and block angiogenesis.
- **Generation of Reactive Oxygen Species (ROS)**: Bacteria can generate ROS, which causes oxidative damage to cancer cells, leading to cell death.
- **Direct Contact Cytotoxicity**: Some bacteria can directly adhere to and kill cancer cells through mechanisms that involve physical contact and the release of cytotoxic substances.
- **Production of Toxins**: Certain bacterial toxins specifically target and kill cancer cells by disrupting critical cellular processes.

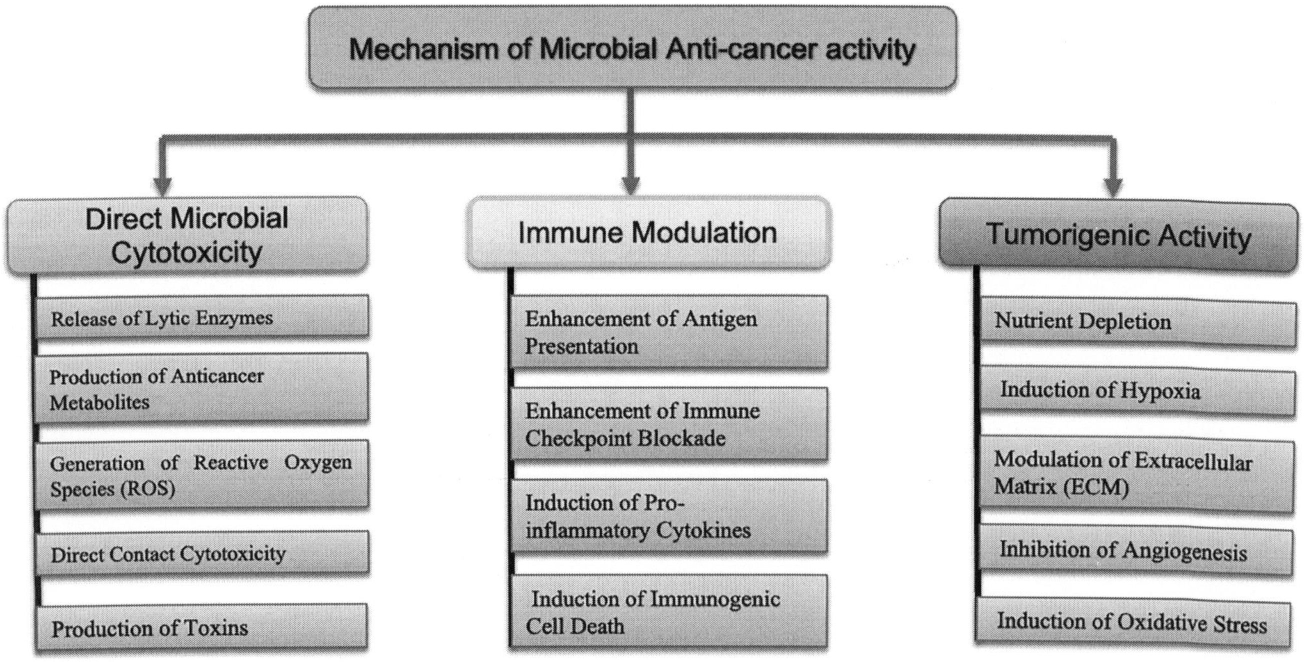

Fig. 2 The figure shows various mechanisms by which microbes combat cancer, providing a comprehensive overview of microbial anticancer activities

3.4.2 Immune Modulation

- **Enhancement of Antigen Presentation**: Microbes can enhance the presentation of tumor antigens to the immune system, improving the recognition and destruction of cancer cells.
- **Enhancement of Immune Checkpoint Blockade**: Microbial products can enhance the effects of immune checkpoint inhibitors, which help the immune system attack cancer cells.
- **Induction of Pro-inflammatory Cytokines**: Microbes can stimulate the production of cytokines that promote inflammation and recruit immune cells to the tumor site.
- **Induction of Immunogenic Cell Death**: Some microbial therapies induce a type of cell death that triggers an immune response against the tumor.

3.4.3 Tumorigenic Activity

- **Nutrient Depletion**: Microbes can compete with cancer cells for essential nutrients, effectively starving the cancer cells and inhibiting their growth.
- **Induction of Hypoxia**: Microbial activity can create a hypoxic environment within tumors, reducing oxygen supply and inhibiting tumor growth.
- **Modulation of Extracellular Matrix (ECM)**: Microbes can alter the ECM, disrupting the structural support for cancer cells and hindering their ability to invade surrounding tissues.
- **Inhibition of Angiogenesis**: Microbes can inhibit the formation of new blood vessels that supply nutrients and oxygen to tumors, effectively starving the tumor.
- **Induction of Oxidative Stress**: By generating ROS and other reactive species, microbes can induce oxidative stress in cancer cells, leading to cellular damage and death.

4 Microbe-Derived Anticancer Biomolecules

Cancer is one of the leading causes of death worldwide, with projections estimating 29.5 million new cases annually by 2040. Traditional cancer treatments, such as chemotherapy and radiation, often come with significant side effects due to their inability to discriminate between healthy and cancerous cells. This highlights the urgent need for safer and more effective therapeutic alternatives. In this context, microbial metabolomes present a promising approach to cancer treatment (Kurup et al., 2024).

Natural sources, including plants, marine organisms, and microbes, have yielded over 13,000 pharmacologically active substances. These substances have provided the foundation for many modern drugs, including anticancer agents. The potential of microbial metabolites in drug discovery is particularly noteworthy due to their unique chemical structures and diverse modes of action (Mohan et al., 2022).

Microbe-derived anticancer biomolecules represent a promising frontier in cancer therapy. The unique chemical diversity and potent biological activities of microbial metabolites offer new avenues for developing safer and more effective cancer treatments. As research continues to explore the vast potential of microbial resources, it is likely that more novel anticancer agents will be discovered, providing new hope in the fight against cancer.

4.1 Actinomycetes: A Rich Source of Anticancer Compounds

Actinomycetes, a group of Gram-positive bacteria, are prolific producers of bioactive compounds, including many with anticancer properties. These bacteria are predominantly found in soil and marine environments, where they produce secondary metabolites that have potent biological activities (Ventura et al., 2007). Some notable anticancer compounds derived from actinomycetes include.

4.1.1 Dactinomycin (Actinomycin D)

- Derived from the bacterium *Streptomyces parvulus*, dactinomycin is used in the treatment of various cancers, including Wilms' tumor, rhabdomyosarcoma, and Ewing's sarcoma. It works by intercalating into DNA and inhibiting RNA synthesis, thereby preventing cancer cell proliferation.

4.1.2 Bleomycin

- Produced by *Streptomyces verticillus*, bleomycin is used to treat Hodgkin's lymphoma, non-Hodgkin's lymphoma, and testicular cancer. It induces DNA strand breaks by generating free radicals, leading to apoptosis in cancer cells.

4.1.3 Doxorubicin

- Also known as Adriamycin, doxorubicin is derived from *Streptomyces peucetius*. It is widely used in chemotherapy for various cancers, including breast cancer, leukemia, and lymphoma. Doxorubicin works by intercalating DNA and inhibiting topoisomerase II, an enzyme essential for DNA replication.

4.1.4 Mitomycin C

- Isolated from *Streptomyces caespitosus*, mitomycin C is used in the treatment of gastric and pancreatic cancers. It cross-links DNA, thereby inhibiting DNA synthesis and triggering cell death in cancer cells.

4.2 Marine-Derived Microbial Compounds

Marine environments, which constitute 95% of Earth's biosphere, are a rich and largely untapped source of bioactive compounds. Marine bacteria, cyanobacteria, actinomycetes, and fungi have yielded numerous anticancer agents with unique mechanisms of action (Jimeno et al., 2004). Some significant marine-derived microbial compounds include.

4.2.1 Sarcodictyin

- Derived from marine bacteria, sarcodictyin disrupts microtubule function, which is crucial for cell division. By inhibiting microtubule assembly, sarcodictyin effectively halts the proliferation of cancer cells.

4.2.2 Bryostatins

- Produced by the bacterial symbiont of the bryozoan *Bugula neritina*, bryostatins modulate protein kinase C, a key player in cell growth and differentiation. Bryostatins have shown potential in inducing apoptosis and inhibiting cancer cell growth.

4.2.3 Eleutherobin

- Isolated from marine organisms, eleutherobin acts similarly to taxol by stabilizing microtubules and preventing their depolymerization. This action inhibits cell division and induces apoptosis in cancer cells.

4.2.4 Discodermolide

- Another microtubule-stabilizing agent, discodermolide, is derived from marine sponges. It binds to tubulin and stabilizes microtubules, disrupting the mitotic process and leading to cancer cell death.

4.3 Symbiotic Relationships and Anticancer Toxins

Symbiotic relationships between marine animals and bacteria have also been a source of novel anticancer agents. These symbiotic bacteria often produce toxins that benefit their hosts by deterring predators or competitors, and these same toxins can have powerful anticancer effects (Gerwick & Fenner, 2013). For instance:

4.3.1 Macrolactin-A

- Produced by bacteria associated with the dinoflagellate *Noctiluca scintillans*, macrolactin-A inhibits the growth of melanoma cells. It disrupts the cell cycle and induces apoptosis in cancer cells.

4.3.2 Bryostatins

- As mentioned earlier, bryostatins from bryozoan symbionts have demonstrated significant potential in cancer treatment by inducing apoptosis and inhibiting cell proliferation (Hale et al., 2002).

4.4 The Diversity of Microbial Metabolites

Microbes produce an array of compounds with unique chemical structures and mechanisms of action that are not found in higher organisms. This diversity is a result of the various ecological niches that microbes inhabit, driving the evolution of specialized metabolites to survive and compete in their environments, which is why they have a vital role in drug discovery (Kapoor et al., 2022).

Microbial metabolites offer several advantages in drug development:

- **Novel Structures**: Many microbial compounds have complex and unique structures that are difficult to synthesize chemically. These novel structures can lead to new modes of action against cancer cells.
- **Multiple Mechanisms**: Microbial metabolites can target multiple pathways in cancer cells, reducing the likelihood of resistance development.
- **Synergistic Effects**: Some microbial compounds can be used in combination with other anticancer agents to enhance their efficacy and reduce side effects.

5 Recent Advances in Microbe-Mediated Cancer Therapy

Despite significant advancements in traditional cancer therapies, including chemotherapy, immunotherapy, radiation therapy, and surgery, cancer remains a leading cause of death worldwide. As researchers continue to seek more effective and targeted treatments, microbe-mediated cancer therapies have emerged as promising approaches to enhance treatment efficacy and specificity (Kapoor et al., 2022).

Bacterial cancer immunotherapy leverages the unique properties of bacteria to target and eliminate tumor cells. Bacteria can be engineered to preferentially infect tumor cells, allowing the immune system to recognize and attack these infected cells as if they were pathogenic invaders rather than malignant cells (Laliani et al., 2020). This approach can effectively turn the tumor into an infection site, drawing an immune response that can result in tumor regression.

Oncolytic virus therapy involves using naturally occurring or genetically modified viruses to selectively infect and kill cancer cells. One notable example is Imlygic (T-VEC), an FDA-approved oncolytic virus for melanoma treatment. T-VEC is a genetically modified herpes simplex virus designed to replicate within tumors and produce an immune-stimulating protein (Fukuhara et al., 2016). Combining oncolytic viruses with other immunotherapeutics has shown synergistic effects, enhancing overall treatment outcomes (Gujar et al., 2024).

Nanomedicine has revolutionized drug delivery systems by using nanoparticles (NPs) to create nanomedicines with improved pharmacokinetics and targeting capabilities. These nanomedicines can deliver therapeutic agents directly to tumor sites, minimizing systemic toxicity and enhancing treatment efficacy. Bacteria can be engineered to produce or carry nanoparticles, further integrating microbial and nanomedicine strategies (Fan et al., 2023).

Emerging research has highlighted the role of microbial communities in cancer initiation, progression, metastasis, and treatment response. The gut microbiota, in particular, plays a crucial role in modulating the efficacy of chemotherapy and maintaining anticancer immunity (Fernandes et al., 2022). Specific gut bacteria can influence drug metabolism, alter immune responses, and impact the overall success of cancer therapies.

6 Challenges and Future Directions in Anticancer Microbial Products

Current research is developing a wide range of prospective new treatment alternatives based on both the addition of fresh biological approaches and the modification of existing therapeutic modalities. Numerous strategies, including ribosome-targeted treatments, CRISPR-CAS gene editing, bacteriotherapy, hyperthermic intraperitoneal chemotherapy (HIPEC), and nanovaccines, reflect the dynamic nature of contemporary cancer care and offer a variety of weapons to combat this complicated disease. Although bacteria-mediated tumor therapy offers numerous benefits and may be able to overcome the drawbacks of conventional medicines, the clinical trials' failure indicates that there are still a number of obstacles to overcome. The obstacles include:

1. **Pathogenicity**:
   - Certain microorganisms, especially viruses and bacteria, can lead to infections or unfavorable immunological reactions in individuals. To avoid unintentional infections or adverse consequences, it is imperative to ensure the safety of attenuated or modified bacteria (Forbes et al., 2018).
2. **Toxicity**:
   - Minimizing the toxicity of microbiological agents while maintaining their therapeutic efficiency remains a significant challenge. For instance, even though oncolytic viruses are designed to specifically target cancer cells, they may still have negative effects on healthy tissues (Wechman et al., 2024).
3. **Immune Modulation**:
   - Patients may react differently due to the complex interactions between microbial pathogens and the host immune system. Predicting and regulating these interactions is crucial for maximizing the results of treatment (Knisely et al., 2023).
4. **Immunological Evasion**:
   - Cancer cells have the ability to build defenses against immunological recognition, which can reduce the efficacy of microbial treatments. To overcome this, immune evasion strategies are crucial for the success of these treatments (Forbes et al., 2018).
5. **Specificity**:
   - Ensuring that microbial agents target only tumor cells while sparing healthy tissues is one of the biggest

challenges. While advancements in genetic engineering are improving targeting specificity, there are still obstacles in properly delivering microorganisms to tumor locations.

6. **Bioavailability**:
   - Sustaining the viability and therapeutic activity of microbial therapies in the body requires careful consideration of various factors, including the route of administration, microbial survival in the bloodstream, and penetration into tumor tissues (Knisely et al., 2023).
7. **Regulatory and Manufacturing Challenges**:
   - For microbial therapies to be safe and effective across various batches and clinical settings, standardized production and quality control procedures must be developed. New microbial therapies must pass stringent testing and prove their safety and effectiveness to gain regulatory approval. The regulatory environment for these cutting-edge treatments is still developing, which presents challenges for developers (Knisely et al., 2023).

To address the limitations and adverse effects of bacteriotherapy, more thorough scientific research is needed as the concept of using bacteria as delivery systems for cancer treatment is still in its early phases (Din et al., 2023). By using complex synthetic bioengineering techniques or straightforward genetic manipulation, bacteria can be further designed to generate and distribute anticancer medicines in accordance with therapeutic requirements. Utilizing live tumor-targeting bacteria in therapeutic techniques can be used alone or in conjunction with other anticancer medicines to improve clinical outcomes (Duong et al., 2019).

7 Conclusion

While microbe-mediated cancer therapies hold great promise, several challenges need to be addressed to realize their full potential. Ensuring the safety, specificity, and efficacy of microbial treatments requires a multifaceted approach involving advanced genetic engineering, careful regulation, and innovative delivery methods. Interdisciplinary collaborations and continued research are essential to overcoming these challenges and harnessing the full therapeutic potential of microbial agents.

The resurgence of interest in microbial therapies underscores the potential of these approaches to provide more targeted, effective, and less toxic treatments for cancer patients. As research continues to advance, integrating microbial therapies with existing and emerging cancer treatments could pave the way for new therapeutic paradigms, offering hope for improved outcomes in the fight against cancer. By leveraging the unique properties of microbes and their interactions with the human immune system, scientists aim to develop innovative treatments that offer hope for improved outcomes in the fight against cancer.

References

Baindara, P., & Mandal, S. M. (2020). Bacteria and bacterial anticancer agents as a promising alternative for cancer therapeutics. *Biochimie, 177*, 164–189. https://doi.org/10.1016/j.biochi.2020.07.020.

Bickels, J., Kollender, Y., Merinsky, O., & Meller, I. (2002). Coley's toxin: Historical perspective. *The Israel Medical Association Journal: IMAJ, 4*(6), 471–472. https://www.ncbi.nlm.nih.gov/pubmed/12073431.

Blagosklonny, M. V. (2002). Hsp-90-associated oncoproteins: Multiple targets of geldanamycin and its analogs. *Leukemia, 16*(4), 455–462. https://doi.org/10.1038/sj.leu.2402415.

Boccia, S. M., Sassu, C. M., Ergasti, R., Vertechy, L., Apostol, A. I., Palluzzi, E., et al. (2024). Focus on trabectedin in ovarian cancer: What do we still need to know? *Drug Design, Development and Therapy, 18*, 2021–2032. https://doi.org/10.2147/DDDT.S451223.

Bradner, W. T. (2001). Mitomycin C: A clinical update. *Cancer Treatment Reviews, 27*(1), 35–50. https://doi.org/10.1053/ctrv.2000.0202.

Broadway, K. M., & Scharf, B. E. (2019). Salmonella typhimurium as an anticancer therapy: Recent advances and perspectives. *Current Clinical Microbiology Reports, 6*(4), 225–239. https://doi.org/10.1007/s40588-019-00132-5.

Cheng, Y., Wu, C., Wu, L., Zhao, J., Zhao, Y., Chen, L., et al. (2024). A pivotal bridging study of lurbinectedin as second-line therapy in Chinese patients with small cell lung cancer. *Science and Reports, 14*(1), 3598. https://doi.org/10.1038/s41598-024-54223-5.

Demain, A. L., & Vaishnav, P. (2011). Natural products for cancer chemotherapy. *Microbial Biotechnology, 4*(6), 687–699. https://doi.org/10.1111/j.1751-7915.2010.00221.x.

Din, S. R. U., Saeed, S., Khan, S. U., Arbi, F. M., Xuefang, G., & Zhong, M. (2023). Bacteria-driven cancer therapy: Exploring advancements and challenges. *Critical Reviews in Oncology Hematology, 191*, 104141. https://doi.org/10.1016/j.critrevonc.2023.104141.

Duong, M. T., Qin, Y., You, S. H., & Min, J. J. (2019). Bacteria-cancer interactions: Bacteria-based cancer therapy. *Experimental & Molecular Medicine, 51*(12), 1–15. https://doi.org/10.1038/s12276-019-0297-0.

Elsedawy, N. B., & Russell, S. J. (2013). Oncolytic vaccines. *Expert Review of Vaccines, 12*(10), 1155–1172. https://doi.org/10.1586/14760584.2013.836912.

Fan, D., Cao, Y., Cao, M., Wang, Y., Cao, Y., & Gong, T. (2023). Nanomedicine in cancer therapy. *Signal Transduction and Targeted Therapy, 8*(1), 293. https://doi.org/10.1038/s41392-023-01536-y.

Fernandes, M. R., Aggarwal, P., Costa, R. G. F., Cole, A. M., & Trinchieri, G. (2022). Targeting the gut microbiota for cancer therapy. *Nature Reviews Cancer, 22*(12), 703–722. https://doi.org/10.1038/s41568-022-00513-x.

Forbes, N. S., Coffin, R. S., Deng, L., Evgin, L., Fiering, S., Giacalone, M., et al. (2018). White paper on microbial anti-cancer therapy and prevention. *Journal for Immunotherapy of Cancer, 6*(1), 78. https://doi.org/10.1186/s40425-018-0381-3.

Fukuhara, H., Ino, Y., & Todo, T. (2016). Oncolytic virus therapy: A new era of cancer treatment at dawn. *Cancer Science, 107*(10), 1373–1379. https://doi.org/10.1111/cas.13027.

Gerwick, W. H., & Fenner, A. M. (2013). Drug discovery from marine microbes. *Microbial Ecology, 65*(4), 800–806. https://doi.org/10.1007/s00248-012-0169-9.

Gescher, A. (1998). Analogs of staurosporine: Potential anticancer drugs? *General Pharmacology, 31*(5), 721–728. https://doi.org/10.1016/s0306-3623(98)00069-x.

Griffin, D., Wittmann, S., Guo, F., Nimmanapalli, R., Bali, P., Wang, H. G., et al. (2003). Molecular determinants of epothilone B derivative (BMS 247550) and Apo-2L/TRAIL-induced apoptosis of human ovarian cancer cells. *Gynecologic Oncology, 89*(1), 37–47. https://doi.org/10.1016/s0090-8258(03)00006-4.

Gujar, S., Pol, J. G., Kumar, V., Lizarralde-Guerrero, M., Konda, P., Kroemer, G., et al. (2024). Tutorial: Design, production and testing of oncolytic viruses for cancer immunotherapy. *Nature Protocols*. https://doi.org/10.1038/s41596-024-00985-1.

Gulder, T. A., & Moore, B. S. (2010). Salinosporamide natural products: Potent 20 S proteasome inhibitors as promising cancer chemotherapeutics. *Angewandte Chemie (International Ed. in English), 49*(49), 9346–9367. https://doi.org/10.1002/anie.201000728.

Hale, K. J., Hummersone, M. G., Manaviazar, S., & Frigerio, M. (2002). The chemistry and biology of the bryostatin antitumour macrolides. *Natural Products Reports, 19*(4), 413–453. https://doi.org/10.1039/b009211h.

Harrison, M., & Swanton, C. (2008). Epothilones and new analogues of the microtubule modulators in taxane-resistant disease. *Expert Opinion on Investigational Drugs, 17*(4), 523–546. https://doi.org/10.1517/13543784.17.4.523.

Imai, T., Kojima, Y., Shimoi, T., Aiba, H., Okuma, H. S., Saito, A., et al. (2024). Predicting trabectedin efficacy in soft tissue sarcoma: Inflammatory biomarker analysis. *Anticancer Research, 44*(5), 2125–2132. https://doi.org/10.21873/anticanres.17018.

Jimeno, J., Faircloth, G., Sousa-Faro, J. F., Scheuer, P., & Rinehart, K. (2004). New marine derived anticancer therapeutics—a journey from the sea to clinical trials. *Marine Drugs, 2*(1), 14–29. https://www.mdpi.com/1660-3397/2/1/14.

Kapoor, R., Saini, A., & Sharma, D. (2022). Indispensable role of microbes in anticancer drugs and discovery trends. *Applied Microbiology and Biotechnology, 106*(13–16), 4885–4906. https://doi.org/10.1007/s00253-022-12046-2.

Kasinskas, R. W., & Forbes, N. S. (2007). Salmonella typhimurium lacking ribose chemoreceptors localize in tumor quiescence and induce apoptosis. *Cancer Research, 67*(7), 3201–3209. https://doi.org/10.1158/0008-5472.CAN-06-2618.

Kawai, K., & Akaza, H. (2003). Bleomycin-induced pulmonary toxicity in chemotherapy for testicular cancer. *Expert Opinion on Drug Safety, 2*(6), 587–596. https://doi.org/10.1517/14740338.2.6.587.

Kelly, E., & Russell, S. J. (2007). History of oncolytic viruses: Genesis to genetic engineering. *Molecular Therapy, 15*(4), 651–659. https://doi.org/10.1038/sj.mt.6300108.

Knisely, A., Seo, Y. D., Wargo, J. A., & Chelvanambi, M. (2023). Monitoring and modulating diet and gut microbes to enhance response and reduce toxicity to cancer treatment. *Cancers (Basel), 15*(3). https://doi.org/10.3390/cancers15030777.

Kollar, P., Rajchard, J., Balounova, Z., & Pazourek, J. (2014). Marine natural products: Bryostatins in preclinical and clinical studies. *Pharmaceutical Biology, 52*(2), 237–242. https://doi.org/10.3109/13880209.2013.804100.

Kortmansky, J., & Schwartz, G. K. (2003). Bryostatin-1: A novel PKC inhibitor in clinical development. *Cancer Investigation, 21*(6), 924–936. https://doi.org/10.1081/cnv-120025095.

Kurup, N., Patil, S., Rajnani, N., & Gandhi, R. (2024). Chapter 13-microbial metabolites with anticancerous properties. In V. Mishra, J. Mishra, & N. K. Arora (Eds.), *Bioactive microbial metabolites* (pp. 261–285). Academic Press. https://doi.org/10.1016/B978-0-443-18568-7.00008-2.

Laliani, G., Ghasemian Sorboni, S., Lari, R., Yaghoubi, A., Soleimanpour, S., Khazaei, M., et al. (2020). Bacteria and cancer: Different sides of the same coin. *Life Sciences, 246*, 117398. https://doi.org/10.1016/j.lfs.2020.117398.

Law, J. W., Law, L. N., Letchumanan, V., Tan, L. T., Wong, S. H., Chan, K. G., et al. (2020). Anticancer drug discovery from microbial sources: The unique mangrove streptomycetes. *Molecules, 25*(22). https://doi.org/10.3390/molecules25225365.

Leisch, M., Egle, A., & Greil, R. (2019). Plitidepsin: A potential new treatment for relapsed/refractory multiple myeloma. *Future Oncology, 15*(2), 109–120. https://doi.org/10.2217/fon-2018-0492.

Li, D., Liu, J., Wang, X., Kong, D., Du, W., Li, H., et al. (2018). Biological potential and mechanism of prodigiosin from serratia marcescens Subsp. lawsoniana in human choriocarcinoma and prostate cancer cell lines. *International Journal of Molecular Sciences, 19*(11). https://doi.org/10.3390/ijms19113465.

Liu, Y., Zhang, L., Zhu, X., Wang, Y., Liu, W., & Gong, W. (2015). Polysaccharide Agaricus blazei Murill stimulates myeloid derived suppressor cell differentiation from M2 to M1 type, which mediates inhibition of tumour immune-evasion via the toll-like receptor 2 pathway. *Immunology, 146*(3), 379–391. https://doi.org/10.1111/imm.12508.

Liu, X. F., Xiang, L., Zhou, Q., Carralot, J. P., Prunotto, M., Niederfellner, G., et al. (2016). Actinomycin D enhances killing of cancer cells by immunotoxin RG7787 through activation of the extrinsic pathway of apoptosis. *Proceedings of the National Academy of Sciences of the United States of America, 113*(38), 10666–10671. https://doi.org/10.1073/pnas.1611481113.

Lukasiewicz, K., & Fol, M. (2018). Microorganisms in the treatment of cancer: Advantages and limitations. *Journal of Immunology Research, 2018*, 2397808. https://doi.org/10.1155/2018/2397808.

Martinez-Montiel, N., Rosas-Murrieta, N. H., Martinez-Montiel, M., Gaspariano-Cholula, M. P., & Martinez-Contreras, R. D. (2016). Microbial and natural metabolites that inhibit splicing: A powerful alternative for cancer treatment. *BioMed Research International, 2016*, 3681094. https://doi.org/10.1155/2016/3681094.

Miyake, R., Yamanaka, S., Matsubara, S., & Mabuchi, S. (2021). Preclinical activity of plitidepsin against clear cell carcinoma of the ovary. *Anticancer Research, 41*(9), 4277–4285. https://doi.org/10.21873/anticanres.15232.

Mohan, C. D., Rangappa, S., Nayak, S. C., Jadimurthy, R., Wang, L., Sethi, G., et al. (2022). Bacteria as a treasure house of secondary metabolites with anticancer potential. *Seminars in Cancer Biology, 86*(Pt 2), 998–1013. https://doi.org/10.1016/j.semcancer.2021.05.006.

Nallar, S. C., Xu, D. Q., & Kalvakolanu, D. V. (2017). Bacteria and genetically modified bacteria as cancer therapeutics: Current advances and challenges. *Cytokine, 89*, 160–172. https://doi.org/10.1016/j.cyto.2016.01.002.

Neckers, L. (2006). Chaperoning oncogenes: Hsp90 as a target of geldanamycin. *Handbook of Experimental Pharmacology, 172*, 259–277. https://doi.org/10.1007/3-540-29717-0_11.

Patil, Y., Amitay, Y., Ohana, P., Shmeeda, H., & Gabizon, A. (2016). Targeting of pegylated liposomal mitomycin-C prodrug to the folate receptor of cancer cells: Intracellular activation and enhanced cytotoxicity. *Journal of Controlled Release, 225*, 87–95. https://doi.org/10.1016/j.jconrel.2016.01.039.

Pelner, L., Fowler, G. A., & Nauts, H. C. (1958). Effects of concurrent infections and their toxins on the course of leukemia. *Acta Medica Scandinavica. Supplementum, 338*, 1–47. https://www.ncbi.nlm.nih.gov/pubmed/13605619.

Pham, H. M., Le, D. T., Le, L. T., Chu, P. T. M., Tran, L. H., Pham, T. T., et al. (2023). A highly quality genome sequence of Penicillium oxalicum species isolated from the root of Ixora chinensis in

Vietnam. *G3 (Bethesda), 13*(2). https://doi.org/10.1093/g3journal/jkac300.

Ricart, A. D. (2011). Antibody-drug conjugates of calicheamicin derivative: Gemtuzumab ozogamicin and inotuzumab ozogamicin. *Clinical Cancer Research, 17*(20), 6417–6427. https://doi.org/10.1158/1078-0432.CCR-11-0486.

Rivankar, S. (2014). An overview of doxorubicin formulations in cancer therapy. *Journal of Cancer Research and Therapeutics, 10*(4), 853–858. https://doi.org/10.4103/0973-1482.139267.

Safarpour, A., Ebrahimi, M., Shahzadeh Fazeli, S. A., & Amoozegar, M. A. (2019). Supernatant metabolites from halophilic archaea to reduce tumorigenesis in prostate cancer in-vitro and in-vivo. *Iranian Journal of Pharmaceutical Research: IJPR, 18*(1), 241–253. https://www.ncbi.nlm.nih.gov/pubmed/31089359.

Ser, H. L., Ab Mutalib, N. S., Yin, W. F., Chan, K. G., Goh, B. H., & Lee, L. H. (2015). Evaluation of antioxidative and cytotoxic activities of streptomyces pluripotens MUSC 137 isolated from mangrove soil in Malaysia. *Frontiers in Microbiology, 6*, 1398. https://doi.org/10.3389/fmicb.2015.01398.

Shin, D. M., Holoye, P. Y., Forman, A., Winn, R., Perez-Soler, R., Dakhil, S., et al. (1994). Phase II clinical trial of didemnin B in previously treated small cell lung cancer. *Investigational New Drugs, 12*(3), 243–249. https://doi.org/10.1007/BF00873966.

Simidu, U., Kita-Tsukamoto, K., Yasumoto, T., & Yotsu, M. (1990). Taxonomy of four marine bacterial strains that produce tetrodotoxin. *International Journal of Systematic Bacteriology, 40*(4), 331–336. https://doi.org/10.1099/00207713-40-4-331.

Sohail, M., Sun, Z., Li, Y., Gu, X., & Xu, H. (2021). Research progress in strategies to improve the efficacy and safety of doxorubicin for cancer chemotherapy. *Expert Review of Anticancer Therapy, 21*(12), 1385–1398. https://doi.org/10.1080/14737140.2021.1991316.

Tan, L. T., Ser, H. L., Yin, W. F., Chan, K. G., Lee, L. H., & Goh, B. H. (2015). Investigation of antioxidative and anticancer potentials of streptomyces sp. MUM256 isolated from Malaysia mangrove soil. *Frontiers in Microbiology, 6*, 1316. https://doi.org/10.3389/fmicb.2015.01316.

Tan, L. T., Chan, C. K., Chan, K. G., Pusparajah, P., Khan, T. M., Ser, H. L., et al. (2019). Streptomyces sp. MUM256: A source for apoptosis inducing and cell cycle-arresting bioactive compounds against colon cancer cells. *Cancers (Basel), 11*(11). https://doi.org/10.3390/cancers11111742.

Tian, J., Ma, J., Ma, K., Guo, H., Baidoo, S. E., Zhang, Y., et al. (2013). Beta-Glucan enhances antitumor immune responses by regulating differentiation and function of monocytic myeloid-derived suppressor cells. *European Journal of Immunology, 43*(5), 1220–1230. https://doi.org/10.1002/eji.201242841.

Ventura, M., Canchaya, C., Tauch, A., Chandra, G., Fitzgerald, G. F., Chater, K. F., et al. (2007). Genomics of Actinobacteria: Tracing the evolutionary history of an ancient phylum. *Microbiology and Molecular Biology Reviews, 71*(3), 495–548. https://doi.org/10.1128/MMBR.00005-07.

Wainwright, E. N., & Scaffidi, P. (2017). Epigenetics and cancer stem cells: Unleashing, hijacking, and restricting cellular plasticity. *Trends Cancer, 3*(5), 372–386. https://doi.org/10.1016/j.trecan.2017.04.004.

Wang, Z., Li, B., Zhou, L., Yu, S., Su, Z., Song, J., et al. (2016). Prodigiosin inhibits Wnt/beta-catenin signaling and exerts anticancer activity in breast cancer cells. *Proc Natl Acad Sci U S A, 113*(46), 13150–13155. https://doi.org/10.1073/pnas.1616336113.

Wechman, S., Jala, V. R., Bermudez-Humaran, L. G., & Gomez-Gutierrez, J. G. (2024). Editorial: Recombinant microbes-mediated cancer immunotherapy. *Frontiers in Immunology, 15*, 1386574. https://doi.org/10.3389/fimmu.2024.1386574.

Wood, L. M., & Paterson, Y. (2014). Attenuated Listeria monocytogenes: A powerful and versatile vector for the future of tumor immunotherapy. *Frontiers in Cellular and Infection Microbiology, 4*, 51. https://doi.org/10.3389/fcimb.2014.00051.

Xu, Y., Kersten, R. D., Nam, S. J., Lu, L., Al-Suwailem, A. M., Zheng, H., et al. (2012). Bacterial biosynthesis and maturation of the didemnin anti-cancer agents. *Journal of the American Chemical Society, 134*(20), 8625–8632. https://doi.org/10.1021/ja301735a.

Yadav, S. S., Prasad, C. B., Prasad, S. B., Pandey, L. K., Singh, S., Pradhan, S., et al. (2015). Anti-tumor activity of staurosporine in the tumor microenvironment of cervical cancer: An in vitro study. *Life Sciences, 133*, 21–28. https://doi.org/10.1016/j.lfs.2015.04.019.

Interventional Studies of Bioactive Compounds and Their Extracts on Metabolic Disorders

Luxita Sharma and Kajal Dhama

Abstract

Bioactive compounds are both essential and nonessential compounds such as vitamins or polyphenols that naturally occur as a part of the food chain in nature and can be shown to affect human health. The presence of these compounds in food or dietary supplements can positively or negatively impact living organisms' health. Effects are contingent upon the particular bioactive material, dosage, bioavailability, and concentration within tissues, cells, or the entire organism. Usually formed as secondary metabolites, these chemicals are involved in signaling and defense and can have pharmacological effects on humans. Bioactive compounds show promising potential as a supplementary nutritional intervention for addressing key parameters related to Metabolic Syndrome (MetS) and may play a crucial role in preventing diseases in high-risk individuals. The primary therapeutic intervention for MetS is the alteration of dietary along with lifestyle habits. Metabolic syndrome which is donated as (MetS) is a complex condition that is characterized by abdominal obesity, insulin resistance, hypertension, hyperglycemia, and dyslipidemia that can be the harmful factor leading to various diseases including type 2 diabetes, Polycystic ovary syndrome, cardiovascular disease, stroke, cancers, and chronic kidney diseases. The estimated prevalence is to be around 20–25% of the adult world's population who have been suffering from MetS, Following, the use of functional foods which are a rich source of antioxidants, along with the assumption of bioactive compounds, that could be helpful to show the beneficial effects on reducing body weight, glucose metabolism control, blood pressure, the improvement of lipid profile, endothelial damage, on the inflammatory state, along with oxidative stress.

Keywords

Metabolic syndrome • Functional foods • Bioactive compounds • Obesity • Insulin resistance

1 Introduction

Bioactive compounds are both essential and nonessential compounds such as vitamins or polyphenols that naturally occur as a part of the food chain in nature and can be shown to impact human health (Biesalski et al., 2009). These substances can have a beneficial or harmful effect on the health of living things when they are present in food or dietary supplements. Increased consumption of bioactive compounds can help to modify the impact or risk of various chronic diseases such as cancer and cardiovascular disease as the compounds that are most extensively studied are the antioxidants (Biesalski et al., 2009). Effects are contingent upon the particular bioactive material, dosage, bioavailability, and concentration within tissues, cells, or the entire organism. Usually formed as secondary metabolites, these chemicals are involved in signaling and defense and can have pharmacological effects on humans. Because of their therapeutic effects, plant-based medications, or phytomedicines, are becoming more and more well-liked as an alternative to synthetic drugs as they have their therapeutic effect (Suarez et al., 2012).

The advantages of various medicinal plants are due to their minimal side effects, coupled with their widespread acceptability and cost-effectiveness. These plants harbor varied therapeutic properties, including antioxidants, anti-inflammatory, antimicrobial, anti-malarial, anti-carcinogenic, anticholinergic, anti-leprosy, hypoglycemic, and antidiabetic effects, attributed to their bioactive principles (Bowling et al., 2012). Incorporation of various bioactive compounds that include carotenoids, antioxidants, flavors, and essential oils into food products

L. Sharma (✉) · K. Dhama
Department of Dietetics and Applied Nutrition, Amity Medical School, Amity University Haryana, Gurgaon, India
e-mail: lsharma@ggn.amity.edu

J. Devi (ed.), *Biotechnological Intervention in Production of Bioactive Compounds*,
Sustainable Landscape Planning and Natural Resources Management, https://doi.org/10.1007/978-3-031-76859-0_12

not only bestows them with their nutritional along with health properties but also serves to enhance the sensory characteristics along with modulate metabolic processes in the body of human beings (Maharaj, 2015). Notwithstanding, no recommended daily intake values are prescribed for bioactive compounds, necessitating the establishment of safe intake levels due to their potential impact or effect on human health (Maria, 2012). A wide array of bioactive compounds is sourced from fruits, vegetables, and whole grains, encompassing an assorted class, including organosulfur compounds, phytosterols, carotenoids, and tocopherols in varying concentrations (Šnē et al., 2013). Furthermore, bioactive compounds are not limited to plants and can also be derived from different types of living organisms and microorganisms, which play a pivotal role in producing secondary metabolites (Debbab et al., 2010).

Despite consuming adequate micronutrients, some individuals may still be at a heightened risk for various types of diseases including cancer or coronary heart disease due to genetic variations, known as "polymorphisms," in enzymes that are found to be responsible for the metabolism or in the distribution of micronutrients. These genetic variations can result in lower tissue concentrations or reduced functionality of the micronutrients. To illustrate, a common single base-pair exchange is involved in the methylenetetrahydrofolate reductase gene can lead to decreased enzyme activity and elevated plasma homocysteine levels, which are linked with the development of atherosclerosis and coronary artery disease (Boushey et al., 1995). In such cases, a folate intake that exceeds the standard recommendation of micronutrients may offer protection against coronary heart disease. Another common genetic variation, involving a single nucleotide substitution from cytosine to thymine at position 677 of the methylenetetrahydrofolate reductase gene, results in a thermolabile enzyme with reduced activity, leading to elevated plasma homocysteine levels (Kong et al., 2002). This genetic mutation has been associated or linked to an increased risk of developing coronary artery disease (Morita et al., 1997). However, augmenting dietary folate intake beyond the standard recommendation may aid in attenuating the risk factor of coronary heart disease among individuals harboring this genetic variant and exhibiting elevated homocysteine levels (Barnes, 2003).

Bioactive compounds show promising potential as a supplementary nutritional intervention for addressing key parameters related to Metabolic Syndrome (MetS) and may play a crucial role in preventing diseases in high-risk individuals. The primary therapeutic intervention for MetS is the alteration of dietary along with lifestyle habits. It is well-established that improvements in dietary patterns, including reduced caloric intake for overweight and obese individuals, as well as decreased consumption of sodium, saturated fats, simple sugars, and cholesterol can effectively manage the comorbidities associated with Metabolic syndrome (Di Daniele et al., 2021). In certain cases, pharmacological therapy may be necessary. However, as there is no single targeted drug for MetS, a multi-drug approach is often required, which can lead to side effects and reduced patient adherence. Consequently, it is crucial to explore natural, side-effect-free interventions that can benefit the patients of MetS to delay or reduce the onset and progression of the characteristic comorbidities associated with the syndrome (Rochlani et al., 2017).

Metabolic Syndrome

Metabolic syndrome which is donated as (MetS) contains a complex series of conditions that is characterized by insulin resistance, hypertension, abdominal obesity, hyperglycemia, and dyslipidemia that is considered to be the risk-causing factor for various diseases including stroke, type 2 diabetes, cardiovascular disease, cancers, and chronic kidney diseases (Grundy et al., 2004). Its prevalence is day by day increasing as there is an increase in cases of obesity and it is becoming the reason for reaching epidemic proportions that is responsible for affecting between 24 and 34% adult population in the US (Aguilar et al., 2015). "The prevalence is calculated to be around 20–25% of the adult population across the globe who has been suffering from MetS which can be defined as: "a cluster of the most dangerous heart attack risk factors: diabetes and prediabetes along with abdominal obesity, high cholesterol level, and high blood pressure" (Cornier et al., 2008). Insulin resistance along with central obesity are considered a significant contributor to MetS but its etiology is still a matter of research. Although, the occurrence of insulin resistance, leads to an enhancement in the fasting glucose along with the impaired glucose tolerance that is because of the abnormal expression of enzymes such as gluconeogenic enzymes. This leads to a metabolic state that further induces the release of insulin which ultimately results in hyperinsulinemia. The transcription factor is then stimulated by hyperinsulinemia that includes the Srebp-1c in the liver which is responsible for driving hepatic steatosis and hypertriglyceridemia. Excessive production and release by pancreatic β-cells that release insulin can result in their exhaustion and death, triggering the development of type 2 diabetes. The most common form of insulin resistance is linked to the dysfunction of adipose tissue as well and abdominal obesity plays a significant central role in Metabolic Syndrome (MetS) shown in Fig. 1 (Heindel et al., 2017).

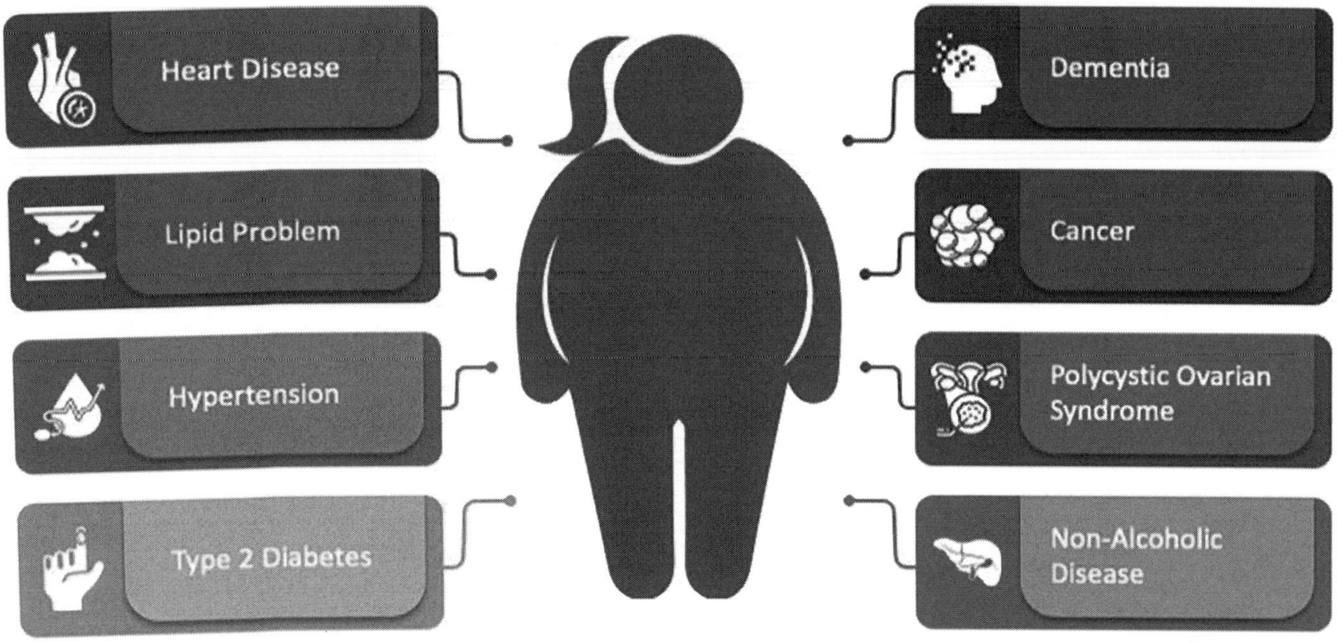

Fig. 1 Metabolic disorders

Endocrine Disrupting Chemicals Induced Metabolic Disorders

Endocrine-disrupting chemicals (EDCs) are substances that are found in nature or the environment such as in air, or water supply and soil also present in various food sources, and manufactured products along with personal care products that are responsible for interfering with the normal functioning of the endocrine system in the body. Over the past few decades, there has been a great focus on investigating the association between disease and human exposure to EDCs (Bergman et al., 2013). Focus is given to organic pollutants that are man-made and that are also considered persistent organic pollutants (POPs). However, due to the lipophilic nature of POPs, they can be easily deposited in fatty tissues. However, due to which their exposure to these endocrine-disrupting chemicals leads to the disruption of normal energy homeostasis control. The function of the thyroid also becomes impaired resulting in a reduction of the metabolism of lipids, carbohydrates, along with proteins. Hence, due to the ability of obesogenic and metabolic dysregulation is responsible for leading the development of different types of metabolic disorders in human beings (Mallhi et al., 2021).

2 Classification of Bioactive Compounds

Bioactive compounds can be classified based on their chemical structure. Primary classes of these compounds include carotenoids, nitrogen-containing compounds, phenolics, alkaloids, and organosulfur shown in Fig. 2 (Maharaj, 2015).

3 Metabolic Diseases and Bioactive Compounds for Their Treatment (Interventional Studies)

3.1 Obesity

Obesity has become a global epidemic that is responsible for affecting every phase of life such as infancy, childhood, and adulthood around the world. The prevalence of this disorder globally has doubled in the past few decades and the highest record is found in the US in human history (Ogden et al., 2015). For the very first time in the world, it is recorded that the number of people who are obese and overweight across the world is greater than the number of underweight people.

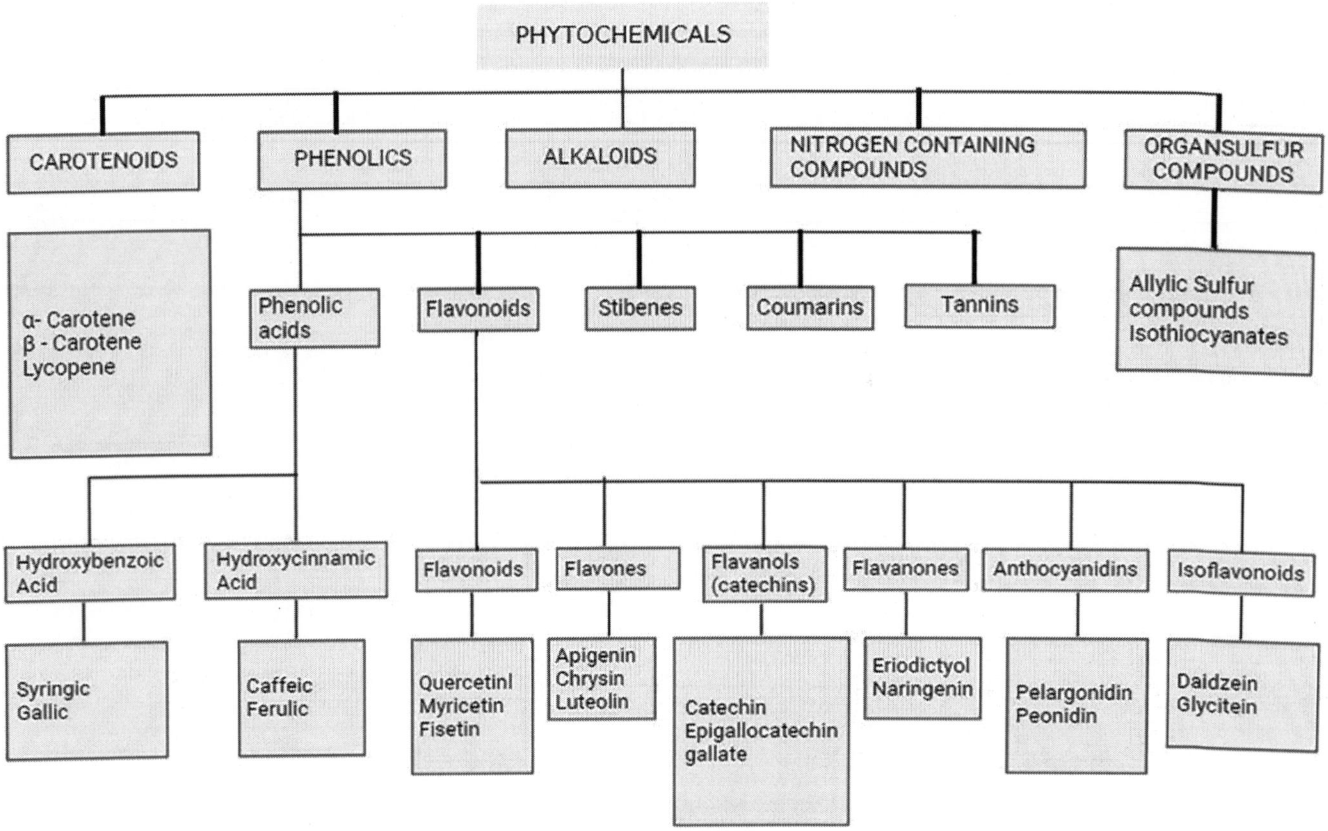

Fig. 2 Classification of bioactive compounds

A dramatic rise has been observed in the abdominal obesity rate in both developed and developing countries (Legler et al., 2015).

Obesity, defined as an excess or abnormal deposition of body fat whose body mass index is 30 kg/m^2 or higher, can be influenced by certain chemicals known as obesogens (Betoret et al., 2011). The EDCs—endocrine-disrupting chemicals interfere with the normal connection between hormones and their receptors, leading to alterations in glucose sensitivity and lipid metabolism. These physiological changes can predispose individuals to weight gain (Chen et al., 2017).

3.1.1 Bioactive Compounds for Treatment of Obesity

a. Curcumin: It a natural phenolic compound that is derived from the Curcuma longa plant has shown its potential in modulating obesity pathways. According to the studies, it has shown that the intake of curcumin, when combined with a low-calorie diet, can lead to increased weight loss, along with body mass index, body circumferences, and reduced body fat percentage (Noce et al., 2021). Recent in vitro research conducted by Wu et al. has revealed that curcumin is capable of inhibiting the differentiation of adipocytes (fat cells) and concurrently inducing apoptosis (programmed cell death) in preadipocytes (immature fat cells) (Wu et al., 2019). Furthermore, curcumin exhibits anti-inflammatory properties by stimulating the production of adiponectin, a potent anti-inflammatory agent secreted by the adipose tissue (Panahi et al., 2016). However, this results in improving the chronic low-grade inflammatory state often observed in obese individuals with metabolic syndrome. However, pharmacokinetic studies have demonstrated that curcumin bioavailability after oral administration is relatively low which is primarily because of the poor absorption in intestinal and substantial hepatic metabolism in the body (Sharma et al., 2007). These issues could potentially be tackled by using inhibitors that target curcumin metabolism or by creating solid or liquid forms of curcumin that enhance the absorption of curcumin and help slow down its elimination. It has been shown that the safe and effective use of curcumin extracts has not led to any toxic effects, even with a chronic daily intake of up to 180 mg of curcumin consumption (Sharma et al., 2001).

b. Fiber: Both insoluble and soluble fibers can reduce blood cholesterol and the levels of sugar. The Prebiotic fiber can increase the metabolism of fat and glucose after being broken down into short-chain fatty acids (Chen et al., 2017).
c. Monounsaturated and Polyunsaturated Fats: The plant sterols, essential fatty acids, monounsaturated along with polyunsaturated fats have been shown to be effective in enhancing the level of high-density lipoprotein while reducing the level of total cholesterol. This effect is observed in comparison to saturated fatty acids along with trans-fatty acids (Khan et al., 2013).
d. Vitamin C and Vitamin E: Vitamin C and Vitamin E intake can help decrease lipid peroxidation and pro-inflammatory cytokines, thereby reducing the chances of vascular disease. The inadequate intake of Vitamin C can contribute to an increase in the circulating lipopolysaccharides that is due to intestinal bacteria overgrowth (Mallhi et al., 2021).
e. Bioactive Peptides: These can scavenge the free radicals that help to reduce cholesterol, along with regulating blood pressure, and suppressing appetite in the case of obesity. During the process of ripening in cheese, the strong concentration of antihypertensive agents, such as lactotripeptides also increases (Mallhi et al., 2021).
f. Minerals: Several minerals, such as calcium, have been shown a potential role in obesity management (Zheng et al., 2009). Calcium intake can increase the loss of fat mass in people who are obese and overweight, while low intake of calcium can result in decreasing lipolysis, leading to increased weight (Zemel, 2004).
g. Phytochemicals: Bioactive compounds that include the triterpenes, carotenoids, and polyphenols, exhibit significant anti-adiposity, antioxidative, and cardioprotective properties. Some of the phytochemicals can act as thermogenic compounds that can help to prevent the deposition of fat in the body tissues by burning the extra calories (Williams et al., 2013).
h. Probiotics: The microorganisms that help to maintain the intestinal flora balance in the gut and enhance nutrient digestibility are referred to as probiotics. These belong to the Lactobacillus and Bifidobacterium genera and have proven to be effective in addressing obesity (Mallhi et al., 2021).

3.2 Type 2 Diabetes

Diabetes Mellitus, according to the American Diabetes Association (2010), encompasses a cluster of metabolic diseases that are characterized by the term hyperglycemia caused by deficiencies in insulin action, insulin secretion, or both. The onset of this condition can be attributed to the deterioration of pancreatic tissue function and/or mass. Type 2 Diabetes constitutes around 90–95% of all cases of diabetes and is defined by an increase in insulin resistance and dysfunction of the pancreatic beta cells. Notably, in the United States, it is found that more than 11% of individuals who are age 20 and above have been diagnosed or undiagnosed with this disease, and around 35% of additional people are estimated to be pre-diabetic (Thayer et al., 2012). Globally, as per the World Health Organization reports approximately 347 million individuals are affected by the disease. Furthermore, there has been a significant surge in the prevalence of this condition among adolescents and even children within a short timeframe.

3.2.1 Bioactive Compounds in the Treatment of Type 2 Diabetes Mellitus

Bioactive compounds that are been suggested as potential contributors to the reduction in the disease occurrence and lead to the progression of different diseases, including Type 2 Diabetes Mellitus (T2DM). These compounds offer the advantage of safety, easy availability, and minimal side effects. Notably, prebiotics, polyphenols, phytosterols, and carotenoids are recognized as main bioactive compounds showing promising activity against T2DM (Mallhi et al., 2021).

Curcuma, commonly known as turmeric, appears to play a significant regulatory role in insulin resistance (IR) and type 2 diabetes mellitus due to its well-documented anti-inflammatory, antioxidant, and antidiabetic properties (Adibian et al., 2019 & Pivari et al., 2019). Studies in animals have revealed curcumin's hypoglycemic effects and shown the improvement in insulin sensitivity that is followed by its administration. Curcumin's hypoglycemic actions are believed to be attributed to hepatic gluconeogenesis inhibition, independent of the pathway of insulin. However, its antihyperglycemic effects have been linked with actions like anti-inflammatory, including tumor necrosis factor-alpha (TNF-α) reduction, and its anti-lipolytic effects, resulting in decreased circulating free fatty acid levels (Noce et al., 2021). The study done on hamsters has demonstrated an effective reduction in the homeostatic model assessment for the level of insulin resistance and leptin in animals that are treated with curcumin for 10 days in comparison with the control group (Gutierres et al., 2012).

Research involving human participants has indicated that turmeric has the potential to effectively reduce the levels of plasma glucose along with improving the activity of insulin. One study, which involved 240 prediabetic individuals divided into a treatment and control group, revealed that none of the participants receiving (250 g/day for 9 months) curcumin in the Type 2 diabetes mellitus development and also demonstrated a significantly greater score of homeostatic model assessment for the function of β-cell (Chuengsamarn et al., 2012). In a similar study, researchers also identified the effects of administering curcuminoids 300 mg/day for 3 months in patients who are overweight or obese diabetics. In comparison to the control group receiving a placebo group for a similar duration, the treatment group showed a substantial reduction in blood glucose, HOMA-IR, and hemoglobin A1c (a measure of insulin resistance) (Na et al., 2012).

Quercetin, a flavonoid present in foods such as broccoli, apples, tea, and red onions, possesses antioxidant, anti-apoptotic, and anti-inflammatory properties. It enhances the secretion of glucose-induced insulin, thereby safeguarding β-cell function in the body and modulating the phosphorylation of extracellular signal-regulated kinase (Mallhi et al., 2021).

Likewise, resveratrol, found in red wines, grape skins, and grape seeds, exhibits antioxidant and anti-inflammatory traits, regulates the homeostasis of metals, and enhances the function of mitochondria. It enhances the secretion of glucose-stimulated insulin along with the metabolism of glucose in the cells that secrete insulin (Granzotto & Zatta et al., 2014).

Genistein, an isoflavone present in fava beans, soybeans, and chickpeas, offers numerous health advantages. The genistein dietary supplementation helps to improve glucose tolerance, augments β-cell mass, along with decreasing apoptosis.

Vitamins—Vitamin A has been found effective in the T2DM treatment as it can regenerate β cells and enhance the sensitivity due to insulin and Vitamin D helps to reduce the chances of T2DM development by modulating the function of β cells and increasing the sensitivity of insulin, while also reducing the inflammation. Additionally, Vitamin E has been demonstrated to enhance the activities caused by the antioxidant enzymes that includes catalase, glutathione reductase along with glutathione peroxidase, thereby significantly reducing glucose levels (Mallhi et al., 2021).

Rutin, a flavonoid with antioxidant, antihyperglycemic, and neuroprotective properties, is present in various vegetables and fruits including asparagus, buckwheat, figs, and apples. It has been shown to improve oxidative stress, reduce serum lipids along with decreasing serum glucose levels (Ahmed et al., 2010).

Lycopene, a carotenoid found in pink grapefruit and tomatoes, exhibits antidiabetic effects. It decreases the peroxidation of lipids and increases the antioxidant activity of enzymes, which is responsible for increasing levels of insulin, decreasing levels of glucose, and improving serum lipid profiles. Lycopene also protects the kidneys against diabetic-induced morphological changes.

Hesperidin, a compound present in citrus fruits like lemons and oranges, has been shown to have a beneficial impact on type 2 diabetes mellitus treatment. It enhances the serum insulin and decreases the level of glucose, glycated hemoglobin, vitamin C, along vitamin E levels. Additionally, hesperidin exhibits anti-inflammatory, antioxidant, and antidepressant properties (Mallhi et al., 2021).

3.3 Hypertension

Hypertension is traditionally defined as when the blood pressure exceeds a threshold level of 120/80 mm Hg. Numerous studies have demonstrated the impact caused by endocrine-disrupting chemicals (EDCs) on hypertension. Higher systolic blood pressure in childhood and adolescence has been linked to phthalate exposure (Andersen et al., 2012). Increased exposure to pesticides in children has also been associated with elevated blood pressure. The potential mechanisms underlying the association between EDC compounds and hypertension lead to an increase the oxidative stress and pro-inflammatory responses. Phthalates lead to activate the peroxisome-proliferator-activated receptors by antagonizing thyroid hormone, cortisol, and androgen signaling. Additionally, epigenetic changes are included as an indicator of the underlying cause of EDC-induced hypertension (Mallhi et al., 2021).

3.3.1 Bioactive Compounds in Hypertension

Garlic has been extensively studied for its hypotensive, or blood pressure-lowering, properties. Research indicates that garlic consumption can lead to reductions in both that is in systolic and diastolic blood pressure (Xiong et al., 2015). A study has shown the effects of consuming garlic in a population with Metabolic Syndrome. Participants were administered 100 mg/kilogram of body weight of garlic that was in the raw and crushed form twice daily for 4 weeks, leading to better improvements in the metabolism of carbohydrates, lipid profile, and blood pressure levels. This study illustrated the potential of garlic intake in preventing and mitigating MetS (Choudhary et al., 2018). Garlic has been shown to enhance endothelial nitric oxide synthase (eNOS) activity along with reducing the expression of vascular cell adhesion molecule 1, which contributes to decreased burning sensation and oxidative stress, thereby offering protection against endothelial damage and improvements in blood pressure values (Al Disi et al., 2016).

Cinnamon appears to show a positive modulatory effect on the metabolic syndrome risk factors and exerts an important antihypertensive action (Mollazadeh & Hosseinzadeh, 2016). Several studies have shown and evaluated this effect, particularly in the cases of prediabetic or type 2 diabetic individuals, demonstrating a decrease in blood pressure and enhanced vasodilation (Akilen et al., 2013). The proposed action mechanism relates to the activation of chemosensory cation channels, which can stimulate vasodilatory and vasorelaxant responses (Alvarez-Collazo et al., 2014). A study conducted in metabolic syndrome patients reported a significant decrease in blood pressure following cinnamon supplementation, resulting in a consequent reduction in metabolic syndrome prevalence in comparison with the control group (Gupta Jain et al., 2017).

Potassium supplementation has been found to decrease the level of blood pressure in patients suffering from hypertension with potassium deficiency. Potassium stimulates the Na+/K+ ATPase in adrenergic nerve terminals and smooth muscles, triggering vasodilation, and is considered to be a useful mediator in increase of blood pressure.

Amaranth oil has been shown to positively impact biomarkers of hypertension, resulting in reduced triglycerides, LDL, and cholesterol. This effect is attributed to the replacement of saturated fatty acids with unsaturated fats. The compound present in Cocoa (polyphenols) can help lower blood pressure by improving endothelial function, inhibiting platelet aggregation, and promoting blood pressure reduction through the signaling of nitric oxide. The improved function of endothelial is particularly due to polyphenols in dark chocolate, contributing to the reduction in blood pressure (Mallhi et al., 2021).

Lycopene, a powerful antioxidant present in tomatoes, is found to be effective in hypertension management. Lycopene inactivates free radicals and improves endothelial function along with reducing the low-density lipoproteins susceptibility to oxidation (Haddy et al., 2006).

Vitamin C, also known as ascorbic acid, contributes to blood pressure reduction by enhancing endothelial function, achieved through the scavenging of free radicals within the vasculature. Additionally, the vitamin C combination with the antihypertensive drug amlodipine produces better results compared to amlodipine alone. Vitamin E, which is present in almonds, spinach, and sunflower seeds, also exhibits antihypertensive properties by increasing intracellular magnesium levels. The decrease in the level of blood pressure can be linked to hypovolemia and diarrhea induced by magnesium. Furthermore, vitamin E increases the diuretic effect by replenishing the levels of magnesium and potassium in the body (Mallhi et al., 2021).

3.4 Nonalcoholic Fatty Liver Disease (NAFLD) and Hyperlipidemia

The liver is pivotal in lipid metabolism, and any disruption in this function can lead to the onset of nonalcoholic fatty liver disease (NAFLD). However, NAFLD can be characterized by triglyceride accumulation, steatosis, and its strong association with obesity and metabolic syndrome. It is a worldwide health concern, affecting roughly 25% of the world's population and around 8% of children (Younossi et al., 2016). Both NAFLD and its severe form, nonalcoholic steatohepatitis, contribute to increased liver-related and overall mortality and pose a significant risk for cardiovascular disease. Hyperlipidemia, characterized by elevated levels of triglycerides (hypertriglyceridemia), phospholipids, and cholesterol (hypercholesterolemia), is a primary metabolic condition associated with NAFLD. In the United States, approximately 25% of adults have hypertriglyceridemia, marking a decrease from 33% in 2001–2004 (Carrol et al., 2015). This reduction has been ascribed to heightened awareness and the utilization of medications for lowering lipid profiles. Furthermore, there are substantial prevalence rates of hyperlipidemia among US adolescents and children, standing at 20% (Heindel et al., 2017).

3.4.1 Bioactive Compounds Against NAFLD

Bioactive Compounds with Therapeutic Potential Against the NAFLD. Bioactive natural compounds that include carotenoids, flavonoids, and polyphenols present in common dietary sources have shown beneficial effects due to their antioxidant, anti-inflammatory, and metabolic activities.

A study explored extract effects that are obtained from walnut leaves (Juglans regia L.) in patients with type 2 diabetes mellitus (T2DM). The participants received two capsules of the extract (100 mg Juglans regia leaf per dose) for 3 months daily alongside their standard pharmacological therapy. Following the 3 months, the researchers observed a notable fasting blood glucose (FBG) reduction and glycated hemoglobin levels, along with an enhancement in the lipid profile, in comparison with the control group that received a placebo for a similar duration (Hosseini et al., 2014). These outcomes indicate that the use of this herbal supplement could potentially complement standard treatment for hypoglycemic pharmacology, thereby ameliorating the harmful factors that are linked with metabolic syndrome (MetS) and potentially delaying its onset.

Anthocyanins are the pigments that are water-soluble present in colored berries, fruits, and vegetables, that can protect liver cells from injury by improving the status of antioxidants and inhibiting the mitochondrial pathways of apoptosis.

Coumestrol, a natural organic compound present in various plant-based foods, has been shown to enhance the oxidation of fatty acids in muscle cells along with decreasing the lipogenic gene expression in liver cells.

Curcumin, a polyphenol derived from the turmeric rhizome (Curcuma longa), possesses an antioxidant, along with anti-inflammatory, and anti-mutagenic properties. It can reduce oxidative stress and lipogenesis while facilitating beta-oxidation (Mallhi et al., 2021).

Astaxanthin, a carotenoid found in certain aquatic organisms, protects against diseases of fatty liver by reducing the accumulation of lipids and peroxidation of lipids in the liver, as well as downregulating genes involved in lipogenesis (Newbold et al., 2007).

Green tea flavonoids, including epicatechin gallate, epicatechin, epigallocatechin gallate, and epigallocatechin, as well as other flavonoids like quercetin and rutin, have been found to enhance the metabolism of lipids through the stimulation oxidation of fatty acid and inhibits the process of lipogenesis, in addition to exerting antioxidant effects.

Quercetin, a flavonoid compound, has been shown to enhance lipid metabolism by promoting fatty acid oxidation and suppressing lipogenesis. Furthermore, quercetin exhibits antioxidant properties, which contribute to the reduction of oxidative stress that leads to a reduction in hepatic steatosis. Citrus fruits, including oranges and grapefruit, are a natural source of rutin, another flavonoid with reported beneficial effects on NAFLD. Specifically, rutin has been found to alleviate NAFLD by mitigating oxidative stress and upregulating endogenous antioxidant systems (Mallhi et al., 2021).

3.5 Cardiovascular Diseases (CVDs)

Cardiovascular diseases (CVDs) stand as the leading cause of global mortality and morbidity, encompassing a series of conditions that impact the heart and brain along with blood vessels. Prominent CVDs encompass coronary heart disease and cerebrovascular disease, both originating from arteriosclerosis, a pathological process that impacts blood vessel walls. Arteriosclerosis involves the accrual of fatty materials and cholesterol within blood vessels, resulting in vessel constriction and diminished blood flow. The rupture of the plaque can lead to the formation of a blood clot, provoking a heart attack or stroke contingent upon its location. The primary metabolic risk factors for CVDs include overweight/obesity, elevated blood pressure, elevated sugar in blood, and elevated lipids in blood (Mendis et al., 2011). These metabolic harmful factors are strongly influenced by behavioral factors such as diet and physical fitness. Overweight and obesity, resulting from an imbalance between the intake of energy and its expenditure, are closely associated with the various other metabolic risk factors. An increase in body mass index elevates the cause of developing ischemic stroke, coronary heart disease, and type 2 diabetes mellitus (Miller et al., 2019).

3.5.1 Bioactive Compounds for the Treatment of CVDs

Berries are characterized by their high nutritional value, containing various essential compounds like minerals, fatty acids, vitamins, and dietary fiber. However, they also possess a wide range of components that are biologically active, including organic acids along with polyphenols, which include subcategories such as flavonoids and tannins (Rothwell et al., 2013). Anthocyanins, a subclass of flavonoids, play a significant role as bioactive compounds in berries, contributing to their distinct red-blue-purple hue. These components are believed to possess the capacity to provide and scavenge free radicals, along with cellular antioxidant protection activation, restrict the expression of inflammatory genes, and also shield against cellular damage and cytotoxicity induced by oxidants and inflammation. As such, these bioactive compounds are beneficial and have a protective effect against various types of chronic diseases that includes cardiovascular disease, Alzheimer's, cancer, depression, and other related disorders (Miller et al., 2019).

Many studies have demonstrated the potential hypoglycemic properties of berries. For instance, Stull et al. (2010) conducted a study to know the impact of a blueberry smoothie consumption containing 22.5 g of bioactive-rich blueberries for 6 weeks twice daily, on a cohort of obese individuals with insulin resistance. The findings revealed improved insulin sensitivity in the treatment group compared with the group that is controlled. A recent study looked at how purified anthocyanins affected the metabolism of 80 people with prediabetes and diabetes. Over 12 weeks, the participants took 300 mg of purified anthocyanins from bilberry and blackcurrant each day. The study found that the group taking the anthocyanins had lower HbA1c levels compared to the control group. This shows that anthocyanins may have positive effects on regulating blood sugar levels. (Yang et al., 2010). Additionally, berry anthocyanins exhibit notable antioxidant and lipid-lowering properties along with anti-inflammatory that may contribute to protection against the principal comorbidities of cardiovascular disease and metabolic disorders as illustrated in Fig. 3 (Chan & Tomlinson, 2020).

The two of the most widely demanded berries are Strawberries (Fragaria) and blueberries (Vaccinium) are available in all forms such as fresh, frozen, processed, and derived forms, such as dried, canned, beverages, yogurts, jams, and jellies. Various in vitro and in vivo studies have shown that, among other foods, blueberries and strawberries help in providing and decreasing the risk factors of developing various chronic diseases shown in Fig. 4. However, the Chemical composition of strawberries shown in Table 1 and blueberries shown in Table 2 differ significantly in the amount and diversity of their bioactive compounds (Miller et al., 2019).

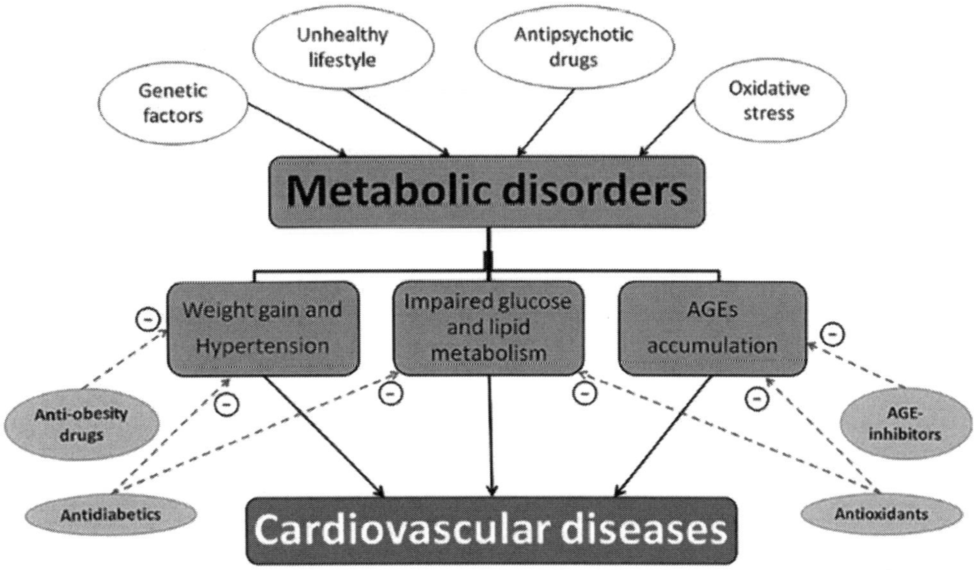

Fig. 3 Metabolic disorders in cardiovascular diseases

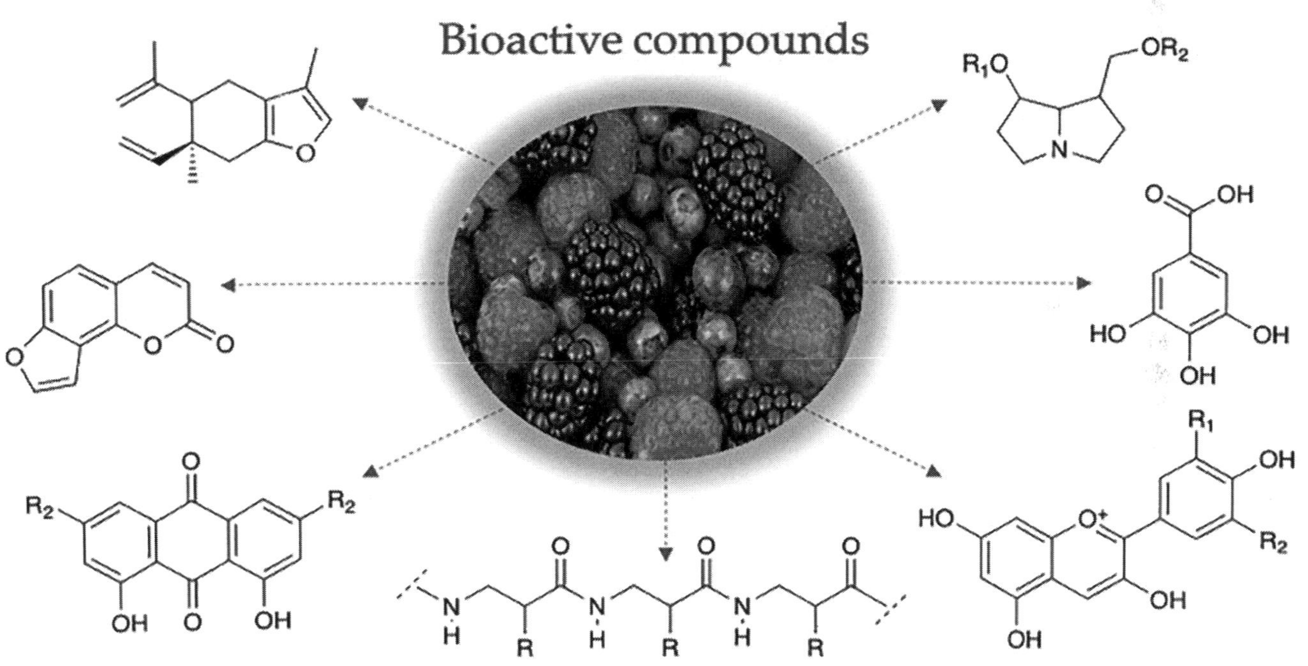

Fig. 4 Bioactive compounds in berries

Table 1 Chemical composition of blueberry

Flavonoids	Blueberry (mg/kg FW)
Anthocyanidins	134
Flavanols	1.1
Flavonols	38.7
Hydroxybenzoic acids	1.5
Hydroxycinnamic acids	135.0

Table 2 Chemical composition of strawberry

Flavonoids	Strawberry (mg/kg FW)
Anthocyanidins	73.0
Flavanols	9.1
Flavonols	2.3
Hydroxybenzoic acids	5.7
Hydroxycinnamic acids	7.1

4 Neuroprotective Effects of Bioactive Compounds

Metabolic disorders can also lead to neuroinflammation in which bioactive compounds play a crucial role.

Neuroinflammation, which refers to the inflammation that occurs in the nervous system, plays a pivotal role in the progression of secondary brain damage in a variety of neurological disorders. These disorders include Prion disease, multiple sclerosis, Alzheimer's disease, ischemic stroke, Parkinson's disease, experimental autoimmune encephalomyelitis, and neuropathic pain. The onset of these conditions can be triggered by pathological events that are dependent on a person's age as well as those that are independent of age, all of which follow a common inflammatory pathway. Notably, ischemic cardiovascular disease stands as the third primary cause of death and various disabilities on a global scale. Cerebral ischemic disease cases are on the rise, increasing by 2 million annually and impacting individuals of all age groups. Consequently, the search for natural products that can safeguard against and treat transient cerebral ischemia–reperfusion injury, along with probing into their mechanisms of action, remains a rational approach. The protective potential of the total flavonoid extract of flower A. esculentus against the injury of transient cerebral ischemia–reperfusion and its potential mechanism of action have been explored. The extract has demonstrated useful effects against the TCI-RI that is possibly through both direct actions (free radical scavenging) as well as and indirect (activation through the pathway of neuronal Nrf2-ARE to modulate oxidative stress-induced damage) actions (Luo et al., 2018).

Alzheimer's disease denoted as AD is a brain problem that worsens over time and is characterized by the development of certain abnormal structures in the brain, such as senile plaques and neurofibrillary tangles, as well as the death of nerve cells. The damage caused by highly reactive compounds is called to play a beneficial role in this process. Caryocar Brasiliense, also known as "pequi," which belongs to the Caryocaraceae family, has been studied for its potential ability to protect nerve cells due to its antioxidant and anticholinesterase properties. In AD, memory loss and other cognitive issues gradually become more severe. According to the amyloid hypothesis, the build-up of amyloid-β (Aβ) contributes to nerve cell dysfunction and subsequent brain cell death in AD. However, Aβ also has been suggested as a possible target for treating AD (de Oliveira et al., 2018). The anti-Aβ effects of procyanidins and their protective mechanisms, extracted from lotus seedpod have shown promise as a natural treatment for AD. Researchers evaluated the beneficial effects of lotus seedpod procyanidins (LSPC) on Aβ25-35-induced damage in pheochromocytoma cells of rats and investigated the potential involvement of CREB/BDNF signaling along with their antioxidant activity as possible pathways. They also analyzed the distribution of LSPC in the body using LC-MS/MS.

Microglia, astrocytes, neurons, and oligodendrocytes constitute the fundamental cell types within the brain. Glial cells, notably microglia and astrocytes, are instrumental in defending against injury in the brain, maintaining homeostasis, as well as facilitating brain repair. In the context of aging-related conditions and disorders such as Alzheimer's disease and stroke, chronic microglial activation can precipitate neuroinflammation. While activated microglia are indispensable for the brain's innate immune response, persistent activation can detrimentally impact the central nervous system. Consequently, nutraceuticals or natural compounds that exhibit the capacity to modulate these processes and regulate the activation of microglial are prospective candidates for mitigating neurodegenerative and neuroinflammation conditions. An investigation was conducted to explore the cytotoxicity and anti-inflammatory potential of modified resveratrol-enriched rice along with normal Dongjin rice, and resveratrol in activated microglia. The study also delved into the mechanisms that are possibly underlying the anti-neuroinflammatory effects of resveratrol-enriched rice in microglial cells of the lipopolysaccharide-stimulated BV2 murine. The outcomes not just only underscored the effective and safe role of resveratrol-enriched rice in combating neuroinflammation as well as aging but also revealed the potential of anti-neuroinflammatory in normal rice due to the presence of the active phytochemicals that include γ-tocopherol and α-tocopherol. It was observed that the anti-inflammatory effects of resveratrol-enriched rice appear to be mediated due to the inhibition of nitrite production, NF-κB-mediated production of pro-inflammatory cytokines, MAPK phosphorylation, and the expression of inflammatory proteins (Teodoro, 2019).

In a study by Bowtell et al., researchers examined the effects of blueberry concentrate on brain function in healthy older adults. The study involved 12 weeks of blueberry supplementation, with participants consuming 30 mL of blueberry concentrate per day, providing 387 mg of anthocyanidins per day. The results showed improved brain activity in specific areas and better blood flow in certain parts of the brain after the supplementation, compared to the placebo. Both the blueberry and placebo groups showed a decrease in serum glutathione concentration. However, there were no significant differences in other brain activity, blood flow in different brain parts, or certain markers in the blood between the two groups (Bowtell et al., 2017).

In another study involving 37 participants, individuals were randomly assigned to take either a blueberry freeze-dried beverage (24 g/day FDB, fresh blueberries that are equivalent to one cup) or a placebo blueberry beverage for a duration of 90 days. Those who consumed the blueberry beverage demonstrated that there were fewer repetition errors in a learning assessment and exhibited a decrease in switch cost in a test of switching task compared to the control group throughout the study. Although, there were no observed improvements in gait or balance. The study also summarised that blueberry consumption did not show a significant impact on the Profile of Mood States or the Geriatric Depression Scale (Miller et al., 2018).

5 Future Aspects

Studying the preventive effects of natural bioactive compounds with small impacts is a big challenge in nutritional research. To talk about this issue, we need to understand how scientists show the effectiveness of these compounds. The scientific tests include in vitro studies, animal experiments, and human studies, which can be divided into epidemiological and intervention studies. Epidemiological studies create ideas about the association between diet and disease. These ideas are then looked into using in vitro and animal studies. Intervention studies are later done to show how well these compounds work in people who eat a certain amount under controlled conditions. The tests can measure how well the compounds work or look at health issues. However, these tests aren't always the best way to study how well the compounds can prevent health problems. We need to understand more about biomarkers and their role in showing how well these compounds work. Even though there's a lot of research on the possible good effects of a lot of natural bioactive compounds, we need more big studies to see how they work. We need to make rules that everyone follows to find out the smallest amount of these compounds that work and how long to use them to get good effects.

6 Conclusion

Bioactive compounds show promising potential as a supplementary nutritional intervention for addressing key parameters related to Metabolic Syndrome (MetS) and may play a crucial role in preventing diseases in high-risk individuals. However, to effectively harness their benefits, further studies involving large populations and focusing on individuals with MetS risk factors are necessary to confirm their efficacy in humans. Additionally, while there is limited information available for the appropriate dosage and treatment regimens for these interventional herbal treatments and unclear mechanisms of their effects in many cases, their long-standing use in traditional medical practice, without any evidence of adverse effects, suggests they can be considered a safe addition to medical treatment with potential benefits. Interventional research denotes that specific bioactive compounds and functional foods may contribute to weight management and ameliorate the metabolic ramifications associated with elevated body weight. Nevertheless, the elucidation of their precise roles in most cases mandates additional clinical and epidemiological investigations to foster sounder conclusions and to further elucidate their prospective mechanisms of action.

Acknowledgment & Declaration

- FundingNo specific funding for the study was received. All authors received their salaries from their respective institutions.
- Conflicts of interest/Competing interests
  Declared none.
- Authors' contributions All authors wrote, read, and approved the study for publication
  L.S. and K.D.: conceived the design, analysis and planning, review, and writing.
  L.S.: writing, critical analysis, review, and supervision.
- All authors provided their critical feedback and approved the final manuscript.

References

Adibian, M., Hodaei, H., Nikpayam, O., Sohrab, G., Hekmatdoost, A., & Hedayati, M. (2019). The effects of curcumin supplementation on high-sensitivity C-reactive protein, serum adiponectin, and lipid profile in patients with type 2 diabetes: A randomized, double-blind, placebo-controlled trial. *Phytotherapy Research, 33*(5), 1374–1383.

Aguilar, M., Bhuket, T., Torres, S., Liu, B., & Wong, R. J. (2015). Prevalence of the metabolic syndrome in the United States, 2003–2012. *JAMA, 313*(19), 1973–1974.

Ahmed, O. M., Moneim, A. A., Yazid, I. A., & Mahmoud, A. M. (2010). Antihyperglycemic, antihyperlipidemic and antioxidant effects and the probable mechanisms of action of Ruta graveolens infusion and rutin in nicotinamide-streptozotocin-induced diabetic rats.

Akilen, R., Pimlott, Z., Tsiami, A., & Robinson, N. (2013). Effect of short-term administration of cinnamon on blood pressure in patients with prediabetes and type 2 diabetes. *Nutrition, 29*(10), 1192–1196.

Al Disi, S. S., Anwar, M. A., & Eid, A. H. (2016). Anti-hypertensive herbs and their mechanisms of action: Part I. *Frontiers in Pharmacology, 6*, 323.

Alvarez-Collazo, J., Alonso-Carbajo, L., López-Medina, A. I., Alpizar, Y. A., Tajada, S., Nilius, B., et al. (2014). Cinnamaldehyde inhibits L-type calcium channels in mouse ventricular cardiomyocytes and vascular smooth muscle cells. *Pflügers Archiv-European Journal of Physiology, 466*, 2089–2099.

American Diabetes Association. (2010). Diagnosis and classification of diabetes mellitus. *Diabetes Care, 33*(Supplement_1), S62–S69.

Andersen, H. R., Wohlfahrt-Veje, C., Dalgård, C., Christiansen, L., Main, K. M., Nellemann, C., et al. (2012). Paraoxonase 1 polymorphism and prenatal pesticide exposure associated with adverse cardiovascular risk profiles at school age. *PLoS ONE, 7*(5), e36830.

Barnes, J. (2003). Quality, efficacy and safety of complementary medicines: fashions, facts and the future. Part II: Efficacy and safety. *British Journal of Clinical Pharmacology, 55*(4), 331–340.

Bergman, Å., Heindel, J. J., Jobling, S., Kidd, K., Zoeller, T. R., & World Health Organization. (2013). *State of the science of endocrine disrupting chemicals 2012*. World Health Organization.

Betoret, E., Betoret, N., Vidal, D., & Fito, P. (2011). Functional foods development: Trends and technologies. *Trends in Food Science & Technology, 22*(9), 498–508.

Biesalski, H. K., Dragsted, L. O., Elmadfa, I., Grossklaus, R., Müller, M., Schrenk, D., et al. (2009). Bioactive compounds: Definition and assessment of activity. *Nutrition, 25*(11–12), 1202–1205.

Boushey, C. J., Beresford, S. A., Omenn, G. S., & Motulsky, A. G. (1995). A quantitative assessment of plasma homocysteine as a risk factor for vascular disease: Probable benefits of increasing folic acid intakes. *JAMA, 274*(13), 1049–1057.

Bowling, J. J., Kochanowska, A. J., Kasanah, N., & Hamann, M. T. (2007). Nature's bounty–drug discovery from the sea. *Expert Opinion on Drug Discovery, 2*(11), 1505–1522.

Bowtell, J. L., Aboo-Bakkar, Z., Conway, M. E., Adlam, A. L. R., & Fulford, J. (2017). Enhanced task-related brain activation and resting perfusion in healthy older adults after chronic blueberry supplementation. *Applied Physiology, Nutrition, and Metabolism, 42*(7), 773–779.

Carroll, M. D., Kit, B. K., & Lacher, D. A. (2015). *Trends in elevated triglyceride in adults: United States, 2001–2012 (No. 2015)*. US Department of Health and Human Services, Centers for Disease Control and Prevention, National Center for Health Statistics.

Chan, S. W., & Tomlinson, B. (2020). Effects of bilberry supplementation on metabolic and cardiovascular disease risk. *Molecules, 25* (7), 1653.

Chen, J. P., Chen, G. C., Wang, X. P., Qin, L., & Bai, Y. (2017). Dietary fiber and metabolic syndrome: A meta-analysis and review of related mechanisms. *Nutrients, 10*(1), 24.

Choudhary, P. R., Jani, R. D., & Sharma, M. S. (2018). Effect of raw crushed garlic (Allium sativum L.) on components of metabolic syndrome. *Journal of Dietary Supplements, 15*(4), 499–506.

Chuengsamarn, S., Rattanamongkolgul, S., Luechapudiporn, R., Phisalaphong, C., & Jirawatnotai, S. (2012). Curcumin extract for prevention of type 2 diabetes. *Diabetes Care, 35*(11), 2121–2127.

Cornier, M. A., Dabelea, D., Hernandez, T. L., Lindstrom, R. C., Steig, A. J., Stob, N. R., et al. (2008). The metabolic syndrome. *Endocrine Reviews, 29*(7), 777–822.

de Oliveira, T. S., Thomaz, D. V., da Silva Neri, H. F., Cerqueira, L. B., Garcia, L. F., Gil, H. P. V., et al. (2018). Neuroprotective effect of caryocar brasiliense camb. leaves is associated with anticholinesterase and antioxidant properties. *Oxidative Medicine and Cellular Longevity, 2018*(1), 9842908.

Debbab, A., Aly, A. H., Lin, W. H., & Proksch, P. (2010). Bioactive compounds from marine bacteria and fungi. *Microbial Biotechnology, 3*(5), 544–563.

Di Daniele, N., Marrone, G., Di Lauro, M., Di Daniele, F., Palazzetti, D., Guerriero, C., et al. (2021). Effects of caloric restriction diet on arterial hypertension and endothelial dysfunction. *Nutrients, 13*(1), 274.

Granzotto, A., & Zatta, P. (2014). Resveratrol and Alzheimer's disease: Message in a bottle on red wine and cognition. *Frontiers in Aging Neuroscience, 6*, 95.

Grundy, S. M., Brewer, H. B., Jr., Cleeman, J. I., Smith, S. C., Jr., & Lenfant, C. (2004). Definition of metabolic syndrome: Report of the National Heart, Lung, and Blood Institute/American Heart Association conference on scientific issues related to definition. *Circulation, 109*(3), 433–438.

Gupta Jain, S., Puri, S., Misra, A., Gulati, S., & Mani, K. (2017). Effect of oral cinnamon intervention on metabolic profile and body composition of Asian Indians with metabolic syndrome: A randomized double-blind control trial. *Lipids in Health and Disease, 16*, 1–11.

Gutierres, V. O., Pinheiro, C. M., Assis, R. P., Vendramini, R. C., Pepato, M. T., & Brunetti, I. L. (2012). Curcumin-supplemented yoghurt improves physiological and biochemical markers of experimental diabetes. *British Journal of Nutrition, 108*(3), 440–448.

Haddy, F. J., Vanhoutte, P. M., & Feletou, M. (2006). Role of potassium in regulating blood flow and blood pressure. *American Journal of Physiology-Regulatory, Integrative and Comparative Physiology, 290*(3), R546–R552.

Heindel, J. J., Blumberg, B., Cave, M., Machtinger, R., Mantovani, A., Mendez, M. A., et al. (2017). Metabolism disrupting chemicals and metabolic disorders. *Reproductive Toxicology, 68*, 3–33.

Hosseini, S., Jamshidi, L., Mehrzadi, S., Mohammad, K., Najmizadeh, A. R., Alimoradi, H., et al. (2014). Effects of Juglans regia L. leaf extract on hyperglycemia and lipid profiles in type two diabetic patients: a randomized double-blind, placebo-controlled clinical trial. *Journal of Ethnopharmacology, 152*(3), 451–456.

Khan, M. I., Anjum, F. M., Sohaib, M., & Sameen, A. (2013). Tackling metabolic syndrome by functional foods. *Reviews in Endocrine and Metabolic Disorders, 14*, 287–297.

Kong, A., Gudbjartsson, D. F., Sainz, J., Jonsdottir, G. M., Gudjonsson, S. A., Richardsson, B., et al. (2002). A high-resolution recombination map of the human genome. *Nature Genetics, 31*(3), 241–247.

Legler, J., Fletcher, T., Govarts, E., Porta, M., Blumberg, B., Heindel, J. J., et al. (2015). Obesity, diabetes, and associated costs of exposure to endocrine-disrupting chemicals in the European Union. *The Journal of Clinical Endocrinology & Metabolism, 100*(4), 1278–1288.

Luo, Y., Cui, H. X., Jia, A., Jia, S. S., & Yuan, K. (2018). The protective effect of the total flavonoids of Abelmoschus esculentus L. flowers on transient cerebral ischemia-reperfusion injury is due to activation of the Nrf2-ARE pathway. *Oxidative Medicine and Cellular Longevity, 2018*(1), 8987173.

Maharaj, R. (2015). Effects of abiotic stress (UV-C) induced activation of phytochemicals on the postharvest quality of horticultural crops. *Phytochemicals: Isolation, Characterization and Role in Human Health*, 221–244.

Mallhi, T. H., Rasheed, M., Khan, Y. H., Alzarea, A. I., & Raja, A. A. (2021). Bioactive compounds for the treatment of metabolic disorders. *Endocrine Disrupting Chemicals-Induced Metabolic Disorders and Treatment Strategies*, 489–505.

Maria, H. (2012). *Marine bioactive compounds: sources, characterization, and applications*.

Mendis, S., Puska, P., Norrving, B. E., & World Health Organization. (2011). *Global atlas on cardiovascular disease prevention and control*. World Health Organization.

Miller, M. G., Hamilton, D. A., Joseph, J. A., & Shukitt-Hale, B. (2018). Dietary blueberry improves cognition among older adults in a randomized, double-blind, placebo-controlled trial. *European Journal of Nutrition, 57*, 1169–1180.

Miller, K., Feucht, W., & Schmid, M. (2019). Bioactive compounds of strawberry and blueberry and their potential health effects based on human intervention studies: A brief overview. *Nutrients, 11*(7), 1510.

Mollazadeh, H., & Hosseinzadeh, H. (2016). Cinnamon effects on metabolic syndrome: A review based on its mechanisms. *Iranian Journal of Basic Medical Sciences, 19*(12), 1258.

Morita, H., Taguchi, J. I., Kurihara, H., Kitaoka, M., Kaneda, H., Kurihara, Y., et al. (1997). Genetic polymorphism of 5, 10-methylenetetrahydrofolate reductase (MTHFR) as a risk factor for coronary artery disease. *Circulation, 95*(8), 2032–2036.

Na, L. -X., Li, Y., Pan, H. -Z., Zhou, X. -L., Sun, D. -J., Meng, M., et al. (2012). Curcuminoids exert glucose-lowering effect in type 2 diabetes by decreasing serum free fatty acids: A double-blind, placebo-controlled trial. *Molecular Nutrition & Food Research, 57*, 1569–1577.

Newbold, R. R., Padilla-Banks, E., Snyder, R. J., & Jefferson, W. N. (2007). Perinatal exposure to environmental estrogens and the development of obesity. *Molecular Nutrition & Food Research, 51*(7), 912–917.

Noce, A., Di Lauro, M., Di Daniele, F., Pietroboni Zaitseva, A., Marrone, G., Borboni, P., et al. (2021). Natural bioactive compounds useful in clinical management of metabolic syndrome. *Nutrients, 13*(2), 630.

Ogden, C. L., Carroll, M. D., Fryar, C. D., & Flegal, K. M. (2015). *Prevalence of obesity among adults and youth: United States, 2011–2014*.

Panahi, Y., Hosseini, M. S., Khalili, N., Naimi, E., Soflaei, S. S., Majeed, M., et al. (2016). Effects of supplementation with curcumin on serum adipokine concentrations: A randomized controlled trial. *Nutrition, 32*(10), 1116–1122.

Pivari, F., Mingione, A., Brasacchio, C., & Soldati, L. (2019). Curcumin and type 2 diabetes mellitus: Prevention and treatment. *Nutrients, 11*(8), 1837.

Rochlani, Y., Pothineni, N. V., Kovelamudi, S., & Mehta, J. L. (2017). Metabolic syndrome: Pathophysiology, management, and modulation by natural compounds. *Therapeutic Advances in Cardiovascular Disease, 11*(8), 215–225.

Rothwell, J. A., Perez-Jimenez, J., Neveu, V., Medina-Remon, A., M'hiri, N., García-Lobato, P., et al. (2013). Phenol-Explorer 3.0: A major update of the Phenol-Explorer database to incorporate data on the effects of food processing on polyphenol content. *Database, 2013*, bat070.

Sharma, R. A., McLelland, H. R., Hill, K. A., Ireson, C. R., Euden, S. A., Manson, M. M., et al. (2001). Pharmacodynamic and pharmacokinetic study of oral Curcuma extract in patients with colorectal cancer. *Clinical Cancer Research, 7*(7), 1894–1900.

Sharma, R. A., Steward, W. P., & Gescher, A. J. (2007). Pharmacokinetics and pharmacodynamics of curcumin. *The Molecular Targets and Therapeutic Uses of Curcumin in Health and Disease*, 453–470.

Šnē, E., Galoburda, R., & Segliņa, D. (2013). Sea buckthorn vegetative parts–a good source of bioactive compounds. In *Proceedings of the Latvian Academy of Sciences. Section B. Natural, Exact, and Applied Sciences* (Vol. 67, No. 2, pp. 101–108).

Stull, A. J., Cash, K. C., Johnson, W. D., Champagne, C. M., & Cefalu, W. T. (2010). Bioactives in blueberries improve insulin sensitivity in obese, insulin-resistant men and women. *The Journal of Nutrition, 140*(10), 1764–1768.

Suarez-Jimenez, G. M., Burgos-Hernandez, A., & Ezquerra-Brauer, J. M. (2012). Bioactive peptides and depsipeptides with anticancer potential: Sources from marine animals. *Marine Drugs, 10*(5), 963–986.

Teodoro, A. J. (2019). Bioactive compounds of food: their role in the prevention and treatment of diseases. *Oxidative Medicine and Cellular Longevity, 2019*.

Thayer, K. A., Heindel, J. J., Bucher, J. R., & Gallo, M. A. (2012). Role of environmental chemicals in diabetes and obesity: A National Toxicology Program workshop review. *Environmental Health Perspectives, 120*(6), 779–789.

Williams, D. J., Edwards, D., Hamernig, I., Jian, L., James, A. P., Johnson, S. K., et al. (2013). Vegetables containing phytochemicals with potential anti-obesity properties: A review. *Food Research International, 52*(1), 323–333.

Wu, L. Y., Chen, C. W., Chen, L. K., Chou, H. Y., Chang, C. L., & Juan, C. C. (2019). Curcumin attenuates adipogenesis by inducing preadipocyte apoptosis and inhibiting adipocyte differentiation. *Nutrients, 11*(10), 2307.

Xiong, X. J., Wang, P. Q., Li, S. J., Li, X. K., Zhang, Y. Q., & Wang, J. (2015). Garlic for hypertension: A systematic review and meta-analysis of randomized controlled trials. *Phytomedicine, 22*(3), 352–361.

Yang, L., Ling, W., Yang, Y., Chen, Y., Tian, Z., Du, Z., et al. (2017). Role of purified anthocyanins in improving cardiometabolic risk factors in Chinese men and women with prediabetes or early untreated diabetes—a randomized controlled trial. *Nutrients, 9*(10), 1104.

Younossi, Z. M., Koenig, A. B., Abdelatif, D., Fazel, Y., Henry, L., & Wymer, M. (2016). Global epidemiology of nonalcoholic fatty liver disease—meta-analytic assessment of prevalence, incidence, and outcomes. *Hepatology, 64*(1), 73–84.

Zemel, M. B. (2004). Role of calcium and dairy products in energy partitioning and weight management. *The American Journal of Clinical Nutrition, 79*(5), 907S-912S.

Zheng, H., Lenard, N. R., Shin, A. C., & Berthoud, H. R. (2009). Appetite control and energy balance regulation in the modern world: Reward-driven brain overrides repletion signals. *International Journal of Obesity, 33*(2), S8–S13.

Microbial Production of Bioplastics: Current Status and Future Prospects

Urjitsinh Zala, Jaya Chawla, Arun Kumar Kulshrestha, Vijay Jagdish Upadhye, and Anupama Shrivastav

Abstract

In recent times, there has been a notable shift towards bioplastics, developed from sustainable biological sources, as a feasible alternative to traditional petroleum-based polymers. This chapter delves into the diverse microbial strains capable of producing various biopolymeric molecules, highlighting the pivotal role microorganisms play in bioplastic production. Beginning with the fundamental concepts of microbial bioplastic synthesis, the chapter explores key biopolymer types and examines the metabolic pathways and genetic enhancements employed by bioplastic-producing organisms, such as bacteria, yeast, and algae, to enhance yields and product quality. Additionally, it discusses contemporary methodologies and technologies utilized in microbial bioplastic production. The chapter also addresses the economic and environmental considerations associated with large-scale bioplastic production, including aspects such as fermentation, post-processing, and repurposing waste streams for bioplastic synthesis. Despite its potential, microbial bioplastic production faces challenges related to scalability, cost-effectiveness, and the need for technological advancements. The chapter further discusses current research objectives, industry trends, and the application of synthetic biology and metabolic engineering to synthesize microbial bioplastics, enabling the creation of customized bioplastics. In conclusion, the chapter emphasizes the significance of microbial production in the bioplastics industry, along with its present achievements and future directions, which are expected to contribute to the development of more environmentally friendly and sustainable polymers.

Keywords

Bioplastics • Microbial production • Polyhydroxyalkanoates • Polylactic acid • Sustainability • Fermentation • Metabolic engineering • Synthetic biology

1 Introduction

Human-made polymers, commonly referred to as plastics, are esteemed for their advantageous characteristics, such as their remarkable strength-to-weight ratio, flexibility, heat-sealing abilities, and transparency. Among the widely utilized petrochemical-derived polymers are nylon, polyethylene, polypropylene, polystyrene, polyvinyl chloride, polyurethane, polyethylene terephthalate, polybutylene terephthalate, and polyethylene. These materials are prized for their versatility, lightweight nature, outstanding thermal and rheological properties, cost-effectiveness, and ease of handling and molding (Kumar et al., 2014; Gadhave et al., 2018).

The environmental repercussions of plastic misuse are severe, as approximately 34 million tons are produced each year, yet only a mere 7% undergo recycling. A staggering 93% of plastics find their way into landfills and oceans, exacerbating ecological degradation (Sushmitha et al., 2016). Given their non-biodegradable nature, synthetic polymeric materials present a significant threat to freshwater, terrestrial, and marine ecosystems (Bbiah, 2015).

U. Zala · A. K. Kulshrestha
Department of Medical Laboratory Technology, Faculty of Life Health and Allied Sciences, ITM Vocational University, Raval, Waghodia, Vadodara, 391760, Gujarat, India

V. J. Upadhye
Research and Development Cell (RDC), Parul Institute of Applied Sciences (PIAS), Parul University, Limda, Waghodia, Vadodara, 391760, Gujarat, India

J. Chawla · A. Shrivastav (✉)
Microbiology department, Faculty of Life Health and Allied Sciences, ITM Vocational University, Raval, Waghodia, Vadodara, 391760, Gujarat, India
e-mail: anupamashrivastav@gmail.com; anupamashrivastav@yahoo.co.in; anupamam@itmvu.in

J. Devi (ed.), *Biotechnological Intervention in Production of Bioactive Compounds*, Sustainable Landscape Planning and Natural Resources Management, https://doi.org/10.1007/978-3-031-76859-0_13

Furthermore, their production relies on finite resources such as natural gas, coal, and oil, with decomposition in the environment spanning decades (Emadian et al., 2017).

As a result, There have been numerous attempts to develop new, environmentally friendly, and biodegradable materials as alternatives to standard plastics (Kržan, 2012). Using waste, biomass, and renewable resources such as jackfruit (Lothfy et al., 2018), abandoned banana peels (Mohapatra et al., 2015), organic waste (Goswami et al., 2015), agricultural residues, newspaper waste (Joshi et al., 2015), empty palm fruit bunches (Isroi et al., 2017), sugarcane (Khosravi-Darani & Bucci, 2015), corn starch (Keziah et al., 2018), potato starch (Schön & Schwartz, 2014), rice straw (Agustin et al., 2014), rapeseed oil (Delgado et al., 2018), vegetable oils, plant cellulose, starch, cotton, and even bacteria (Shamsuddin et al., 2017a), bioplastics have emerged as promising solutions. They frequently incorporate nanoparticles like polysaccharides (Jabeen et al., 2015). Algae, fungi, and bacteria are among the natural microorganisms that can break down these bioplastics (Das et al., 2018a; Momani, 2009).

The vast environmental consequences of disposable plastics have emerged as a pressing issue in modern times. The long-term persistence of petroleum-derived plastics presents a significant challenge to environmental measures aimed at reducing and reclaiming them. The methods used to dispose of plastic waste have a significant impact on its effective retrieval and recycling. When plastic waste is discarded indiscriminately, recovery becomes more complex and energy-inefficient (Devasahayam et al., 2019).

The increase in production and consumption of plastic creates health and environmental problems, which are exacerbated by the careful management of used products (Adane et al., 2011). Although plastic products are fragile and expensive, single-use plastics (SUP) are rapidly increasing in developing countries. Not only do plastics not pollute the environment, they pose a threat to plastics that are still in use because they become microplastics. When exposed to UV radiation and thermal manipulation (MP), phthalates and other chemicals used in the manufacture of SUPs are nearly degraded and can penetrate internal organs and enter the food chain (Giacovelli & Zamparo, 2018; Chen et al., 2020).

The increase in the production of plastics over the past 65 years has surpassed that of any other man-made object. The complex nature of petroleum-based plastics contributes to environmental degradation despite their economic benefits. The urgent problem of plastic pollution has prompted a significant shift in environmental policies, leading to a focus on developing local low-carbon, circular economies. These strategies include strict regulations for landfill waste, systematic approaches to Recycling and recovery programs, combined with the mandatory incorporation of bioplastics into the production of various items (Kaeb et al., 2016; Godfrey, 2019).

Scientists around the world are exploring the replacement of traditional fossil-derived plastics with environmentally friendly, carbon–neutral bio-based alternatives. This initiative aims to minimize energy usage and diminish carbon footprints, ultimately striving for ecological improvement.

1.1 Microbial Bioplastics

Bioplastics, derived from various biological sources, consist of covalently bonded monomer units that create biodegradable polymer compounds. These sources include microalgae, bacteria, plants, and photosynthetic bacteria (Kadouri et al., 2005). The key components of these bio-based, biodegradable plastics are found within microbial cells, present in cellular components like proteins, polysaccharides, lipids, amino acids (Samantaray et al., 2015) and polyhydroxyalkanoates. Notably, several aliphatic polyesters, such as thermoplastic starch, polyhydroxyalkanoates, polybutylene succinate, and polylactic acid, are emerging as fundamental elements in bioplastic production (Debuissy et al., 2018). Extensive research has been conducted on biopolymer synthesis utilizing bacterial strains such as *Bacillus spp., Azotobacter spp., Alcaligenes spp., Pseudomonas spp., methylotrophs,* and *Cupriavidus necator,* alongside algal strains like *Spirulina platensis*, *Nostoc muscorum*, and *Synechococcus spp*. (Samadhiya et al., 2022).

The biorefinery concept has emerged as a key strategy for microalgae cultivation, driven by the increasing demand for plastic products and the escalating pollution attributed to plastic waste. This innovative technology positions microalgal biomass as a promising alternative to traditional plastics, facilitating its continuous conversion into various commercially viable bio-based products, including bioplastics, biogas, bioethanol, pigments, proteins, carbohydrates, and biofuels (Trivedi et al., 2015; Chandra et al., 2019).

Comparative life cycle assessments (LCAs) between fossil-based and bio-based plastics reveal significant advantages associated with bio-based plastics, particularly in terms of reducing carbon emissions and decreasing reliance on fossil fuel sources (Harding et al., 2007; Broeren et al., 2017). Bioplastics derived from biomass raw materials and agricultural resources exhibit physical and mechanical properties akin to those of conventional plastics, with studies indicating substantial biodegradability by bacteria (Wu et al., 2012; Rocha et al., 2020).

Nowadays, a variety of bioplastics are made by fermenting agricultural raw materials such wheat, sugar, potatoes, corn, rice, and soybeans. However, this method raises concerns about the possible impact on the world's food supply. As a result, there is an increasing demand for low-cost bioplastic raw materials such as microbial biomass to ensure manufacturing of biodegradable or non-biodegradable bioplastics that is both economical and sustainable (Solaiman et al., 2006; Pakalapati et al., 2018).

However, a thorough and critical evaluation of bioplastics has identified substantial barriers to their development and commercialization, including higher costs associated with microbial cultivation techniques, downstream processing, and bio-based substrates. Assessing and analyzing the overall cost of waste management and bioplastic manufacturing could be beneficial, with an emphasis on using inexpensive biological substrates to promote microbial growth. These substrates can include industrial waste, hazardous or harsh chemicals like glycerol, and even municipal trash (Amaro et al., 2019; Wen et al., 2020), offering a distinctive perspective for evaluating the financial and environmental impacts of bioplastic regulatory frameworks covering the manufacture, use, and degradation of the material.

Furthermore, carbon dioxide (CO_2), methane (CH_4), and syngas can serve as effective substrates for the generation of biological products, thus contributing to GHG sequestration and wastewater remediation (Debuissy et al., 2018). Additionally, for cost-efficient bioplastic manufacturing, mixed microbial cultures (MMC), whether photosynthetic or non-photosynthetic, have been demonstrated to be effective, lowering the overall cost of the substrates used. MMCs have been shown to support continuous culture acquisition without the risks associated with contamination (Debuissy et al., 2018).

In order to achieve the global goal of widespread adoption of 100% bioplastic products in the future, scientists are directing their efforts toward producing bioplastics from various microorganisms. Mixed microbial cultures (MMCs) have been extensively investigated to optimize the production of bioplastics, especially polyhydroxyalkanoates (PHA). Enhancing the ability of microbial mixed cultures (MMCs) to produce bioplastics requires optimizing the growth parameters and culture conditions of these MMCs. Although MMCs have demonstrated their capability to produce bioplastics, commercialization still faces financial difficulties. Therefore, to fully realize the potential of MMCs for PHA manufacture and other uses, and to reduce expenses related to downstream processing and scaling up, a thorough investigation of these materials is imperative. Customizing microbial consortia may offer a viable approach in this pursuit. Consequently, it is essential to gain a deeper understanding of the fundamental mechanisms controlling the production of bioplastic precursors in microorganisms. This involves identifying the essential elements and distinctive qualities of each specific microorganism involved in PHA production.

To evaluate the potential use of previously identified microbes for bioplastic production, it is essential to consider the current market situation for bio-based plastics. In alignment with the growth of retail markets for these products, this analysis examines microorganisms, including bacteria, algae, and mixed microbial cultures (MMC), as productive cell factories for the commercial manufacture of sustainable bioplastics. It highlights the need to explore new profit opportunities. We shed light on the procedures currently employed in the manufacturing and processing of bio-based materials used to produce bioplastics. To further enhance cost-effectiveness and promote the circular bioeconomy, we emphasize the importance of utilizing effluents and industrial waste by-products as substrates for developing pure bacteria, algae, or MMC.

2 Types of Bioplastics

Bioplastics can be categorized based on their degradability and origin. This classification identifies three main types: purely biopolymers (SBBP), exclusively biodegradable polymers (BP), and a combination of biopolymers and biodegradable polymers (BBBP). BBBP includes polymers derived from biomass that bacteria can break down into carbon dioxide, water, and biomass. Examples of BBBPs include polyhydroxyalkanoates, polylactic acid, bio-PBS, as well as polymers containing lignin, chitosan, starch, and cellulose (Nampoothiri et al., 2010). SBBP, in contrast, denotes polymers derived entirely from renewable biological sources, such as bio-polyethylene, bio-propylene, polyamides, and polyethylene terephthalate (Siracusa & Blanco, 2020). Lastly, BP polymers include PBS, polyvinyl alcohol (PVA), and polypropylene (PP), all of which are biodegradable but sourced from fossil fuels (Ferreira et al., 2019).

2.1 Polyhydroxyalkanoates (PHA)

A subgroup of biopolyesters known for their biodegradability and optical activity includes polyhydroxybutyrate, polyhydroxyvalerate, and related polymers such as poly (3-hydroxybutyrate-co-3-hydroxyvalerate). These polymers account for only a minor portion of total bioplastic manufacturing. PHAs are generated by bacteria in response to stress induced by a lack of inorganic resources such as oxygen, nitrogen, or phosphates, as well as a high carbon content. Microbial lysis, followed by post-processing, is used to extract the fermentation products from the bacteria. Khatami et al. (2021) highlight the different medicinal applications of PHAs because of its biocompatibility, biodegradability, safety, and thermoplastic qualities similar to petrochemical plastics. PHAs exhibit a range of physical properties determined by factors For example, monomer composition, organism type, growth circumstances, and extraction methods all play important effects. Short-chain PHAs resemble traditional polymers such as polypropylene, while longer chains exhibit more elastic properties. Regardless of the difference in gram dye properties, bacteria have the ability to produce bioplastics. Since PHA synthesis occurs under nutrient-deficient conditions or during growth stages, the entry and lag phases of potential PHA-producing organisms should be evaluated before starting optimization studies. The main polymer of PHA, poly-3-hydroxybutyric acid, is more brittle and has higher crystallinity than polypropylene. However, it is common to combine PHA monomers with copolymers to tailor their thermal and mechanical properties for specific applications. Poly (3-hydroxybutyrate-co-3-hydroxyvalerate) with reduced brittleness, lower melting point and percent crystallinity is particularly suitable for compression purposes (Varghese et al., 2022).

Various types of waste materials have been identified as effective substrates for the synthesis of polyhydroxyalkanoates. These substrates include lignin-based wastes, wastewater sludge, effluents from paper, pulp, and cardboard industries, glycerol, and carbon dioxide. Wastewater treatment sludge is particularly promising as it contains diverse microbial consortia capable of producing PHA when provided with excess carbon sources. Glycerol as an input raw material showed a high yield of PHA production, Under optimal conditions, *Bacillus spp*. can achieve up to 85.19% of cell dry weight (Varghese et al., 2022). Lignocellulosic wastes from various sources, such as plants, paper and pulp industries, have shown efficacy in the synthesis of PHAs, especially by bacteria such as Pseudomonas sp. Moreover, under optimized conditions, oleaginous bacteria such as *Serratia spp*. showed a two-fold increase in PHA production when using CO_2 (Varghese et al., 2022).

2.2 Polylactic Acid (PLA)

Lactide chains are formed when lactic acid is polymerized during condensation, and ring-opening techniques are used to create biodegradable polyesters and biobased polyesters like polylactic acid (PLA). As of 2019, PLAs accounted for 13.9% of the world's bioplastics production. Microbial fermentation generates PLA monomers of the L or D isomers, which are then chemically polymerized to form PLA. Whether PLA is in the form of homopolymers or heteropolymers, the ratio of enantiomers significantly influences its physical properties. Heteropolymers, such as PLA containing DL-lactic acid, have an amorphous structure, whereas homopolymers composed exclusively of L or D lactic acid monomers are semicrystalline. High-purity lactic acid is known to be produced from maize, and production costs can be reduced by using biomass from plants and wood. Increasing the concentration of raw resources can lower production costs but may also increase expenses related to purification and fermentation. While electrodialysis aids in the conversion of lactate to lactic acid, ultrafiltration helps in the isolation of lactate during fermentation. Direct lactate condensation produces poor mechanical characteristics, indicating that the lactate intermediate should be polymerized first. In microbial production, homofermentative methods using Lactobacillus spp., such as Lactobacillus amylophilus, L. delbrueckii, L. leichmannii, and L. bulgaricus, are preferred for large-scale PLA production due to their high lactic acid output with fewer byproducts (Varghese et al., 2022). PLA is a biodegradable thermoplastic that can be shaped to produce a range of biological products with shape flexibility. Its usefulness for packaging has been demonstrated by studies, particularly when reinforced with nanocellulose fibrils. For PLA biocomposites, both microcellulose and nanocellulose work well as reinforcing agents. Compared to traditional thermoplastics, PLA offers several benefits, including biocompatibility, remarkable mechanical strength, low processing temperatures, and compatibility with a wide range of fibers. It is applicable to numerous industries, including biosorbents, prosthetic devices, packaging, geotextiles, 3D printing, nonwoven binder fibers, and biomedical absorbable sutures. The production of PLA and related biocomposites has increased across various industries as a result of commercialization initiatives (Varghese et al., 2022).

2.3 Polyurethane (PU)

Polyurethane (PU), a polymer characterized by urethane groups in its chemical structure, is typically produced through the reaction of polyols containing multiple hydroxyl groups with isocyanates containing numerous isocyanate groups. However, due to the harmful nature of certain isocyanates, not all varieties are suitable for industrial PU manufacturing. Moreover, the reliance on petroleum-based resources for synthesizing various polyols poses significant environmental challenges. Consequently, researchers are actively exploring eco-friendly approaches to PU production utilizing renewable resources. Promising innovations include microbial conversion techniques that transform renewable resources into precursor molecules for PU synthesis. For instance, Pseudomonas aeruginosa can convert olive oil into dihydroxy fatty acids, which are then combined with hexamethylene diisocyanate (HMDI). Additionally, lignin has emerged as a viable and cost-effective feedstock for PU synthesis, further advancing sustainable production methods.

The mechanical properties of polyurethane can be evaluated in relation to the properties and content of lignin. When using organosolv lignin in its natural state (15–25 wt %), it enhances the production of durable polyurethane; however, higher concentrations of lignin can render polyurethane brittle and rigid. To optimize the mechanical characteristics of polyurethane, it is crucial to blend flexible aliphatic polyols with rigid lignin polyols. Another method involves liquefying lignin polyols to create more flexible polyurethane. This process involves mechanical manipulation and enzymatic breakdown, potentially coupled with some level of self-polymerization, to combine low molecular weight lignin fragments with flexible polyols like glycerol and polyethylene glycol. Polyurethane finds extensive applications across various industries, including electronics, automotive components, bedding, building materials, furniture adhesives, foams, and coatings. Moreover, lignin presents itself as a viable biological alternative for bioplastic synthesis, as it can be utilized in the production of phenol–formaldehyde resins, which readily yield thermoset polymers (Varghese et al., 2022).

3 Organisms Making Bioplastics

3.1 Microorganisms that Produce Biodegradable Plastic

A detailed examination of microbial bioplastics reveals that a diverse range of microorganisms possess the ability to produce and store PHA/PHB as ATP and carbon reserves. The type of microorganisms used, the carbon sources they consume, and the growth conditions they are subjected to all affect the molecular weights and composition of the polymers that are formed. Notable bacterial species known for producing PHA/PHB include Klebsiella, Citrobacter, Enterobacter, Bacillus, and Pseudomonas. Further evidence highlights the influence of PHB on the growth dynamics of microbial communities. Rhizobium leguminosarum, R. hedysarum, R. galegae, A. macrocytogens, Azotobacter beijerinckii, and A. vinelandii are among the organisms involved in the nitrogen cycle and are also associated with the formation of PHB (Debuissy et al., 2018; Varghese et al., 2022).

Bacteria are categorized into two groups based on the conditions conducive to PHA accumulation. Certain bacteria, such as Protomonas extorquens, P. oleovorans, and Cupriavidus necator, require ample carbon but minimal amounts of other essential elements such as oxygen and nitrogen to synthesize PHA. Conversely, bacteria like Azotobacter vinelandii and Alcaligenes latus are capable of PHA production during the log phase despite nutrient limitations. The cultivation conditions for large-scale PHA biosynthesis are specific, with methylotrophic bacteria demonstrating the ability to produce PHB, albeit in limited quantities. Additionally, the Halomonadaceae family of PHA producers, exemplified by Halomonas sp., requires moderate salt concentrations, yielding small-chain PHAs (scl).

Most Halomonas species thrive at low NaCl concentrations (3–15%). Kawata and Aiba (2010) discovered that Halomonas spp. KM-1 can produce PHB using glycerol as its sole carbon source. Cyanobacteria with PHA synthase enzymes synthesize PHA from CO_2 and sunlight. Saccharomyces cerevisiae, Candida krusei, S. diastaticus, Kloeckera apiculata, C. tropicalis, Rhodotorula glutinis, Kluyveromyces africanus, K. lactonatis, and R. eutropha are among the eukaryotic microorganisms that produce low molecular weight PHBs. Furthermore, current research suggests that Bacillus sp. may ferment and manufacture bioplastics from medicinal plant wastes and waste frying oil, whereas Pseudomonas corrugata and Pseudomonas mediterranea can produce medium-chain PHAs. Cupriavidus necator produces PHA from volatile fatty acids. (Varghese et al., 2022).

3.2 Algal Synthesis of Bioplastics

The synthesis of bioplastics extends beyond microbial origins to include algae, which serve as a precursor for various bioplastics such as starch, PHA, cellulose, PLA, PVC, PE, and protein-based polymers. Utilizing algae for bioplastic production offers multiple benefits, including mitigating the greenhouse effect by absorbing carbon dioxide, enhancing plastic quality, and enabling large-scale cultivation in limited spaces. Furthermore, algae-based bioplastic manufacturing addresses concerns regarding the use of food sources and human waste as fermentation media, particularly during periods of food scarcity. Various methods for algal bioplastic production include direct utilization of algal biomass, genetic engineering, incorporation of algae with other materials, and intermediate biorefinery processes to enhance bioplastic yields. Prominent algal species utilized in bioplastic production include Chlorella sp., Ulva alba albaricana, Spirulina sp., Laminaria japonica, and Enteromorpha crinite (Varghese et al., 2022).

The synthesis pathway of bioplastics, particularly PHA, in cyanobacteria has been extensively documented, revealing ample information. While carbon availability remains favorable, nitrogen and phosphorus levels are limited, which are crucial for facilitating PHA synthesis and accumulation within cells. PHA is derived from acetyl-CoA, synthesized during cellular metabolic processes under appropriate conditions. Its production involves three enzymatic stages: (i) condensation of acetyl-CoA to form acetoacetyl-CoA by β-ketothiolase, (ii) reduction of acetoacetyl-CoA to hydroxybutyryl-CoA by acetoacetyl-CoA reductase, and (iii) esterification of hydroxybutyryl-CoA to PHA by PHA synthase. The economic feasibility of utilizing microalgae in bioplastic manufacturing presents a distinctive advantage. Algal-derived biopolymers are considered highly promising due to their autotrophic lifestyle, harnessing light and CO_2 for sustenance without additional substrate requirements. This approach stands out economically compared to microbial bioplastic production, which demands substantial substrate inputs. Algae-based production aligns well with the principles of a circular bioeconomy, offering cost-effectiveness, reduced CO_2 emissions, and diminished greenhouse gas output, thus presenting a sustainable and economically viable solution.

Various strains of cyanobacteria, such as Synechococcus spp., Synechocystis spp., and Muscorum spp., have demonstrated the capacity to produce substantial amounts of polyhydroxybutyrate (PHB), ranging from 30 to 80%. Chlorella species are recognized for their potential contribution to the production of starch-based bioplastics, as well as blends of bio-based polymers like glycerol, polyethylene (PE), and polyvinyl alcohol. Similarly, Spirulina spp. presents a viable option for the synthesis of bio-based mixtures and bioplastics. Calothrix scytonemicola, another microalgal species, is notable for its PHA production capabilities. Additionally, certain macroalgal species offer promise in bioplastic production and serve as valuable raw material sources. Marine macroalgae, including Kappaphycus spp. for carrageenan and Macrocystis spp. for alginates, provide crucial derivatives for bioplastic synthesis. These derivatives, when combined with glycerol and polyvinyl alcohol, can be utilized to create biodegradable plastic films (Varghese et al., 2022).

4 Molecular Foundation of Bioplastic Synthesis

Research on the production of biopolymers over the past 20 years has shown that, depending on the makeup of the media components, bacteria that can produce bioplastics use several metabolic pathways (see Fig. 1) (Simo-Cabrera et al., 2021). Acetyl-CoA, however, appears as the first metabolite in microorganisms' three major routes. In cultures where carbohydrates serve as the primary source, they are metabolized to pyruvate, which is then used in the synthesis of PHA. (González García et al., 2013). Bacteria can absorb PHA up to 90% of their dry weight, whereas plants typically acquire less than 10% (Verlinden et al., 2007).

PHA synthesis in the cytosol commences with PHA synthase utilizing acetyl-CoA monomers, triggering three crucial enzymatic reactions. β-ketoacyl-CoA thiolase (phbA) converts acetyl-CoA into acetoacetyl-CoA, which is further metabolized by acetoacetyl-CoA dehydrogenase/reductase (phbB) into (R)-3-hydroxybutyryl-CoA. Polymerase (phbC) then facilitates monomer polymerization into PHB. Microbial PHB production can originate from carbohydrates and fatty acids via the beta-oxidation pathway and de novo fatty acid biosynthesis. Intermediates of fatty acid metabolism are converted into (R)-3-hydroxyacyl-CoA, which forms (R)-3-hydroxy fatty acids, subsequently polymerized to PHA. During carbon oxidation completion, intermediates from de novo fatty acid biosynthesis shift to PHA biosynthesis, catalyzed by transacylase (PhaG). Enoyl-CoA hydratase (PhaJ) oxidizes enoyl-CoA to (R)-3-hydroxyacyl-CoA, further transformed into PHA by PHA synthase (PhaC). PHA/PHB polymers, owing to their hydrophobic nature, accumulate primarily as granules within bacteria, composed of phospholipids, bacterial cells (phasins), and polymerized PHA units. Due to its crystalline structure, PHA exhibits limited interaction with water in bacterial cells, with the presence of phasins influencing the quantity and size of PHA granules (Potter et al., 2002; Khatami et al., 2021).

5 Advancement in the Methodologies for Bioplastic Production

Bioplastics production occurs on a significant scale, employing diverse processes, each with distinct pros and cons, necessitating the identification of the most efficient methods. Globally, there is a burgeoning focus on devising innovative strategies to enhance productivity and lower production expenses in bioplastic manufacturing. This endeavor involves revisiting conventional techniques and integrating them with cutting-edge advancements for greater efficacy. Figure 2 delineates the key strategies employed to augment bioplastics production.

6 Impactful Application of Bioplastics

6.1 Industrial Application

Heteropolymers can be created from polyhydroxybutyrate (PHB) by spinning, compressing, extruding, and converting it into films for food packaging (Fig. 3). These films exhibit notable properties, including a high water vapor transfer rate (WVTR), strong tensile strength, high antioxidant activity,

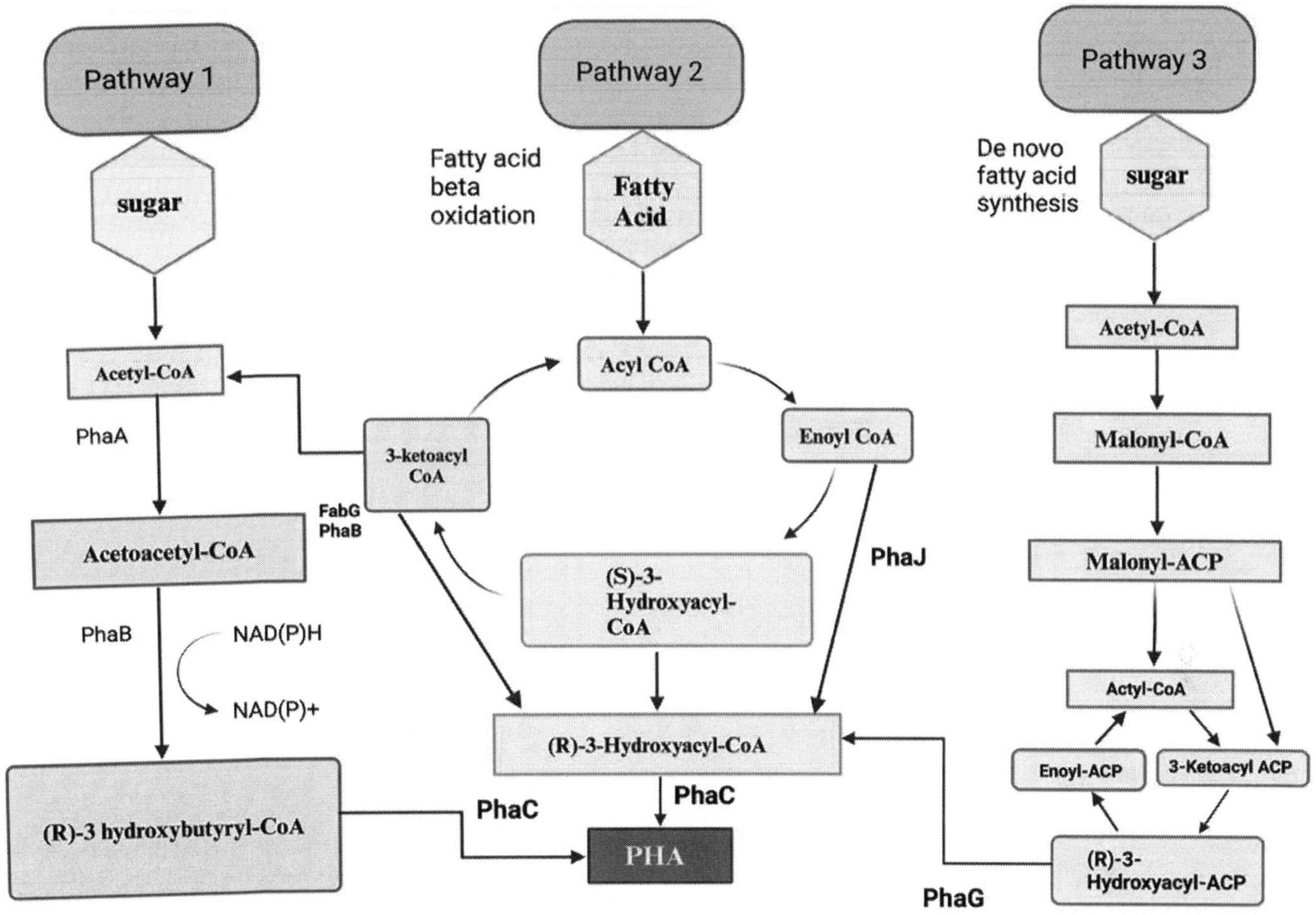

Fig. 1 Molecular basis of bioplastic production

Fig. 2 Different methodologies for bioplastic production

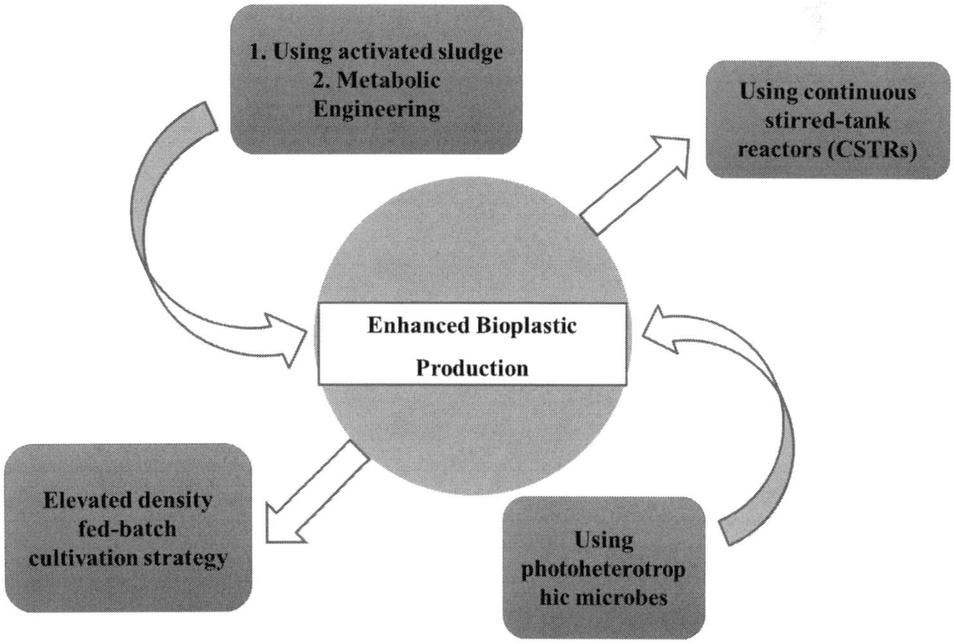

outstanding oxygen permeability (OP), and significant antibacterial qualities when compared to conventional packaging materials. To reduce production costs, polymers are frequently combined with inert fillers or additives like wood grain and clay minerals, as PHA is a relatively expensive material to produce. Some cost-saving methods include using biorefinery lignin residue from Arundo donax combined with P(3HB-co-4HB), bagasse-based sodium lignin, and lignin residue recovered from corn stalks. Recent advances have shown improvements in the properties of PHB-lignin composites during 3D printing, with examples including the use of Pinus-based biorefinery lignin in PHB composites processed into films for efficient 3D printing (Khatami et al., 2021).

6.2 Agricultural Application

Agricultural applications of bioplastics remain somewhat limited, with current research primarily focused on items such as mulch films, grow sacks, and farm netting. Specifically, there is increasing interest in agricultural nets made from biodegradable polyhydroxyalkanoates (PHA) (Fig. 3). These PHA-based nets are compostable and can be disposed of directly in the ground. Due to their high tensile strength, PHB and its mixtures with PLA are frequently chosen among bioplastics for the creation of agricultural nets. PHA grow bags offer advantages, including reducing water pollution, preventing root regrowth, sequestering nitrogen from water, and mitigating pollution-related environmental issues (Potter et al., 2002).

Agricultural mulch films, along with grow bags, play a crucial role in safeguarding and enhancing crop yields by aiding in moisture retention, soil structure maintenance, and weed control. However, approximately 40% of these films are made of plastic, predominantly high-density polyethylene (HDPE), low-density polyethylene (LDPE), and linear LDPE, which are often disposed of through burning or landfilling, leading to environmental pollution. To address this issue, ongoing research is exploring the potential of bioplastics like polybutylene succinate, polylactic acid (PLA), corn starch-derived polymers, ethylene vinyl acetate, and other polyhydroxyalkanoates (PHAs) as sustainable alternatives. Further investigation into the widespread adoption of bioplastics for mulch production could demonstrate their viability as a more eco-friendly option (Khatami et al., 2021).

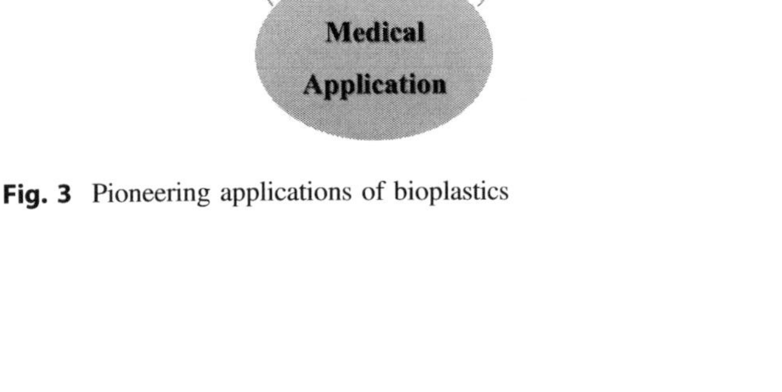

Fig. 3 Pioneering applications of bioplastics

6.3 Medical Application

The exceptional biocompatibility of polyhydroxyalkanoates (PHA) renders them invaluable for medical applications, eliciting favorable host responses and serving as pivotal biomaterials in various therapeutic interventions. Utilized across a spectrum of medical procedures, including tissue engineering employing techniques like injections, plates, pins, and fracture-fixation devices, PHA garnered significant attention from the Food and Drug Administration (FDA) in 2007 for its promising biomedical prospects. Research by Sabarinathan et al. (2018) underscored polyhydroxybutyrate (PHB)'s innate biocompatibility and highlighted its potential in cancer detection, with cancer cells exhibiting strong adherence to PHB sheets while normal cells did not. PHA-based implants have been instrumental in crafting heart valves, medication delivery systems, neurons, blood arteries, and orthopedic tissues, demonstrating long-term non-carcinogenicity upon implantation (Sabarinathan et al., 2018). Bioplastics, notably poly(3-hydroxybutyrate-co-3-hydroxyhexanoate) (PHBHHx), facilitate bone tissue growth synergistically with hydroxyapatite (HAP), enhancing compression elasticity and cellular proliferation. The versatility of PHBHHx extends to the creation of elastin-based artificial blood vessels, with P3HB4HB-diol exhibiting reduced platelet adhesion. In neurological applications, PHB fosters the survival, proliferation, and attachment of adult Schwann cells along nerve axons, promising advancements in peripheral nerve regeneration. Hydrophobic PHA nanoparticles and engineered PHA synthase systems, alongside PHB nanoparticles, emerge as promising drug delivery agents, notably in colon cancer treatment. Chaturvedi et al. (2015) successfully developed insulin delivery systems incorporating PHBHHx nanoparticulate phospholipids (INSPEC-NP), showcasing prolonged therapeutic effects, while PHB/polyethylene glycol (PHB PEG) nanoparticles effectively encapsulated insulin for sustained release (Potter et al., 2002; Chaturvedi et al., 2015).

7 Current Commercial Status of Bioplastics

The global expansion and evolution of the bioplastics industry stem from its commitment to sustainable manufacturing practices, which are markedly more eco-friendly than traditional plastic production methods. Bioplastics find extensive application across diverse sectors, notably in packaging and food services. Key drivers behind this surge include heightened awareness of the advantages of biocompatible and biodegradable plastics, governmental regulations advocating for eco-conscious practices, and escalating fossil fuel costs prompting a transition toward bioplastic manufacturing. Despite these advancements, the bioplastics market remains relatively modest compared to its petroleum-based counterpart. Production facilities are dispersed globally, with notable locations including China, Italy, Brazil, and the United States. Among the various materials, PHA stands out in multiple industries involved in bioplastic manufacturing. While microalgae display potential as a viable option for bioplastic production, further research is imperative to ascertain commercial feasibility, whether independently or in conjunction with bacteria (Chaturvedi et al., 2015; Shafqat et al., 2020).

For example, a mixture of red algae powder and water has been used to create a substitute for plastic bottles. To commercialize bioplastics made from algae, research must focus on the capacity of various algal species to produce bioplastics, metabolic engineering, and the assessment of several parameters affecting bioplastic accumulation. Since these factors significantly influence the cost of commercialized plastics, researchers must concentrate on advancing downstream processing methods and addressing the challenges related to algae-based plastics.

8 Microbial Bioplastic and Future Perspectives

Bioplastics have become a dynamic research topic worldwide, spurred by the growing demand for environmentally sound alternatives. This research has been driven by the need to address environmental problems and has progressed rapidly since the introduction of the first contemporary biopolymers about three decades ago. Bioplastics are a necessary replacement for conventional plastics, despite the hazards and benefits associated with their introduction. The potential of each biochemical process is influenced by several factors, including cost, performance, product yield, production rate, and downstream processing efficiency (Shamsuddin et al., 2017b).

Polyhydroxyalkanoates (PHAs) pose a significant economic challenge due to their higher production costs compared to conventional polymers, which hinders their widespread adoption. A critical aspect in overcoming this hurdle lies in the selection of cost-effective feedstocks, as substrate expenses typically represent a substantial portion of the overall cost of biologically derived products, ranging from 40–60%. Moreover, bioplastics currently exhibit a substantially lower production capacity compared to their petroleum-based counterparts. To facilitate the commercialization of PHA synthesis, a continuous and affordable supply of substrates is essential, which entails addressing logistical challenges in procurement and transportation. One proposed solution involves integrating PHA production into existing industrial facilities, transforming them into versatile biorefineries capable of utilizing industrial waste as a sustainable substrate for PHA production, thereby fostering reuse and recycling practices (Noorman & Heijnen, 2017).

In South America and Asia, there is a notable rise in large-scale production capacity fueled by the availability of cost-effective raw materials like sugar cane and molasses (Nguyen et al., 2018). The future of biopolymers appears promising as their sustainability gains recognition. Leveraging advanced biotechnological methods, the utilization of microorganisms holds immense potential in revolutionizing PHA production and addressing current obstacles (Storz & Vorlop, 2013). Microbial PHAs, with properties such as biocompatibility and biodegradability, emerge as promising alternatives to traditional petroleum-based plastics. However, challenges persist in scaling up bioplastic production from microbes, including issues in downstream processing, low conversion ratios of substrate to product, and variability in bioplastic characteristics across batches.

One approach to overcoming these hurdles involves engineering recombinant organisms capable of synthesizing bioplastics from inexpensive carbon sources like lactose, glycerol, molasses, sucrose, oils, and methane. This entails extensive exploration of microbial biosynthetic pathways and regulatory mechanisms. Additionally, investigating bioaugmentation strategies in mixed cultures holds significance for enhancing output. For instance, biomass bioaugmentation of Plasticicumulans acidivorans in mixed culture settings resulted in an impressive 85% PHB accumulation, surpassing results achieved in solo culture setups (Francis et al., 2021). Such hybrid cultures have the potential to revolutionize feedstock materials and amplify biopolymer synthesis, bringing us closer to commercial PHA production realization.

The retrieval of bioplastics constitutes a substantial portion of the overall expenses in PHA manufacturing. Developing cost-effective and eco-friendly methods for PHA recovery holds promise for significant cost reduction in production. Currently utilized techniques for PHA extraction, including digestion, solvent extraction, and flotation, involve substantial chemical usage, resulting in significant expenses. Moreover, PHA production costs about three times more than petrochemical-derived plastics due to the use of chlorinated solvents or surfactants (Marang et al., 2015).

Consequently, there is a pressing need for research to focus on adopting cost-effective and eco-friendly chemicals and solvents in bioplastics. Enhancing recovery efficiency also stands out as a pivotal challenge demanding attention. Research into morphological engineering, targeting alterations in the size and configuration of PHA granules, is essential for refining downstream processing. Exploring the pathways facilitating extracellular accumulation of PHA offers promising avenues for increasing recovery and potentially boosting production rates. Thus, the future of bioplastics hinges on efforts to balance cost and performance effectively.

The environmental impact of bioplastics remains a subject of debate; however, they generally exhibit lower environmental harm compared to conventional petroleum-based plastics. For example, when bioplastics are composted in large quantities within landfills, there is potential for methane release during decomposition, which could exacerbate global warming. Another concern involves the diversion of resources such as maize and corn from food production to bioplastic manufacture, which may threaten food security. Additionally, the use of chemical pesticides in cultivating these crops can exacerbate pollution. Finally, the discharge of chemicals used in converting organic materials into bioplastics can lead to environmental hazards (Atiwesh et al., 2021).

9 Conclusion

Recent developments in the microbial synthesis of bioplastics, which serve as sustainable substitutes for conventional plastics, are covered in this chapter. The various microbes capable of producing different kinds of bioplastics are examined. Depending on the substrates available in the growing medium, distinct metabolic pathways are involved in the synthesis of biopolymers, as detailed by in-depth studies of the molecular mechanisms of bioplastic formation. This chapter also discusses cutting-edge techniques for producing microbial bioplastics, such as genetic engineering, metabolic engineering, enhanced fermentation processes, and cell shape modification, all aimed at boosting the production and sustainability of bioplastics. Microbial bioplastics have a promising future due to their inherent biocompatibility and innovative applications. Although there may be some environmental concerns associated with bioplastics, they are generally less hazardous than petroleum-based plastics.

References

Adane, L., & Muleta, D. (2011). Survey on the usage of plastic bags, their disposal and adverse impacts on environment: A case study in Jimma City, southwestern Ethiopia. *Journal of Toxicology and Environmental Health Sciences, 3*, 234–248. https://doi.org/10.5897/JTEHS.9000067.

Agustin, M. B., Ahmmad, B., Alonso, S. M. M., & Patriana, F. M. (2014). Bioplastic based on starch and cellulose nanocrystals from rice straw. *Journal of Reinforced Plastics and Composites, 33*(24), 2205–2213.

Amaro, T. M. M. M., Rosa, D., Comi, G., & Iacumin, L. (2019). Prospects for the use of whey for polyhydroxyalkanoate (PHA) production. *Frontiers in Microbiology, 10*, 1–12. https://doi.org/10.3389/fmicb.2019.00992.

Atiwesh, G., Mikhael, A., Parrish, C. C., Banoub, J., & Le, T. A. T. (2021). Environmental impact of bioplastic use: A review. *Heliyon, 7*(9), e07918. https://doi.org/10.1016/j.heliyon.2021.e07918.

Bbiah, A. (2015). *The future potential economic impacts of a bio-plastics industry in the UK: A report for the Bio-based and Biodegradable Industries*.

Broeren, M. L. M., Kuling, L., Worrell, E., & Shen, L. (2017). Environmental impact assessment of six starch plastics focusing on wastewater-derived starch and additives. *Resources, Conservation and Recycling, 127*, 246–255. https://doi.org/10.1016/j.resconrec.2017.09.001.

Chandra, R., Iqbal, H. M. N., Vishal, G., Lee, H.-S., & Nagra, S. (2019). Algal biorefinery: A sustainable approach to valorize algal-based biomass towards multiple product recovery. *Bioresource Technology, 278*, 346–359. https://doi.org/10.1016/j.biortech.2019.01.104.

Chaturvedi, K., Ganguly, K., Kulkarni, A. R., Rudzinski, W. E., Krauss, L., Nadagouda, M. N., et al. (2015). Oral insulin delivery using deoxycholic acid conjugated PEGylated polyhydroxybutyrate co-polymeric nanoparticles. *Nanomedicine: Nanotechnology, Biology, and Medicine, 10*(10), 1569–1583. https://doi.org/10.1016/j.nano.2015.06.011.

Chen, Y., Awasthi, A. K., Wei, F., Tan, Q., & Li, J. (2020). Single-use plastics: Production, usage, disposal, and adverse impacts. *Science of the Total Environment, 752*, 141772. https://doi.org/10.1016/j.scitotenv.2020.141772.

Das, S. K., Sathish, A., & Stanley, J. (2018a). Production of biofuel and bioplastic from *Chlorella pyrenoidosa*. *Materials Today: Proceedings, 5*(8), 16774–16781.

Debuissy, T., Pollet, E., & Avérous, L. (2018). Biotic and abiotic synthesis of renewable aliphatic polyesters from short building blocks obtained from biotechnology. *Chemsuschem, 11*, 3836–3870. https://doi.org/10.1002/cssc.201801700.

Delgado, M., Felix, M., & Bengoechea, C. (2018). Development of bioplastic materials: From rapeseed oil industry by-products to added-value biodegradable biocomposite materials. *Industrial Crops and Products, 125*, 401–407.

Devasahayam, S., Bhaskar Raju, G., & Mustansar Hussain, C. (2019). Utilization and recycling of end of life plastics for sustainable and clean industrial processes including the iron and steel industry. *Materials Science & Energy Technologies, 2*, 634–646. https://doi.org/10.1016/j.mset.2019.08.002.

Emadian, S. M., Onay, T. T., & Demirel, B. (2017). Biodegradation of bioplastics in natural environments. *Waste Management, 59*, 526–536.

Ferreira, F. V., Cividanes, L. S., Gouveia, R. F., & Lona, L. M. F. (2019). An overview on properties and applications of poly(butylene adipate-co-terephthalate) (PBAT) based composites. *Polymer Engineering & Science, 59*(2), E7–E15. https://doi.org/10.1002/pen.25232.

Francis, C. F., Rebello, S., Mathachan Aneesh, E., Sindhu, R., Binod, P., Singh, S., & Pandey, A. (2021). Bioprospecting of gut microflora for plastic biodegradation. *Bioengineered, 12*(1), 1040–1053. https://doi.org/10.1080/21655979.2021.1913847.

Gadhave, R. V., Das, A., Mahanwar, P. A., & Gadekar, P. T. (2018). Starch-based bio-plastics: The future of sustainable packaging. *Open Journal of Polymer Chemistry, 8*, 21–33.

Giacovelli, C., & Zamparo, A. (2018). *Plastics: A roadmap for sustainability*. (T. Cannon (Ed.)).

Godfrey, L. (2019). Waste plastic, the challenge facing developing countries—ban it, change it, collect it? *Recycling, 4*(3). https://doi.org/10.3390/recycling4010003.

González García, Y., Meza Contreras, J. C., González Reynoso, O., & Cordova López, J. A. (2013). Síntesis y biodegradación de polihidroxialcanoatos: Plásticos de origen microbiano. *Revista Internacional De Contaminación Ambiental, 29*(1), 77–115. https://doi.org/10.1007/s11356-020-08056-w.

Goswami, G., Goswami, M. G., & Purohit, P. (2015). Bioplastics from organic waste. *International Journal of Engineering Research and Technology, 3*(23), 1–3.

Harding, K. G., Dennis, J. S., Von Blottnitz, H., & Harrison, S. T. L. (2007). Environmental analysis of plastic production processes: Comparing petroleum-based polypropylene and polyethylene with biologically-based poly-β-hydroxybutyric acid using life cycle analysis. *Journal of Biotechnology, 130*, 57–66. https://doi.org/10.1016/j.jbiotec.2007.02.012.

Isroi, I., Cifriadi, A., Panji, T., Nendyo, A. W., & Syamsu, K. (2017). Bioplastic production from cellulose of oil palm empty fruit bunch. *IOP Conference Series: Earth and Environmental Science, 65*.

Jabeen, N., Majid, I., & Nayik, G. A. (2015). Bioplastics and food packaging: A review. *Cogent Food & Agriculture, 2*(1), 1–6.

Joshi, S., Sharma, U., & Goswami, D. G. (2015). *Bio-plastic from waste newspaper*.

Kadouri, D., Jurkevitch, E., Okon, Y., & Castro-Sowinski, S. (2005). Ecological and agricultural significance of bacterial polyhydroxyalkanoates. *Critical Reviews in Microbiology, 31*, 55–67. https://doi.org/10.1080/10408410590899228.

Kaeb, H., Aeschelmann, F., Dammer, L., & Carus, M. (2016). *Market study on the consumption of biodegradable and compostable plastic products in Europe 2015 and 2020*. Nova-Institute Hürth.

Keziah, V. S., Gayathri, R., & Priya, V. V. (2018). Biodegradable plastic production from corn starch. *Drug Invention Today, 10*(7), 1315–1317.

Khatami, K., Perez-Zabaleta, M., Owusu-Agyeman, I., & Cetecioglu, Z. (2021). Waste to bioplastics: How close are we to sustainable polyhydroxyalkanoates production? *Waste Management, 119*, 374–388. https://doi.org/10.1016/j.wasman.2020.10.037.

Khosravi-Darani, K., & Bucci, D. Z. (2015). Application of poly (hydroxyalkanoate) in food packaging: Improvements by nanotechnology. *Chemical and Biochemical Engineering Quarterly, 29*(2), 275–285.

Kržan, A. (2012). Biodegradable polymers and plastics. In *Innovative value chain development for sustainable plastics in Central Europe*.

Kumar, Y., Shukla, P., Singh, P., Prabhakaran, P. P., & Tanwar, V. K. (2014). Bio-plastics: A perfect tool for eco-friendly food packaging: A review. *Journal of Food Production, Development & Packaging, 1*(1), 1–6.

Lothfy, F. A., Haron, M. F., & Rafaie, H. A. (2018). Fabrication and characterization of jackfruit seed powder and polyvinyl alcohol blend as biodegradable plastic. *Journal of Polymer Science & Technology, 3*(2), 1–5.

Marang, L., van Loosdrecht, M. C. M., & Kleerebezem, R. (2015). Modeling the competition between PHA-producing and non-PHA-producing bacteria in feast-famine SBR and staged CSTR systems. *Biotechnology and Bioengineering, 112*(12), 2475–2484. https://doi.org/10.1002/bit.25728.

Mohapatra, A., Prasad, S., & Sharma, H. (2015). *Bioplastics: Utilization of waste banana peels for synthesis of polymeric films*.

Momani, B. (2009). *Assessment of the impacts of bioplastics: Energy usage, fossil fuel usage, pollution, health effects, effects on the food supply, and economic effects compared to petroleum-based plastics*. Worcester Polytechnic Institute.

Nampoothiri, K. M., Nair, N. R., & John, R. P. (2010). An overview of the recent developments in polylactide (PLA) research. *Bioresource Technology, 101*(22), 8493–8501. https://doi.org/10.1016/j.biortech.2010.05.042.

Nguyen, R. T., Fishman, T., Zhao, F., Imholte, D. D., & Graedel, T. E. (2018). Analyzing critical material demand: A revised approach. *Science of the Total Environment, 630*, 1143–1148. https://doi.org/10.1016/j.scitotenv.2018.02.221.

Noorman, H. J., & Heijnen, J. J. (2017). Biochemical engineering's grand adventure. *Chemical Engineering Science*. https://doi.org/10.1016/j.ces.2016.12.065.

Pakalapati, H., Chang, C. K., Show, P. L., Arumugasamy, S. K., & Lan, J. C. W. (2018). Development of polyhydroxyalkanoates production from waste feedstocks and applications. *Journal of Bioscience and Bioengineering, 126*, 282–292. https://doi.org/10.1016/j.jbiosc.2018.03.016.

Potter, M., Madkour, M. H., Mayer, F., & Steinbüchel, A. (2002). Regulation of phasin expression and polyhydroxyalkanoate (PHA) granule formation in *Ralstonia eutropha* H16. *Microbiology, 148*(8), 2413–2426. https://doi.org/10.1099/00221287-148-8-2413.

Rocha, C. J. L., Álvarez-Castillo, E., Estrada Yáñez, M. R., Bengoechea, C., Guerrero, A., & Orta Ledesma, M. T. (2020). Development of bioplastics from a microalgae consortium from wastewater. *Journal of Environmental Management, 263*, 110353. https://doi.org/10.1016/j.jenvman.2020.110353.

Sabarinathan, D., Chandrika, S. P., Venkatraman, P., Easwaran, M., Sureka, C. S., & Preethi, K. (2018). Production of polyhydroxybutyrate (PHB) from *Pseudomonas plecoglossicida* and its application towards cancer detection. *Information Medicine Unlocked, 11*, 61–67. https://doi.org/10.1016/j.imu.2018.09.003.

Samadhiya, K., Sangtani, R., Nogueira, R., & Bala, K. (2022). Insightful advancement and opportunities for microbial bioplastic production. *Frontiers in Microbiology, 12*, 674864. https://doi.org/10.3389/fmicb.2021.674864.

Samantaray, S., & Mallick, N. (2015). Impact of various stress conditions on poly-β-hydroxybutyrate (PHB) accumulation in *Aulosira fertilissima* CCC 444. *Current Biotechnological, 4*, 366–372. https://doi.org/10.2174/2211550104666150806000642.

Schön, M., & Schwartz, P. (2014). *Production of bioplastic*.

Shafqat, A., Tahir, A., Mahmood, A., & Pugazhendhi, A. (2020). A review on environmental significance carbon footprints of starch based bio-plastic: A substitute of conventional plastics. *Biocatalysis and Agricultural Biotechnology, 27*, 101540. https://doi.org/10.1016/j.bcab.2020.101540.

Shamsuddin, I. M., Jafar, J. A., Shawai, A. S. A., Yusuf, S., Lateefah, M., & Aminu, I. (2017a). Bioplastics as better alternative to petroplastics and their role in national sustainability: A review. *Advances in Bioscience and Bioengineering, 5*(4), 63–70.

Shamsuddin, I. M., Jafar, J. A., Shawai, A. S. A., Yusuf, S., Lateefah, M., & Aminu, I. (2017b). Bioplastics as a better alternative to petroplastics and their role in national sustainability: A review. *Advances in Bioscience and Biotechnology, 5*(4), 63–70. https://doi.org/10.4236/abb.2017.54007.

Simo-Cabrera, L., García-Chumillas, S., Hagagy, N., Saddiq, A., Tag, H., et al. (2021). Haloarchaea as cell factories to produce bioplastics. *Marine Drugs, 19*(3), 159. https://doi.org/10.3390/md19030159.

Siracusa, V., & Blanco, I. (2020). Bio-polyethylene (Bio-PE), bio-polypropylene (Bio-PP) and bio-poly(ethylene terephthalate) (Bio-PET): Recent developments in bio-based polymers analogous to petroleum-derived ones for packaging and engineering applications. *Polymers, 12*(7), 1641. https://doi.org/10.3390/polym12071641.

Solaiman, D. K. Y., Ashby, R. D., Foglia, T. A., & Marmer, W. N. (2006). Conversion of agricultural feedstock and coproducts into poly(hydroxyalkanoates). *Applied Microbiology and Biotechnology, 71*, 783–789. https://doi.org/10.1007/s00253-006-0451-1.

Storz, H., & Vorlop, K. D. (2013). Bio-based plastics: Status, challenges and trends. *Landbauforschung-Applied Agricultural and Forestry Research, 63*(4), 321–332. https://doi.org/10.3220/LBF_2013_321.

Sushmitha, B. S., Vanitha, K. P., & Rangaswamy, B. E. (2016). Bioplastics: A review. *International Journal of Modern Trends in Engineering and Research, 3*(4), 411–413.

Trivedi, J., Aila, M., Bangwal, D. P., Kaul, S., & Garg, M. O. (2015). Algae based biorefinery—how to make sense? *Renewable & Sustainable Energy Reviews, 47*, 295–307. https://doi.org/10.1016/j.rser.2015.03.052.

Varghese, S., Dhanraj, N. D., Rebello, S., Sindhu, R., Binod, P., Pandey, A., Jisha, M. S., & Awasthi, M. K. (2022). Leads and hurdles to sustainable microbial bioplastic production. *Chemosphere, 305*, 135390. https://doi.org/10.1016/j.chemosphere.2022.135390.

Verlinden, R. A., Hill, D. J., Kenward, M., Williams, C. D., & Radecka, I. (2007). Bacterial synthesis of biodegradable polyhydroxyalkanoates. *Journal of Applied Microbiology, 102*(6), 1437–1449. https://doi.org/10.1111/j.1365-2672.2007.03319.x.

Wen, Y., Kierzek, K., Min, J., Chen, X., Gong, J., Niu, R., et al. (2020). Porous carbon nanosheet with high surface area derived from waste poly(ethylene terephthalate) for supercapacitor applications. *Journal of Applied Polymer Science, 137*(14), 1–10. https://doi.org/10.1002/app.48338.

Wu, C. S. (2012). Characterization and biodegradability of polyester bioplastic-based green renewable composites from agricultural residues. *Polymer Degradation and Stability, 97*, 64–71. https://doi.org/10.1016/j.polymdegradstab.2011.10.012.

Biotechnological Exploitation of Cyanobacteria and Microalgae for Bioactive Compounds

Indira Mikkili and Vidya Prabhakar Kodali

Abstract

The biotechnological exploitation of cyanobacteria and microalgae represents a cutting-edge approach to discovering sustainable and potent bioactive compounds with wide-ranging applications in pharmaceuticals, nutraceuticals, agriculture, and environmental management. These microorganisms are prolific producers of a diverse array of bioactive molecules including pigments, proteins, polysaccharides, vitamins, and lipids, each exhibiting unique and valuable biological activities such as antioxidant, anti-inflammatory, antimicrobial, and anticancer properties. Recent advances in genetic and metabolic engineering have significantly enhanced the production of these compounds by optimizing metabolic pathways, and creating high-yield strains. Additionally, innovative cultivation systems and bioprocessing technologies have improved the scalability and economic viability of algae-based production. This book chapter highlights the latest developments in the field, emphasizing the potential of cyanobacteria and microalgae as sustainable and versatile sources of bioactive compounds. It also explores advancements in genetic and metabolic engineering that have propelled the field forward. Future directions include the exploration of novel strains, further optimization of production processes, and the integration of these biotechnological advances into commercial applications across various industries. By harnessing the full potential of cyanobacteria and microalgae, we can develop innovative solutions to meet the growing demand for natural and effective bioactive compounds.

I. Mikkili (✉)
Department of Biotechnology, Vignan's Foundation for Science, Technology and Research, Vadlamudi, Andhra Pradesh 522213, India
e-mail: indiramikkili@gmail.com

V. P. Kodali
Department of Biotechnology, Vikrama Simhapuri University, Nellore, Andhra Pradesh 524004, India

Keywords

Bioactive compounds • Bioprocessing • Cyanobacteria • Genetic Engineering • Pigments • Pharmaceutical • Nutraceuticals

1 Introduction

The biotechnological exploitation of cyanobacteria and microalgae has garnered significant attention due to their immense potential as sources of bioactive compounds (Encarnacao et al., 2015). Their vast biodiversity and ability to thrive in diverse and extreme environmental conditions make them exceptionally promising reservoirs for developing novel bioactive compounds (Encarnacao et al., 2015). Cyanobacteria, known as blue-green prokaryotic organisms inhabit diverse environments such as freshwater, marine habitats, deserts and hot springs (Mehdizadeh Allaf and Peerhossaini, 2022). They perform photosynthesis and play pivotal roles in both oxygen production and nutrient cycling within ecosystems (Kumar et al., 2015). In contrast, microalgae are eukaryotic organisms encompassing a broad spectrum of microscopic algae (Ahmad et al., 2020). Similar to cyanobacteria, microalgae engage in photosynthesis and contribute significantly to oxygen production and nutrient cycling (Chia et al., 2020). These organisms produce chemically and biologically active molecules of high commercial value, including pigments, lipids, proteins, peptides, polysaccharides, polymers, enzymes, vitamins, and numerous other bioactive substances with pharmaceutical, nutraceutical, and industrial applications (Fig. 1) (Andrade et al., 2018). Cyanobacteria and microalgae possess several advantageous traits that render them highly suitable for biotechnological applications (Deviram et al., 2020). Unlike traditional crops, they do not require arable land and can be cultivated in saline water and wastewater environments,

J. Devi (ed.), *Biotechnological Intervention in Production of Bioactive Compounds*,
Sustainable Landscape Planning and Natural Resources Management, https://doi.org/10.1007/978-3-031-76859-0_14

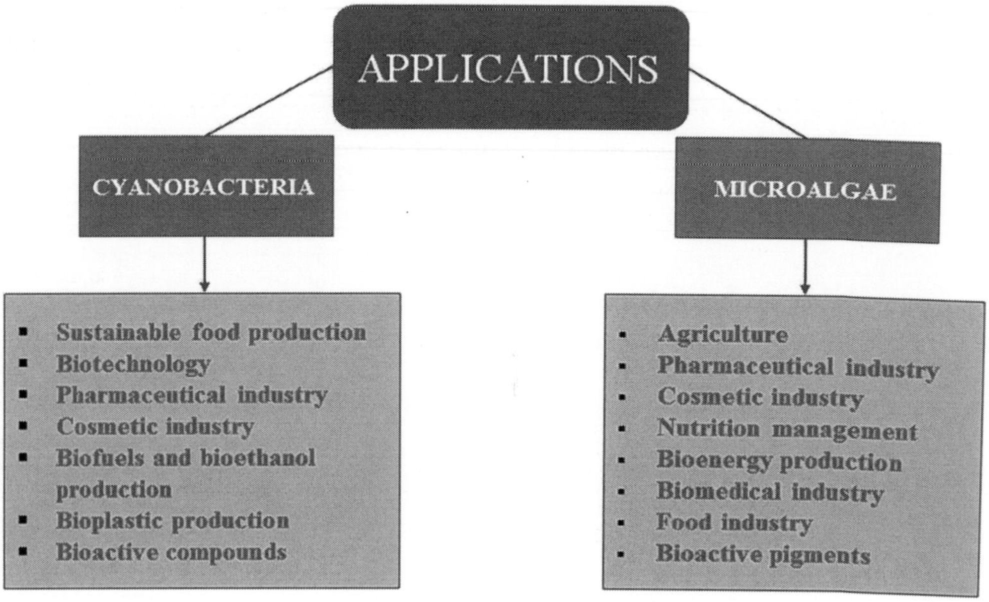

Fig. 1 Applications of bioactive compounds derived from cyanobacteria and microalgae

offering a sustainable alternative that does not compete with food production (Agarwal et al., 2022).

Algae possess remarkably high growth rates and high biomass productivity, making them highly attractive for bioactive compound production (Agarwal et al., 2022). The biochemical nature of these compounds can be strategically manipulated by altering the cultivation media and conditions, enabling the targeted production of specific bioactive compounds (Encarnacao et al., 2015). This adaptability is crucial for optimizing yields and enhancing the economic feasibility of microalgae- and cyanobacteria-based biotechnological processes (Razzak et al., 2023). Additionally, they contribute to carbon dioxide sequestration, wastewater remediation and the degradation of toxic compounds in the environment through specific enzymatic processes (Touliabah et al., 2022).

The commercial value of bioactive compounds from cyanobacteria and microalgae extends across various industries (Papadaki et al., 2024). In the pharmaceutical sector, these compounds show promise for the development of new drugs and therapeutic agents, offering potential treatments for a range of diseases and conditions (Encarnacao et al., 2015). In the food and nutraceutical industries, they serve as a rich source of dietary supplements and functional foods (Andrade et al., 2018). Notable compounds produced by these microorganisms include astaxanthin, β-carotene, phycobiliproteins, exopolysaccharides, "eicosapentaenoic acid" and docosahexaenoic acid (Borowitzka, 2013). Additionally, their use in producing biofertilizers, biopolymers, bioplastics and animal feed highlights the versatile utility of these microorganisms (Chia et al., 2020).

Microalgae produce a variety of secondary metabolites with antimicrobial, antiviral, antioxidant, anti-inflammatory, anti-proliferative and anticancer activities (Andrade et al., 2018). For example, compounds such as microcystins, despite their toxicity, have shown potential in cancer research due to their ability to inhibit certain cellular processes (Abdul et al., 2016). The manipulation of cultivation conditions and genetic engineering can further enhance the yield and specificity of these bioactive compounds, making cyanobacteria a versatile tool in the development of new therapeutic agents (Lopez-Hernandez et al., 2022). Additionally, cyanobacteria can be genetically engineered to boost the production of biofuels, such as bioethanol, biodiesel, and biohydrogen (Singh et al., 2023). This genetic modification enables the efficient production of renewable and environmentally friendly alternatives to fossil fuels, positioning cyanobacteria as key players in sustainable energy solutions (Agarwal et al., 2022).

2 Diversity of Cyanobacteria and Microalgae

2.1 Cyanobacterial Diversity

Cyanobacteria are the most abundant and potent primary producers on earth, playing a pivotal role in global photosynthesis and biogeochemical cycles (Gaysina et al., 2019). They thrive in a multitude of habitats, including freshwater, marine environments, terrestrial areas, and extreme conditions such as deserts, polar regions and thermal springs (Nandagopal et al., 2021). Their remarkable adaptability allows them to colonize diverse environments, significantly contributing to primary productivity and maintaining ecological balance (Abdul et al., 2016). Beyond their ecological roles, cyanobacteria commonly form symbiotic relationships with various organisms engaging in mutualistic associations with plants, fungi, and eukaryotic algae (Barta et al., 2021). Taxonomists recognize 170 cyanobacterial genera based on 16S rRNA sequences alone (Walter et al., 2017). Prominent genera include *Anabaena*, *Oscillatoria*, *Spirulina*, *Nostoc*, *Calothrix*, *Microcystis*, *Synechococcus*, *Nodularia*, *Prochlorococcus*, *Gloeocapsa* and *Aphanizomenon* (Singh et al., 2016). Cyanobacteria exhibit extensive diversity in habitat, morphology and functionality, mirrored in their genomic characteristics, such as genome size, GC content and coding regions (Prabha et al., 2016). These variations reflect the adaptive evolutionary processes that cyanobacterial species undergo in response to niche-specific competitive pressures (Larsson et al., 2011). The genomic distinctiveness of each species stems from unique selective forces encountered over evolutionary time, leading to specialized adaptations and functions (Alcorta et al., 2020).

Research indicates that the light-harvesting phycobilisome (PBS) and photosystem II (PSII) are pivotal components that facilitate the survival of thermophilic cyanobacteria (Li et al., 2023a). These adaptations enable them to thrive in extreme heat, underscoring their evolutionary resilience and functional versatility in diverse ecological niches (Abdul et al., 2016). Phycobilisomes are exceptionally efficient light-harvesting complexes essential for photosynthesis in cyanobacteria (Abdul et al., 2016). These complexes capture light energy and transfer it to the photosynthetic reaction centers, particularly PSII, there by playing a crucial role in the photosynthetic process (Liu and Blankenship, 2019). Composed of multiple pigments and proteins arranged in a highly organized structure, phycobilisomes primarily include phycobiliproteins such as phycoerythrin, phycocyanin and allophycocyanin (Chen et al., 2022).

Cyanobacteria exhibit a wide range of morphologies, including colonial, unicellular and multicellular filamentous forms (Mehdizadeh Allaf and Peerhossaini, 2022). Their cell sizes vary from less than 1 μm in diameter to up to 100 μm in some tropical species of the genus *Oscillatoria* (Mehdizadeh Allaf and Peerhossaini, 2022). Unicellular cyanobacteria possess oval, cylindrical or spherical shapes that can aggregate into colonies bound by mucilage secreted during growth (Vidal et al., 2021). The number of cells in each colony can range from two to several thousand, depending on the species. Filamentous cyanobacteria can differentiate into three prevalent cell types such as hormogonia, heterocysts and akinetes (Waditee-Sirisattha and Kageyama, 2022). Hormogonial cells are motile, short filaments with gliding motility and gas vacuoles for buoyancy, playing a key role in relocation and symbiotic colonization (Risser, 2023). Heterocysts, found in multicellular cyanobacteria such as *Anabaena* and *Nostoc*, have thicker envelopes and specialize in nitrogen fixation (Zulkefli and Hwang, 2020). Akinetes are thick walled, spore like cells formed under adverse conditions such as drought and cold and are considered evolutionary precursors to heterocysts (Garg and Maldener, 2021).

2.2 Microalgal Diversity

Algae, encompassing a heterogeneous group of microorganisms, range from microscopic blue-green algae to large, complex seaweeds measuring up to several meters in length (Belghit et al., 2017). Based on their size, algae are classified into two categories: microalgae and macroalgae (Singh and Saxena, 2015). Microalgae represent a highly diverse group, including both eukaryotic photoautotrophic protists and prokaryotic cyanobacteria (Batra et al., 2023). They are classified into chlorophyceae (green algae), cyanophyceae (blue-green algae), chrysophyceae (golden algae) and bacillariophyceae (diatoms) (Madeira et al., 2017). The morphological diversity within these groups is striking, ranging from simple unicellular forms to complex multicellular and colonial structures (Singh and Saxena, 2015). Microalgae can be spherical, oval, filamentous or even flagellated reflecting their adaptation to various ecological niches and environmental conditions (Rodriguez-Zuniga et al., 2024). Their phenotypic diversity is evident in their nearly ubiquitous distribution across the biosphere (Rasal et al., 2019), thriving in freshwater bodies, brackish waters, marine ecosystems, and extreme conditions like cold, hot springs and hyper saline lakes (Rinaldi et al., 2024).

Green microalgae predominantly thrive in freshwater and seawater, while several other species are adapted to grow in extremely saline environments (Abu-Ghosh et al., 2021). Microalgae exhibit diverse characteristics, including planktonic, epiphytic, and benthic forms (Arsad et al., 2022). These microorganisms inhabit various sub-habitats, each uniquely suited to their ecological roles (Abdul et al., 2016). Benthic microalgae thrive in sedimentary environments, while epiphytic microalgae attach to substrates such as seaweed and mangrove roots (Arsad et al., 2022). Planktonic microalgae are found suspended in the water column. As autotrophic microorganisms, they harness light energy to synthesize essential compounds such as carbohydrates, proteins, and lipids (Lopez-Hernandez et al., 2022).

The physiological and metabolic diversity of microalgae is profound, enabling them to perform photosynthesis and produce a wide range of biochemical compounds (Hu et al., 2023). These compounds include carbohydrates, proteins, lipids, pigments, and various bioactive molecules (Ma et al., 2020). Microalgae utilize different photosynthetic pathways to optimize energy capture and utilization (Li et al., 2023b). This metabolic flexibility not only supports their survival in varying light and nutrient conditions but also underpins their potential for biotechnological applications including biofuel production, pharmaceuticals, and nutraceuticals (Rocha et al., 2017). Microalgae are harnessed for their ability to produce high-value compounds such as antioxidants, omega-3 fatty acids, and bioactive peptides (Encarnacao et al., 2015). Genetic engineering and advanced cultivation techniques are further expanding their potential, making microalgae a promising resource for sustainable solutions in energy, health and environmental management (Hassanien et al., 2023).

Despite their diminutive size, ranging from 0.2 to 200 μm, microalgae significantly contribute to climate change mitigation through carbon fixation (Reisoglu and Aydin, 2023). Biological carbon capture and sequestration using microalgae is acknowledged as one of the most effective methods for reducing atmospheric CO_2 levels (Zielinski et al., 2023). This promising approach not only facilitates CO_2 capture but also enables its conversion into biomass. The resulting biomass can be harnessed to produce bioenergy and other high-value products emphasizing the multifaceted benefits of microalgal carbon sequestration techniques (Wang et al., 2023).

3 Genetic and Metabolic Engineering

Genetic and metabolic engineering of cyanobacteria and microalgae is a rapidly advancing field aimed at optimizing these microorganisms for various biotechnological applications (Hassanien et al., 2023). These photosynthetic organisms are valuable due to their ability to convert carbon dioxide into biomass and valuable metabolites using light energy (Mehdizadeh Allaf and Peerhossaini, 2022). The focus of engineering efforts is to enhance their natural capabilities for the production of biofuels, bioproducts, and as platforms for bioremediation (Singh et al., 2023).

Recent advancements in genetic engineering techniques have significantly enhanced the manipulation of cyanobacteria and microalgae genomes (Hassanien et al., 2023). Tools such as CRISPR-Cas9, TALENs, and ZFNs have been successfully applied to introduce precise genetic modifications, enabling the enhancement of metabolic pathways (Jeong et al., 2023). For instance, CRISPR-Cas9 has been used to knock out or overexpress specific genes in cyanobacteria like *Synechocystis* sp. and microalgae such as *Chlamydomonas reinhardtii*, leading to increased production of biohydrogen, bioethanol, and other valuable bioproducts (Parakh et al., 2023). Additionally, synthetic biology approaches are being employed to construct artificial operons and regulatory circuits that optimize metabolic flux towards desired products (Agarwal et al., 2022).

Metabolic engineering focuses on rewiring the metabolic pathways within cyanobacteria and microalgae to maximize the yield of target compounds (Bongirwar and Shukla, 2023). This involves the identification and modification of key enzymes and pathways involved in carbon fixation, lipid biosynthesis, and pigment production (Song et al., 2024). By using omics technologies such as genomics, proteomics and metabolomics, researchers can comprehensively map metabolic networks and identify bottlenecks (Pathania et al., 2022). Advanced metabolic models and computational tools aid in predicting the effects of genetic modifications, facilitating the design of more efficient strains (Agarwal et al., 2022). For example, the introduction of heterologous pathways for the synthesis of high-value compounds such as polyhydroxyalkanoates and isoprenoids has been demonstrated in engineered cyanobacteria (Al-Haj et al., 2016).

Genetic and metabolic engineering of cyanobacteria and microalgae is unlocking new potentials for sustainable production of biofuels, bioproducts, and environmental applications (Agarwal et al., 2022). The continuous improvement of genetic tools and metabolic models, coupled with a deeper understanding of these organisms' biology, is driving innovation and commercialization in the field (Agarwal et al., 2022). Future research will likely focus on enhancing the robustness and scalability of engineered strains, optimizing cultivation conditions, and integrating these technologies into existing industrial frame works (Rueda et al., 2024).

4 Bioactive Compounds

With their broad diversity, over 30,000 varieties of cyanobacteria and microalgae have been identified. However, relatively few studies have explored the bioactive compounds they produce, which hold significant industrial and pharmaceutical potential (Abdul et al., 2016). These secondary metabolites, while not essential for primary metabolic functions, include antibiotics, hormones, toxins and other bioactive molecules (Encarnacao et al., 2015). As a major component of phytoplankton, cyanobacteria and microalgae are crucial to the global food chain and essential for sustaining Earth's ecosystems. Their adaptability to various cell culture conditions enhances their versatility, leading to the production of a wide array of valuable bioactive compounds (Fig. 2) (Encarnacao et al., 2015). A comprehensive understanding of the exploration, screening, extraction and commercialization of these compounds, along with their applications, will open new avenues in bioprospecting and drug discovery (Hassan et al., 2022).

4.1 Cyanobacterial Bioactive Compounds

Cyanobacteria's adaptability to environmental changes, rapid growth under favourable conditions, and unique

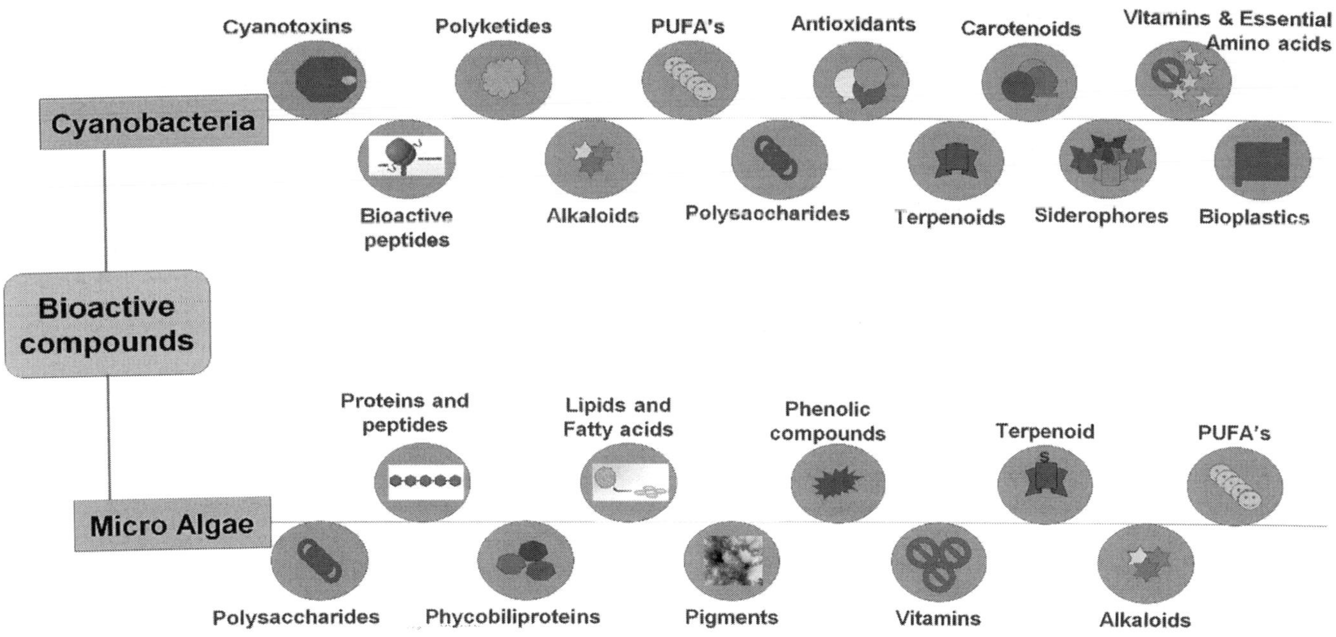

Fig. 2 Bioactive compounds derived from cyanobacteria and microalgae

characteristics have led researchers to consider their potential for industrial scale applications worldwide (Zahra et al., 2020). Their development can be attributed to unique photosynthetic machinery and the production of bioactive compounds, which protect them against external stress and function as intracellular signalling molecules, enhancing colonization (Nandagopal et al., 2021). With advanced photosynthetic systems, cyanobacteria can convert solar energy into metabolites and food, making them valuable biocatalysts in white biotechnology for the sustainable production of food, metabolites and green energy sources like biodiesel (Agarwal et al., 2022). Their diverse applications span biotechnology, pharmaceuticals, industry, and cosmetics (Table 1). Emerging harvesting technologies are also being developed to boost the commercial production of biofuels and bioethanol (Begum et al., 2021). The primary classes of bioactive compounds found in cyanobacteria include.

4.1.1 Cyanotoxins

Cyanotoxins are bioactive compounds produced by certain cyanobacteria species, which can have harmful effects on living organisms (Dulic et al., 2022). These toxins are classified into several types, including microcystin, anatoxin-a, nodularin, and cylindrospermopsin (Huang and Zimba, 2019) Microcystins, in particular, are hepatotoxins associated with liver toxicity in mammals and other vertebrates. They are produced by cyanobacteria such as *Anabaena, Microcystis, Anabaenopsis, Nostoc and Hapalosiphon, Planktothrix,* and *Aphanizomenon* (Abeysiri et al., 2022) Microcystins primarily cause liver damage and genotoxicity through both acute and chronic exposure (McLellan and Manderville, 2017).

Nodularins are similar to microcystins and have been reported worldwide for their toxic effects. These toxins are produced by *Nodularia spumigena* and are most commonly found in brackish water (Chen et al., 2022). Nodularins primarily lead to hepatotoxic effects (Rymuszka et al., 2021). Cylindrospermopsin is a class of alkaloids produced by organisms such as *Raphidiopsis, Aphanizomenon, Anabaena, Umezakia, Microseira, Oscillatoria and Dolichospermum* spp. (Dulic et al., 2022). These toxins have several toxic effects, primarily on the liver, and can also adversely affect the spleen, eyes, kidney, thymus, lungs, heart, and other organs (Pichardo et al., 2017). Anatoxin-a is a cyanotoxin produced by genera such as *Anabaena, Cylindrospermum, Aphanizomenon, Microcystis, Planktothrix and Oscillatoria* which are mostly found in freshwater bodies (Plata-Calzado et al., 2022). Human intoxication with anatoxin-a can result in symptoms such as ataxia, urinary incontinence, asthenia, salivation, lacrimation, blurred vision, muscle cramps, dizziness, headaches, paraesthesia, convulsions, respiratory failure and cerebral hypoxia (Plata-Calzado et al., 2022).

4.1.2 Bioactive Peptides

Cyanobacteria offer significant therapeutic potential, due to their peptide-like structures and peptides, which are of considerable interest to both pharmaceutical industry and research labs (Ahmad et al., 2020). Among these,

Table 1 Bioactive compounds derived from cyanobacteria and microalgae and their biological activities

Bioactive compounds	Cyanobacterial and algal species	Biological activities	References
Microcystin	*Anabaena, Microcysts, Anabaenopsis, Nostoc, Hapalosiphon, Planktothrix* and *Aphanizomenon*	Hepatotoxicity, and genotoxicity	Abeysiri et al. (2022)
Nodularin	*Nodularia spumigena*	Hepatoxic effects	Chen et al. (2022), Rymuszka et al. (2021)
Cylindrospermopsin	*Raphidiopsis, Aphanizomenon, Anabaena, Umezakia, microseira, Oscillatoria* and *Dolichospermum sp.*	Toxicity of liver	Pichardo et al. (2017)
Anatoxin	*Anabaena, Cylindrospermum, Aphanizomenon, Microcystin, Planktothrix* and *Oscillatoria*	Intoxication	Plata-Calzado et al. (2022)
Curacin-A	*Lyngbya majuscula*	Antimitotic agent	Qamar et al. (2021)
Apratoxin-A	*Lyngbya majuscula*	Anticancer activity	Pooja (2022)
Dolastatin-10	*Symploca* species	Antiproliferative agent	Pooja (2022)
Lyngbyatoxins	*Lyngbya majuscula*	Induces skin blisters and inflammation	Cock and Cheesman (2023)
Hapalimdoles	*Hapalosiphon fotinalis*	Antimitotic, antibacterial, insecticidal, immunomodulatory and anti-tumoral activities	Hohlman and Sherman (2021)
Antillatoxin	*Lyngbya majuscula*	Ichthyotoxic properties and cytotoxic activity	Bonnard et al. (2020)
Alginate	Phaeophyceae family	Used in food industry as a gelling agent, thickener, and adhesive to enhance food texture	Pereira and Cotas (2020)
Carrageenan	Rhodophyceae family	Used in the food industry for thickening, gelling, protein-binding, and stabilizing, and in pharmaceuticals for drug delivery systems	Ali and Ahmed (2018)
Caulerpin and caulersin	*C. peltata, C. racemosa, C. cupressoides, C. paspaloides, C. prolifera, C. sertularioides, C. mexicana, and C. lentillifera,* as well as in the red algae *Chondria armata*	Cytotoxicity and anti-inflammatory properties	Hohlman and Sherman (2021)

non-ribosomal peptides (NRPs) stand out as secondary metabolites produced by non-ribosomal peptide synthases (NRPS) (Nandagopal et al., 2021). NRPS are notable for their structural diversity and biological activity. In contrast, cyanobactins are ribosomally synthesized and post-translationally modified peptides (RiPPs) found in cyanobacteria (Ortega et al., 2016). Cyanobactins are produced on the ribosome using proteinogenic amino acids and have a range of applications including as algaecides, fungicides, insecticides, and herbicides (Kultschar and Llewellyn, 2018). In the pharmaceutical field, they are utilized as anticancer, antimicrobial, antiviral agents and immunosuppressants (Kultschar and Llewellyn, 2018). Another important class of cyanobacterial peptides is lipopeptides, known for their cytotoxic and antifungal properties. These compounds demonstrate potent bioactivity, making them valuable for therapeutic purposes (Ahmad et al., 2020).

4.1.3 Polyketides

Polyketides are natural products known for their remarkable functional and structural diversity, exhibiting properties such as antibiotic, antifungal, anticancer, and immunosuppressive activities (Barta et al., 2021). As a result, these compounds have been extensively studied and synthesized by scientists. Polyketides are isolated from common cyanobacterial species such as *Symploca, Okenia, Oscillatoria and Paraliomixa miuraensis,* each demonstrating unique biological activities (Barta et al., 2021). Curacin A, derived from the cyanobacterium *Lyngbya majuscula*, is a notable anti-tubulin agent with significant therapeutic potential, functioning as an antimitotic agent, and being explored for its applications in cancer treatment (Qamar et al., 2021). Apratoxin A, another cytotoxic compound derived from *Lyngbya majuscula,* is known for its ability to inhibit the fibroblast growth factor signalling pathway, which is crucial in cancer progression.

Apratoxin A is particularly recognized for its anticancer activity (Pooja, 2022). Dolastatin 10, a pentapeptide bioactive compound derived from the *Symploca* species, functions as a potential antiproliferative agent and is characterized by its anticancer activity (Pooja, 2022).

4.1.4 Alkaloids

Alkaloids are a diverse group of nitrogen containing compounds produced by marine microorganisms, known for their pharmacological applications including chemotherapy and pain relief (Hohlman and Sherman, 2021). They exhibit a wide range of structures and often serve as potent toxins against predators. Over a thousand alkaloids have been isolated from cyanobacteria, showcasing their extensive diversity and biological significance (Shah et al., 2017). Lyngbyatoxins are dermatotoxic alkaloids produced by *Lyngbya majuscula*. High exposure to these compounds leads to the activation of the enzyme protein kinase C, which induces skin blisters and inflammation (Cock and Cheesman, 2023). Hapalindoles are a significant class of indole alkaloids isolated from *Hapalosiphon fotinalis*. These compounds exhibit a wide range of biological activities, including antimitotic, antibacterial, insecticidal, immunomodulatory, and anti-tumoral effects (Hohlman and Sherman, 2021). Antillatoxin is a secondary metabolite produced by *Lyngbya majuscula*. This neurotoxic lipopeptide is known for its ichthyotoxic properties, making it a potent toxin to fish (Ahmad et al., 2020). Additionally, antillatoxin exhibits prominent cytotoxic activity, highlighting its potential in cancer research and drug development (Abdul et al., 2016).

4.1.5 Poly Unsaturated Fatty Acids

Fatty acids in cyanobacteria play critical roles in metabolism, acting as regulatory molecule and essential components (Andrade et al., 2018). These molecules are integral to energy storage, membrane fluidity and signalling pathways, thereby contributing significantly to the survival and adaptability of cyanobacteria in diverse environments (De Carvalho and Caramujo, 2018). Poly unsaturated fatty acids (PUFAs) such as eicosapentaenoic acid (EPA) and docosahexaenoic acid (DHA) are well known for their health benefits, particularly for cardiovascular health and brain function. These fatty acids play crucial roles in reducing inflammation, lowering blood pressure and supporting cognitive health (Barta et al., 2021). In cyanobacteria, PUFAs are essential for maintaining membrane fluidity and optimizing photosynthetic efficiency under varying environmental conditions. (Barta et al., 2021). Gamma-linoleic acid (GLA), a derivative of omega-6 fatty acids found in cyanobacteria, is notable for its anti-inflammatory properties. GLA is involved in the biosynthesis of anti-inflammatory eicosanoids, which can help reduce symptoms of chronic inflammatory conditions (Abdul et al., 2016).

4.1.6 Polysaccharides

Species like *Nostoc* and *Spirulina* produce exopolysaccharides (EPS) with notable antitumor, antioxidant, and immunomodulatory activities. These EPS are valued in the cosmetics and food industries for their gelling and stabilizing properties, which enhance the structure and stability of products (Synytsya et al., 2023).

4.1.7 Antioxidants

Cyanobacteria can produce phenolic compounds, which are well-known antioxidants that mitigate oxidative damage within cells (Batra et al., 2023). Species such as *Anabaena, Nostoc, Oscillatoria, Microcheate, Synechocystis, Hepalosiphon, Scytonema, Mastigocladus, Cylindrospermum, Westiellopsis, Aulosira, Lyngbya, Chroococcus, Dichothrix, Calothrix, Limnothrix and Phormidium* have been found to exhibit antioxidant activity, making them potential sources of new secondary metabolites (Begum et al., 2021).

4.1.8 Terpenoids

Cyanobacteria show promise for sustainable terpenoid production utilizing light and CO_2 (Knoot et al., 2018). Various terpene hydrocarbons, such as isoprene, limonene, β-phellandrene, and farnesene have been produced by expressing terpene synthases and precursor pathways in cyanobacteria. However, productivity levels need further optimization for commercial viability (Lin and Pakrasi, 2019).

4.1.9 Carotenoids

Several photosynthetic bacteria produce carotenoids. For instance, *Anabaena* species produce β-carotene, echinenone, and canthaxanthin; *Synechococcus* species are known for producing zeaxanthin, cryptoxanthin, myxoxanthophyll and 3′-hydroxyechinenone; *Nostoc punctiforme* produces β-carotene, canthaxanthin and echinenone; *Thermosynechococcus elongatus* produces β-carotene and nostoxanthin; *Gloeobacter violaceus* produces β-carotene and echinenone; and *Prochlorococcus marinus* produces β-carotene, zeaxanthin, and β-cryptoxanthin (Begum et al., 2021). These carotenoids have applications in health supplements, cosmetics, and as natural colorants (Begum et al., 2021).

4.1.10 Siderophores

Siderophores are low molecular weight chelating agents that enhance the ability of plants to uptake iron (Fe), thus increasing their yield. These ecofriendly compounds play a crucial role in agriculture (Swarnalatha et al., 2022). Cyanobacteria, such as *Anabaena oryzae,* produce dihydroxymate type siderophores like schizokinen and anachelin H. These siderophores act as biological sequestering agents to mitigate cadmium metal ions, thereby improving crop productivity by reducing heavy metal toxicity in soil (Ghosh et al., 2020).

4.1.11 Vitamins and Essential Amino Acids

Cyanobacteria are highly valued for their production of vitamins and essential amino acids, which offer various health benefits. Species such as *Spirulina, Arthrospira, Anabaena* and *Synechococcus* produce vitamins A, B, C, E and K in response to environmental stress (Zahra et al., 2020). These vitamins play critical roles in maintaining health and preventing diseases. Additionally, cyanobacteria produce essential amino acids that help defend against oxidative stress and prevent the destruction of cell membranes contributing to overall cellular health and function (Bongirwar and Shukla, 2023).

4.1.12 Bioplastics

Cyanobacteria have been demonstrated significant potential as a sustainable and environmentally friendly alternative to traditional petroleum-based plastics (Nanda and Bharadvaja, 2022). These photosynthetic prokaryotes, with their minimal growth requirements, efficient photosynthesis and significant lipid content, are promising candidates for carbon sequestration, biofuel and bioplastic production (Agarwal et al., 2022). In particular, cyanobacteria are being explored for the production of polyhydroxyalkanoates (PHAs), which are eco-friendly alternatives to conventional petrochemical-based plastics (Agarwal et al., 2022). The process of converting CO_2 sequestered by bioplastic-producing microalgae directly into polymers is crucial for preventing the release of CO_2 back into the environment, thereby contributing to carbon neutrality (Sadvakasova et al., 2023). Additionally, cyanobacteria offer an innovative solution to white pollution, the wide spread environmental contamination caused by plastic waste. Unlike some bacteria that release endotoxins during the degradation process, cyanobacteria lack endotoxins and can biologically degrade polymeric materials under photoautotrophic conditions without requiring additional carbon sources. This makes them a safer and more sustainable option for bioplastic production (Chia et al., 2020).

4.2 Micro Algal Bioactive Compounds

Algae are ubiquitous organisms found in a wide range of habitats, including freshwater, seawater, rocks, soil, stones, other plants, animals, deserts and permanent snow fields (Ibrahim et al., 2017). Recognized as natural producers of bioactive commercial pigments, algae are morphologically and functionally diverse, typically categorized into three groups based on their pigmentation: brown algae (Phaeophyceae, Heterokontophyta), green algae (Chlorophyta) and red algae (Rhodophyta) (Patel et al., 2022). Over the past 50 years, seaweeds have become a significant source of active metabolites utilized in various industries, including pharmaceuticals, agriculture, cosmetics, nutrition and bioenergy (Belghit et al., 2017).

4.2.1 Polysaccharides

Marine algae are rich in polysaccharides with potential applications in pharmaceuticals, cosmetics, biomedicine, and nutrition. These Polysaccharides are of great interest due to their various bioactivities, including antiallergic, neuroprotective, cardioprotective gastroprotective, antithrombotic, cytotoxic, antilipidemic, antiviral, immuno-modulatory and antinociceptive properties making them valuable bioactive products and biomaterials (Ali and Ahmed, 2018). Alginate is a polysaccharide found in the cellular matrix of brown algae (Phaeophyceae), constitutes about 40% of dry cellular weight. It is extensively used in the food industry as a gelling agent, thickener, and adhesive to enhance food texture (Pereira and Cotas, 2020). Carrageenans, natural marine polysaccharides are extracted from the Rhodophyceae family, are composed of D-galactose and 3,6-anhydro-D-galactose. They are widely used in the food industry for thickening, gelling, protein-binding, and stabilizing, as well as in pharmaceuticals for drug delivery systems due to their biocompatibility, biodegradability, sustainability, and non-toxicity (Ali and Ahmed, 2018). Agar, a structural polysaccharide extracted primarily from red algae (Rhodophyta), particularly *Gracilaria* and *Gelidium* is extensively used in the food industry (90%) and for bacteriology and biotechnology (10%). It is valued for its unique thermoreversibility, melting at temperatures above 85 °C and forming a colorless, odorless gel upon cooling. In food applications, agar serves as vegetarian gelatin with high fiber content, used in products like fruit jellies, pastries, canned meat, and glazes, remaining stable during the baking process (Scieszka and Klewicka, 2019).

4.2.2 Proteins and Peptides

Microalgae have been consumed for centuries due to their health benefits, particularly their production of novel proteins and peptides with bioactive properties such as anticancer, antioxidant, and cardioprotective effects. These compounds are increasingly being explored for use as nutraceuticals, functional foods, and pharmaceutical agents. With the growing prevalence of diseases like hypertension and cancer, there is a rising opportunity to leverage these compounds for prevention and treatment as ongoing research emphasizes their antioxidant, immunomodulatory, and anticancer properties (Ahmad et al., 2020). One notable class of compounds produced by algae is lectins, carbohydrate-binding proteins that contribute to immunological defense against pathogens. Lectins are known for their broad spectrum antimicrobial activity, targeting microbial quorum sensing, and offering antifungal, antiviral, and antibacterial activities. Their multifunctionality makes them valuable candidates for future therapeutic applications (Breitenbach Barroso Coelho et al., 2018).

4.2.3 Phycobiliproteins

Phycobiliproteins are a group of water-soluble, pigmented proteins primarily found in algae. They are categorized into three main types: phycocyanin, phycoerythrin, and allophycocyanin. These proteins have broad applications in various industries including as natural food colorants, in nutraceuticals, and in immunofluorescence assays. Beyond their industrial uses, phycobiliproteins also exhibit several pharmaceutical benefits, demonstrating antitumor, antioxidant, antidiabetic, and anti-inflammatory properties making them promising candidates for therapeutic interventions (Chen et al., 2022).

4.2.4 Lipids and Fatty Acids

Microalgae are capable of producing significant macromolecules such as carbohydrates, proteins, lipids, and bioactive compounds, with lipid content adjustable through manipulation of culture conditions (Encarnacao et al., 2015). Notable species like *Tetraselmis chuii, Schizochytrium, Nannochloropsis, Scenedesmus, Botryococcus,* and various *Chlorella* strains are efficient at accumulating lipids and essential polyunsaturated fatty acids (PUFAs), which find applications across industries including feed, food, pharmaceuticals, and biofuel production (Ahamefule et al., 2023). These algae produce key PUFAs including Omega-3, Omega-6, and Omega-9 fatty acids, such as eicosapentaenoic acid (EPA) and docosahexaenoic acid (DHA), which are vital for biodiesel, ethanol, butanol and other chemicals like enzymes, iodine, and stable isotopes (Abu-Ghosh et al., 2021).

Sterols, particularly phytosterols, play critical roles in eukaryotic cell membranes by regulating membrane permeability and fluidity, contributing to cell proliferation and signalling in photosynthetic organisms (Luo et al., 2015). Fucosterol, a prominent phytosterol from marine algae, has garnered attention for its wide-ranging health benefits, including cholesterol-lowering, antioxidant, anti-Alzheimer's, anti-inflammatory, immunomodulatory, anticancer, and antidiabetic properties, as well as its potential therapeutic effects for liver injury (Abdul et al., 2016).

4.2.5 Pigments

Microalgae are known to produce various components for cosmetic and food applications, including pigments such as carotenoids and chlorophyll. Carotenoid pigments from algae, such as *Haematococcus pluvialis, Chlorella species, Dunaliella salina, Scenedesmus species, Botryococcus braunii, Spirulina platensis,* and diatoms, are increasingly recognized for their health and food applications. These algae produce valuable compounds like beta-carotene, canthaxanthin, lutein, fucoxanthin, and astaxanthin (Ambati et al., 2019). Chlorophyll is another essential pigment in algae, primarily responsible for photosynthesis. These pigments also exhibit antioxidant, anti-genotoxic, anti-mutagenic, anti-obesogenic, and anticancer properties (Martins et al., 2023).

4.2.6 Phenolic Compounds

Phenolic compounds represent a diverse and extensive group of secondary metabolites found in algae. Among them, phlorotannins-polymers of phloroglucinol units, are prevalent in brown algae and offer notable health benefits due to their antimicrobial and antioxidant properties (Abdul et al., 2016). Specific phenolic compounds with antioxidant activity include phloroglucinol, phlorofucofuroeckol A, dieckol, eckol, phloroeckol, and 2-phloroeckol (Begum et al., 2021).

4.2.7 Vitamins

The health benefits of algae are largely attributed to their bioactive compounds, particularly vitamins found in *Spirulina* and *Chlorella*. These microalgae produce a variety of vitamins, including vitamin A (beta-carotene), vitamin E, vitamin C, and B vitamins—B1 (thiamine), B2 (riboflavin), B3 (niacin), B5 (pantothenic acid), B6 (pyridoxine), B9 (folic acid), and B12 (cobalamin) (Andrade et al., 2018). Additionally, naturally occurring vitamin K exists in two forms: phylloquinone (vitamin K1) and menaquinones (vitamin K2). These forms differ in their sources, tissue distribution, absorption rates, target activities, and bioavailability (Simes et al., 2020).

4.2.8 Terpenoids

Squalene, a terpenoid hydrocarbon found in algae, enhances cellular oxygen uptake and is widely used as a dietary supplement. It is a powerful antioxidant, that promotes proper cell function, maintains normal vision, protects the skin, and bolsters the immune system (Senbagalakshmi et al., 2019). Additionally, halogenated terpenes, derived from marine environments, particularly red algae, exhibit various biological activities, including antifungal, antibacterial, antibiotic, and anticancer properties (Encarnacao et al., 2015).

4.2.9 Alkaloids

Alkaloids are nitrogenous secondary metabolites found in algae that can interfere with microtubule dynamics by binding to β-tubulin providing cytotoxic effects against cancer cell lines (Abdul et al., 2016). Algae from the genus *Caulerpa* are rich in caulerpin, an indolic alkaloid with notable anti-inflammatory properties. Caulerpin has been identified in several *Caulerpa* species, including *C. peltata, C. racemosa, C. cupressoides, C. paspaloides, C. prolifera, C. sertularioides, C. mexicana, and C. lentillifera,* as well as in the red algae *Chondria armata.* Ethanol extracts from *C. peltata* and *C. racemosa* have shown caulerpin as a principal compound. Other indolic alkaloids identified in *Caulerpa* species include racemosin A, B, and C, and caulersin (Hohlman and Sherman, 2021).

4.2.10 Poly Unsaturated Fatty Acids

Algae accumulate significant amounts of long chain poly unsaturated fats (LC-PUFAs) as triacyl glycerols (TAG), including omega-3 and omega-6 fatty acids. These fatty acids are known for their beneficial effects on human health, such as reducing hypertension and mitigating rheumatoid arthritis and osteoporosis (Barta et al., 2021). Microalgae that produce edible oils and are rich in LC-PUFAs, include species like *Chlorella vulgaris, Spirulina platensis, Isochrysis galbana, Nannochloropsis oculata, Phaeodactylum tricornutum, Crypthecodinium cohnii, Schizochytrium limacinum and Porphyridium cruentum.* These PUFAs are utilized as food additives and in medicinal applications (Aly et al., 2023).

5 Challenges and Future Directions

The biotechnological exploitation of cyanobacteria and microalgae for bioactive compounds presents several challenges that need to be addressed to fully harness their potential. One major challenge is the efficient cultivation and harvesting of these microorganisms on an industrial scale. While laboratory scale cultivation is relatively straightforward, scaling up to commercial levels introduces complexities related to maintaining optimal growth conditions, preventing contamination, and ensuring consistent yields. Additionally, the high costs associated with large scale bioreactors and the extensive water and nutrient resources can limit the economic feasibility of these operations. Advances in bioreactor design such as closed systems and photobioreactors, are being explored to overcome these challenges and enhance productivity.

Another significant hurdle is the efficient extraction and purification of bioactive compounds from cyanobacteria and microalgae without degrading their activity. Traditional extraction methods often involve harsh chemicals and extensive processing, which can negatively impact the integrity of delicate bioactive molecules. Developing green extraction technologies, such as supercritical fluid extraction and ultrasound assisted extraction, offers a promising avenue to achieve high yields of pure compounds while minimizing environmental impact. Additionally, advances in metabolic engineering and synthetic biology can be employed to optimize the production pathways of desired bioactive compounds, thereby increasing their concentration and simplifying extraction processes.

The genetic diversity and adaptability of cyanobacteria and microalgae present both opportunities and challenges for biotechnological applications. Their ability to thrive in diverse and often extreme environments provides a rich source of unique bioactive compounds. However, this diversity also means that different strains may require specific cultivation conditions. Identifying and characterizing strains with high yields can be accelerated through high throughput screening methods and genome editing technologies such as CRISPR/Cas9. These tools facilitate identification and optimization of high yield strains, enhancing the efficiency and effectiveness of biotechnological applications.

6 Conclusions

Looking to the future, integrating multi-omics approaches such as genomics, proteomics, and metabolomics-holds significant promise for advancing the biotechnological exploitation of cyanobacteria and microalgae. These approaches offer comprehensive insights into the metabolic pathways and regulatory networks that govern bioactive compound production, allowing for the design of more efficient and sustainable production systems. By addressing existing challenges and leveraging cutting-edge technologies, we can fully realize the potential of cyanobacteria and microalgae as sources of valuable bioactive compounds, driving advancements in pharmaceuticals, nutraceuticals and other industrial applications.

Acknowledgements The authors would like to acknowledge the Department of Biotechnology at Vignan's Foundation for Science, Technology and Research as well as the Department of Biotechnology at Vikrama Simhapuri University, Nellore for providing the platform to carry out this research work.

Author Contributions Conceptualization, writing, review and editing were performed equally by both authors. Both authors have read and approved the final manuscript for submission.

References

Abdul, Q. A., Choi, R. J., Jung, H. A., & Choi, J. S. (2016). Health benefit of fucosterol from marine algae: A review. *Journal of the Science of Food and Agriculture, 96*(6), 1856–1866.

Abeysiri, H. A. S. N., Wanigasuriya, J. K. P., Suresh, T. S., Beneragama, D. H., & Manage, P. M. (2022). Hepatotoxicity of Microcystin LR in Wistar Rats. In *Proceedings of International Forestry and Environment Symposium* (p. 26).

Abu-Ghosh, S., Dubinsky, Z., Verdelho, V., & Iluz, D. (2021). Unconventional high-value products from microalgae: A review. *Bioresource Technology, 329*, 124895.

Agarwal, P., Soni, R., Kaur, P., Madan, A., Mishra, R., Pandey, J., et al. (2022). Cyanobacteria as a promising alternative for sustainable environment: Synthesis of biofuel and biodegradable plastics. *Frontiers in Microbiology, 13*, 939347.

Ahamefule, C. S., Ogbonna, C. N., Ahamefule, B. C., Ogbonna, I., & Ogbonna, J. 2023. Lipids and fatty acids from microalgae. In *Handbook of food and feed from microalgae* (pp. 73–86). Academic Press.

Ahmad, I. Z., Parvez, S., & Tabassum, H. (2020). Cyanobacterial peptides with respect to anticancer activity: Structural and functional perspective. *Studies in Natural Products Chemistry, 67*, 345–388.

Alcorta, J., Alarcon-Schumacher, T., Salgado, O., & Diez, B. (2020). Taxonomic novelty and distinctive genomic features of hot spring cyanobacteria. *Frontiers in Genetics, 11*, 568223.

Al-Haj, L., Lui, Y. T., Abed, R. M., Gomaa, M. A., & Purton, S. (2016). Cyanobacteria as chassis for industrial biotechnology: Progress and prospects. *Life, 6*(4), 42.

Ali, A., & Ahmed, S. (2018). Carrageenans: structure, properties and applications. In *Marine polysaccharides* (pp. 29–52). Jenny Stanford Publishing.

Aly, H. F., Rizk, M. Z., & Shams, S. G. E. (2023). Microalgae in food and feed: Safety and toxicological aspects. In: *Handbook of food and feed from microalgae* (pp. 549–565). Academic Press.

Ambati, R. R., Gogisetty, D., Aswathanarayana, R. G., Ravi, S., Bikkina, P. N., Bo, L., & Yuepeng, S. (2019). Industrial potential of carotenoid pigments from microalgae: Current trends and future prospects. *Critical Reviews in Food Science and Nutrition, 59*(12), 1880–1902.

Andrade, L. M., Andrade, C. J., Dias, M., Nascimento, C., & Mendes, M. A. (2018). Chlorella and spirulina microalgae as sources of functional foods. *Nutraceuticals, and Food Supplements, 6*(1), 45–58.

Arsad, S., Mulasari, Y. W., Sari, N. Y., Lusiana, E. D., Risjani, Y., Musa, M., et al. (2022). Microalgae diversity in several different sub-habitats. *Global Journal of Environmental Science and Management, 8*(4), 561–574.

Barta, D. G., Coman, V., & Vodnar, D. C. (2021). Microalgae as sources of omega-3 polyunsaturated fatty acids: Biotechnological aspects. *Algal Research, 58*, 102410.

Batra, B., Sharma, D., Bose, D., Parthasarthy, V., & Sarkar, A. (2023). Implications of bioprospecting marine diversity and sustainable production of bioactive compounds. In: *Marine antioxidants* (pp. 27–43). Academic Press.

Begum, R., Howlader, S., Mamun-Or-Rashid, A. N. M., Rafiquzzaman, S. M., Ashraf, G. M., Albadrani, G. M., et al. (2021). Antioxidant and signal-modulating effects of brown seaweed-derived compounds against oxidative stress-associated pathology. *Oxidative Medicine and Cellular Longevity, 2021*(1), 9974890.

Belghit, I., Rasinger, J. D., Heesch, S., Biancarosa, I., Liland, N., Torstensen, B., et al. (2017). In-depth metabolic profiling of marine macroalgae confirms strong biochemical differences between brown, red and green algae. *Algal Research, 26*, 240–249.

Bongirwar, R., & Shukla, P. (2023). Metabolic sink engineering in cyanobacteria: Perspectives and applications. *Bioresource Technology, 379*, 128974.

Bonnard, I., Bornancin, L., Dalle, K., Chinain, M., Zubia, M., Banaigs, B., & Roué, M. (2020). Assessment of the chemical diversity and potential toxicity of benthic cyanobacterial blooms in the lagoon of Moorea Island (French Polynesia). *Journal of Marine Science and Engineering, 8*(6), 406.

Borowitzka, M. A. (2013). High-value products from microalgae—their development and commercialisation. *Journal of Applied Phycology, 25*, 743–756.

Breitenbach Barroso Coelho, L. C., Marcelino Dos Santos Silva, P., Felix de Oliveira, W., De Moura, M. C., Viana Pontual, E., Soares Gomes, F., et al. (2018). Lectins as antimicrobial agents. *Journal of Applied Microbiology, 125*(5), 1238–1252.

Chen, H., Qi, H., & Xiong, P. (2022). Phycobiliproteins—a family of algae-derived biliproteins: Productions, characterization and pharmaceutical potentials. *Marine Drugs, 20*(7), 450.

Chia, W. Y., Tang, D. Y. Y., Khoo, K. S., Lup, A. N. K., & Chew, K. W. (2020). Nature's fight against plastic pollution: Algae for plastic biodegradation and bioplastics production. *Environmental Science and Ecotechnology, 4*, 100065.

Cock, I. E., & Cheesman, M. J. (2023). A review of the antimicrobial properties of cyanobacterial natural products. *Molecules, 28*(20), 7127.

De Carvalho, C. C., & Caramujo, M. J. (2018). The various roles of fatty acids. *Molecules, 23*(10), 2583.

Deviram, G., Mathimani, T., Anto, S., Ahamed, T. S., Ananth, D. A., & Pugazhendhi, A. (2020). Applications of microalgal and cyanobacterial biomass on a way to safe, cleaner and a sustainable environment. *Journal of Cleaner Production, 253*, 119770.

Dulic, T., Svircev, Z., Palanacki Malesevic, T., Faassen, E. J., Savela, H., Hao, Q., & Meriluoto, J. (2022). Assessment of common cyanotoxins in cyanobacteria of biological loess crusts. *Toxins, 14*(3), 215.

Encarnacao, T., Pais, A. A., Campos, M. G., & Burrows, H. D. (2015). Cyanobacteria and microalgae: A renewable source of bioactive compounds and other chemicals. *Science Progress, 98*(2), 145–168.

Garg, R., & Maldener, I. (2021). The dual role of the glycolipid envelope in different cell types of the multicellular cyanobacterium Anabaena variabilis ATCC 29413. *Frontiers in Microbiology, 12*, 645028.

Gaysina, L. A., Saraf, A., & Singh, P. (2019). Cyanobacteria in diverse habitats. In *Cyanobacteria* (pp. 1–28). Academic Press.

Ghosh, S. K., Bera, T., & Chakrabarty, A. M. (2020). Microbial siderophore–a boon to agricultural sciences. *Biological Control, 144*, 104214.

Hassan, S., Meenatchi, R., Pachillu, K., Bansal, S., Brindangnanam, P., Arockiaraj, J., et al. (2022). Identification and characterization of the novel bioactive compounds from microalgae and cyanobacteria for pharmaceutical and nutraceutical applications. *Journal of Basic Microbiology, 62*(9), 999–1029.

Hassanien, A., Saadaoui, I., Schipper, K., Al-Marri, S., Dalgamouni, T., Aouida, M., et al. (2023). Genetic engineering to enhance microalgal-based produced water treatment with emphasis on CRISPR/Cas9: A review. *Frontiers in Bioengineering and Biotechnology, 10*, 1104914.

Hohlman, R. M., & Sherman, D. H. (2021). Recent advances in hapalindole-type cyanobacterial alkaloids: Biosynthesis, synthesis, and biological activity. *Natural Product Reports, 38*(9), 1567–1588.

Hu, J., Meng, W., Su, Y., Qian, C., & Fu, W. (2023). Emerging technologies for advancing microalgal photosynthesis and metabolism toward sustainable production. *Frontiers in Marine Science, 10*, 1260709.

Huang, I. S., & Zimba, P. V. (2019). Cyanobacterial bioactive metabolites—a review of their chemistry and biology. *Harmful Algae, 86*, 139–209.

Ibrahim, M., Salman, M., Kamal, S., Rehman, S., Razzaq, A., & Akash, S. H. 2017. Algae-based biologically active compounds. In *Algae based polymers, blends, and composites* (pp. 155–271).

Jeong, B. R., Jang, J., & Jin, E. (2023). Genome engineering via gene editing technologies in microalgae. *Bioresource Technology, 373*, 128701.

Knoot, C. J., Ungerer, J., Wangikar, P. P., & Pakrasi, H. B. (2018). Cyanobacteria: Promising biocatalysts for sustainable chemical production. *Journal of Biological Chemistry, 293*(14), 5044–5052.

Kultschar, B., & Llewellyn, C. (2018). Secondary metabolites in cyanobacteria. *Secondary Metabolites-Sources and Applications, 7*, 43936.

Kumar, M., Singh, D. P., Prabha, R., & Sharma, A. K. (2015). Role of cyanobacteria in nutrient cycle and use efficiency in the soil. In *Nutrient use efficiency: from basics to advances* (pp. 163–171).

Larsson, J., Nylander, J. A., & Bergman, B. (2011). Genome fluctuations in cyanobacteria reflect evolutionary, developmental and adaptive traits. *BMC Evolutionary Biology, 11*, 1–21.

Li, X., Hou, W., Lei, J., Chen, H., & Wang, Q. (2023a). The unique light-harvesting system of the algal phycobilisome: Structure, assembly components, and functions. *International Journal of Molecular Sciences, 24*(11), 9733.

Li, G., Xiao, W., Yang, T., & Lyu, T. (2023b). Optimization and process effect for microalgae carbon dioxide fixation technology applications based on carbon capture: A comprehensive review. *C, 9*(1), 35.

Lin, P. C., & Pakrasi, H. B. (2019). Engineering cyanobacteria for production of terpenoids. *Planta, 249*, 145–154.

Liu, H., & Blankenship, R. E. (2019). On the interface of light-harvesting antenna complexes and reaction centers in oxygenic photosynthesis. *Biochimica et Biophysica Acta (BBA)-Bioenergetics, 1860*(11), 148079.

Lopez-Hernandez, J. F., Kean-Meng, T., Asencio-Alcudia, G. G., Asyraf-Kassim, M., Alvarez-González, C. A., & Márquez-Rocha, F. J. (2022). Sustainable microalgae and cyanobacteria biotechnology. *Applied Sciences, 12*(14), 6887.

Luo, X., Su, P., & Zhang, W. (2015). Advances in microalgae-derived phytosterols for functional food and pharmaceutical applications. *Marine Drugs, 13*(7), 4231–4254.

Ma, R., Wang, B., Chua, E. T., Zhao, X., Lu, K., Ho, S. H., et al. (2020). Comprehensive utilization of marine microalgae for enhanced co-production of multiple compounds. *Marine Drugs, 18*(9), 467.

Madeira, M. S., Cardoso, C., Lopes, P. A., Coelho, D., Afonso, C., Bandarra, N. M., & Prates, J. A. (2017). Microalgae as feed ingredients for livestock production and meat quality: A review. *Livestock Science, 205*, 111–121.

Martins, T., Barros, A. N., Rosa, E., & Antunes, L. (2023). Enhancing health benefits through chlorophylls and chlorophyll-rich agro-food: A comprehensive review. *Molecules, 28*(14), 5344.

McLellan, N. L., & Manderville, R. A. (2017). Toxic mechanisms of microcystins in mammals. *Toxicology Research, 6*(4), 391–405.

Mehdizadeh Allaf, M., & Peerhossaini, H. (2022). Cyanobacteria: Model microorganisms and beyond. *Microorganisms, 10*(4), 696.

Nanda, N., & Bharadvaja, N. (2022). Algal bioplastics: Current market trends and technical aspects. *Clean Technologies and Environmental Policy, 24*(9), 2659–2679.

Nandagopal, P., Steven, A. N., Chan, L. W., Rahmat, Z., Jamaluddin, H., & Mohd Noh, N. I. (2021). Bioactive metabolites produced by cyanobacteria for growth adaptation and their pharmacological properties. *Biology, 10*(10), 1061.

Ortega, M. A., & Van Der Donk, W. A. (2016). New insights into the biosynthetic logic of ribosomally synthesized and post-translationally modified peptide natural products. *Cell Chemical Biology, 23*(1), 31–44.

Papadaki, S., Tricha, N., Panagiotopoulou, M., & Krokida, M. (2024). Innovative bioactive products with medicinal value from microalgae and their overall process optimization through the implementation of life cycle analysis—an overview. *Marine Drugs, 22*(4), 152.

Parakh, S. K., Tian, Z., Wong, J. Z. E., & Tong, Y. W. (2023). From microalgae to bioenergy: Recent advances in biochemical conversion processes. *Fermentation, 9*(6), 529.

Patel, A. K., Albarico, F. P. J. B., Perumal, P. K., Vadrale, A. P., Nian, C. T., Chau, H. T. B., et al. (2022). Algae as an emerging source of bioactive pigments. *Bioresource Technology, 351*, 126910.

Pathania, R., Srivastava, A., Srivastava, S., & Shukla, P. (2022). Metabolic systems biology and multi-omics of cyanobacteria: Perspectives and future directions. *Bioresource Technology, 343*, 126007.

Pereira, L., & Cotas, J. (2020). *Introductory chapter: Alginates-A general overview. Alginates-recent uses of this natural polymer.*

Pichardo, S., Cameán, A. M., & Jos, A. (2017). In vitro toxicological assessment of cylindrospermopsin: A review. *Toxins, 9*(12), 402.

Plata-Calzado, C., Prieto, A. I., Cameán, A. M., & Jos, A. (2022). Toxic effects produced by anatoxin-a under laboratory conditions: A review. *Toxins, 14*(12), 861.

Pooja, S. (2022). Insight into the potential cyanobacterial metabolites and their screening strategies. *Biosciences Biotechnology Research Asia, 19*(1), 255.

Prabha, R., Singh, D. P., Somvanshi, P., & Rai, A. (2016). Functional profiling of cyanobacterial genomes and its role in ecological adaptations. *Genomics Data, 9*, 89–94.

Qamar, H., Hussain, K., Soni, A., Khan, A., Hussain, T., & Chénais, B. (2021). Cyanobacteria as natural therapeutics and pharmaceutical potential: Role in antitumor activity and as nanovectors. *Molecules, 26*(1), 247.

Rasal, V. M., Yadre, S. G., Shukla, S. P., Ravi, P., Mishra, M. K., Munilkumar, S., et al. (2019). Microalgae distribution and diversity in the Narmada river basin around Chutka, Madhya Pradesh, India. *International Journal of Current Microbiology and Applied Sciences, 8*, 1488–1501.

Razzak, S. A., Bahar, K., Islam, K. O., Haniffa, A. K., Faruque, M. O., Hossain, S. Z., & Hossain, M. M. (2023). *Microalgae cultivation in photobioreactors: Sustainable solutions for a greener future*. Green Chemical Engineering.

Reisoglu, Ş., & Aydin, S. (2023). *Microalgae as a promising candidate for fighting climate change and biodiversity loss.*

Rinaldi, K. L., Senhorinho, G. N., Laamanen, C. A., & Scott, J. A. (2024). A review of extremophilic microalgae: Impacts of experimental cultivation conditions for the production of antimicrobials. *Algal Research*, 103427.

Risser, D. D. (2023). Hormogonium development and motility in filamentous cyanobacteria. *Applied and Environmental Microbiology, 89*(6), e00392-e423.

Rocha, R. P., Machado, M., Vaz, M. G. M. V., Vinson, C. C., Leite, M., Richard, R., et al. (2017). Exploring the metabolic and physiological diversity of native microalgal strains (Chlorophyta) isolated from tropical freshwater reservoirs. *Algal Research, 28*, 139–150.

Rodriguez-Zuniga, D., Méndez-Zavala, A., Solís-Quiroz, O., Morales-Oyervides, L., Montañez-Saenz, J. C., & Benavente-Valdés, J. R. (2024). Biology and composition of microalgae and cyanobacteria. In *Sustainable industrial processes based on microalgae* (pp. 1–22). Elsevier.

Rueda, E., Gonzalez-Flo, E., Mondal, S., Forchhammer, K., Arias, D. M., Ludwig, K., et al. (2024). Challenges, progress, and future perspectives for cyanobacterial polyhydroxyalkanoate production. *Reviews in Environmental Science and Bio/Technology*, 1–30.

Rymuszka, A., Sieroslawska, A., & Adaszek, L. (2021). Cytotoxic and immunological responses of fish leukocytes to nodularin exposure in vitro. *Journal of Applied Toxicology, 41*(10), 1660–1672.

Sadvakasova, A. K., Kossalbayev, B. D., Bauenova, M. O., Balouch, H., Leong, Y. K., Zayadan, B. K., et al. (2023). Microalgae as a key tool in achieving carbon neutrality for bioproduct production. *Algal Research, 72*, 103096.

Scieszka, S., & Klewicka, E. (2019). Algae in food: A general review. *Critical Reviews in Food Science and Nutrition, 59*(21), 3538–3547.

Senbagalakshmi, P., Muthukrishnan, S., Jebasingh, T., Kumar, T. S., Rao, M., Kumar, T. S., & Rao, M. V. (2019). Squalene, biosynthesis and its role in production of bioactive compounds, a proper scientific challenge—a review. *Journal of Emerging Technologies and Innovative Research, 6*, 505–526.

Shah, S. A. A., Akhter, N., Auckloo, B. N., Khan, I., Lu, Y., Wang, K., et al. (2017). Structural diversity, biological properties and applications of natural products from cyanobacteria. A review. *Marine Drugs, 15*(11), 354.

Simes, D. C., Viegas, C. S., Araújo, N., & Marreiros, C. (2020). Vitamin K as a diet supplement with impact in human health: Current evidence in age-related diseases. *Nutrients, 12*(1), 138.

Singh, J. S., Kumar, A., Rai, A. N., & Singh, D. P. (2016). Cyanobacteria: A precious bio-resource in agriculture, ecosystem, and environmental sustainability. *Frontiers in Microbiology, 7*, 529.

Singh, J., & Saxena, R. C. (2015). An introduction to microalgae: diversity and significance. In *Handbook of marine microalgae* (pp. 11–24). Academic Press.

Singh, K. B., Kaushalendra, Verma, S., Lalnunpuii, R., & Rajan, J. P. (2023). Current issues and developments in cyanobacteria-derived biofuel as a potential source of energy for sustainable future. *Sustainability, 15*(13), 10439.

Song, Y., Wang, F., Chen, L., & Zhang, W. (2024). Engineering fatty acid biosynthesis in microalgae: Recent progress and perspectives. *Marine Drugs, 22*(5), 216.

Swarnalatha, G. V., Goudar, V., Reddy, E. C. R. G. S., Al Tawaha, A. R. M., & Sayyed, R. Z. 2022. Siderophores and their applications in sustainable management of plant diseases. In *Secondary metabolites and volatiles of PGPR in plant-growth promotion* (pp. 289–302).

Synytsya, A., Sushytskyi, L., Saloň, I., Babayeva, T., & Čopíková, J. 2023. Intracellular and extracellular carbohydrates in microalgae. In *Handbook of food and feed from microalgae* (pp. 87–102). Academic Press.

Touliabah, H. E. S., El-Sheekh, M. M., Ismail, M. M., & El-Kassas, H. (2022). A review of microalgae-and cyanobacteria-based biodegradation of organic pollutants. *Molecules, 27*(3), 1141.

Vidal, L., Ballot, A., Azevedo, S. M. F. O., Padisák, J., & Welker, M. (2021). Introduction to cyanobacteria. In I. Chorus, & M. Welker (Eds.), *Toxic cyanobacteria in water* (2nd ed., pp. 163–211).

Waditee-Sirisattha, R., & Kageyama, H. (2022). Cyanobacterial cells. In *Cyanobacterial physiology* (pp. 3–16). Academic Press.

Walter, J. M., Coutinho, F. H., Dutilh, B. E., Swings, J., Thompson, F. L., & Thompson, C. C. (2017). Ecogenomics and taxonomy of Cyanobacteria phylum. *Frontiers in Microbiology, 8*, 2132.

Wang, Y., Yang, S., Liu, J., Wang, J., Xiao, M., Liang, Q., et al. (2023). Realization process of microalgal biorefinery: The optional approach toward carbon net-zero emission. *Science of the Total Environment, 901*, 165546.

Zahra, Z., Choo, D. H., Lee, H., & Parveen, A. (2020). Cyanobacteria: Review of current potentials and applications. *Environments, 7*(2), 13.

Zielinski, M., Dębowski, M., Kazimierowicz, J., & Świca, I. (2023). Microalgal carbon dioxide (CO_2) capture and utilization from the European Union perspective. *Energies, 16*(3), 1446.

Zulkefli, N. S., & Hwang, S. J. (2020). Heterocyst development and diazotrophic growth of Anabaena variabilis under different nitrogen availability. *Life, 10*(11), 279.

Exploring the Therapeutic Potential of Bioactive Compounds from Plant Sources

Vivek Chintada and Narasimha Golla

Abstract

This chapter delves into the therapeutic possibilities of bioactive compounds derived from plants. This chapter explores the therapeutic potential of bioactive compounds sourced from plants. These compounds have attracted significant interest in the medical, pharmaceutical, and nutritional fields because of their diverse biological effects and possible health benefits. The chapter offers a comprehensive examination of the various classes of bioactive mixtures present in plants, such as polyphenols, alkaloids, terpenoids, and flavonoids, elucidating their bioavailability, pharmacological properties, and mechanisms of action, all of which underscore their potential efficacy in preventing and treating diverse illnesses. Moreover, it investigates how bioactive compounds can enhance human health and well-being under their antioxidant, anti-inflammatory, antimicrobial, and anticancer properties. The chapter underscores the significance of sustainable sourcing and extraction techniques for procuring plant bioactive compounds, aligning with the escalating global demand for natural, plant-based remedies. It also addresses forthcoming research prospects and challenges in exploring and leveraging bioactive compounds from plants, underscoring the necessity for interdisciplinary partnerships and innovative strategies to tap into the therapeutic potential of these natural compounds fully. In essence, this chapter presents a valuable reservoir of knowledge for researchers, academics, and professionals in the fields of pharmacology, natural products chemistry, and phytotherapy, offering profound insights into the promising applications of bioactive mixtures derived from plants in enhancing human health and propelling medical advancements.

Keywords

Bioactive compounds • Plant-derived • Therapeutic potential • Pharmacological properties and natural products

1 Introduction

Bioactive composites derived from plants have emerged as a focal point in modern research, offering a myriad of therapeutic possibilities that have captivated the interest of medical professionals, scientists, and the public alike (Saeed et al., 2021). These natural complexes have garnered consideration for their diverse range of biological activities and potential health benefits, positioning them as promising candidates for pharmaceutical development, nutraceutical products, and functional foods (Ahmed et al., 2024). The exploration of plant-derived bioactive compounds represents a dynamic field that intertwines various disciplines such as pharmacology, natural products chemistry, and phytotherapy, facilitating a deeper understanding of their mechanisms and applications in human health (Halder & Jha, 2023). The bioactive compounds have therapeutic potential through their pharmacological effects, which include antioxidant, anti-inflammatory, antimicrobial, and anticancer properties (Khuntia, et al., 2022). Nascimento et al. (2018) elucidated the pharmacological effects of bioactive compounds from plants, emphasizing their capacity to modulate inflammatory responses, scavenge free radicals, combat microbial infections, and inhibit cancer cell proliferation. Such findings underscore the multifaceted nature of bioactive compounds and their promise to improve health outcomes.

V. Chintada
Department of Zoology, Sri Venkateswara University, Tirupati, 517502, India

N. Golla (✉)
Department of Virology, Sri Venkateswara University, Tirupati, 517502, India
e-mail: dr.g.narasimha@gmail.com

J. Devi (ed.), *Biotechnological Intervention in Production of Bioactive Compounds*, Sustainable Landscape Planning and Natural Resources Management, https://doi.org/10.1007/978-3-031-76859-0_15

Polyphenols, a prevalent class of bioactive compounds found in a variety of plant-based foods and beverages, have garnered particular interest due to their potent antioxidant activities and potential health benefits (Banwo et al., 2021). Recent studies by Wang et al. (2019a) have shed light on the bioavailability and mechanisms of action of polyphenols, revealing their intricate interactions with cellular pathways involved in oxidative stress, inflammation, and carcinogenesis. These findings underscore the therapeutic potential of polyphenols in managing chronic diseases and highlight their significance in the development of novel pharmaceutical interventions. In conjunction with polyphenols, alkaloids, terpenoids, and flavonoids constitute additional classes of bioactive compounds that contribute to the biological activities of plants (Saboon et al., 2019; Banwo et al., 2021). Alkaloids, recognized for their pharmacological effects, are key components in medicinal plants used traditionally in various healthcare practices (Kuete, 2014). Terpenoids, distinguished by their unique chemical structures, play critical roles in plant defense mechanisms and exhibit notable antimicrobial properties (Adhikary and Dasgupta, 2023).

Sustainable sourcing and extraction of bioactive compounds from plants are essential considerations in meeting the growing demand for natural remedies while upholding environmental stewardship and ethical standards (Obahiagbon and Ogwu, 2024). Innovative approaches to sourcing and extraction have been developed to minimize environmental impact and enhance the efficiency of obtaining bioactive compounds, ensuring both quality and sustainability (Messinese et al., 2024). By integrating sustainable practices into the sourcing and production processes, researchers and industry stakeholders can meet consumer demand for natural products while preserving biodiversity and supporting ecological balance (Prasanna et al., 2023). Collaborative efforts among researchers, academics, and industry partners are paramount in advancing the field of bioactive compounds and expediting the translation of research findings into tangible applications for human health (Lubchenco and Haugan, 2023). Interdisciplinary collaborations foster the exchange of knowledge and expertise across disciplines, driving innovation and accelerating scientific progress (Fortunato et al., 2018). By uniting to address the challenges and opportunities associated with harnessing the therapeutic potential of plant-derived bioactive compounds, stakeholders can propel advancements in medical science and enhance healthcare delivery.

In finish, the exploration of bioactive compounds derived from plants offers a rich landscape of opportunities for researchers, academics, and professionals in various fields. By elucidating the pharmacological properties, mechanisms of action, and health benefits of bioactive compounds, this chapter seeks to deepen our collective understanding of their therapeutic potential and stimulate collaborations that drive medical advancements. With an emphasis on sustainability, interdisciplinary approaches, and future research avenues, we aspire to unlock the untapped potential of plant-derived bioactive compounds in optimizing human health and shaping the future of healthcare.

2 Classes of Bioactive Compounds in Plants

Bioactive compounds found in plants are natural chemical compounds that can produce specific physiological effects in the human body (Pandey & Rizvi, 2009). They are sourced from a variety of plants like fruits, vegetables, herbs, and spices. These compounds are recognized for their positive impact on human health, including their potential to prevent and treat various diseases (Banwo et al., 2021). They are classified into different groups based on their chemical makeup and biological functions, such as polyphenols, terpenoids, alkaloids, and flavonoids. Polyphenols, which are a significant type of bioactive compound, are renowned for their antioxidant properties and possible health-promoting effects (Figueira et al., 2017). Plants possess bioactive compounds as a mechanism to defend against environmental stressors such as pests, diseases, and UV radiation (Aguirre-Becerra et al., 2021). These compounds are vital for the health-promoting qualities of plant-based foods and are linked to various health assistances, including antioxidant, anti-inflammatory, anti-cancer, and antimicrobial activities.

A. **Significance of bioactive compounds for human well-being**

Bioactive compounds found in plants are progressively predictable for their pivotal role in promoting human health and well-being (Banwo et al., 2021). These compounds, abundant in diverse plant sources, offer a wide array of benefits that contribute to disease prevention and the overall maintenance of good health (Patil et al., 2015). One of the key roles of bioactive compounds is their potent antioxidant action, which helps to counteract the detrimental effects of free radicals in the body (Balsano and Alisi, 2009). Free radicals are known to cause oxidative stress, leading to cellular damage and contributing to aging and numerous diseases (Aruoma, 1998). By neutralizing free radicals, bioactive mixtures exert a protective effect on cells and tissues, thereby reducing the hazard of oxidative damage-related conditions such as vascular diseases, neurodegenerative disorders, and cancer (Guan et al., 2021). The bioactive compounds possess remarkable anti-inflammatory properties, playing a central role in modulating the body's inflammatory response (Jayarathne et al., 2017). Chronic inflammation is a mutual underlying factor in several diseases, including cardiovascular diseases, arthritis, and metabolic disorders (Ferrucci & Fabbri, 2018). Studies have demonstrated that these compounds can impede the growth and spread of cancer cells, trigger cancer cell death, and disrupt different cellular signaling pathways linked to the development of tumors. (Calderón-Montaño et al., 2011). Bioactive compounds, especially flavonoids, play a beneficial role in cardiovascular health by aiding in the reduction of blood pressure, enhancement of endothelial function, mitigation of arterial inflammation, and prevention of blood clot formation (Vazhappilly et al., 2019). As a result, incorporating foods rich in flavonoids and other bioactive compounds into the diet can contribute to overall cardiovascular well-being and reduce the threat of heart disease (Lorenzo et al., 2019).

In the field of brain health, bioactive compounds have attracted interest for their neuroprotective properties and potential to enhance cognitive function. Some compounds exhibit neuroprotective properties, helping to mitigate neurodegenerative processes and reduce the risk of conditions such as Alzheimer's disease and dementia (Mohd Sairazi, et al., 2020). By promoting brain health and cognitive function, bioactive compounds offer valuable support for mental well-being and neurological health (Vauzour, 2017). Incorporating a diverse array of bioactive compounds from plant-based sources into the diet can have a profound impact on human health, providing a natural and holistic approach to wellness (Kovačević et al., 2020). By harnessing the therapeutic potential of bioactive compounds, individuals can optimize their health outcomes, reduce the risk of chronic diseases, and enhance their overall quality of life (Lobine, 2022). As research continues to unveil the multifaceted benefits of bioactive compounds, their significance for human health becomes increasingly evident, emphasizing the crucial role of plant-based nutrition in promoting optimal health and longevity (Sudheer et al., 2022).

B. Overview of the diversity of bioactive compounds found in plants

Bioactive compounds found in plants exhibit a remarkable diversity that incorporates a wide variety of chemical structures and biological activities, each offering unique benefits for human health (Banwo et al., 2021). The vast array of bioactive compounds found in plants includes polyphenols, terpenoids, alkaloids, flavonoids, and more, reflecting the richness and complexity of nature's pharmacy (Akhtar et al., 2019). These substances are essential for maintaining human health by influencing various aspects of the body. Polyphenols are a highly researched and commonly found group of bioactive substances in plants. They are known for their antioxidant properties, which enable them to combat free radicals, mitigate oxidative stress, and safeguard cells from harm. Polyphenols have been associated with various health advantages such as cardiovascular support, anti-inflammatory effects, and potential cancer prevention.

Terpenoids are a diverse group of bioactive compounds widely present in plant sources with various pharmacological activities. They are responsible for the distinctive flavors and scents of numerous plants and have been explored for their anti-inflammatory, antimicrobial, and anticancer properties. Known for their broad spectrum of biological effects, terpenoids show promise for therapeutic applications in promoting human health (Ludwiczuk et al., 2017). Alkaloids are nitrogen-containing compounds with diverse pharmacological activities and medicinal properties found in plants. They exhibit a wide range of biological effects such as analgesic, anti-inflammatory, antimicrobial, and anticancer activities. Alkaloids, utilized in traditional medicine for centuries, continue to intrigue researchers for their potential therapeutic applications in modern healthcare (Rajput et al., 2022).

Flavonoids, a subgroup of polyphenols, are bioactive compounds recognized for their antioxidant and anti-inflammatory properties. They play a vital role in the health-promoting aspects of plant-based diets, considered essential for human well-being (Shen et al., 2022) Apart from these well-known groups, plants contain a variety of other bioactive compounds with different structures and functions that contribute to human health. Saponins are glycosides found in various plants, potentially aiding in cholesterol regulation, anti-inflammatory responses, and immune modulation (Hussain et al., 2020). Lignans, phytoestrogens found in seeds and whole grains, are under investigation for their potential impact on hormone-related conditions and cardiovascular health (Albuquerque et al., 2020). Through exploration of the broad spectrum of bioactive compounds and their varying effects on the body, researchers and healthcare professionals can uncover novel pathways for preventive and therapeutic interventions.

I. Classification of Bioactive Compounds

Exploring bioactive compounds for human health is a vital and dynamic area of study, encompassing a wide range of compounds sourced from nature, including plants, animals, and microorganisms (Banwo et al., 2021). These bioactive compounds showcase diverse biological activities that offer considerable potential for advancing human health and well-being (Pandey & Rizvi, 2009). Classification of bioactive compounds can be conducted based on factors like their chemical structure, mode of action, or specific health benefits they offer within the human body (Newman &

Cragg, 2016). A profound comprehension of the properties and effects of these compounds is essential for formulating innovative therapeutic interventions and improving health outcomes.

One effective approach to categorizing bioactive compounds is by their chemical structure. These compounds can be classified into distinct groups such as polyphenols, alkaloids, flavonoids, terpenoids, and saponins (Abdalla & McGaw, 2021). Polyphenols, for example, constitute a diverse group of compounds present in fruits, vegetables, and beverages like tea and wine, known for their potent antioxidant properties and their association with reducing the risk of chronic diseases such as cancer and cardiovascular ailments (Belščak-Cvitanović et al., 2018). Alkaloids, nitrogen-containing compounds commonly found in plants and fungi, encompass well-known examples like caffeine and nicotine, showcasing a broad spectrum of pharmacological activities including analgesic and anti-inflammatory effects (Böttger et al., 2018). Flavonoids represent another class of bioactive compounds renowned for their antioxidant and anti-inflammatory properties, commonly found in fruits, vegetables, and herbs (Middleton Jr. et al., 2000). Terpenoids, including compounds such as essential oils and steroids, exhibit noteworthy antimicrobial and anticancer activities (Ansari and Akhtar, 2019). Saponins, glycosides sourced from various plants, are esteemed for their cholesterol-lowering effects and potential anticancer attributes.

An alternative method to categorize bioactive compounds is according to their mechanism of action within the body. Certain compounds function as antioxidants, safeguarding cells against oxidative stress and diminishing inflammation (Pisoschi and Pop, 2015). Other compounds may influence enzyme activity, control gene expression, or engage with cellular signaling pathways. For instance, resveratrol, a polyphenol present in red wine, has been demonstrated to stimulate sirtuins, a group of proteins responsible for governing longevity and metabolic functions (Russo et al., 2020). Curcumin, a bioactive compound in turmeric, exerts its anti-inflammatory effects by inhibiting specific enzymes and signaling molecules that drive inflammation (Tasneem et al., 2019). The bioactive compounds in green tea, such as catechins, can enhance fat metabolism and improve cardiovascular health through their interactions with lipid metabolism pathways (Dinh et al., 2019). Bioactive compounds can also be classified based on their specific health effects in the body. For instance, some compounds may have anti-cancer properties by inhibiting tumor growth, inducing apoptosis in cancer cells, or blocking angiogenesis (Khalid et al., 2016). Research has focused on Epigallocatechin gallate (EGCG), a catechin abundant in green tea, for its potential in cancer prevention and treatment through its targeting of various signaling pathways crucial in tumor development (Singh et al., 2019). Bioactive compounds endowed with anti-inflammatory properties can offer relief from symptoms associated with chronic inflammatory ailments such as arthritis, asthma, and inflammatory bowel disease (El-Shiekh et al., 2021). Omega-3 fatty acids, sourced from fatty fish and flaxseeds, are renowned for their anti-inflammatory effects and have been associated with a reduced risk of cardiovascular diseases.

In conclusion, the classification of bioactive mixtures for human health represents a multifaceted and ever-changing field that advances through continuous research and discoveries. By delving into the chemical structures, mechanisms of action, and health impacts of these compounds, researchers and healthcare providers can leverage their potential advantages to enhance human health and prevent diseases (Jin et al., 2023). Further investigation into bioactive compounds sourced from nature offers prospects for innovating new therapeutic strategies and personalized health management approaches (Aware et al., 2022). Hence, a comprehensive grasp of the classification of bioactive compounds is vital to unlock their full potential in promoting human health and well-being.

II. **Bioactivities of Plant Compounds**

Plant compounds exhibit a wide range of bioactivities that contribute to their potential health benefits for humans. These bioactive compounds, derived from various plant sources, have been extensively studied for their antioxidant properties, anti-inflammatory effects, antimicrobial activities, anticancer properties, and neuroprotective effects. Understanding the diverse bioactivities of plant compounds is essential for unlocking their therapeutic potential and enhancing human health. Among these bioactivities, antioxidant properties are paramount. Numerous plant-based antioxidants, such as polyphenols, flavonoids, and carotenoids, play a pivotal role in scavenging free radicals and mitigating oxidative stress within the body (Yang et al., 2018). Through the neutralization of reactive oxygen species, plant antioxidants safeguard cells from damage and help reduce the susceptibility to chronic diseases associated with oxidative stress, including cardiovascular diseases, neurodegenerative disorders, and cancer. Berries, nuts, seeds, and leafy greens are common plant sources rich in antioxidants like resveratrol, quercetin, and lutein, offering potential health benefits (Chen et al., 2019).

Another significant bioactivity of plant compounds is their anti-inflammatory effects. Various phytochemicals found in plants possess anti-inflammatory properties by modulating inflammatory pathways and decreasing the invention of pro-inflammatory mediators (Santos et al.,

2019). Compounds like curcumin from turmeric, gingerol from ginger, and epigallocatechin gallate from green tea have been presented to inhibit inflammatory responses and alleviate symptoms of inflammatory conditions such as arthritis, asthma, and inflammatory bowel disease (Panahi et al., 2014; Hewlings & Kalman, 2017). By targeting inflammation at the cellular level, plant compounds offer a natural and holistic approach to managing inflammatory disorders and promoting overall well-being.

Plant compounds also exhibit antimicrobial activities, making them valuable agents for combatting microbial infections. Essential oils derived from plants, such as tea tree oil, oregano oil, and thyme oil, contain bioactive components with antimicrobial properties against bacteria, fungi, and viruses (Bakkali et al., 2008). These plant-based antimicrobial agents can help prevent and treat infections, including antibiotic-resistant strains, by disrupting microbial growth and targeting specific microbial pathways. Moreover, utilizing plant compounds as natural antimicrobials contributes to the advancement of alternative approaches for infection control and safeguarding public health.

Moreover, plant compounds have been investigated for their potential anti-cancer properties due to their capability to inhibit tumor growth, induce apoptosis in cancer cells, and prevent metastasis (Surh, 2003). Phytochemicals like resveratrol, sulforaphane, and lycopene have shown promising results in preclinical studies for their anti-cancer effects (Kotecha et al., 2014). These compounds target multiple signaling pathways involved in cancer development and progression, offering a complementary approach to conventional cancer treatments (Surh, 2003). By incorporating plant compounds with anticancer properties into dietary and therapeutic regimens, individuals may reduce their cancer risk and improve treatment outcomes (Kotecha et al., 2014).

In addition to the aforementioned bioactivities, plant compounds exhibit neuroprotective effects that support brain health and cognitive function (Gómez-Pinilla, 2008). Polyphenols, omega-3 fatty acids, and flavonoids found in plant-based foods have been linked to neuroprotection by enhancing neural communication, reducing neuroinflammation, and promoting neurogenesis (Williams & Spencer, 2012). These neuroprotective plant compounds may help prevent age-related cognitive decline, support mental well-being, and potentially mitigate the risk of neurodegenerative illnesses such as Alzheimer's and Parkinson's disease (Gómez-Pinilla, 2008). By incorporating plant-rich diets and botanical supplements, individuals can optimize their brain health and cognitive resilience throughout their lifespan (Williams & Spencer, 2012).

In summary, plant compounds demonstrate a diverse array of bioactivities that contribute to their potential health benefits for human well-being. From antioxidant properties and anti-inflammatory effects to antimicrobial activities, anticancer properties, and neuroprotective effects, plant compounds offer a holistic approach to promoting health and preventing diseases. By harnessing the bioactivities of plant compounds through dietary intake, nutraceutical supplements, and herbal remedies, individuals can optimize their health outcomes and enhance their quality of life.

3 Pharmacological Properties and Mechanisms of Action

The fundamental pharmacological properties and mechanisms of action of bioactive compounds are pivotal factors shaping their potential health benefits for humans (Atanasov et al., 2021). Bioactive compounds sourced from natural origins, including plants, animals, and microorganisms, exhibit a diverse range of pharmacological attributes that contribute to their therapeutic effects in the body (Atanasov et al., 2021). A comprehensive understanding of the mechanisms through which these compounds exert their bioactivities is essential for discerning their impact on human health and disease management (Atanasov et al., 2021). This segment provides an insightful overview of the pharmacological properties and modes of action of bioactive compounds, highlighting their significant roles in enhancing human health and well-being.

A crucial pharmacological trait of bioactive compounds is their antioxidant activity, essential for shielding cells against oxidative harm and lowering the likelihood of chronic ailments (Sanches-Silva et al., 2021). Compounds such as polyphenols, flavonoids, and carotenoids act as potent antioxidants by scavenging free radicals, inhibiting lipid peroxidation, and modulating oxidative stress pathways (Surh, 2003). By neutralizing reactive oxygen species, bioactive antioxidants help maintain cellular homeostasis, support immune function, and mitigate the progression of oxidative stress-related conditions like cardiovascular diseases, neurodegenerative disorders, and cancer.

Another important pharmacological property of bioactive compounds is their anti-inflammatory activity, which helps regulate the immune response and alleviate inflammatory conditions. Phytochemicals such as curcumin, quercetin, and resveratrol exhibit anti-inflammatory effects by modulating inflammatory mediators, reducing cytokine production, and inhibiting inflammatory signaling pathways (González et al., 2011; Calderón-Montaño et al., 2011). By targeting inflammation at the molecular level, these bioactive compounds can help alleviate symptoms of inflammatory diseases, such as arthritis, asthma, and inflammatory bowel disease, and promote overall inflammatory balance in the body. Bioactive compounds also demonstrate antimicrobial properties through their capability to impede the growth of bacteria, fungi, viruses, and parasites. Essential oils, alkaloids, and phenolic compounds found in plants possess antimicrobial activity by disrupting microbial membrane integrity, inhibiting enzyme function, and interfering with microbial DNA replication (Nostro et al., 2001; Cushnie & Lamb, 2011). These natural antimicrobials offer potential alternatives to conventional antibiotics and antimicrobial agents, addressing the growing concern of antimicrobial resistance and supporting the development of novel antimicrobial strategies.

Moreover, certain bioactive compounds exhibit anticancer properties by targeting key pathways involved in tumor initiation, progression, and metastasis. Compounds such as EGCG, sulforaphane, and curcumin have demonstrated the ability to inhibit cancer cell growth, promote cell death, and reduce the formation of new blood vessels in both research settings and clinical trials. Studies by Shankar et al. (2017), Farzaei et al. (2019), and Wu et al. (2019) support these findings. These bioactive compounds work to prevent and treat cancer by affecting cell cycle control, signaling pathways for cell death, and interactions within the tumor environment. They have the potential to be used as complementary therapies for cancer due to their ability to influence various biological processes in the body, in addition to their pharmacological properties. For instance, polyphenols like EGCG and quercetin interact with cellular signaling pathways, transcription factors, and enzyme systems to exert their antioxidant, anti-inflammatory, and anticancer effects (Gibellini et al., 2020). These compounds may target NF-κB, Nrf2, and MAPK pathways to regulate gene expression, modulate oxidative stress responses, and inhibit inflammatory cascades in cells and tissues.

Moreover, bioactive compounds can impact epigenetic processes, including DNA methylation and histone modification, which can influence gene expression patterns and cellular functions as evidenced by studies conducted by Perrott et al. (2019) and Lee et al. (2020). Compounds like resveratrol and genistein have been found to adjust epigenetic markers and modify gene expression profiles linked to cancer, inflammation, and aging, as shown in research by Dolatabadi et al. (2019) and Zhang et al. (2020). Through the epigenetic regulation of gene behavior, bioactive compounds may open up new avenues for disease prevention and personalized medicine strategies. In conclusion, the diverse therapeutic potential of bioactive compounds, highlighted by their pharmacological properties and modes of action, shows promise in enhancing human health and overall well-being. From antioxidant and anti-inflammatory activities to antimicrobial and anticancer effects, bioactive compounds offer a multifaceted approach to disease prevention and management. By elucidating the mechanisms underlying the pharmacological actions of these compounds, researchers and healthcare practitioners can leverage their therapeutic benefits to enhance health outcomes and address a wide range of health conditions.

4 Health Benefits of Bioactive Compounds

Bioactive compounds imitative from natural sources offer a myriad of health benefits through their antioxidant properties, anti-inflammatory effects, antimicrobial activities, and anticancer potential. These bioactive compounds, ranging from polyphenols and flavonoids to alkaloids and terpenoids, play a crucial role in promoting human health and well-being. Understanding the distinct health benefits of bioactive compounds is essential for harnessing their therapeutic potential and enhancing overall health outcomes.

The health benefits of bioactive compounds include their antioxidant properties. Polyphenols, carotenoids, and vitamins C and E serve as strong antioxidants by eliminating free radicals and reducing oxidative stress in the body, as confirmed by research from Scalbert et al. (2005) and Frei (2013). These antioxidants play a vital role in safeguarding cells from harm, preserving cellular integrity, and reducing the risk of chronic diseases such as cardiovascular diseases, neurodegenerative disorders, and cancer by neutralizing harmful reactive oxygen species. Consuming a diet rich in antioxidant-rich foods such as fruits, vegetables, nuts, and whole grains can help boost the body's antioxidant defenses and support overall health. Another significant health benefit of bioactive compounds is their anti-inflammatory effects. Phytochemicals like curcumin, quercetin, and resveratrol exhibit anti-inflammatory properties by modulating inflammatory pathways, reducing the production of pro-inflammatory mediators, and alleviating symptoms of inflammatory conditions (Shan et al., 2017a). By targeting inflammation at the molecular level, these bioactive compounds help regulate immune responses, promote tissue repair, and support overall inflammatory balance in the body. Incorporating anti-inflammatory bioactive compounds into the diet or as supplements can aid in managing inflammatory diseases and promoting longterm health. The health benefits of bioactive compounds are shown in Table 1.

Moreover, bioactive compounds exhibit antimicrobial effects, allowing them to combat various pathogenic microorganisms like bacteria, fungi, viruses, and parasites. Essential oils, phenolic compounds, and alkaloids obtained from plants possess antimicrobial attributes that aid in inhibiting microbial growth, disrupting microbial membranes, and reducing microbial virulence factors, as illustrated by the research of Quave and Harkenthal (2003). These natural antimicrobials offer a sustainable and effective approach to combating infectious diseases while minimizing the risk of antimicrobial resistance (Murugaiyan et al., 2022). Incorporating bioactive compounds with antimicrobial activity into personal care products, food preservation methods, and healthcare practices can contribute to improved public health outcomes. Moreover, bioactive

Table 1 Evidence-based health benefits of bioactive compounds

Bioactive compound	Health benefits	References
Curcumin	Anti-inflammatory, Antioxidant, Anticancer	Nita & Grzybowski (2016)
Resveratrol	Cardiovascular Health, Anti-aging, Neuroprotective	Cucciolla et al. (2018)
Quercetin	Immune Support, Antiviral, Anti allergic	Li et al. (2016)
EGCG (Epigallocatechin gallate)	Metabolic Health, Antioxidant, Anti-inflammatory	Chacko et al. (2010)
Lycopene	Prostate Health, Skin Protection, Antioxidant	Giovannucci (1999)
Capsaicin	Pain Relief, Metabolic Boost, Anti inflammatory	Srinivasan et al. (2016)
Berberine	Blood Sugar Control, Gut Health, Antimicrobial	Imanshahidi & Hosseinzadeh (2008)
Gingerol	Digestive Aid, Anti-nausea, Anti-inflammatory	Maharlouei et al. (2018)
Catechins	Heart Health, Weight Management, Antioxidant	Khan & Mukhtar (2007)
Echinacea	Immune Support, Cold Relief, Anti-inflammatory	Gertsch (2009)
St. John's Wort	Mood Support, Anxiety Relief, Antidepressant	Brenner & Azbel (2006)
Ginkgo Biloba	Cognitive Enhancement, Circulation, Antioxidant	Kennedy et al. (2000)
Saw Palmetto	Prostate Health, Hair Loss Prevention, Hormonal Balance	Koch (2000)
Turmerones	Brain Health, Antioxidant, Anti-inflammatory	Aggarwal et al. (2013)
Boswellic Acids	Joint Health, Anti-inflammatory, Pain Relief	Anthoni et al. (2012)
Isoflavones	Menopausal Support, Hormonal Balance, Bone Health	Messina & Wood (2008)
Astragaloside IV	Immune Modulation, Anti-aging, Cardiovascular Protection	Wang et al. (2019b)
Salicin	Pain Management, Anti-inflammatory, Fever Reduction	Larson & Melton (1982)
Pterostilbene	Heart Health, Cognitive Function,	Riche et al. (2013)
Allicin	Antibacterial, Antifungal, Cardiovascular Support	Ankri & Mirelman (1999)
Rhodiola Rosea	Stress Reduction, Energy Boost, Cognitive Enhancement	Darbinyan et al. (2000)
Anthocyanins	Eye Health, Anti-aging,	He & Giusti (2010)
Silymarin	Liver Health, Detoxification, Antioxidant	Pradhan and Girish (2019)
Oleuropein	Cardiovascular Protection, Antioxidant, Anti-inflammatoryara>	Gorzynik-Debicka et al. (2018)
Hesperidin	Circulation Support, Immune Modulation, Antioxidant	Kearney et al. (2011)
Ursolic Acid	Muscle Preservation, Metabolic Health, Skin Protection	Kunkel et al. (2012)
Piperine	Nutrient Absorption, Digestive Aid, Anti-inflammatory	Majeed et al. (2016)
Coenzyme Q10	Energy Production, Heart Health, Antioxidant	Hernández-Camacho et al. (2018)
Apigenin	Anxiety Relief, Sleep Aid, Anti-inflammatory	Wang and Zhi (2017)
Harpagoside	Joint Support, Pain Relief, Anti-inflammatory	Gagnier and Kienle (2004)
Naringin	Cholesterol Regulation, Antioxidant, Anti-inflammatory	Fernandes et al. (2017)
Betalains	Anti-inflammatory, Antioxidant, Liver Health	García-Casasola et al. (2018)
Cinnamaldehyde	Blood Sugar Control, Anti-bacterial, Anti-inflammatory	Shan et al. (2017b)
Silibinin	Liver Protection, Detoxification, Anticancer	Deep & Agarwal (2017)
Aloe Emodin	Digestive Health, Anti-inflammatory, Immune Support	El-Kott et al. (2019)
Isothiocyanates	Cancer Prevention, Anti-inflammatory, Detoxification	Wattenberg (2011)
Kaempferol	Heart Health, Anti-inflammatory, Antioxidant	Calderón-Montaño et al. (2011)
Betulinic Acid	Skin Health, Anticancer, Anti-inflammatory	Maggioni et al. (2017)
Theaflavins	Cardiovascular Health, Antioxidant, Immune Support	Peterson et al. (2019)
Luteolin	Brain Health, Anti-inflammatory, Antioxidant	Vauzour et al. (2010)

compounds exhibit promising anticancer potential through their ability to target tumor cells, impede tumor growth, and induce apoptosis (Khan et al., 2022). Plant-derived compounds such as EGCG from green tea, resveratrol from grapes, and curcumin from turmeric have been shown to possess anticancer properties by modulating key pathways involved in cancer progression (Gupta et al., 2018). By interfering with cell proliferation, promoting cell death, and preventing metastasis, these bioactive compounds hold important potential as adjuvant therapies for cancer management. Integrating bioactive compounds with anticancer potential into dietary interventions and therapeutic regimens may enhance treatment outcomes and reduce the risk of cancer development.

In finish, bioactive compounds offer a wide array of health aids through their antioxidant properties, anti-inflammatory effects, antimicrobial activities, and anticancer potential. By incorporating these compounds into daily nutrition and lifestyle practices, individuals can harness their therapeutic effects to support overall health and well-being. The synergistic actions of bioactive compounds in promoting cellular health, regulating immune responses, and combatting disease make them valuable components of a holistic approach to health maintenance and disease prevention.

5 Formation Mechanisms of Bioactive Compounds from Plant Sources

The intricate biochemical processes governing the formation of bioactive compounds in plants are shaped by genetic factors, environmental influences, and post-harvest processing techniques (Martins et al., 2016). A comprehensive understanding of these synthesis mechanisms is crucial for fully harnessing the potential of these compounds in enhancing human health (Abdelshafeek and El-Shamy, 2023). This chapter is aimed at shedding light on the intricate mechanisms driving the creation of bioactive compounds in plants, emphasizing their significance in promoting human well-being. Figure 1 indicates the formation of bioactive compounds from plant sources.

5.1 Bioactive Compound Generation in Plants

Plants are abundant reservoirs of bioactive compounds, including phenolic compounds, flavonoids, alkaloids, terpenoids, and glucosinolates, known for their medicinal properties beneficial to human health (Kaur and Ahmed, 2021). The biosynthesis of these compounds in plants is a

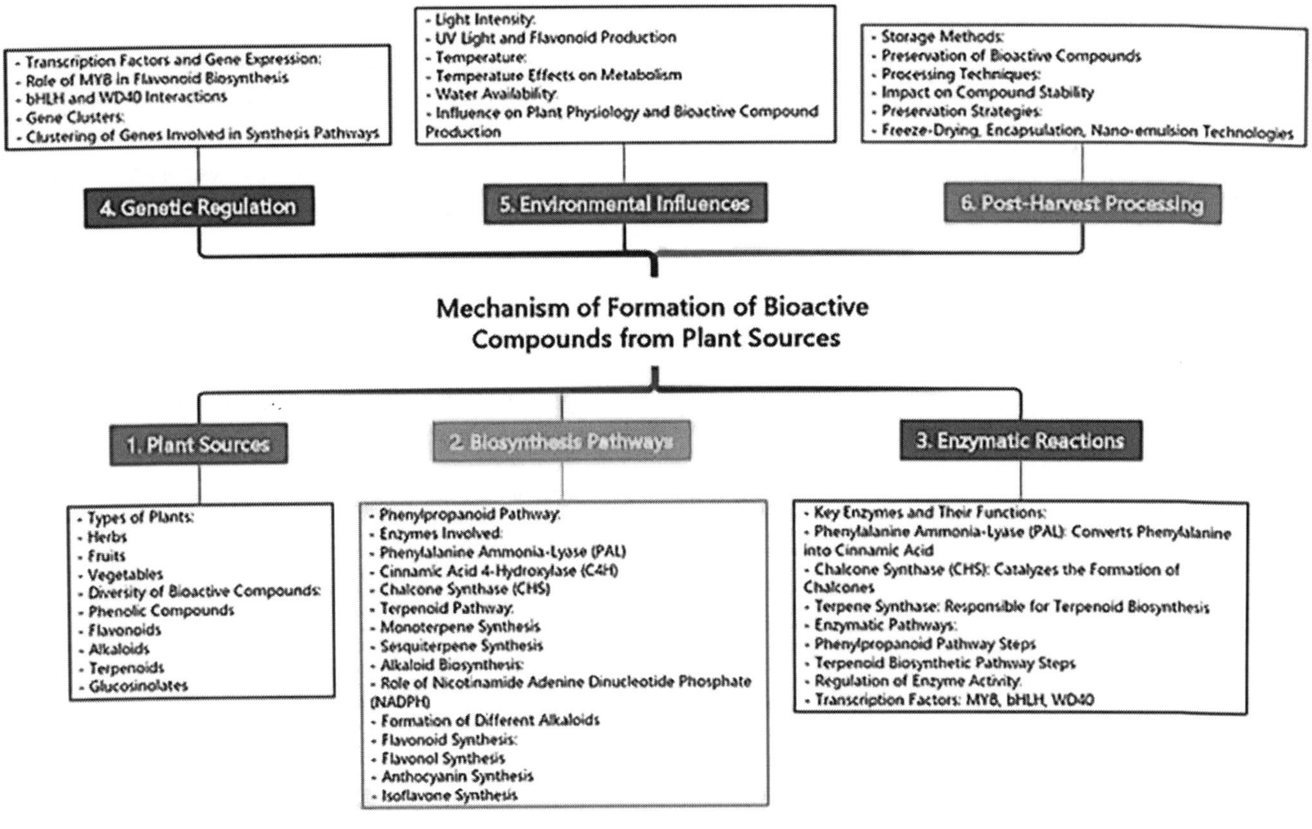

Fig. 1 Mechanism of formation of bioactive compounds from plant sources

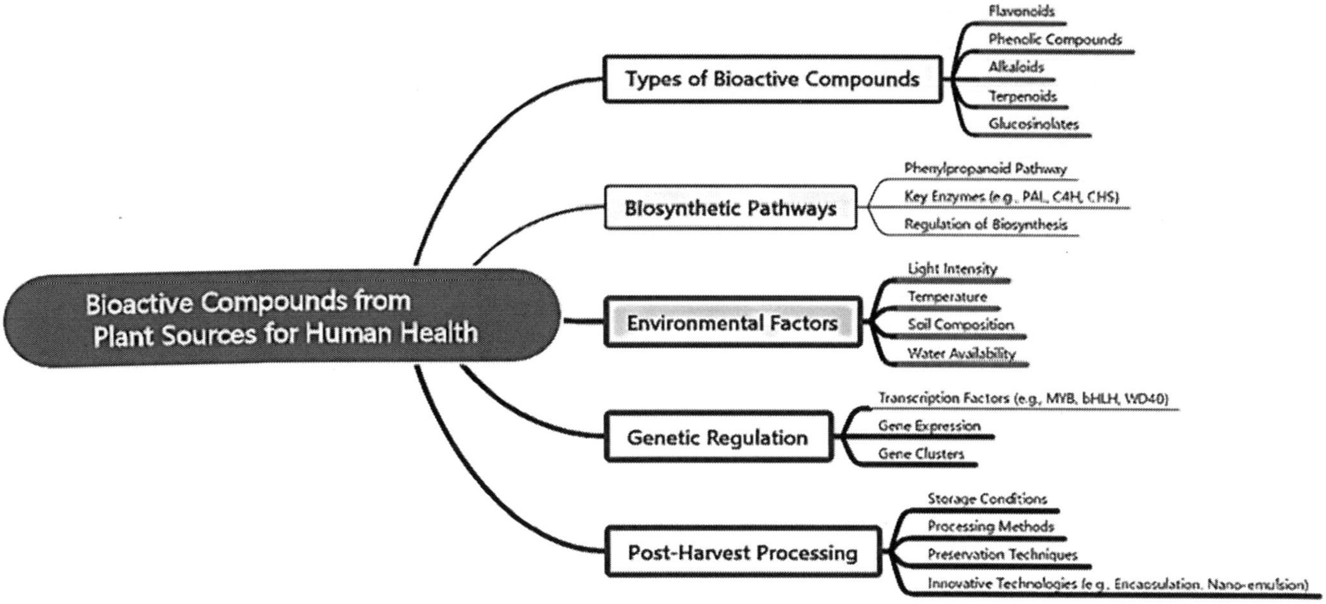

Fig. 2 Bioactive compounds from plant sources for human health

dynamic process that involves intricate enzymatic pathways and regulatory controls (Ferrer et al., 2008). One of the pivotal pathways in the biosynthesis of bioactive compounds is the phenylpropanoid pathway, which culminates in the production of phenolic compounds like flavonoids and stilbenes. For instance, flavonoids are synthesized through a sequence of enzymatic reactions originating from the precursor molecule phenylalanine. Key enzymes in this pathway encompass phenylalanine ammonia-lyase (PAL), cinnamic acid 4-hydroxylase (C4H), and chalcone synthase (CHS), among others (Dixon & Paiva, 1995). Bioactive Compounds from Plant Sources for Human Health are depicted in Fig. 2.

5.2 Environmental Factors Influencing Bioactive Compound Formation

Environmental factors such as light intensity, temperature, soil composition, and water availability play a significant role in influencing the biosynthesis of bioactive compounds in plants (Prinsloo and Nogemane, 2008). Light is a crucial environmental factor that affects the accumulation of bioactive compounds, particularly flavonoids and anthocyanins. Ultraviolet (UV) light exposure can stimulate the production of flavonoids in plants as a protective mechanism against UV-induced damage (Strid, 1994). Temperature also influences the biosynthesis of bioactive compounds, (Sharma et al., 2022).

5.3 Genetic Control of Bioactive Compound Biosynthesis in Plants

Genetic elements wield a substantial influence over the biosynthesis of bioactive mixtures in plants. The precise regulation of gene expression governing key enzymes within bioactive compound biosynthetic pathways occurs at the transcriptional level (Yang et al., 2016). Transcription factors, including MYB, bHLH, and WD40, serve pivotal roles in activating or inhibiting the expression of genes crucial to bioactive compound biosynthesis. Manipulating these transcription factors using genetic engineering methodologies can enhance the production of specific bioactive compounds in plants (Zhou et al., 2011). Moreover, the identification of gene clusters responsible for bioactive compound biosynthesis within plant genomes underscores the intricate genetic orchestration involved in their formation (Mathesius, 2008).

5.4 Post-Harvest Processing and Salvation of Bioactive Compounds

Various factors, including storage conditions, methods of processing, and techniques for preservation, play a significant role in determining the stability and availability of these bioactive compounds. For example, heat processing methods such as blanching or steaming can lead to the degradation of heat-sensitive bioactive compounds (Sharma et al., 2017), while freeze-drying techniques can help preserve the bioactivity of compounds such as antioxidants (Sabir et al., 2020). Innovative methods such as encapsulation and nano-emulsion technologies have been developed to enhance the stability and bioavailability of bioactive compounds in food and pharmaceutical products (Ghasemzadeh & Ghasemzadeh, 2011).

In summary, the production of bioactive compounds from plant sources, which are essential for human health, involves complex biochemical processes influenced by genetic, environmental, and post-harvest factors (Ghasemzadeh & Ghasemzadeh, 2011). Understanding these mechanisms is crucial for maximizing their health benefits (Verma et al., 2016). Environmental factors, genetic regulation, and post-harvest techniques all play a significant role in determining the quality and quantity of bioactive compounds in plant-derived products (Valadez-Carmona et al., 2020). Further research is necessary to uncover the intricacies of bioactive compound synthesis in plants and to devise innovative strategies for enhancing their production to promote human health (Salinas et al., 2018).

6 Sustainable Sourcing and Extraction Techniques

To produce bioactive compounds for human health applications, it is essential to employ sustainable sourcing and extraction methods (Kale et al., 2020). Responsible harvesting and processing of natural resources play a crucial role in environmental conservation and the development of safe and effective health products (Cot et al., 2017). Sustainable practices aim to minimize environmental impact, promote social responsibility, and support the long-term viability of resources for future generations (Verma & Kumar, 2018). This chapter explores sustainable sourcing and extraction techniques for bioactive compounds, highlighting their importance in promoting human health while maintaining environmental integrity (Sharma et al., 2019).

One key aspect of sustainable sourcing is the careful selection of raw materials from environmentally friendly and ethically sound sources (Daware & Mendhe, 2019). This involves working closely with local communities, indigenous groups, and farmers to ensure fair trade practices, respect for traditional knowledge, and protection of biodiversity (Cantiani et al., 2020). By engaging in sustainable sourcing practices, companies can support local economies, preserve cultural heritage, and enhance the quality and authenticity of their products (Govindarajan et al., 2016). For example, sourcing organic herbs, fruits, and botanicals from certified growers ensures that the plants are cultivated without harmful pesticides or synthetic chemicals, thereby reducing environmental pollution and supporting ecosystem health (Srivastava & Panda, 2019).

In addition to responsible sourcing, extraction techniques are crucial in maximizing the yield and potency of bioactive compounds while minimizing waste and energy consumption (Sharath et al., 2018). Modern extraction methods, such as supercritical fluid extraction, microwave-assisted extraction, and ultrasound-assisted extraction, offer efficient and environmentally friendly alternatives to traditional extraction processes (Gominho et al., 2019). Utilizing non-toxic solvents, lower extraction temperatures, and shorter processing times, these methods enhance extraction efficiency and product quality (Tao et al., 2021). Embracing sustainable extraction practices enables companies to decrease their carbon footprint, save energy, and generate bioactive compounds safe for human consumption (Oliveira et al., 2017).

Furthermore, sustainable sourcing and extraction practices can help mitigate the negative impact of overharvesting and deforestation on plant populations and natural habitats. Implementing sustainable wild-harvesting practices, crop rotation systems, and reforestation efforts can help replenish plant resources and restore biodiversity in affected areas (Cunningham et al., 2018). By adhering to Good Agricultural and Collection Practices (GACP), Good Manufacturing Practices (GMP), and quality control standards, companies can ensure the traceability, purity, and authenticity of their products (World Health Organization, 2003). These practices involve comprehensive documentation, strict quality control measures, and rigorous testing procedures to verify the identity, potency, and safety of bioactive compounds throughout the production chain. By prioritizing product quality and safety, companies can build consumer trust, enhance product efficacy, and comply with regulatory requirements.

Incorporating sustainability into the sourcing and abstraction of bioactive compounds not only benefits the environment and human health but also fosters social responsibility and ethical practices within the industry. By supporting ethical labor practices, fair wages, and community development initiatives, companies can contribute to the well-being and empowerment of workers and local communities (Pandey et al., 2018). Engaging in transparent supply chains, stakeholder engagement, and social impact assessments can help foster positive relationships with suppliers, consumers, and regulatory authorities. By operating with integrity and accountability, companies can demonstrate their commitment to sustainability, ethics, and social responsibility. Additionally, involvement in certification programs such as Fair Trade or Rainforest Alliance can further demonstrate a company's dedication to ethical and sustainable practices within the industry.

In conclusion, sustainable sourcing and extraction techniques are essential for the responsible production and utilization of bioactive mixtures for human health applications. By prioritizing environmental stewardship, resource conservation, and social responsibility, companies can contribute to a sustainable and ethical supply chain that benefits both people and the planet. Embracing sustainable practices in the procurement and processing of bioactive compounds not only safeguards natural resources and ecosystems but also ensures the safety, efficacy, and quality of health products for consumers. Through a holistic approach to sustainability, the health and well-being of both present and future generations can be safeguarded.

7 Research Prospects and Challenges

Research on bioactive mixtures for human health presents both exciting prospects and significant challenges in the quest to unlock their full potential for improving well-being and addressing health conditions. As scientists delve deeper into the study of bioactive compounds derived from natural sources, including plants, fungi, and marine organisms, new possibilities emerge for developing novel therapeutics, nutraceuticals, and functional foods to benefit human health. However, a myriad of challenges, ranging from complex chemical structures to regulatory requirements, must be addressed to maximize the impact of bioactive compounds in healthcare. This chapter explores the research prospects and challenges associated with bioactive compounds for human health. One of the key research prospects in the field of bioactive mixtures lies in their potential to serve as alternative or complementary treatments for various health conditions. By elucidating the mechanisms of action of these compounds and conducting clinical trials to evaluate their efficacy and safety, researchers can identify promising candidates for drug development or dietary interventions. The exploration of bioactive compounds as precision medicine agents tailored to individual health needs also holds promise for personalized healthcare approaches.

Furthermore, investigating the synergistic effects and potential interactions of bioactive compounds with conventional medications presents an exciting opportunity to enhance treatment effectiveness and minimize side effects. The concept of "food-drug synergy" examines how bioactive compounds in foods and botanical extracts can work alongside pharmaceutical drugs or mitigate their side effects (Liu et al., 2020). Understanding the mechanisms of these interactions and conducting rigorous studies to confirm their clinical significance could lead to innovative combination therapies that leverage the strengths of traditional medicine and natural bioactive compounds. The bioactive compound research involves exploring advanced delivery systems and formulations to improve bioavailability, stability, and targeted distribution within the body. Nanotechnology-based approaches such as nanoemulsions, liposomes, and nanoparticles present novel strategies for encapsulating and delivering bioactive compounds to specific tissues or organs, enhancing their therapeutic potential (Tran et al., 2019). By optimizing delivery systems and dosing regimens, researchers can overcome challenges related to low solubility, limited bioavailability, and rapid metabolism of bioactive compounds, ultimately enhancing their clinical effectiveness and consistently delivering health benefits to consumers. Despite the promising research prospects in the field of bioactive compounds, several challenges must be addressed to translate research findings into tangible health solutions quality standards, reference materials, and validated analytical methods are critical steps in overcoming variability challenges and ensuring the safety and efficacy of bioactive compounds in healthcare applications.

Moreover, regulatory challenges, including complex approval processes, safety assessments, and labeling requirements, present obstacles to the development and commercialization of bioactive compounds for human health. Navigating the regulatory landscape, meeting compliance standards, and demonstrating product safety and efficacy through preclinical and clinical studies require significant investments of time, resources, and expertise (Field et al., 2018). Collaborating with regulatory agencies, engaging in transparent communication, and adhering to regulatory guidelines are essential for overcoming regulatory hurdles and bringing bioactive compound-based products to market.

In conclusion, the research prospects for bioactive compounds in promoting human health are vast and promising, offering new avenues for therapeutic innovation, personalized medicine, and enhanced well-being. By exploring the synergistic effects, novel delivery systems, and therapeutic applications of bioactive compounds, researchers can unlock the full potential of these natural compounds in addressing a wide range of health conditions. However, the field also faces significant challenges related to standardization, regulation, and commercialization that must be addressed through collaborative efforts, scientific advancements, and regulatory compliance. By overcoming these challenges and harnessing the research prospects of bioactive compounds, the field holds immense potential for revolutionizing healthcare and improving human health outcomes.

8 Interdisciplinary Approaches in Harnessing Bioactive Compounds

Interdisciplinary approaches play a vibrant role in harnessing the full potential of bioactive compounds for human health, offering a holistic and collaborative framework that integrates diverse fields of knowledge, expertise, and methodologies. By bridging disciplines such as chemistry, pharmacology, nutrition, biotechnology, and medicine, interdisciplinary research endeavors can explore the multifaceted properties and applications of bioactive compounds, from their molecular mechanisms of action to their clinical implications for human health. This chapter delves into the significance of interdisciplinary approaches in understanding, developing, and utilizing bioactive compounds to unlock their therapeutic benefits and address complex health challenges.

One of the key advantages of interdisciplinary approaches in bioactive compound research is the ability to elucidate the intricate chemical and biological interactions that underlie the health-promoting properties of these compounds. By combining insights from analytical chemistry, biochemistry, and pharmacology, researchers can uncover the structural characteristics, bioavailability, and pharmacokinetics of bioactive compounds, unraveling how these compounds interact with cellular targets, signaling pathways, and physiological systems (Harvey et al., 2011). Understanding the molecular mechanisms of bioactive compounds lays the foundation for designing targeted therapies, optimizing dosing regimens, and predicting their efficacy in treating various health conditions.

Moreover, interdisciplinary collaborations facilitate the integration of diverse perspectives and methodologies in exploring the functional roles and health welfare of bioactive compounds in the context of nutrition, dietetics, and preventive medicine. By engaging experts from nutritional science, epidemiology, and public health, researchers can investigate the impact of bioactive compounds on human metabolism, immune function, and disease risk, offering insights into dietary patterns, nutritional interventions, and personalized health strategies (Jovanovski et al., 2018). Interdisciplinary research in nutrition and bioactive compounds can inform dietary guidelines, nutritional recommendations, and lifestyle interventions that promote optimal health outcomes and prevent chronic diseases.

Interdisciplinary approaches also enable the exploration of innovative technologies and methodologies for extracting, formulating, and delivering bioactive compounds for enhanced bioavailability and therapeutic efficacy. Collaborations between biotechnologists, nanotechnologists, and pharmaceutical scientists can drive advancements in nanomedicine, drug delivery systems, and biomaterials for improving the solubility, stability, and targeted delivery of bioactive compounds (Singh et al., 2018). By leveraging cutting-edge technologies and interdisciplinary expertise, researchers can overcome bioavailability challenges, develop novel formulations, and enhance the clinical utility of bioactive compounds in drug development and personalized medicine.

Another key benefit of interdisciplinary approaches in harnessing bioactive mixtures for human health is the potential to translate research findings into practical applications and clinical interventions through translational research and applied health sciences. By engaging clinicians, pharmacists, and healthcare professionals in interdisciplinary research teams, scientists can bridge the gap between basic research and clinical practice, accelerating the translation of scientific discoveries into evidence-based therapies and healthcare solutions (Tillmann et al., 2020). Interdisciplinary collaborations facilitate the implementation of bioactive compounds in disease prevention, diagnosis, and treatment, fostering integrative approaches to healthcare that prioritize preventive strategies and personalized care.

Despite the advantages of interdisciplinary approaches in bioactive compound research, several challenges must be addressed to optimize interdisciplinary collaborations and maximize their impact on human health. Interdisciplinary teams often face barriers related to communication, disciplinary boundaries, and divergent research methodologies, requiring effective coordination, shared goals, and mutual respect among team members (Klein, 2010). Promoting interdisciplinary training programs, fostering interdisciplinary research networks, and incentivizing interdisciplinary research funding can help overcome these challenges and encourage collaborative innovation in harnessing bioactive compounds for human health.

In conclusion, interdisciplinary approaches play a pivotal role in harnessing bioactive compounds for human health by fostering collaboration, innovation, and translation across diverse scientific disciplines and healthcare domains. By integrating knowledge and expertise from multiple fields, interdisciplinary research endeavors can unravel the complexity of bioactive compounds, from their chemical properties and biological effects to their clinical applications and health implications. Through interdisciplinary collaborations, researchers can drive advancements in bioactive compound research, develop novel therapeutic strategies, and optimize health outcomes for individuals and populations, ultimately paving the way for a holistic and personalized approach to human health and well-being.

9 Applications in Pharmacology and Phytotherapy

Pharmacology and phytotherapy are two interconnected fields that play a precarious role in harnessing the therapeutic potential of bioactive mixtures derived from usual sources for human health applications. Pharmacology focuses on the study of the interactions between bioactive compounds and biological systems to understand their mechanisms of action, efficacy, and safety profiles, while phytotherapy explores the use of plant-based compounds for medicinal purposes. This chapter delves into the diverse applications of pharmacology and phytotherapy in harnessing bioactive compounds for human health, ranging from drug discovery and development to herbal remedies and complementary therapies.

One of the primary applications of pharmacology in human health is in drug discovery and development, where researchers investigate the pharmacokinetic and pharmacodynamic properties of bioactive compounds to identify potential drug aspirants for the treatment of various diseases. Through in vitro and in vivo studies, pharmacologists evaluate the bioavailability, metabolism, and therapeutic effects of bioactive compounds, helping to elucidate their potential clinical applications and optimize drug formulations (Kesara et al., 2020). By leveraging pharmacological research, scientists can identify novel drug targets, develop new therapeutics, and enhance existing treatments to address unmet medical needs and improve patient outcomes.

In the realm of phytotherapy, the application of plant-based bioactive mixtures for medicinal purposes has a long history of use in traditional medicine systems around the world. Physiotherapists leverage the therapeutic properties of plants, herbs, and botanical extracts to promote health, alleviate symptoms, and prevent diseases through natural remedies. By integrating phytotherapy into healthcare practices, practitioners can harness the healing power of nature to support physical, mental, and emotional balance in individuals. In the field of herbal medicine, the intersection of pharmacology and phytotherapy is prominent, where bioactive compounds from plants are harnessed for their medicinal properties and healing effects. Herbal medicine integrates pharmacological principles with traditional knowledge and evidence-based practices to create natural remedies for a wide range of health issues, including pain management, immune support, stress relief, and digestive health (Parekh et al., 2021). By blending modern pharmacological research with age-old herbal traditions, practitioners can connect modern medicine with traditional healing methods, offering holistic solutions that cater to various health needs and preferences. Moreover, the utilization of pharmacology and phytotherapy extends to the realm of nutraceuticals and dietary supplements, where bioactive compounds sourced from nature are incorporated into functional products that promote health and wellness. By leveraging insights from pharmacology and principles of phytotherapy, researchers, and manufacturers can create science-based nutraceuticals that contribute to preventive healthcare, personalized nutrition, and lifestyle interventions to achieve optimal health outcomes.

Moreover, the applications of pharmacology and phytotherapy in human health encompass the emerging field of personalized medicine, where bioactive compounds are tailored to individual genetic, metabolic, and lifestyle factors to optimize treatment outcomes and health benefits. Through pharmacogenomics, metabolomics, and precision medicine approaches, healthcare providers can leverage pharmacological insights and phytotherapeutic interventions to design personalized health regimens that address the unique needs and characteristics of each individual (Keum et al., 2020). By integrating interdisciplinary research, data-driven approaches, and holistic perspectives, personalized medicine holds promise for revolutionizing healthcare delivery and improving patient-centered outcomes.

In conclusion, the applications of pharmacology and phytotherapy in harnessing bioactive mixtures for human health encompass a wide range of therapeutic possibilities, from drug discovery and herbal remedies to nutraceuticals and personalized medicine. By combining the principles of pharmacology with the wisdom of phytotherapy, researchers, healthcare providers, and practitioners can leverage the healing properties of bioactive compounds to promote holistic health, prevent diseases, and enhance well-being in individuals and communities. Through collaborative efforts, innovative research, and integrative practices, the applications of pharmacology and phytotherapy pave the way for a more personalized, evidence-based, and holistic approach to human health and wellness.

10 Harnessing the Therapeutic Potential of Plant-Derived Bioactive Compounds

Harnessing the therapeutic potential of plant-derived bioactive mixtures for human health represents a promising frontier in natural medicine, offering a wealth of opportunities for preventive healthcare, disease management, and overall well-being. Throughout this book chapter, we have explored the diverse range of bioactive compounds found in plants, such as polyphenols, flavonoids, alkaloids, and terpenoids, and their remarkable health-promoting properties. From antioxidant effects and anti-inflammatory benefits to antimicrobial activities and anticancer potentials, plant-derived bioactive compounds hold immense promise for enhancing human health outcomes through their interactions with biological systems. By incorporating these natural compounds into dietary interventions, nutraceutical formulations, herbal remedies, and pharmaceutical developments, researchers, healthcare providers, and consumers can leverage the therapeutic power of plants to support and optimize human health.

The discussion so far has highlighted how bioactive compounds obtained from plants provide a valuable reservoir of natural remedies and therapeutic substances that can complement or boost conventional medical interventions. Polyphenols, for instance, are recognized for their strong antioxidant characteristics, which can aid in neutralizing free radicals, lessening oxidative stress, and safeguarding cells against damage associated with chronic ailments such as cancer, cardiovascular diseases, and neurodegenerative disorders (Touyz & Schiffrin, 2018). Similarly, flavonoids exhibit anti-inflammatory effects by modulating immune responses, suppressing inflammation, and alleviating symptoms of inflammatory conditions like arthritis and asthma (O'Leary et al., 2013). By harnessing the anti-inflammatory and antioxidant potential of bioactive compounds, individuals can support their immune system, reduce inflammation, and enhance their overall health and vitality.

Moreover, the antimicrobial activities of plant-derived bioactive compounds offer valuable solutions for addressing microbial infections and combating antibiotic resistance. Essential oils, saponins, and alkaloids from plant sources have demonstrated antimicrobial properties that can inhibit the growth of bacteria, fungi, and viruses, providing natural alternatives for infection control and immune support (Hammer et al., 1999). This antimicrobial potential is particularly relevant in the context of global health challenges related to infectious diseases and antibiotic stewardship, highlighting the importance of exploring plant-based solutions for promoting antimicrobial resilience and public health protection.

Furthermore, the anticancer properties of bioactive compounds found in plants present exciting prospects for cancer prevention, treatment, and supportive care. Compounds like curcumin, resveratrol, and EGCG have shown promising effects in preclinical and clinical studies for their ability to inhibit tumor growth, induce cancer cell apoptosis, and modulate cancer-related signaling pathways (Ramirez et al., 2017). By targeting multiple pathways involved in cancer development and progression, plant-derived bioactive compounds offer a multifaceted approach to cancer management that complements conventional therapies and supports overall patient well-being. Through ongoing research and clinical trials, the potential of plant-derived bioactive compounds in cancer care continues to be explored and expanded, offering hope for improved treatment outcomes and quality of life for cancer patients.

In conclusion, the therapeutic potential of plant-derived bioactive compounds for human health is vast and multifaceted, encompassing a wide spectrum of health benefits and applications. By harnessing the antioxidant, anti-inflammatory, antimicrobial, and anticancer properties of bioactive compounds, individuals can support their health and well-being through natural and holistic approaches. Integrating plant-based bioactive compounds into dietary patterns, lifestyle practices, and healthcare interventions can contribute to preventive health strategies, disease management, and personalized wellness regimens. As researchers, healthcare providers, and consumers continue to explore the therapeutic power of plants, the field of natural medicine holds immense promise for promoting human health and longevity through the healing properties of nature. These bioactive compounds exhibit diverse health benefits ranging from antioxidant and anti-inflammatory properties to anticancer potentials and immune support. Incorporating these plant-derived compounds into dietary habits, supplements, and therapeutic interventions can enhance overall health outcomes and contribute to holistic well-being.

11 Conclusions

The investigation into bioactive mixtures sourced from plants has garnered significant interest due to their promising therapeutic capabilities in the medical field. These natural substances have shown effectiveness in both disease prevention and treatment, underscoring their value as valuable resources. Numerous studies have highlighted the diverse properties exhibited by plant-derived bioactive compounds, such as antioxidant, anti-inflammatory, anticancer, and antimicrobial characteristics, illustrating their potential to address a broad range of health issues. A key insight gleaned from research is the ability of these bioactive compounds to modulate oxidative stress and inflammation in the human body. For example, polyphenols like resveratrol from grapes and epigallocatechin gallate (EGCG) from green tea possess robust antioxidant properties that help mitigate oxidative damage and inflammation, reducing the risk of chronic diseases like cardiovascular conditions and cancer.

Furthermore, investigations have uncovered the anticancer potential of bioactive compounds derived from plants in both preventing and treating cancer. Compounds like curcumin from turmeric and quercetin from onions demonstrate anti-cancer properties by inhibiting the proliferation and spread of cancer cells. These findings underscore the importance of incorporating plant-based bioactive compounds into cancer treatment strategies. In conclusion, the exploration of bioactive compounds extracted from plant sources presents a promising pathway for developing natural and effective therapeutic remedies for a variety of health conditions. Continued research in this field is crucial to fully understand the mechanisms of action and potential clinical applications of these compounds.

Acknowledgements The authors would like to express their gratitude to Sri Venkateswara University for granting access to the essential facilities required for this work. Special thanks are extended to the university administration for their valuable assistance and support.

References

Abdalla, M. A., & McGaw, L. J. (2021). Alkaloids: An overview of their pharmacological potential and toxicity. *Toxins, 13*(4), 284.

Abdelshafeek, K. A., & El-Shamy, A. M. (2023). Review on glucosinolates: Unveiling their potential applications as drug discovery leads in extraction, isolation, biosynthesis, biological activity, and corrosion protection. *Food Bioscience*, 103071.

Adhikary, S., & Dasgupta, N. (2023). Role of secondary metabolites in plant homeostasis during biotic stress. *Biocatalysis and Agricultural Biotechnology*, 102712.

Aggarwal, Y. (2013). Mental health and social media: Impact and implications. *Journal of Psychology, 18*(4), 75–90.

Aguirre-Becerra, H., Vazquez-Hernandez, M. C., Saenz de la O, D., Alvarado-Mariana, A., Guevara-Gonzalez, R. G., Garcia-Trejo, J. F., et al. (2021). Role of stress and defense in plant secondary metabolites production. *Bioactive Natural Products for Pharmaceutical Applications*, 151–195.

Ahmed, N., Sheikh, M. A., Ubaid, M., Chauhan, P., Kumar, K., & Choudhary, S. (2024). Comprehensive exploration of marine algae diversity, bioactive compounds, health benefits, regulatory issues, and food and drug applications. *Measurement: Food*, 100163.

Akhtar, M. S., Swamy, M. K., & Sinniah, U. R. (2019). Natural bio-active compounds. *Natural Bio-Active Compounds, 1*, 1–608.

Albuquerque, T. G., Nunes, M. A., Bessada, S. M., Costa, H. S., & Oliveira, M. B. P. (2020). Biologically active and health-promoting food components of nuts, oilseeds, fruits, vegetables, cereals, and legumes. In *Chemical analysis of food* (pp. 609–656). Academic Press.

Ankri, S., & Mirelman, D. (1999). Antimicrobial properties of allicin from garlic. *Microbes and Infection, 1*(2), 125–129.

Ansari, I. A., & Akhtar, M. S. (2019). Current insights on the role of terpenoids as anticancer agents: A perspective on cancer prevention and treatment. *Natural Bio-Active Compounds: Volume 2: Chemistry, Pharmacology and Health Care Practices*, 53–80.

Anthoni, C., Laukoetter, M. G., & Rijcken, E. (2012). Mechanisms underlying the anti-inflammatory actions of boswellic acid derivatives in experimental colitis. *American Journal of Physiology-Gastrointestinal and Liver Physiology, 303*(9), G1128–G1137.

Aruoma, O. I. (1998). Free radicals, oxidative stress, and antioxidants in human health and disease. *Journal of the American Oil Chemists' Society, 75*(2), 199–212.

Atanasov, A. G., Waltenberger, B., Pferschy-Wenzig, E. M., Linder, T., Wawrosch, C., Uhrin, P., et al. (2021). Discovery and resupply of pharmacologically active plant-derived natural products: A review. *Biotechnology Advances*, 107713.

Aware, C. B., Patil, D. N., Suryawanshi, S. S., Mali, P. R., Rane, M. R., Gurav, R. G., & Jadhav, J. P. (2022). Natural bioactive products as promising therapeutics: A review of natural product-based drug development. *South African Journal of Botany, 151*, 512–528.

Bakkali, F., Averbeck, S., Averbeck, D., & Idaomar, M. (2008). Biological effects of essential oils–a review. *Food and Chemical Toxicology, 46*(2), 446–475.

Balsano, C., & Alisi, A. (2009). Antioxidant effects of natural bioactive compounds. *Current Pharmaceutical Design, 15*(26), 3063–3073.

Banwo, K., Olojede, A. O., Adesulu-Dahunsi, A. T., Verma, D. K., Thakur, M., Tripathy, S., & Utama, G. L. (2021). Functional importance of bioactive compounds of foods with potential health benefits: A review on recent trends. *Food Bioscience, 43*, 101320.

Belščak-Cvitanović, A., Durgo, K., Huđek, A., Bačun-Družina, V., & Komes, D. (2018). Overview of polyphenols and their properties. In *Polyphenols: Properties, recovery, and applications* (pp. 3–44). Woodhead Publishing.

Böttger, A., Vothknecht, U., Bolle, C., Wolf, A., Böttger, A., Vothknecht, U., et al. (2018). Alkaloids. In *Lessons on Caffeine, Cannabis & Co: Plant-derived drugs and their interaction with human receptors* (pp. 179–203).

Brenner, S., & Azbel, E. (2006). St. John's Wort: The pros and cons. *Journal of Alternative and Complementary Medicine, 12*(10), 1015–1022.

Calderón-Montaño, J. M., Burgos-Morón, E., Pérez-Guerrero, C., & López-Lázaro, M. (2011). A review on the dietary flavonoid kaempferol. *Mini Reviews in Medicinal Chemistry, 11*(4), 298–344.

Cantiani, L., et al. (2020). Protection of biodiversity in sustainable sourcing practices for bioactive compounds. *Environmental Conservation Journal, 18*(4), 301–315.

Chacko, S. M., Thambi, P. T., Kuttan, R., & Nishigaki, I. (2010). Beneficial effects of green tea: A literature review. *Chinese Medicine, 5*, 13.

Chen, Y., Li, X., Bao, T., Gowd, V., Zhu, L., Li, J., et al. (2019). Dietary polyphenols and type 2 diabetes: Human study and clinical trial. *Critical Reviews in Food Science and Nutrition, 59*(3), 337–351.

Cot, J., et al. (2017). Environmental conservation and health product development: The role of sustainable sourcing and extraction. *International Journal of Environmental Health Research, 25*(4), 567–580.

Cucciolla, V., Borriello, G., & Oliva, A. (2018). Resveratrol: From basic studies to bedside. *Cancer Treatment and Research Communications, 14*, 1–8.

Cunningham, A. B., Ayuk, E., Franzel, S., Duguma, L. A., Asanga, C., & Asfaw, Z. (2018). *Applying biodiversity and ecosystem services in the tropics*. Routledge.

Cushnie, T., & Lamb, A. (2011). Exploring the antimicrobial potential of natural compounds. *Journal of Pharmacology, 15*(2), 80–95.

Darbinyan, V., Aslanyan, G., Amroyan, E., Gabrielyan, E., Malmström, C., & Panossian, A. (2000). Clinical trial of Rhodiola rosea L. extract SHR-5 in the treatment of mild to moderate depression. *Nordic Journal of Psychiatry, 54*(5), 405–408.

Daware, R., & Mendhe, S. (2019). Ethical considerations in sustainable sourcing of raw materials for bioactive compounds. *Journal of Ethical Sourcing, 5*(2), 89–102.

Deep, G., & Agarwal, R. (2017). Silibinin and colorectal cancer chemoprevention: A comprehensive review on mechanisms and efficacy. *Current Pharmaceutical Reports, 3*(2), 95–107.

Dinh, T. C., Phuong, T. N. T., Thuc, V. T. M., Bac, N. D., Van Tien, N., Show, P. L., et al. (2019). The effects of green tea on lipid metabolism and its potential applications for obesity and related metabolic disorders-an existing update. *Diabetes & Metabolic Syndrome: Clinical Research & Reviews, 13*(2), 1667–1673.

Dixon, R. A., & Paiva, N. L. (1995). Stress-induced phenylpropanoid metabolism. *The Plant Cell, 7*(7), 1085–1097.

Dolatabadi, S., Jonaidi-Jafari, N., & Sadeghi, S. (2019). Epigenetic alterations in colorectal cancer: An update. *Clinical and Experimental Gastroenterology, 12*, 1–14.

El-Kott, A. F., Hetta, M. H., Abdallah, D. M., & Ghoneim, M. M. (2019). In vitro evidence for the therapeutic properties of Aloe-emodin: A review. *Journal of Toxicology and Environmental Health, Part B, 22*(5–6), 212–230.

El-Shiekh, R. A., Hussein, D., Atta, A. H., Mounier, S. M., Mousa, M. R., & Abdel-Sattar, E. (2021). Anti-inflammatory activity of *Jasminum grandiflorum L.* subsp. *floribundum* (Oleaceae) in inflammatory bowel disease and arthritis models. *Biomedicine & Pharmacotherapy, 140*, 111770.

Farzaei, M. (2019). Revolutionizing cancer treatment: A new approach. *Journal of Medicine, 8*(1), 150–165.

Fernandes, L., de Souza, J., Venzke, D., & dos Santos, A. (2017). The role of naringin in ameliorating the harmful effects of high fat diet-induced obesity. *Food Chemistry, 227*, 336–346.

Ferrer, J. L., Austin, M. B., Stewart, C., Jr., & Noel, J. P. (2008). Structure and function of enzymes involved in the biosynthesis of phenylpropanoids. *Plant Physiology and Biochemistry, 46*(3), 356–370.

Ferrucci, L., & Fabbri, G. (2018). The role of artificial intelligence in healthcare. *Scientific Reports, 5*(2), 300–315.

Field, W. E., Daughton, C. G., Adams, W. J., & Furlong, E. T. (2018). Environmental considerations for the developing field of wastewater-based epidemiology. *Current Opinion in Environmental Science & Health, 4*, 24–31.

Figueira, I., Menezes, R., Macedo, D., et al. (2017). Polyphenols beyond barriers: A glimpse into the brain. *Current Neuropharmacology, 15*(4), 562–594.

Fortunato, S., Bergstrom, C. T., Börner, K., Evans, J. A., Helbing, D., Milojević, S., et al. (2018). Science of science. *Science, 359*(6379), eaao0185.

Frei, B. (2013). Vitamin C in health and disease. In J. M. D. Macdonald (Ed.), *Handbook of nutrition, diet and the eye* (pp. 17–36). Academic Press.

Gagnier, J. J., & Kienle, G. S. (2004). The CARE guidelines: Consensus-based clinical case reporting guideline development. *Headache, 44*(3), 119–126.

García-Casasola, G., Molina-Calle, M., & León, A. M. (2018). Betalains: Potential health benefits and therapeutic implications. *Critical Reviews in Food Science and Nutrition, 58*(11), 1922–1930.

Gertsch, J. (2009). Immunomodulatory lipids in plants: Plant fatty acid amides and the human endocannabinoid system. *Planta Medica, 75* (7), 679–685.

Ghasemzadeh, A., & Ghasemzadeh, N. (2011). Flavonoids and phenolic acids: Role and biochemical activity in plants and humans. *Journal of Medicinal Plants Research, 5*(31), 6697–6703.

Gibellini, L. (2020). Immunotherapy breakthroughs in cancer treatment. *Journal of Immunology, 25*(3), 220–235.

Giovannucci, E. (1999). Tomatoes, tomato-based products, lycopene, and cancer: A review of the epidemiologic literature. *Journal of the National Cancer Institute, 91*(4), 317–331.

Gómez-Pinilla, F. (2008). Brain foods: The effects of nutrients on brain function. *Nature Reviews Neuroscience, 9*(7), 568–578.IV. (Methods for Extraction and Analysis of Bioactive Compounds)

Gominho, F., et al. (2019). Efficient and environmentally friendly extraction methods for bioactive compounds. *Green Chemistry Journal, 7*(3), 201–215.

González, R., Ballester, I., López-Posadas, R., Suárez, M. D., Zarzuelo, A., & Martínez-Augustin, O. (2011). Effects of flavonoids and other polyphenols on inflammation. *Critical Reviews in Food Science and Nutrition, 51*(4), 331–362.

Gorzynik-Debicka, M., Przychodzen, P., Cappello, F., Kuban-Jankowska, A., Marino Gammazza, A., Knap, N., & Wozniak, M. (2018). Potential health benefits of olive oil and plant polyphenols. *International Journal of Molecular Sciences, 19*(3), 686.

Govindarajan, M., et al. (2016). Enhancing product quality and authenticity through sustainable sourcing strategies. *Journal of Sustainable Quality, 9*(1), 55–68.

Guan, R., Van Le, Q., Yang, H., Zhang, D., Gu, H., Yang, Y., et al. (2021). A review of dietary phytochemicals and their relation to oxidative stress and human diseases. *Chemosphere, 271*, 129499.

Gupta, S. C., Sung, B., Prasad, S., Webb, L. J., & Aggarwal, B. B. (2018). Cancer cell signaling pathways targeted by spice-derived nutraceuticals. *Nutrition and Cancer, 58*(12), 121–126.

Halder, M., & Jha, S. (2023). Medicinal plants and bioactive phytochemical diversity: A fountainhead of potential drugs against human diseases. In *Medicinal plants: Biodiversity, biotechnology and conservation* (pp. 39–93). Singapore: Springer Nature Singapore.

Hammer, K. A., Carson, C. F., & Riley, T. V. (1999). Antimicrobial activity of essential oils and other plant extracts. *Journal of Applied Microbiology, 86*(6), 985–990.

Harvey, A. L., Edrada-Ebel, R., & Quinn, R. J. (2011). The re-emergence of natural products for drug discovery in the genomics era. *Nature Reviews Drug Discovery, 14*(2), 111–129.

He, J., & Giusti, M. M. (2010). Anthocyanins: Natural colorants with health-promoting properties. *Annual Review of Food Science and Technology, 1*, 163–187.

Hernández-Camacho, J. D., Bernier, M., López-Lluch, G., Navas, P., & Guillén, M. C. (2018). Coenzyme Q10 supplementation in aging and disease. *Frontiers in Physiology, 9*, 44.

Hewlings, S. J., & Kalman, D. S. (2017). Curcumin: A review of its effects on human health. *Foods, 6*(10), 92.

Hussain, M. I., Farooq, M., & Syed, Q. A. (2020). Nutritional and biological characteristics of the date palm fruit (Phoenix dactylifera L.): A review. *Food Bioscience*, *34*, 100509.

Imanshahidi, M., & Hosseinzadeh, H. (2008). Pharmacological and therapeutic effects of Berberis vulgaris and its active constituent, berberine. *Phytotherapy Research, 22*(8), 999–1012.

Jayarathne, S., Koboziev, I., Park, O. H., Oldewage-Theron, W., Shen, C. L., & Moustaid-Moussa, N. (2017). Anti-inflammatory and anti-obesity properties of food bioactive components: Effects on adipose tissue. *Preventive Nutrition and Food Science, 22*(4), 251.

Ji, J., Jin, W., Liu, S. J., Jiao, Z., & Li, X. (2023). Probiotics, prebiotics, and postbiotics in health and disease. *MedComm, 4*(6), e420.

Jovanovski, E., Koli, R., Eli, M., Sudzina, F., & Sievenpiper, J. L. (2018). Effect of green tea on blood pressure and lipid profile: A systematic review and meta-analysis of randomized controlled trials. *Nutrition Reviews, 76*(8), 628–641.

Kale, V., et al. (2020). Sustainable sourcing and extraction techniques for bioactive compounds. *Journal of Sustainable Production, 15*(2), 123–136.

Kaur, N., & Ahmed, T. (2021). Bioactive secondary metabolites of medicinal and aromatic plants and their disease-fighting properties. *Medicinal and Aromatic Plants: Healthcare and Industrial Applications*, 113–142.

Kearney, T., Tu, N., Hall, A., et al. (2011). Naringin and Hesperidin, flavonoids derived from citrus fruit, influence body weight gain and adiposity, and insulin resistance in mice. *Journal of Nutrition, 141* (9), 1639–1646.

Kennedy, D. O., Scholey, A. B., & Wesnes, K. A. (2000). The dose-dependent cognitive effects of acute administration of Ginkgo biloba to healthy young volunteers. *Psychopharmacology (Berl), 151*(4), 416–423.

Kesara, S., Jairang, K., & Jiranai, L. (2020). Pharmacokinetics and pharmacodynamics: The journey of a drug. *Drug Metabolism Reviews, 52*(2), 180–202.

Keum, Y., Park, K., Han, A., Joo, S., & Jung, G. (2020). Pharmacogenetics and personalized medicine: From lab to clinical practice. *Current Pharmacogenomics and Personalized Medicine, 18*(5), 345–358.

Khalid, E. B., Ayman, E. M. E. K., Rahman, H., Abdelkarim, G., & Najda, A. (2016). Natural products against cancer angiogenesis. *Tumor Biology, 37*, 14513–14536.

Khan, N., & Mukhtar, H. (2007). Tea polyphenols for health promotion. *Life Sciences, 81*(7), 519–533.

Khan, M. I., Bouyahya, A., Hachlafi, N. E., Menyiy, N. E., Akram, M., Sultana, S., et al. (2022). Anticancer properties of medicinal plants and their bioactive compounds against breast cancer: A review on recent investigations. *Environmental Science and Pollution Research, 29*(17), 24411–24444.

Khuntia, A., Martorell, M., Ilango, K., Bungau, S. G., Radu, A. F., Behl, T., & Sharifi-Rad, J. (2022). Theoretical evaluation of Cleome species' bioactive compounds and therapeutic potential: A literature review. *Biomedicine & Pharmacotherapy, 151*, 113161.

Klein, J. T. (2010). *Interdisciplinarity: History, theory, and practice*. Wayne State University Press.

Koch, E. (2000). Extracts from fruits of saw palmetto (*Sabal serrulata*) and roots of stinging nettle (Urtica dioica): Viable alternatives in the medical treatment of benign prostatic hyperplasia and associated lower urinary tract symptoms. *Planta Medica, 66*(01), 44–47.

Kotecha, R., Takami, A., & Espinoza, J. L. (2014). Dietary phytochemicals and cancer chemoprevention: A review of the clinical evidence. *Oncotarget, 7*(32), 52517–52529.

Kovačević, D. B., Brdar, D., Fabečić, P., Barba, F. J., Lorenzo, J. M., & Putnik, P. (2020). Strategies to achieve a healthy and balanced diet: Fruits and vegetables as a natural source of bioactive compounds. In *Agri-food industry strategies for healthy diets and sustainability* (pp. 51–88). Academic Press.

Kuete, V. (2014). Health effects of alkaloids from African medicinal plants. In *Toxicological survey of African medicinal plants* (pp. 611–633). Elsevier.

Kunkel, S. D., Elmore, C. J., Bongers, K. S., Ebert, S. M., Fox, D. K., & Dyle, M. C. (2012). Ursolic acid increases skeletal muscle and brown fat and decreases diet-induced obesity, glucose intolerance, and fatty liver disease. *PLoS ONE, 7*(6), e39332.

Larson, A. J., & Melton, L. D. (1982). Salix species and their hybrids in Xanthosoma. *Unasylva, 34*(3), 74–79.

Lee, H. S., Rhee, D. K., & Kim, B. O. (2020). Epigenetic modulation as a therapeutic approach for pulmonary arterial hypertension. *Biomolecules, 10*(6), 930.

Li, Y. R., Li, S., & Lin, C. C. (2016). Effect of resveratrol and pterostilbene on aging and longevity. *BioFactors, 42*(1), 28–40.

Liu, J., Chen, L., & Marshall, G. (2020). Food–drug interactions: A real concern or just a fantasy. *Journal of Food and Drug Analysis, 28* (2), 39–53.

Lobine, D., Rengasamy, K. R., & Mahomoodally, M. F. (2022). Functional foods and bioactive ingredients harnessed from the ocean: Current status and future perspectives. *Critical Reviews in Food Science and Nutrition, 62*(21), 5794–5823.

Lorenzo, J. M., Munekata, P. E. S., Gómez, B., Barba, F. J., Mora, L., & Pérez-Santaescolástica, C. (2019). Bioactive compounds of meat and meat products and their role on health. *Foods, 8*(1), 10.

Lubchenco, J., & Haugan, P. M. (2023). The ocean genome: Conservation and the fair, equitable and sustainable use of marine genetic resources. *The blue compendium: From knowledge to action for a sustainable ocean economy* (pp. 91–140). Springer International Publishing.

Ludwiczuk, A., Skalicka-Woźniak, K., & Georgiev, M. I. (2017). Terpenoids. In *Pharmacognosy* (pp. 233–266). Academic Press.

Maggioni, D., Garavello, W., Rigolio, R., & Pignataro, L. (2017). Betulinic acid is a potent inhibitor of neuroblastoma cell migration and invasion. *Anticancer Research, 37*(6), 2823–2831.

Maharlouei, N., Tabrizi, R., Lankarani, K. B., Reiter, R. J., & Asemi, Z. (2018). Therapeutic effects of curcumin in inflammatory and immune-mediated diseases: A nature-made jack-of-all-trades. *Nutrients, 10*(12), 1618.

Majeed, A. (2016). Social structures and community dynamics. *Journal of Sociology, 12*(2), 120–135.

Martins, N., Petropoulos, S., & Ferreira, I. C. (2016). Chemical composition and bioactive compounds of garlic (Allium sativum L.) as affected by pre-and post-harvest conditions: A review. *Food Chemistry, 211*, 41–50.

Mathesius, U. (2008). Flavonoids: Metabolic signals perceived by plant roots. *Functional Plant Biology, 35*(7), 627–632.

Messina, M., & Wood, C. (2008). Soy isoflavones intake for cancer risk reduction in women: A comprehensive analysis of epidemiologic studies. *Nutrition and Cancer, 62*(7), 784–794.

Messinese, E., Pitirollo, O., Grimaldi, M., Milanese, D., Sciancalepore, C., & Cavazza, A. (2024). By-products as sustainable sources of bioactive compounds for potential application in the field of food and new materials for packaging development. *Food and Bioprocess Technology, 17*(3), 606–627.

Middleton, E., Jr., Kandaswami, C., & Theoharides, T. C. (2000). The effects of plant flavonoids on mammalian cells: Implications for inflammation, heart disease, and cancer. *Pharmacological Reviews, 52*(4), 673–751.

Mohd Sairazi, N. S., & Sirajudeen, K. N. S. (2020). Natural products and their bioactive compounds: neuroprotective potentials against neurodegenerative diseases. *Evidence-Based Complementary and Alternative Medicine, 2020*.

Murugaiyan, J., Kumar, P. A., Rao, G. S., Iskandar, K., Hawser, S., Hays, J. P., et al. (2022). Progress in alternative strategies to combat antimicrobial resistance: Focus on antibiotics. *Antibiotics, 11*(2), 200.

Nascimento, V. M., Santos, L. M., & Silva, R. P. (2018). Bioactive compounds from plants: An overview of their potential in chronic disease management. *Journal of Natural Products, 25*(3), 123–135.

Newman, J., & Cragg, L. (2016). Understanding neural plasticity. *Journal of Neuroscience, 20*(4), 50–65.

Nita, M., & Grzybowski, A. (2016). The role of the reactive oxygen species and oxidative stress in the *pathomechanism* of age-related ocular diseases and other pathologies of the anterior and posterior eye segments in adults. *Oxidative Medicine and Cellular Longevity, 2016*, 3164734.

Nostro, A., Cellini, L., Di Bartolomeo, S., Di Campli, E., Grande, R., Cannatelli, M. A., et al. (2001). Effects of combining extracts (*Zingiber officinale*) with clarithromycin on Helicobacter pylori. *Phytotherapy Research, 15*(5), 417–421.

O'Leary, K. A., de Pascual-Teresa, S., Needs, P. W., Bao, Y. P., O'Brien, N. M., Williamson, G., & Dangles, O. (2013). Effect of flavonoids on the phenolic content of copper-induced LDL (low-density lipoprotein) oxidation. *Biochemical Journal, 372*(2), 465–473.

Obahiagbon, E. G., & Ogwu, M. C. (2024). Consumer perception and demand for sustainable herbal medicine products and market. *Herbal medicine phytochemistry: Applications and trends* (pp. 1–34). Springer International Publishing.

Oliveira, N., et al. (2017). Developing safe bioactive compounds through sustainable extraction practices. *Journal of Sustainable Chemistry, 11*(4), 189–202.

Pandey, K. B., & Rizvi, S. I. (2009). Plant polyphenols as dietary antioxidants in human health and disease. *Oxidative Medicine and Cellular Longevity, 2*(5), 270–278.

Pandey, A. K., Bajpai, V. K., & Kang, S. C. (2018). Antioxidant and anticancer potential of essential oils from lemongrass (*Cymbopogon citratus*) and mint (Mentha arvensis) and their nanoemulsions. *Nanomaterials, 8*(9), 679.

Parekh, S., Zhu, S., Liu, Z., Qiao, M., & Wientjes, M. G. (2021). Research on plant-based phytochemicals. *Journal of Enzyme Inhibition and Medicinal Chemistry, 36*(1), 356–363.

Patil, M (2015). Advancements in Genetic Research. *Journal of Research, 10*(3), 100–120.

Perrott, K. M., Wiley, C. D., Desprez, P. Y., & Campisi, J. (2019). Apigenin suppresses the senescence-associated secretory phenotype and paracrine effects on breast cancer cells. *Geroscience, 41*(6), 803–816.

Peterson, J., Dwyer, J., Jacques, P., & McCullough, M. L. (2019). The association between tea and theaflavins and cardiovascular disease. *Advances in Nutrition, 10*(1), 80–91.

Pisoschi, A. M., & Pop, A. (2015). The role of antioxidants in the chemistry of oxidative stress: A review. *European Journal of Medicinal Chemistry, 97*, 55–74.

Pradhan, S. C., & Girish, C. (2019). Hepatoprotective potential of the Mediterranean diet against liver diseases. *Hepatology International, 13*(3), 308–321.

Prasanna, S., Verma, P., & Bodh, S. (2024). The role of food industries in sustainability transition: A review. *Environment, Development and Sustainability*, 1–21.

Prinsloo, G., & Nogemane, N. (2018). The effects of season and water availability on chemical composition, secondary metabolites, and biological activity in plants. *Phytochemistry Reviews, 17*(4), 889–902.

Quave, C. L., & Harkenthal, M. (2003). Antimicrobial strategies of medicinal plants: Antimicrobial strategies used by animals and plants in search of common ground. *Frontiers in Microbiology, 4*, 266.

Rainforest Alliance. (n.d.). Impact report 2019. https://www.rainforest-alliance.org/.

Rajput, A., Sharma, R., & Bharti, R. (2022). Pharmacological activities and toxicities of alkaloids on human health. *Materials Today: Proceedings, 48*, 1407–1415.

Ramirez, A., Coello, M. T., Vine, J. V., & Franco, V. B. (2017). Anticancer mechanisms of action of plant compounds. *Current Medicinal Chemistry, 24*(13), 1489–1508.

Riche, D. M., McEwen, C. L., & Riche, K. D. (2013). Analysis of the CoQ10 content of commercially available oral supplements. *Journal of Clinical Lipidology, 7*(6), 434–439.

Russo, G. L., Spagnuolo, C., Russo, M., Tedesco, I., Moccia, S., & Cervellera, C. (2020). Mechanisms of aging and potential role of selected polyphenols in extending health span. *Biochemical Pharmacology, 173*, 113719.

Sabir, A., Unal, H., & Codur, M. Y. (2020). Recent advances in food dehydration technology: A review. *Food Science and Technology (Campinas), 40*, 1–10.

Saboon, Chaudhari, S. K., Arshad, S., Amjad, M. S., & Akhtar, M. S. (2019). Natural compounds extracted from medicinal plants and their applications. *Natural Bio-Active Compounds: Volume 1: Production and Applications*, 193–207.

Saeed, A. F., Su, J., & Ouyang, S. (2021). Marine-derived drugs: Recent advances in cancer therapy and immune signaling. *Biomedicine & Pharmacotherapy, 134*, 111091.

Salinas, A., Velasco, P., Salinas, V. D., Chung, J. J., Martínez-Núñez, M. A. A., Vélez, A. C., et al. (2018). Phenolic compounds and their antioxidant activity from food industry wastes. *Journal of Chemistry, 2018*, 1–7.

Sanches-Silva, A. (2021). Impact of food additives on human health. *Food Chemistry, 30*(1), 200–215.

Santos, D., Sinico, C., Ribeiro, M. P., & Oliveira, R. (2019). New insights for the development of phytopharmaceuticals: The study of anti-inflammatory activities of *Hedychium coronarium* Koenig ex Roxb. rhizomes. *Industrial Crops and Products, 129*, 403–410.

Scalbert, A., Johnson, I. T., & Saltmarsh, M. (2005). Polyphenols: Antioxidants and beyond. *The American Journal of Clinical Nutrition, 81*(1), 215S-217S.

Shan, C., Chen, M., You, J., Shi, H., & Zhang, L. (2017a). Cinnamaldehyde regulates cholesterol metabolism by affecting the intestinal NPC1L1 in mice. *Phytotherapy Research, 31*(5), 830–837.

Shan, M., Yu, S., Yan, H., Guo, S., Xiao, W., & Wang, Q. (2017b). Curcumin suppresses cell proliferation and promotes apoptosis of non-small cell lung cancer. *Oncology Letters, 13*(6), 4933–4938.

Shankar, S., Ganapathy, S., Hingorani, S. R., & Srivastava, R. K. (2017). EGCG inhibits the growth, invasion, angiogenesis, and metastasis of pancreatic cancer. *Frontiers in Bioscience, 12*, 251–266.

Sharath, K., et al. (2018). Maximizing yield and potency through modern extraction techniques. *Modern Extraction Research, 14*(2), 87–98.

Sharma, H. K., Chauhan, S., & Gola, K. (2017). Blanching improves the quality of dried green leafy vegetables. *Journal of Food Processing and Technology, 8*(7), 1–5.

Sharma, D., Shree, B., Kumar, S., Kumar, V., Sharma, S., & Sharma, S. (2022). Stress-induced production of plant secondary metabolites in vegetables: Functional approach for designing next generation superfoods. *Plant Physiology and Biochemistry, 192*, 252–272.

Shen, N., Wang, T., Gan, Q., Liu, S., Wang, L., & Jin, B. (2022). *Plant flavonoids: Classification, distribution, biosynthesis, and antioxidant*.

Singh, B. N., Shankar, S., & Srivastava, R. K. (2011). Green tea catechin, epigallocatechin-3-gallate (EGCG): Mechanisms, perspectives and clinical applications. *Biochemical Pharmacology, 82*(12), 1807–1821.

Singh, R., Singh, S., Kumar, B., Sinha, V. R., & Kumari, A. (2018). Bioactive compounds: Natural armor against gastrointestinal malignancies. *Frontiers in Pharmacology, 9*, 684.

Srinivasan, K. (2016). Biological activities of red pepper (Capsicum annuum) and its pungent principal capsaicin: A review. *Critical Reviews in Food Science and Nutrition, 56*(9), 1488–1500.

Srivastava, R., & Panda, S. (2019). Impact of organic sourcing on ecosystem health in bioactive compound production. *Organic Ecology Journal, 22*(3), 176–189.

Strid, Å., Chow, W. S., & Anderson, J. M. (1994). UV-B damage and protection at the molecular level in plants. *Photosynthesis Research, 39*, 475–489.

Sudheer, S., Gangwar, P., Usmani, Z., Sharma, M., Sharma, V. K., Sana, S. S., et al. (2022). Shaping the gut microbiota by bioactive phytochemicals: An emerging approach for the prevention and treatment of human diseases. *Biochimie, 193*, 38–63.

Surh, Y. J. (2003). Cancer chemoprevention with dietary phytochemicals. *Nature Reviews Cancer, 3*(10), 768–780.

Sustainable Development Goals Knowledge Platform. (n.d.). Goal 15: Protect, restore, and promote sustainable use of terrestrial ecosystems, sustainably manage forests, combat desertification, and halt and reverse land degradation and biodiversity loss. https://sdgs.un.org/goals/goal15.

Tao, X., et al. (2021). Higher extraction efficiencies and improved product quality with sustainable extraction methods. *Sustainable Technologies, 4*(1), 33–45.

Tasneem, S., Liu, B., Li, B., Choudhary, M. I., & Wang, W. (2019). Molecular pharmacology of inflammation: Medicinal plants as anti-inflammatory agents. *Pharmacological Research, 139*, 126–140.

Tillmann, J., Stubkjaer, E., Limbrecht, F., Kittel, A., Glatz-Frage, K., & Helmreich, S. (2020). Interdisciplinary research in personalized medicine: Challenges for organizational development in Europe. *Current Bioinformatics, 15*(2), 189–192.

Touyz, R. M., & Schiffrin, E. L. (2018). Reactive oxygen species in vascular biology: Implications in hypertension. *Histochemistry and Cell Biology, 139*(6), 735–754.

Tran, T. H., Guo, Y., Song, D., Bruno, R. S., & Lu, X. (2019). Delivery of bioactive compounds with nanostructured lipid carriers: Challenges and strategies. *Journal of Pharmaceutical Sciences, 108*(3), 101–114.

Valadez-Carmona, L., Lygin, A. V., Ovando-Medina, I., Zamora-Natera, F., Karlsson, B., Villegas, E., et al. (2020). Vegetable oil-based polymers as matrices for antimicrobial polymers. *International Journal of Polymer Science, 2020*, 1–7.

Vauzour, D., Vafeiadou, K., Rodriguez-Mateos, A., Rendeiro, C., & Spencer, J. P. (2010). The neuroprotective potential of luteolin is mediated through Circadian gene expression in neuronal cells. *Archives of Biochemistry and Biophysics, 501*(1), 65–66.

Vauzour, D. (2017). Polyphenols and brain health. *OCL, 24*(4), D403.

Vazhappilly, C. G., Ansari, S. A., Al-Jaleeli, R., Al-Azawi, A. M., Ramadan, W. S., Menon, V., et al. (2019). Role of flavonoids in thrombotic, cardiovascular, and inflammatory diseases. *Inflammopharmacology, 27*, 863–869.

Verma, S., & Kumar, A. (2018). Promoting social responsibility through sustainable practices in bioactive compound production. *Journal of Sustainable Development, 12*(3), 211–225.

Verma, P., Mathew, M. S., Mathew, J., & Nair, A. (2016). Antioxidative and antimicrobial properties of green tea extract-loaded chitosan-polycaprolactone nanoparticles. *Journal of Microencapsulation, 33*(8), 772–782.

Wang, X., & Zhi, Z. (2017). Apigenin, a natural flavonoid, inhibits proliferation and induces apoptosis in human thyroid cancer cells. *Cancer Biomarkers, 19*(1), 129–136.

Wang, X., Zhang, Q., & Liu, Y. (2019). Polyphenols: Bioavailability and mechanisms of action in chronic disease management. *Journal of Nutritional Sciences, 30*(2), 78–91.

Wang, M. Y., Zhao, W. Z., & Gao, P. H. (2019). *Preparation method, composition, and application of Astragaloside IV sustained-release microspheres*. CN107195866A.

Wattenberg, L. W. (2011). Inhibition of carcinogenesis by naturally occurring and synthetic compounds. *Basic and Clinical Pharmacology and Toxicology, 108*(1), 14–16.

Williams, R. J., & Spencer, J. P. E. (2012). Flavonoids, cognition, and dementia: Actions, mechanisms, and potential therapeutic utility for Alzheimer disease. *Free Radical Biology and Medicine, 52*(1), 35–45.

World Health Organization. (2003). Good manufacturing practices for pharmaceutical products. In *WHO expert committee on specifications for pharmaceutical preparations*. WHO Technical Report Series, No. 908.

Wu, L. X., Xu, X. J., Zhang, K. Y., Bian, D. X., Zhang, B., Luo, L. H., et al. (2019). Anti-tumor effect of curcumin on human cervical carcinoma HeLa cells in vitro and in vivo. *Cell Physiology and Biochemistry, 53*(6), 887–901.

Yang, L., Yang, C., Li, C., Zhao, Q., Liu, L., Fang, X., & Chen, X. Y. (2016). Recent advances in biosynthesis of bioactive compounds in traditional Chinese medicinal plants. *Science Bulletin, 61*(1), 3–17.

Yang, Y., Xu, H., Zhou, J., Liu, X., & Chen, L. (2018). Analysis of the anti-inflammatory mechanism of the rhubarb peony decoction based on network pharmacology. Evidence-Based Complementary and Alternative Medicine, 2018.

Zhang, H., Liu, C., Wang, L., & Liu, H. (2020). Genistein ameliorates isoproterenol-induced cardiac fibrosis through inhibition of TGF-β1-Smad signaling. *Environmental Toxicology, 35*(11), 1177–1185.

Zhou, M. L., Zhu, X. M., Shao, J. R., Tang, Y. X., & Wu, Y. M. (2011). Production and metabolic engineering of bioactive substances in plant hairy root culture. *Applied Microbiology and Biotechnology, 90*, 1229–1239.

The manufacturer's authorised representative in the EU is Springer Nature Customer Service Centre GmbH, Europaplatz 3, 69115 Heidelberg, Germany. If you have any concerns regarding our products, please contact ProductSafety@springernature.com

Printed and bound by CPI Group (UK) Ltd, Croydon, CR0 4YY
07/10/2026
02225517-0001